Cancer Drug Discovery and Development

Series Editor

Beverly A. Teicher, Bethesda, USA

Cancer Drug Discovery and Development is the definitive book series in cancer research and oncology, providing comprehensive coverage of specific topics in the field. The aim of the series is to cover the process of drug discovery, preclinical models in cancer research, specific drug target groups and experimental and approved therapeutic agents across volumes.

Volumes are current and timely, anticipating areas where experimental agents are reaching FDA approval. Each volume is edited by an expert in the related field and chapters are authored by renowned scientists and physicians in their fields of interest. All volumes undergo single blind peer review and review by the series editor prior to acceptance and publication in the series. Cancer Drug Discovery & Development is indexed in SCOPUS.

Akash Sabarwal • Saba Tabasum • Soumitro Pal
Editors

Receptor Tyrosine Kinases in Cancer

Humana Press

Editors
Akash Sabarwal
Division of Nephrology, Department of Pediatrics
Boston Children's Hospital and Harvard Medical School
Boston, MA, USA

Saba Tabasum
Department of Medical Oncology
Dana-Farber Cancer Institute and Harvard Medical School
Boston, MA, USA

Soumitro Pal
Division of Nephrology, Department of Pediatrics
Boston Children's Hospital and Harvard Medical School
Boston, MA, USA

ISSN 2196-9906 ISSN 2196-9914 (electronic)
Cancer Drug Discovery and Development
ISBN 978-3-031-93893-1 ISBN 978-3-031-93894-8 (eBook)
https://doi.org/10.1007/978-3-031-93894-8

This Springer imprint is published by the registered company Springer Nature Switzerland AG
The registered company address is: Gewerbestrasse 11, 6330 Cham, Switzerland

In loving memory of my late parents, Swarn Sabharwal and Naresh Kumar Sabharwal, whose love and wisdom have shaped my life and keep guiding me. This book is also dedicated to all those affected by cancer—may their strength inspire hope and keep driving scientists in the pursuit of better treatments.
—Akash Sabarwal

Foreword

Cancer biology has experienced extraordinary advancements in recent decades, with receptor tyrosine kinases (RTKs) emerging as pivotal players in tumor progression and therapy resistance. Their significance cannot be overstated, as these signaling molecules are integral to key cellular processes that drive malignancies. This edited volume, *Receptor Tyrosine Kinases in Cancer*, to be published by Springer as part of the esteemed *Cancer Drug Discovery and Development* series, offers a comprehensive exploration of RTKs. This series, edited by Beverly Teicher, is indexed in SCOPUS and is renowned for its contributions to advancing cancer research. This volume highlights the central role of RTKs in cancer biology while identifying novel therapeutic opportunities.

The book brings together contributions from leading experts, each offering valuable insights into different aspects of RTK biology. From a foundational understanding of RTKs' involvement in signal transduction and cancer progression, as introduced in the opening chapter, to in-depth analyses of specific RTK families such as c-MET, AXL, VEGFR, and EGFR, this volume provides a thorough and expansive perspective. The work also tackles contemporary challenges in the field, such as acquired therapeutic resistance, the interaction between RTKs and mitochondria, and their influence on the immune microenvironment. Moreover, the book features emerging therapeutic approaches—from natural compounds to combination therapies—highlighting the translational potential of RTK-focused research.

This volume owes much of its depth and richness to the collective expertise of its distinguished editors: Dr. Akash Sabarwal, Dr. Saba Tabasum, and Dr. Soumitro Pal. Dr. Sabarwal's excellent work on RTK-driven signaling pathways in cancer, Dr. Tabasum's pioneering research in angiogenesis and immune regulation, and Dr. Pal's contributions to renal cancer and therapeutic resistance have each shaped this book into a critical resource. Together, their expertise offers unique perspectives that will be invaluable to researchers, clinicians, and students alike in their pursuit of innovative scientific approaches to combat cancer.

By bridging foundational concepts with cutting-edge discoveries, this book serves as a platform to deepen our understanding of RTKs and their therapeutic potential. This volume will be a valuable reference, inspiring future research and advancing transformative breakthroughs in cancer therapy.

With great enthusiasm, I invite readers to immerse themselves in this comprehensive collection of knowledge and insights. May it ignite new ideas, foster collaborations, and contribute to the ongoing battle against cancer.

With best wishes,

Sujit Basu
Jack C. Geer Endowed Professor,
Professor of Medical Oncology,
Department of Internal Medicine,
Professor of Pathology & Director
of Translational Research
Ohio State University,
Columbus, OH, USA

Preface

Cancer research has witnessed remarkable progress in recent decades, moving from conventional chemotherapy to more targeted therapies, with small-molecule kinase inhibitors playing a pivotal role in this evolution. Alongside this, our understanding of immune evasion mechanisms in cancer cells has also deepened, with immune checkpoint blockade emerging as a promising strategy for enhancing cancer treatment. Immune checkpoint inhibitors (ICIs) have demonstrated significant clinical success across a variety of solid tumors, improving treatment outcomes and offering new therapeutic possibilities.

Among the key molecular drivers of cancer, receptor tyrosine kinases (RTKs) have emerged as central regulators of vital processes like cell growth, differentiation, metastasis, metabolism, and immune suppression. When RTKs are aberrantly activated—whether through mutations, overexpression, or dysregulated signaling—this contributes directly to cancer progression and the development of therapy resistance. As a result, RTKs have become critical targets in the quest to develop more effective cancer therapies.

The advent of small-molecule kinase inhibitors marked a transformative shift in cancer treatment, offering strategies that often outperform traditional therapies. Imatinib (Gleevec), the first small-molecule kinase inhibitor approved for the treatment of chronic myeloid leukemia (CML) and gastrointestinal stromal tumors (GIST), represents a landmark achievement in oncology. Its success not only provided life-extending benefits to patients but also sparked the development of a new generation of kinase inhibitors targeting a range of RTKs, further revolutionizing cancer care.

Ongoing research into RTK signaling pathways continues to deepen our understanding of their role in cancer biology. This research has gained increasing importance, with several small-molecule RTK inhibitors now approved by regulatory agencies such as the US FDA. Pharmaceutical companies, too, are investing heavily in the development of novel RTK-targeting molecules to address unmet medical needs in oncology.

This book aims to provide a comprehensive overview of the latest advancements in RTK research and their implications for cancer treatment. It brings together contributions from leading experts worldwide to explore various aspects of RTK biology and its therapeutic potential.

Structured into 12 chapters, the book covers diverse topics, including the roles of RTKs such as EGFR, c-Met, AXL, and VEGFR in pro-tumorigenic signaling, therapeutic resistance, and the development of targeted small-molecule inhibitors for different cancer types. The content strikes a balance between fundamental research and clinical applications, making it a valuable resource for a broad audience.

We would like to express our deepest gratitude to the contributing authors for their exceptional chapters and to the peer reviewers for their invaluable insights and constructive feedback. Our heartfelt thanks also go to the Division of Nephrology, Boston Children's Hospital, for their ongoing support, and to Professor Friedhelm Hildebrandt, Chair of the Division, for his unwavering leadership and encouragement throughout this project. His visionary guidance and commitment to fostering a collaborative research environment have been instrumental in bringing this work to success.

This book is designed to benefit researchers, oncologists, and graduate students alike, offering new perspectives and actionable knowledge in cancer biology. We hope it serves as a source of inspiration, promoting fresh ideas and collaborations that drive the continued advancement of cancer research and therapy.

Sincerely,

Boston, MA, USA Akash Sabarwal
Boston, MA, USA Saba Tabasum
Boston, MA, USA Soumitro Pal

Contents

Editors and Contributors

About the Editors

Akash Sabarwal is a researcher and educator, currently serving as a faculty member in the Division of Nephrology, at Boston Children's Hospital, and as an Instructor in Pediatrics at Harvard Medical School, both located in Boston, Massachusetts, USA. With a profound focus on unraveling the complexities of cancer biology, his research interest is in understanding the intricate role of Receptor Tyrosine Kinases in protumorigenic signaling pathways and acquired therapeutic resistance. He earned his PhD degree from the School of Life Sciences at the Central University of Gujarat, Gandhinagar, India. His doctoral research, investigating the effects of Fisetin on gastric cancer cells, showcased his early commitment to advancing cancer therapeutics. With over a decade of research experience, both as a research fellow and now as a faculty member, he has contributed significantly to the field of cancer biology.

His scholarly work is reflected in numerous research articles published in esteemed, high-impact journals, along with reviews and book chapters. Additionally, his expertise is recognized globally, as he serves as an expert reviewer for various reputed international journals, showcasing his deep understanding of cancer biology and associated signaling pathways. He specializes in elucidating these pathways' modulation using different small molecule inhibitors.

In recognition of his exceptional contributions to cancer research, he has been honored with prestigious awards, including the esteemed Dana-Farber/Harvard Cancer Center (DF/HCC) Kidney Cancer, Specialized Program of Research Excellence (SPORE) Award. Driven by a passion for scientific discovery and a commitment to advancing cancer therapeutics, he continues to make profound strides in the field, leaving an indelible mark on cancer research.

Saba Tabasum is a distinguished Scientist within the Department of Medical Oncology at the Dana-Farber Cancer Institute and Harvard Medical School, located in Boston, Massachusetts, USA. Her research focuses on unraveling the intricate role of angiogenesis in immune regulation and homing, while also identifying and validating innovative biologically based therapeutic targets. Specifically, she delves into understanding the targeting of VEGFA and TGF-β signaling pathways within tumor endothelial cells to modulate the tumor immune microenvironment. Her pioneering work aims to enhance our comprehension of the dynamic interplay between angiogenesis and immune checkpoint blockade, alongside other immune therapy interventions. Her academic journey led her to earn a PhD from the prestigious School of Life Sciences at Jawaharlal Nehru University, New Delhi, India. During her doctoral research, she conducted studies on the anti-metastatic properties of the phytochemical fisetin, elucidating its mechanisms of action through targeted inhibition of TGF-β signaling in non-small cell lung carcinoma. Thus, for over a decade, she has dedicated her efforts to comprehending and targeting tyrosine kinases in cancer. Her scholarly contributions have been disseminated through esteemed publications such as the *Journal of Clinical Investigation*, *Nature Communications*, and *Cancer Immunology Research*. Additionally, she has demonstrated her innovative prowess by filing patents for her groundbreaking work. Furthermore, her expertise extends beyond her own research endeavors. She actively engages in peer reviewing for reputable journals and holds numerous leadership positions where she serves as a voice to address and represent the research community. In recognition of her scholarly achievements and dedication to advancing the field of cancer immunology, she has been awarded the prestigious MacNaught fellowship, which supports her ongoing research efforts at the Dana-Farber Cancer Institute. Through her dedication and innovative research, she continues to advance our understanding of cancer immunology and pave the way for new therapeutic approaches.

Soumitro Pal is an Associate Professor in the Division of Nephrology, Boston Children's Hospital and Harvard Medical School, Boston, Massachusetts, USA. He has received his PhD in Physiology from the University of Calcutta. He did his post-doctoral research in the field of angiogenesis at Beth Israel Deaconess Medical Center and Harvard Medical School. Dr. Pal's research is focused on studying the mechanism of renal injury, renal inflammation, and renal cancer, with particular importance paid to the signal transduction pathways regulating gene expression. One of the main focus of his work is to identify molecular mechanisms of cancer growth in patients receiving organ transplants. For instance, it has been shown that although calcineurin inhibitors (CNI) are very good immunosuppressive agents, they can promote cancer, while the mTOR inhibitor rapamycin (RAPA) can inhibit cancer growth. His work has demonstrated that treatment with CNI can activate the Ras-Raf-Nrf2-HO-1 signaling cascade, which plays a critical role in renal cancer growth. He has developed a sophisticated murine model to study the mechanism of post-transplantation cancer growth. His research aims to identify molecular targets to develop novel drugs for the treatments of renal inflammation and cancer. His studies are also focused on exploring novel combination therapies to overcome therapeutic resistance in cancer.

Contributors

Hina Agraval Department of Medicine, National Jewish Health, Denver, CO, USA

Ryuhjin Ahn David H. Koch Institute for Integrative Cancer Research, Massachusetts Institute of Technology (MIT), Cambridge, MA, USA

Department of Biological Engineering, MIT, Cambridge, MA, USA

Shreya Ahuja Renal Division, Department of Medicine, Brigham and Women's Hospital, Boston, MA, USA

Department of Neuroscience and Biology, Brandeis University, Waltham, MA, USA

Amrendra Kumar Ajay Renal Division, Department of Medicine, Brigham and Women's Hospital, Boston, MA, USA

Department of Medicine, Harvard Medical School, Boston, MA, USA

Dhiraj Bhatia Department of Biological Sciences and Engineering, Indian Institute of Technology, Gandhinagar, Gujarat, India

Jui Bhiwankar Renal Division, Department of Medicine, Brigham and Women's Hospital, Boston, MA, USA

Debashish Chakraborty School of Biological Sciences, National Institute of Science Education and Research (NISER) Bhubaneswar, An OCC of Homi Bhabha National Institute, Khurda, Odisha, India

Alicia D'Souza MIT-Harvard Health Sciences and Technology, Cambridge, MA, USA

David H. Koch Institute for Integrative Cancer Research, Massachusetts Institute of Technology (MIT), Cambridge, MA, USA

Diana Drago-Garcia Stem Cell and Cancer Biology Laboratory, The Francis Crick Institute, London, UK

Liwu Fu State Key Laboratory of Oncology in South China, Collaborative Innovation Center for Cancer Medicine, Guangdong Esophageal Cancer Institute, Sun Yat-sen University Cancer Center, Guangzhou, People's Republic of China

Kunj Bihari Gupta Georgia Cancer Center, Augusta University, Augusta, GA, USA

Sonu Kumar Gupta Department of Physiology and Endocrinology, Medical College of Georgia, Augusta University, Augusta, GA, USA

Won Sik Ham Department of Urology and Urological Science Institute, Yonsei University College of Medicine, Seoul, South Korea

Vrinda Heda Renal Division, Department of Medicine, Brigham and Women's Hospital, Boston, MA, USA

Sanvi Kishore Renal Division, Department of Medicine, Brigham and Women's Hospital, Boston, MA, USA

Bhavya Kumar Renal Division, Department of Medicine, Brigham and Women's Hospital, Boston, MA, USA

Kunal Kumar Department of Biochemistry and Molecular Biology, Medical College of Georgia, Augusta University, Augusta, GA, USA

Rahul Kumar Dr. B. R. A.-Institute Rotary Cancer Hospital, All India Institute of Medical Sciences, New Delhi, India

Sanam Kumari Department of Mathematics, Sciences and Technology, Paine College, Augusta, GA, USA

Ji Eun Lee Department of Urology and Urological Science Institute, Yonsei University College of Medicine, Seoul, South Korea

Myung Eun Lee Department of Urology and Urological Science Institute, Yonsei University College of Medicine, Seoul, South Korea

Min Luo State Key Laboratory of Oncology in South China, Collaborative Innovation Center for Cancer Medicine, Guangdong Esophageal Cancer Institute, Sun Yat-sen University Cancer Center, Guangzhou, People's Republic of China

Ilaria Marrocco Department of Life Sciences and Public Health, Università Cattolica del Sacro Cuore, Rome, Italy

Satyabrata Nayak Cancer Biology Laboratory, School of Life Sciences, Jawaharlal Nehru University, New Delhi, India

Ramendra Pati Pandey Department of Biotechnology and Microbiology, SRM University, Delhi-NCR, Sonepat, Haryana, India

Jee Soo Park Department of Urology and Urological Science Institute, Yonsei University College of Medicine, Seoul, South Korea

Yeong Ji Park Department of Urology and Urological Science Institute, Yonsei University College of Medicine, Seoul, South Korea

Rohit Patel Department of Neurology, Brigham and Women's Hospital, Harvard Medical School, Boston, MA, USA

Ayanna Rohil Renal Division, Department of Medicine, Brigham and Women's Hospital, Boston, MA, USA

Donatella Romaniello Department of Medical and Surgical Sciences, University of Bologna, Bologna, Italy

IRCCS Azienda Ospedaliero-Universitaria di Bologna, Bologna, Italy

Miss Ruby Department of Biotechnology and Microbiology, SRM University, Delhi-NCR, Sonepat, Haryana, India

Akansha Sharma Department of Microbiology and Immunology, Thomas Jefferson University, Philadelphia, PA, USA

Jiten Sharma Division of Molecular Medicine and Gene Therapy, Lund Stem Cell Center, Lund University, Lund, Sweden

Rana P. Singh Cancer Biology Laboratory, School of Life Sciences, Jawaharlal Nehru University, New Delhi, India

Department of Pharmaceutical Sciences, Skaggs School of Pharmacy and Pharmaceutical Sciences, University of Colorado, Anschutz Medical Campus, Aurora, CO, USA

Gautam Buddha University, Greater Noida, UP, India

Raghu Solanki Department of Biological Sciences and Engineering, Indian Institute of Technology, Gandhinagar, Gujarat, India

Pranay Tanwar Dr B. R. A. Institute Rotary Cancer Hospital, All India Institute of Medical Sciences, New Delhi, India

Muthuswamy Thangaraju Department of Biochemistry and Molecular Biology, Medical College of Georgia, Augusta University, Augusta, GA, USA

Mary Luz Uribe Instituto de Neurociencias, Consejo Superior de Investigaciones Científicas (CSIC), and Universidad Miguel Hernández (UMH), Sant Joan d'Alacant, Spain

Johnna Francis Varghese Department of Neurology, Brigham and Women's Hospital, Harvard Medical School, Boston, MA, USA

Randolph S. Watnick Vascular Biology Program, Boston Children's Hospital, Harvard Medical School, Boston, MA, USA

Monika Yadav Laboratory of Nanotechnology and Chemical Biology, Regional Centre for Biotechnology, Faridabad, India

Umesh C. S. Yadav Special Center for Molecular Medicine, Jawaharlal Nehru University, New Delhi, India

An Introduction to Receptor Tyrosine Kinases: Signal Transduction and Cancer Progression

Miss Ruby, Rahul Kumar, Jui Bhiwankar, Sanvi Kishore, Bhavya Kumar, Vrinda Heda, Ayanna Rohil, Shreya Ahuja, Ramendra Pati Pandey, and Amrendra Kumar Ajay

1 Introduction

The discovery of epidermal growth factor (EGF) and nerve growth factor (NGF) in the 1960s marked a significant breakthrough in cellular biology. Since then, the search for proteins and their mediated signaling in cells has led to the identification of receptor tyrosine kinases (RTKs) [1, 2]. RTKs constitute a superfamily of transmembrane proteins that facilitate cellular communication and regulate critical biological processes, including cell growth, proliferation, differentiation, and survival [3].

Authors Miss Ruby and Rahul Kumar have equally contributed to this chapter.

M. Ruby · R. P. Pandey
Department of Biotechnology and Microbiology, SRM University, Delhi-NCR, Sonepat, Haryana, India

R. Kumar
Dr. B. R. A.-Institute Rotary Cancer Hospital, All India Institute of Medical Sciences, New Delhi, India

J. Bhiwankar · S. Kishore · B. Kumar · V. Heda · A. Rohil
Renal Division, Department of Medicine, Brigham and Women's Hospital, Boston, MA, USA

S. Ahuja
Renal Division, Department of Medicine, Brigham and Women's Hospital, Boston, MA, USA

Department of Neuroscience and Biology, Brandeis University, Waltham, MA, USA
e-mail: shreyaahuja@brandeis.edu

A. K. Ajay (✉)
Renal Division, Department of Medicine, Brigham and Women's Hospital, Boston, MA, USA

Department of Medicine, Harvard Medical School, Boston, MA, USA
e-mail: akajay@bwh.harvard.edu

A. Sabarwal et al. (eds.), *Receptor Tyrosine Kinases in Cancer*, Cancer Drug Discovery and Development, https://doi.org/10.1007/978-3-031-93894-8_1

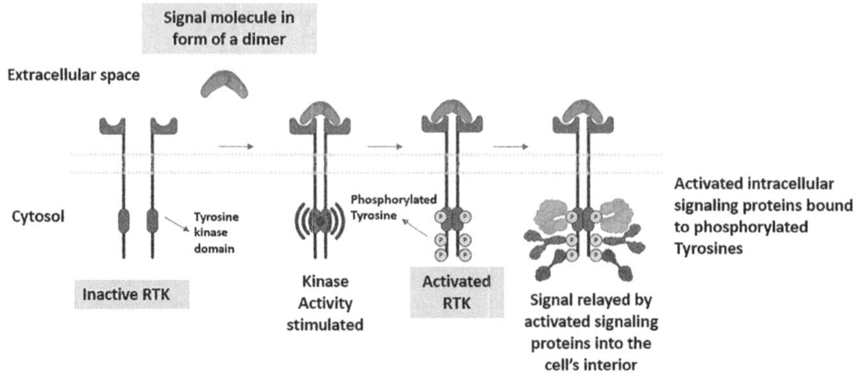

Fig. 1 Activation and signal transduction mechanisms of RTKs

The RTK family includes a diverse array of cell surface receptors that mediate essential physiological and metabolic signaling pathways in response to hormones, growth factors, and cytokines. The human genome encodes 58 RTKs, categorized into 20 subfamilies. These family members vary in size, with some containing only a few members, while others, such as the PDGF subfamily (5 members), the ERBB and FGF subfamilies (4 members each), and the ephrin receptor subfamily (14 members), represent the largest groups [4, 5]. The extracellular domain (ECD) dynamics play a crucial role in defining RTK family function and cellular responses through subsequent signaling cascades. Structurally, RTKs share a conserved structural organization consisting of an extracellular binding domain, a single transmembrane helix or transmembrane domain (TMD), and a cytoplasmic region housing the protein tyrosine kinase domain (TKD), along with an additional carboxy (C-) terminal and juxtamembrane regulatory domains [6, 7]. Given their regulatory importance, the fundamental structure, activation mechanism, and intracellular signaling pathways of RTKs have been largely conserved throughout evolution, from the nematode *Caenorhabditis elegans* to humans [7].

RTK activation is initiated when a ligand binds to the ECD, triggering receptor dimerization or oligomerization, which leads to autophosphorylation of tyrosine residues within the cytoplasmic domain. This phosphorylation event serves as a docking site for intracellular signaling proteins, activating downstream pathways such as the RAS/MAPK, PI3K/Akt, Src, and JAK/STAT. These pathways are pivotal in regulating cell proliferation, differentiation, and survival, as depicted in Fig. 1. However, dysregulation of RTK signaling, often due to mutations or overexpression, has been strongly linked in various diseases, particularly cancers [8].

To ensure controlled RTK activation, several feedback mechanisms maintain receptor activity. These include cellular proteins mobilized upon RTK activation, such as adaptor proteins, enzymes, and signaling molecules, which modulate signal transduction. Negative feedback loops further regulate receptor activity by inducing protein phosphorylation, thereby preventing excessive signaling and maintaining homeostasis [4, 9, 10].

2 Activation Mechanisms of RTKs in Cancer

In normal cells, RTKs are tightly regulated, ensuring precise signal transduction to control key cellular processes such as growth, proliferation, and survival. However, in cancer cells, dysregulation of RTKs leads to aberrant signaling, activating tumor-promoting pathways that drive uncontrolled cell division, angiogenesis, and metastasis. Several mechanisms that contribute to this dysregulation are discussed below.

2.1 Gene Amplification and Overactivation of RTKs in Cancer

Gene amplification is a primary mechanism leading to the overactivation of RTKs in cancer. This process results in an increased number of RTK genes within the genome, leading to the overproduction of receptor proteins on the cell surface. As a result, even with normal ligand levels, the amplified receptors enhance signaling, driving oncogenic pathways [11, 12]. RTK overexpression has been observed in various human malignancies, such as of EGFR in glioblastoma multiforme (GBM) [13] and lung [14] and esophageal cancers [15], as well HER2/ErbB2 in the lung [16], breast [17], and gastric cancers [18, 19].

2.2 Gain-of-Function Mutations in RTKs and their Role in Cancer

Gain-of-function (GOF) mutations in RTKs provide a selective growth advantage and result in constitutive activation, driving uncontrolled cell growth and tumor progression. These mutations can cause ligand-independent activation, where structural alterations keep the receptor active even in the absence of ligand binding, as seen in EGFR L858R mutations in non-small cell lung cancer (NSCLC) [20]. Other mutations occur within the kinase domain, enhancing enzymatic activity and leading to persistent autophosphorylation, such as in the KIT D816V mutations in gastrointestinal stromal tumors (GIST) [21]. Additionally, mutations may disrupt negative regulatory mechanisms, preventing receptor degradation and prolonging signaling, as exemplified by MET exon 14 skipping mutations [22]. Some mutations even increase ligand sensitivity, amplifying downstream signaling even at low concentrations, as observed in FGFR2 mutations in endometrial cancer [23]. These alterations lead to persistent activation of oncogenic pathways, such as MAPK and PI3K-AKT, which promote tumor proliferation, resistance to apoptosis, and metastatic progression.

2.3 Chromosomal Rearrangements and Gene Fusions in RTKs

Chromosomal rearrangements, particularly gene fusions in RTKs, represent another major mechanism of oncogene activation in various cancers. These rearrangements result in the formation of chimeric proteins, often leading to constitutive activation of RTK signaling. This bypasses normal regulatory mechanisms and drives uncontrolled cell growth, survival, and metastasis. One of the most well-characterized RTK gene fusions is the ALK fusion (e.g., EML4-ALK, NPM1-ALK), which occurs in non-small cell lung cancer (NSCLC) and anaplastic large cell lymphoma (ALCL). These fusions are caused by inversions on chromosome 2 or chromosomal translocations such as t(2;5), leading to constitutive ALK kinase activation [24–26]. Similarly, RET fusions (e.g., RET/PTC1, RET/PTC3) in papillary thyroid carcinoma (PTC) arise from chromosomal aberrations like Inv [10](q11) or t(10;17), resulting in ligand-independent RET activation [27, 28].

2.4 Autocrine Activation of RTK Signaling

Autocrine activation is another mechanism by which RTK signaling becomes dysregulated. In this process, a cell produces and secretes its own ligand, which then binds and activates its own receptor. This creates a self-stimulating loop that results in sustained, uncontrolled signaling, independent of external stimuli, and promotes unchecked cell proliferation [29]. For example, in glioblastomas, the overexpression of EGFR and its ligand TGF-α leads to constitutive EGFR activation, driving persistent MAPK and PI3K/Akt pathway signaling, which promotes tumor progression [30, 31].

2.5 Loss of Negative Feedback in RTK Signaling

Under normal conditions, RTK activation is tightly regulated by negative feedback loops that involve protein phosphatases, ubiquitin-mediated degradation, and inhibitory signaling proteins, all of which help terminate or dampen excessive signaling. However, in cancer, genetic mutations, epigenetic alterations, or post-translational modifications can disrupt these feedback mechanisms, leading to persistent RTK activation and hyperactive downstream signaling [32]. For instance, mutations in sprouty (SPRY) proteins, which typically inhibit RAS/MAPK signaling, can result in sustained ERK activation, promoting unchecked tumor growth [33, 34].

RTK-mediated signaling plays a central role in cancer progression by activating multiple downstream pathways that regulate key hallmarks of malignancy. Upon activation, RTKs stimulate RAS, which functions as a central molecular switch to

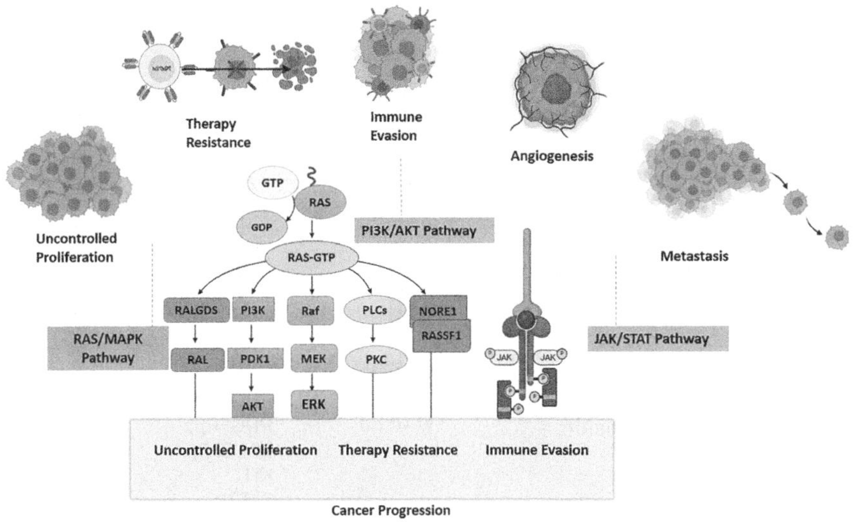

Fig. 2 The roles of the RAS/MAPK, PI3K/AKT, and JAK/STAT pathways in cancer progression. The RAS/MAPK pathway drives uncontrolled proliferation by promoting cell division. The PI3K/AKT pathway enables therapy resistance through anti-apoptotic mechanisms. The JAK/STAT pathway facilitates immune evasion by suppressing immune responses. Each pathway contributes to key cancer hallmarks, as indicated by the color-coded legend

activate the RAS/MAPK, PI3K/AKT, and JAK/STAT pathways. The RAS/MAPK pathway promotes uncontrolled cell proliferation and contributes to therapy resistance by enhancing cell survival mechanisms. The PI3K/AKT pathway supports cell growth, survival, and immune evasion, while the JAK/STAT pathway plays a critical role in immune modulation and metastasis. Together, these pathways drive several aspects of tumor biology, including angiogenesis, evasion of immune surveillance, resistance to treatment, and metastasis. The integration of these signals creates a self-sustaining environment for cancer cells, making RTK signaling a pivotal target for therapeutic intervention (Fig. 2).

3 Role of Notable RTKs in Cancer

RTKs are a diverse family of cell surface receptors that play critical roles in various cellular processes. As previously discussed, abnormal activation or dysregulation of RTKs is closely associated with oncogenesis and other diseases. Below, we highlight some of the most notable RTKs and their contributions to cancer development and progression (Fig. 3).

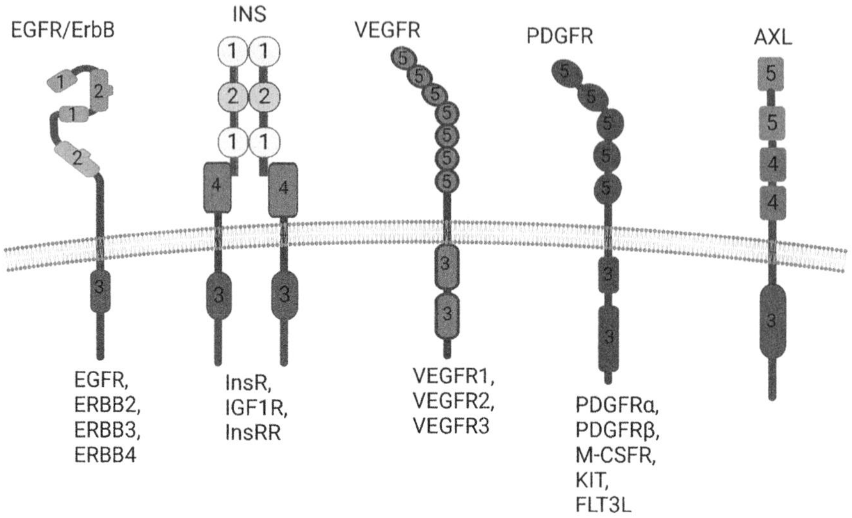

Fig. 3 Key structural features of receptor tyrosine kinase subfamilies

3.1 Epidermal Growth Factor Receptor and the ERBB Family in Cancer

The epidermal growth factor receptor (EGFR), a member of the ErbB receptor tyrosine kinase (RTK) family, is activated by binding to peptide growth factors from the EGF family [35]. Upon activation, EGFR phosphorylates intracellular substrates, triggering signaling pathways that promote cell proliferation, DNA synthesis, and expression of oncogenes such as Fos and Jun. EGFR is frequently overexpressed and mutated in various cancers, playing a pivotal role in tumorigenesis [36, 37]. The EGFR gene can undergo amplification, and mutations in its tyrosine kinase domain have been linked to cancer progression [35, 37, 38]. In addition to EGFR, other members of the ERBB family, HER2, HER3, and HER4, are also involved in cancer pathogenesis [39–41].

HER2 does not bind ligands directly but forms heterodimers with other ERBB receptors, amplifying signaling cascades involved in cell survival and proliferation. HER2 exhibits the highest catalytic activity among the ERBB receptors, making it central to cancer progression.

HER3, lacking intrinsic kinase activity, relies on heterodimerization with other ERBB family members to drive signaling. Dysregulation of HER3 can lead to uncontrolled cell growth and metabolic changes, contributing to cancer progression [42].

HER4, often associated with growth inhibition, activates pathways like MAPK and PI3K/AKT upon ligand binding. Despite its role in growth regulation, dysregulation of HER4 also contributes to cancer development [43]. Figure 4 summarizes

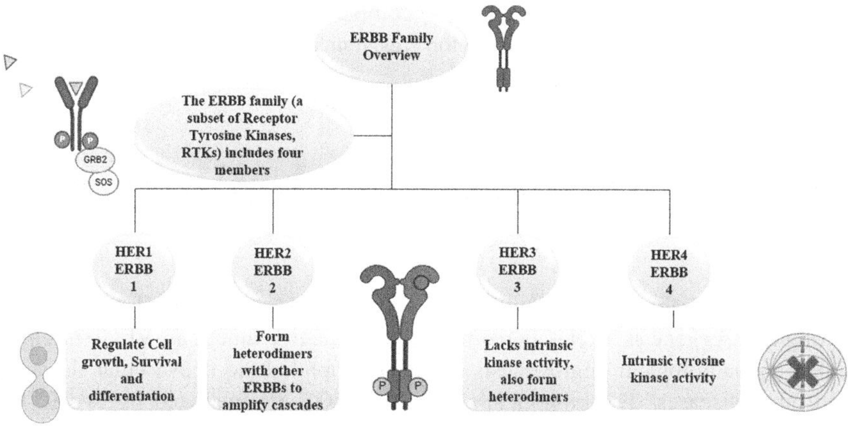

Fig. 4 The overview of the ERBB family (ErbB receptors), comprising four members—the human epidermal growth factor receptors ERBB1–ERBB4—and their function in cancer. Abbreviation: human epidermal growth factor receptor (HER); ErbB receptors, also known as epidermal growth factor (EGF) receptors (ERBB)

the roles of the HER family receptors. Overall, dysregulation through overexpression, mutation, or persistent activation of EGFR and other ERBB receptors drive tumor growth, survival, and metastasis [44].

3.2 *Insulin Receptor (IR) and the IGF-1R Axis in Cancer*

RTKs are synthesized as single-pass transmembrane proteins that primarily bind polypeptide growth factors. With the exception of the insulin receptor (IR) family, most RTKs are activated through ligand-induced receptor dimerization [45]. In pancreatic β cells, insulin secretion is regulated by channel-mediated depolarization, cytoskeletal remodeling, and vesicle tethering at the cell membrane, all of which are influenced by cell surface receptors. RTKs play a crucial role in increasing β-cell mass and function by promoting development and postnatal signaling. However, the activation can also lead to exocytosis [46]. Beyond their physiological roles, RTKs are frequently dysregulated in cancer, where aberrant activation drives uncontrolled cell proliferation, survival, angiogenesis, and metastasis. The insulin-like growth factor-1 receptor (IGF-1R), a key RTK, is a major activator of the phosphatidylinositol 3-kinase-AKT (PI3K/AKT) signaling cascade. IGF-1R is essential for oncogenic transformation and carcinogenesis, making it a significant target for anticancer drug development. These insights have driven efforts to develop biological and small-molecule IGF-1R inhibitors. However, an emerging challenge in targeted

RTK therapy is the ability of one RTK to compensate for another, allowing tumor cells to sustain viability and develop resistance to RTK-specific anticancer drugs [47].

3.3 Vascular Endothelial Growth Factor Receptor (VEGFR) in Cancer

Vascular endothelial growth factor (VEGF) regulates physiological angiogenesis during embryogenesis, skeletal growth, and reproductive functions. VEGF has been linked to pathological angiogenesis in tumors, ophthalmic neovascular diseases, and other diseases. The biological effects of VEGF are mediated by two RTKs, VEGFR-1 and VEGFR-2, whose signaling characteristics differ significantly [48]. Vascular endothelial growth factors (VEGFs) are a family including VEGF-A, VEGF-B, VEGF-C, VEGF-D, and VEGF-E and placental growth factor (PlGF) that control blood vessel (angiogenesis) and lymphatic vessel (lymphangiogenesis) formation. VEGFs bind to three primary RTKs (VEGFR-1, VEGFR-2, and VEGFR-3) and coreceptors like neuropilins and heparan sulfate proteoglycans (HSPG). VEGFRs are activated by ligand-induced dimerization via the extracellular domain (ECD) [49].

3.4 Platelet-Derived Growth Factor Receptor (PDGFR) in Cancer

Platelet-derived growth factor receptor (PDGFR) belongs to the class III RTK family. It is characterized by an extracellular region consisting of five immunoglobulin-like domains (D1–D5) and a split intracellular catalytic tyrosine kinase domain (TK1-TK2). The PDGFR family comprises two receptors: PDGFRα and PDGFRβ. Upon binding of one of the five dimeric platelet-derived growth factor (PDGF) ligands, PDGF-AA, PDGF-BB, PDGF-CC, PDGF-DD, or PDGF-AB, to the extracellular membrane, PDGFRs undergo dimerization. This results in the formation of PDGFRα homodimers, PDGFRα/β heterodimers, or PDGFRβ homodimers. While ligand binding is essential for PDGFR activation, the ligand-receptor complex is further stabilized by direct interactions between the dimerized receptors [50, 51]. Dysregulation of PDGFR signaling has been implicated in various cancers, where aberrant activation promotes tumor growth, angiogenesis, and metastasis. Mutations or overexpression of PDGFR has been observed in malignancies such as glioblastoma, sarcomas, and certain leukemias, making it a promising therapeutic target in oncology [52–54].

3.5 AXL in Cancer

AXL, a member of the TAM (TYRO3, AXL, and MERTK) family, is frequently overexpressed in cancers and is associated with adverse clinicopathological features and poor prognosis [55, 56]. It interacts with its high-affinity ligand, growth arrest-specific 6 (Gas6), a vitamin K-dependent protein, to activate signaling pathways linked to cancer progression, metastasis, immune evasion, and treatment resistance [57]. Notably, studies have shown that AXL is highly expressed in melanoma cell lines, particularly those with mutant NRAS. In these cells, AXL activation via autocrine and paracrine Gas6 signaling drives invasive and metastatic behaviors [58].

AXL can be activated through both Gas6-dependent and ligand-independent mechanisms. Upon binding Gas6, AXL dimerizes and undergoes trans-autophosphorylation, which triggers downstream pathways that promote tumor progression. Autocrine and paracrine loops further reinforce AXL activity, creating a self-sustaining network of survival and growth signals. Moreover, AXL plays a pivotal role in epithelial-to-mesenchymal transition (EMT) by activating transcription factors such as Snail and Slug, and this results in the downregulation of epithelial markers like E-cadherin and upregulation of mesenchymal markers, facilitating cell invasion and migration. The involvement of AXL in cytoskeletal reorganization and matrix degradation through metalloproteinase activation further highlights its crucial role in EMT and cancer metastasis. Figure 5 illustrates AXL signaling pathways and resistance mechanisms [56].

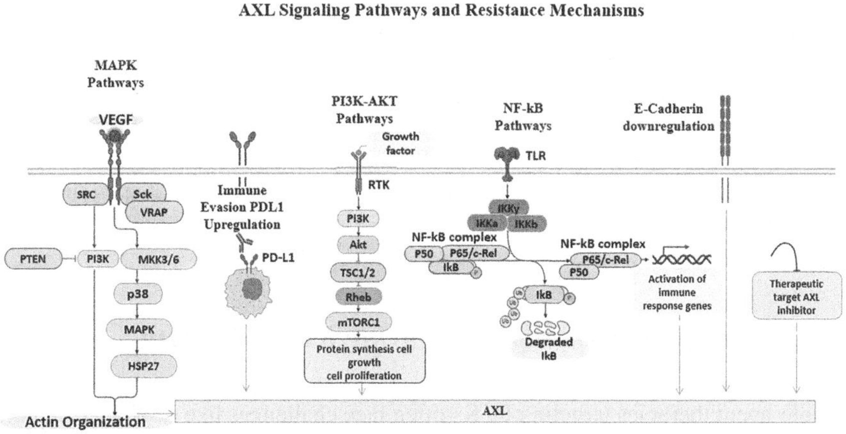

Fig. 5 AXL plays a crucial role downstream of MAPK and NF-κB signaling pathways, regulating E-cadherin and PI3K-AKT pathways. Deregulation of AXL leads to resistance and uncontrolled cell proliferation. The resistance mechanism develops against therapeutic targets of AXL and involves bypassing RTK signaling and dysregulation of immune checkpoint inhibitors

4 RTK as Therapeutic Targets

RTKs have become critical therapeutic targets in cancer due to their central role in regulating essential cellular processes, including proliferation, survival, angiogenesis, migration, and metabolism. In many cancers, RTKs are dysregulated through mechanisms such as mutations, amplifications, overexpression, or chromosomal rearrangements, resulting in the constitutive activation of downstream signaling pathways like RAS/MAPK, PI3K/AKT, and JAK/STAT. This persistent activation of signaling cascades contributes to tumor initiation, progression, and resistance to conventional therapies. Targeting RTKs offers promising strategies for interrupting these pathways, offering potential for more effective cancer treatments.

4.1 *The Role of Small-Molecule Inhibitors and Monoclonal Antibodies*

Small-molecule inhibitors have revolutionized cancer treatment, particularly in cancers driven by mutated RTKs. For example, imatinib, a BCR-ABL kinase inhibitor, has significantly improved the treatment of chronic myeloid leukemia [59]. In NSCLC, osimertinib and gefitinib target EGFR mutations, demonstrating the potential of precision medicine in targeting specific RTK [60]. However, due to compensatory signaling and secondary mutations that can lead to resistance, next-generation inhibitors that target allosteric regions of RTKs are being developed to overcome these challenges [61]. Monoclonal antibodies (mAbs) have also emerged as a powerful strategy in cancer treatment. For instance, trastuzumab, targeting HER2 in HER2-positive breast cancer, prevents ligand binding and triggers immune-mediated cancer cell death [62]. Advancements in antibody engineering have led to the development of bispecific antibodies and antibody-drug conjugates (ADCs). ADCs, such as trastuzumab emtansine, deliver toxic agents directly to cancer cells, minimizing systemic toxicity while maximizing therapeutic efficacy [63]. These targeted therapies highlight the growing potential of RTK inhibition in improving cancer treatment outcomes. Table 1 summarizes the specific RTK inhibition strategies employed in various cancers.

4.2 *Combination Therapies Involving RTK Inhibitors*

Single-agent therapies targeting RTKs often face challenges like resistance and off-target toxicities. Combination therapies have emerged as a promising strategy to enhance treatment efficacy by targeting multiple pathways simultaneously [95]. One approach combines RTK inhibitors (RTKIs) with DNA damage response (DDR) modulators to disrupt cancer cell DNA repair mechanisms, increasing

Table 1 Examples of RTKIs, their target receptors, and associated cancer types

RTKI	Primary target(s)	Cancer type	Reference(s)
Erlotinib	EGFR	Non-small cell lung cancer (NSCLC)	[64]
Gefitinib	EGFR	NSCLC	[65]
Afatinib	EGFR, HER2	NSCLC, metastatic breast cancer	[66–68]
Osimertinib	EGFR (T790M mutation)	NSCLC	[69]
Lapatinib	HER2, EGFR	HER2+ breast cancer	[70]
Dabrafenib	BRAF	Melanoma, NSCLC	[71, 72]
Vemurafenib	BRAF	Melanoma	[73]
Sorafenib	VEGFR, PDGFR, RAF kinase	Renal cell carcinoma (RCC), hepatocellular carcinoma (HCC)	[74, 75]
Sunitinib	VEGFR, PDGFR, KIT	RCC, gastrointestinal stromal tumors (GIST)	[76, 77]
Pazopanib	VEGFR, PDGFR	RCC	[78]
Axitinib	VEGFR	RCC	[79]
Tivantinib	c-MET	HCC	[80–82]
Larotrectinib	NTRK	Solid tumors with NTRK fusions	[83]
Crizotinib	ALK, ROS1, MET	NSCLC	[84, 85]
Selpercatinib	RET	Thyroid cancer, NSCLC	[86]
Vandetanib	VEGFR, EGFR, RET	Medullary thyroid carcinoma	[87]
Imatinib	BCR-ABL, KIT, PDGFR	Chronic myeloid leukemia, GIST	[88–90]
Midostaurin	FLT3, KIT, PDGFR	Acute myeloid leukemia	[91]
Cabozantinib	MET, VEGFR	RCC, thyroid cancer	[92, 93]
Neratinib	HER2	HER2+ breast cancer	[94]

sensitivity to genotoxic therapies. For example, RTKIs with PARP inhibitor combinations interfere with DNA repair pathways, improving therapeutic responses. Similarly, EGFR inhibitors enhance the efficacy of platinum-based chemotherapies by preventing the repair of DNA crosslinks, leading to increased cancer cell death [96]. Integrating RTKIs with immune checkpoint inhibitors represents another transformative strategy. Tumors often exploit RTK-driven oncogenic signaling to suppress immune responses, promoting immune evasion [97]. By combining RTKIs with checkpoint inhibitors like anti-PD-1 antibodies, anti-tumor immunity is enhanced, improving response rates and overcoming resistance [98–100]. Targeting angiogenesis through RTK inhibition is also a crucial therapeutic avenue. VEGFR-targeting agents like bevacizumab, when combined with EGFR inhibitors, suppress tumor angiogenesis and improve therapeutic outcomes [100]. VEGF inhibitors not only disrupt tumor vasculature but also modulate the tumor microenvironment (TME), enhancing immune cell infiltration and reversing immunosuppression. The combination of anti-VEGF agents with PD-1/PD-L1 inhibitors has demonstrated promising clinical benefits in metastatic cancers such as renal cell carcinoma (RCC) and hepatocellular carcinoma (HCC). For example, the combination of atezolizumab (anti-PD-L1) and bevacizumab (anti-VEGF) demonstrated superior survival benefits in HCC patients compared to sorafenib, establishing a new standard of care [101, 102]. These combination strategies highlight the importance of multi-targeted approaches in overcoming therapy resistance and improving cancer treatment outcomes.

4.3 Natural Compounds as RTK Inhibitors

Natural compounds have gained attention as potential RTK inhibitors due to their bioactivity, multi-target interactions, and lower toxicity profiles. Flavonoids such as quercetin, genistein, and fisetin have demonstrated inhibitory effects on EGFR and VEGFR, disrupting angiogenesis and tumor growth [103, 104]. Alkaloids like berberine and vinca alkaloids target c-MET and AXL signaling, reducing metastatic potential and reversing drug resistance [105, 106].

Moreover, natural compounds often have lower toxicity profiles and can enhance the efficacy of conventional therapies [107, 108]. Curcumin, a well-studied natural compound, has shown potential in inhibiting tumor cell proliferation in various cancer models. It enhances the therapeutic effects of temozolomide in glioblastoma by targeting NF-κB and STAT3 pathways [109]. The integration of natural compounds into combination regimens offers a broader therapeutic window and improves patient outcomes. For example, COX-2 activation is associated with Akt pathway activation and poor survival in ER-negative, HER2-positive breast cancer, suggesting that COX-2 inhibitors, potentially combined with Akt inhibitors, could improve treatment outcomes. Similar results are also seen when curcumin suppresses OPN-induced VEGF expression and VEGF-dependent tumor angiogenesis in breast cancer, showing its potential as an anti-angiogenic agent [106, 110–112]. These findings emphasize the growing interest in natural compounds as promising adjuncts in RTK-targeted therapies.

5 Molecular Mechanisms of Therapeutic Resistance

Tyrosine kinase inhibitors (TKIs), including small-molecule inhibitors, have revolutionized cancer treatment by targeting and inhibiting tumor-promoting signaling pathways. However, the emergence of resistance, like acquired resistance, poses a significant challenge to the long-term efficacy of TKIs. Understanding the molecular mechanisms of resistance is essential for developing next-generation therapies to improve patient outcomes [113].

TKI therapeutic resistance can be categorized into intrinsic/primary and acquired resistance [114]. Intrinsic resistance is mediated by pre-existing tumor factors that render the therapy ineffective, leading to minimal or no response. On the other hand, acquired resistance develops after an initial therapeutic response, driven by adaptive alterations within the tumor that allow it to evade treatment and progress [115]. The changes can be genetic, epigenetic, or phenotypic.

The genetic alterations contributing to acquired resistance to TKIs are secondary mutations, gene amplifications, and genomic deletions. Secondary mutations, such as c-KIT, can reduce the binding affinity of TKIs, diminishing their efficacy [116]. Gene amplifications are classified by the overexpression of oncogenic proteins, compensating for inhibited pathways and enabling tumor survival despite treatment [117, 118]. Lastly, genomic deletions in miRNA-encoding regions lead to RTK overexpression, facilitating resistance by activating compensatory signaling pathways such as epithelial-to-mesenchymal transitions and drug resistance [119, 120]. The schematic for the mechanism of TKI resistance is shown in Fig. 6.

Epigenetic changes, such as DNA methylation, histone modifications, and alterations in non-coding RNAs, like miRNAs, can regulate gene expression without changing the genetic sequence [121]. Epigenetic mechanisms facilitate phenotype switching and drug tolerance. Hypoxia-induced upregulation of HIF-1α and epigenetic silencing of tumor suppressor genes like PTEN contribute to TKI resistance [122]. Phenotypic changes include cellular plasticity and epithelial-mesenchymal transition (EMT) methods. Cellular plasticity allows tumor cells to dynamically adapt to environmental or therapeutic stress, enabling cells to transition between proliferative and drug-tolerant states. EMTs facilitate resistance by enabling tumor cells to evade therapeutic targets and survive treatment [123]. This underscores the complexity of resistance and the need for approaches that target dynamic tumor states.

There have been many intrinsic methods of resistance that have proven to be effective. Some examples include overexpression of efflux transporters like

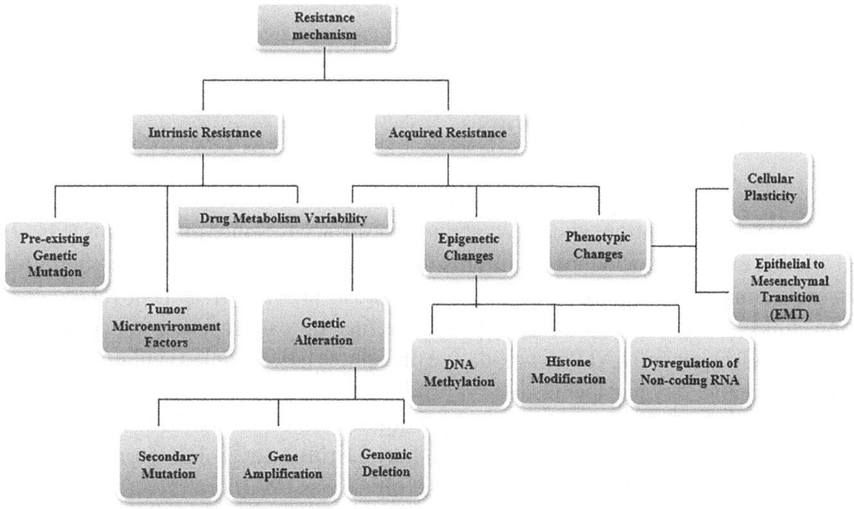

Fig. 6 The role of intrinsic and acquired resistance pathways of RTKs in cancer progression. Common mechanisms involve secondary mutations, gene amplification, and genomic deletions

P-glycoproteins (P-gp) and variations in drug metabolism pathways [123]. Utilizing these mechanisms presents a promising strategy to overcome resistance by reducing intracellular drug concentrations and altering bioavailability.

Overcoming resistance through RTK-targeted therapies mainly involves bypass signaling via alternative RTKs. Secondary signaling pathways, such as c-MET or AXL, can be activated by tumor cells to evade the inhibitory effects of TKIs [124]. Treatment outcomes and overcoming resistance can be improved by targeting these bypass pathways in combination with primary RTK inhibitors.

The tumor microenvironment (TME) influences tumor progression and therapy resistance. Mesenchymal and hematopoietic cells are components of the TME and support tumor growth and metastasis. TKIs can influence the TME by normalizing its dysregulated elements, enhancing their therapeutic impact [125]. Also, combinations of TKIs with chemotherapy or immunotherapy offer synergistic treatment benefits and show promise in targeting both cancer cells and the supportive TME.

Immune resistance is another challenge, with tumors evading detection through immunoediting and upregulation of immune checkpoint molecules like PD-L1. This resistance complicates the efficacy of immune-based therapies, including checkpoint inhibitors [126]. Understanding and counteracting these mechanisms is key to advancing immunotherapy and enhancing patient outcomes.

Emerging approaches, such as combination therapies integrating TKIs with chemotherapy, radiation, immune checkpoint inhibitors, and vaccines, aim to overcome resistance and achieve durable responses. Genetic testing and patient-derived models uncover novel resistance mechanisms and enable the development of next-generation TKIs with improved selectivity and efficacy [126]. These advancements drive personalized medicine and set the stage for transformative cancer care.

6 The Role of RTK Signaling in Mediating Mitochondrial Functions

Receptor tyrosine kinases (RTKs) are primarily known for regulating cellular proliferation, differentiation, and survival through surface receptor interactions (Fig. 7). However, emerging evidence highlights their role within mitochondria, where they influence metabolism, bioenergetics, and apoptosis [127, 128]. This novel localization links RTKs to metabolic reprogramming in cancer and other diseases. Mitochondria generate ATP via oxidative phosphorylation while regulating apoptosis and reactive oxygen species (ROS) production. RTKs such as ErbB2 and EGFR localize to mitochondria, where they modulate mitochondrial bioenergetics and ROS balance [129]. For example, mitochondrial ErbB2 (mtErbB2) promotes a shift from oxidative phosphorylation to aerobic glycolysis (Warburg effect), supporting cancer cell survival in hypoxic conditions and contributing to tumor progression and therapy resistance [129–131].

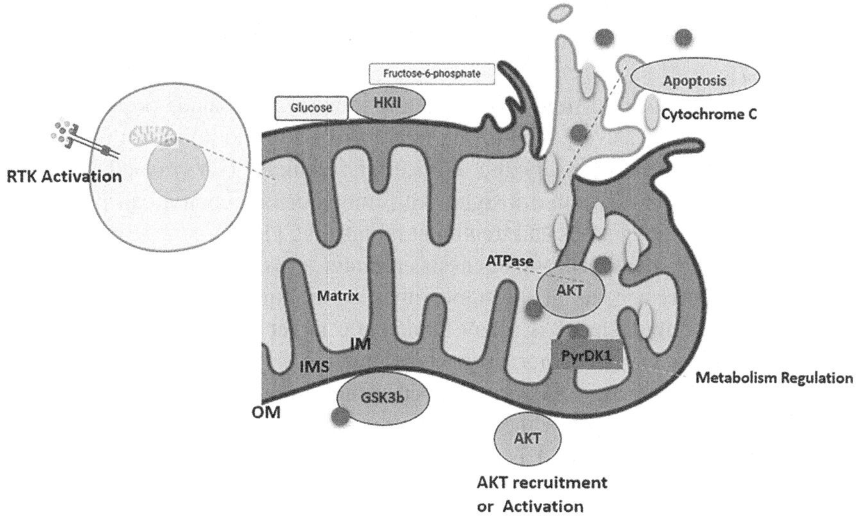

Fig. 7 The effects of RTK activation on mitochondrial function, metabolism, and apoptosis

Beyond cancer, mitochondrial RTKs are implicated in energy metabolism dysregulation and oxidative stress in neurodegenerative diseases and cardiovascular disorders. Src kinase family members are attributed to play a crucial role in mitochondrial tyrosine phosphorylation [132]. Various members of this family, such as Fgr, Fyn, Lyn, and c-Src, are present in the internal structure of mitochondria, as well as Csk, an important enzyme in the regulation of the activity of this family [127, 128, 133, 134]. It has been seen that mitochondrial Src kinase and tyrosine phosphorylation significantly influence mitochondrial respiratory complexes, connecting RTK activity to mitochondrial health.

Targeting mitochondrial RTKs provides a valuable opportunity for therapeutic use. Small-molecule RTK inhibitors that penetrate mitochondria can disrupt metabolic reprogramming and restore apoptotic sensitivity in cancer cells. Combination therapies targeting RTKs and mitochondrial bioenergetics are being explored to enhance therapeutic efficacy [38, 132].

In summary, mitochondrial RTKs are critical regulators of cellular metabolism, apoptosis, and stress responses. Understanding their roles in mitochondrial function offers insights into innovative therapeutic strategies for diseases driven by metabolic dysfunction.

7 Future Directions

RTK-associated biomarkers, including EGFR mutations and HER2 amplification, are essential for patient classification in precision oncology [135]. Proteomics and CRISPR and single-cell RNA sequencing are identifying novel biomarkers that can help direct individualized treatment regimens and forecast therapy responses [136].

Research is also being done on combining RTK inhibitors with immunotherapies and chemotherapies to counteract resistance mechanisms and improve outcomes [137].

In cancer patients, real-time RTK signaling monitoring is made possible by liquid biopsies. Non-invasively monitoring tumor evolution and treatment response is possible for clinicians by analyzing extracellular vehicles (EVs) or ctDNA. This method, which offers dynamic insights into tumor heterogeneity, can potentially transform cancer diagnostics and treatment monitoring [138].

RTK signaling regulates angiogenesis, immune evasion, and stromal interactions, among other aspects of the tumor microenvironment (TME). For instance, VEGFR-driven angiogenesis is a key therapeutic target in cancers like renal cell carcinoma and promotes tumor growth and metastasis [139]. RTK interactions within these intricate networks are mapped using systems biology and computational methods to identify new therapeutic targets [140].

CRISPR-Cas9 has emerged as a valuable tool for understanding RTK activities and discovering vulnerabilities in cancer cells. Functional genomics screens employing CRISPR have revealed new dependencies in RTK-driven malignancies, paving the path for novel treatment approaches [136, 138]. Challenges, including tumor heterogeneity and off-target effects, persist, but continual developments in CRISPR technology are helping to overcome these concerns [141].

8 Conclusion

RTKs play a crucial role in cellular communication and regulation by activating key signaling pathways such as MAPK, PI3K, and JAK/STAT, essential for cell growth and survival. RTK dysregulation, especially in the EGFR, INS, VEGFR, PDGFR, and AXL families, promotes unchecked growth, avoidance of regulatory feedback systems, and increased signaling because of mutation or overexpression. These occurrences greatly accelerate the development of cancer. TKs are essential regulators of cancer progression, with their dysregulation contributing to tumor growth, metastasis, and therapeutic resistance. Though TKIs have transformed cancer treatment, overcoming acquired and intrinsic resistance remains a significant challenge, calling for developing next-generation therapies and combination strategies to improve patient outcomes. Advances in RTK-focused therapies, including inhibitors and combination strategies with immune checkpoint inhibitors, demonstrate their potential to overcome resistance mechanisms and improve patient outcomes in cancer treatment. These advances have paved the way for innovative approaches in precision oncology, offering promising avenues for overcoming resistance and improving patient outcomes.

Acknowledgments The figures are created with Biorender.com.

References

1. Cohen S. Isolation of a mouse submaxillary gland protein accelerating incisor eruption and eyelid opening in the new-born animal. J Biol Chem. 1962;237:1555–62.
2. Levi-Montalcini R, Booker B. Excessive growth of the sympathetic ganglia evoked by a protein isolated from mouse salivary glands. Proc Natl Acad Sci USA. 1960;46(3):373–84.
3. Manning G, Whyte DB, Martinez R, Hunter T, Sudarsanam S. The protein kinase complement of the human genome. Science. 2002;298(5600):1912–34.
4. Salokas K, Liu X, Ohman T, Chowdhury I, Gawriyski L, Keskitalo S, et al. Physical and functional interactome atlas of human receptor tyrosine kinases. EMBO Rep. 2022;23(6):e54041.
5. Kumar R, Goel H, Solanki R, Rawat L, Tabasum S, Tanwar P, et al. Recent developments in receptor tyrosine kinase inhibitors: a promising mainstay in targeted cancer therapy. Med Drug Discovery. 2024;23:100195.
6. Hubbard JB. The role of conformational energetic disorder in the catalytic activity of immobilized enzymes. Biophys Chem. 1999;81(2):81–92.
7. Lemmon MA, Schlessinger J. Cell signaling by receptor tyrosine kinases. Cell. 2010;141(7):1117–34.
8. Rozengurt E, Soares HP, Sinnet-Smith J. Suppression of feedback loops mediated by PI3K/mTOR induces multiple overactivation of compensatory pathways: an unintended consequence leading to drug resistance. Mol Cancer Ther. 2014;13(11):2477–88.
9. Arias-Romero LE, Saha S, Villamar-Cruz O, Yip SC, Ethier SP, Zhang ZY, et al. Activation of Src by protein tyrosine phosphatase 1B is required for ErbB2 transformation of human breast epithelial cells. Cancer Res. 2009;69(11):4582–8.
10. Arteaga CL. ERBB receptors in cancer: signaling from the inside. Breast Cancer Res. 2011;13(2):304.
11. Tomuleasa C, Tigu AB, Munteanu R, Moldovan CS, Kegyes D, Onaciu A, et al. Therapeutic advances of targeting receptor tyrosine kinases in cancer. Signal Transduct Target Ther. 2024;9(1):201.
12. Zhang N, Li Y. Receptor tyrosine kinases: biological functions and anticancer targeted therapy. MedComm (2020). 2023;4(6):e446.
13. Lopez-Gines C, Gil-Benso R, Ferrer-Luna R, Benito R, Serna E, Gonzalez-Darder J, et al. New pattern of EGFR amplification in glioblastoma and the relationship of gene copy number with gene expression profile. Mod Pathol. 2010;23(6):856–65.
14. Selvaggi G, Novello S, Torri V, Leonardo E, De Giuli P, Borasio P, et al. Epidermal growth factor receptor overexpression correlates with a poor prognosis in completely resected non-small-cell lung cancer. Ann Oncol. 2004;15(1):28–32.
15. Hanawa M, Suzuki S, Dobashi Y, Yamane T, Kono K, Enomoto N, et al. EGFR protein overexpression and gene amplification in squamous cell carcinomas of the esophagus. Int J Cancer. 2006;118(5):1173–80.
16. Hirsch FR, Varella-Garcia M, Cappuzzo F. Predictive value of EGFR and HER2 overexpression in advanced non-small-cell lung cancer. Oncogene. 2009;28(Suppl 1):S32–7.
17. Yaziji H, Goldstein LC, Barry TS, Werling R, Hwang H, Ellis GK, et al. HER-2 testing in breast cancer using parallel tissue-based methods. JAMA. 2004;291(16):1972–7.
18. Kim KC, Koh YW, Chang HM, Kim TH, Yook JH, Kim BS, et al. Evaluation of HER2 protein expression in gastric carcinomas: comparative analysis of 1,414 cases of whole-tissue sections and 595 cases of tissue microarrays. Ann Surg Oncol. 2011;18(10):2833–40.
19. Park DI, Yun JW, Park JH, Oh SJ, Kim HJ, Cho YK, et al. HER-2/neu amplification is an independent prognostic factor in gastric cancer. Dig Dis Sci. 2006;51(8):1371–9.
20. Marrocco I, Giri S, Simoni-Nieves A, Gupta N, Rudnitsky A, Haga Y, et al. L858R emerges as a potential biomarker predicting response of lung cancer models to anti-EGFR antibodies: comparison of osimertinib vs. cetuximab. Cell Rep Med. 2023;4(8):101142.
21. Ke H, Kazi JU, Zhao H, Sun J. Germline mutations of KIT in gastrointestinal stromal tumor (GIST) and mastocytosis. Cell Biosci. 2016;6:55.

22. Socinski MA, Pennell NA, Davies KD. MET exon 14 skipping mutations in non-small-cell lung cancer: an overview of biology, clinical outcomes, and testing considerations. JCO Precis Oncol. 2021;5:PO.20.00516.
23. Jeske YW, Ali S, Byron SA, Gao F, Mannel RS, Ghebre RG, et al. FGFR2 mutations are associated with poor outcomes in endometrioid endometrial cancer: an NRG oncology/gynecologic oncology group study. Gynecol Oncol. 2017;145(2):366–73.
24. Camidge DR, Dziadziuszko R, Peters S, Mok T, Noe J, Nowicka M, et al. Updated efficacy and safety data and impact of the EML4-ALK fusion variant on the efficacy of Alectinib in untreated ALK-positive advanced non-small cell lung cancer in the global phase III ALEX study. J Thorac Oncol. 2019;14(7):1233–43.
25. Ladanyi M, Cavalchire G, Morris SW, Downing J, Filippa DA. Reverse transcriptase polymerase chain reaction for the Ki-1 anaplastic large cell lymphoma-associated t(2;5) translocation in Hodgkin's disease. Am J Pathol. 1994;145(6):1296–300.
26. Andraos E, Dignac J, Meggetto F. NPM-ALK: a driver of lymphoma pathogenesis and a therapeutic target. Cancers (Basel). 2021;13(1):144.
27. Zehir A, Benayed R, Shah RH, Syed A, Middha S, Kim HR, et al. Mutational landscape of metastatic cancer revealed from prospective clinical sequencing of 10,000 patients. Nat Med. 2017;23(6):703–13.
28. Landa I, Ibrahimpasic T, Boucai L, Sinha R, Knauf JA, Shah RH, et al. Genomic and transcriptomic hallmarks of poorly differentiated and anaplastic thyroid cancers. J Clin Invest. 2016;126(3):1052–66.
29. Walsh JH, Karnes WE, Cuttitta F, Walker A. Autocrine growth factors and solid tumor malignancy. West J Med. 1991;155(2):152–63.
30. Tang P, Steck PA, Yung WK. The autocrine loop of TGF-alpha/EGFR and brain tumors. J Neuro-Oncol. 1997;35(3):303–14.
31. Ciardiello F, Tortora G. A novel approach in the treatment of cancer: targeting the epidermal growth factor receptor. Clin Cancer Res. 2001;7(10):2958–70.
32. Maiti GP, Mondal P, Mukherjee N, Ghosh A, Ghosh S, Dey S, et al. Overexpression of EGFR in head and neck squamous cell carcinoma is associated with inactivation of SH3GL2 and CDC25A genes. PLoS One. 2013;8(5):e63440.
33. Gross I, Armant O, Benosman S, de Aguilar JL, Freund JN, Kedinger M, et al. Sprouty2 inhibits BDNF-induced signaling and modulates neuronal differentiation and survival. Cell Death Differ. 2007;14(10):1802–12.
34. Mason JM, Morrison DJ, Bassit B, Dimri M, Band H, Licht JD, et al. Tyrosine phosphorylation of Sprouty proteins regulates their ability to inhibit growth factor signaling: a dual feedback loop. Mol Biol Cell. 2004;15(5):2176–88.
35. Normanno N, De Luca A, Bianco C, Strizzi L, Mancino M, Maiello MR, et al. Epidermal growth factor receptor (EGFR) signaling in cancer. Gene. 2006;366(1):2–16.
36. Voldborg BR, Damstrup L, Spang-Thomsen M, Poulsen HS. Epidermal growth factor receptor (EGFR) and EGFR mutations, function and possible role in clinical trials. Ann Oncol. 1997;8(12):1197–206.
37. Sabbah DA, Hajjo R, Sweidan K. Review on epidermal growth factor receptor (EGFR) structure, signaling pathways, interactions, and recent updates of EGFR inhibitors. Curr Top Med Chem. 2020;20(10):815–34.
38. Hamid O, Robert C, Daud A, Hodi FS, Hwu WJ, Kefford R, et al. Safety and tumor responses with lambrolizumab (anti-PD-1) in melanoma. N Engl J Med. 2013;369(2):134–44.
39. Slack-Davis JK, Eblen ST, Zecevic M, Boerner SA, Tarcsafalvi A, Diaz HB, et al. PAK1 phosphorylation of MEK1 regulates fibronectin-stimulated MAPK activation. J Cell Biol. 2003;162(2):281–91.
40. Dalby KN, Morrice N, Caudwell FB, Avruch J, Cohen P. Identification of regulatory phosphorylation sites in mitogen-activated protein kinase (MAPK)-activated protein kinase-1a/p90rsk that are inducible by MAPK. J Biol Chem. 1998;273(3):1496–505.
41. Pinet L, Wang YH, Deville C, Lescop E, Guerlesquin F, Badache A, et al. Structural and dynamic characterization of the C-terminal tail of ErbB2: disordered but not random. Biophys J. 2021;120(10):1869–82.

42. Shaw RJ, Cantley LC. Ras, PI(3)K and mTOR signalling controls tumour cell growth. Nature. 2006;441(7092):424–30.
43. Sangwan V, Park M. Receptor tyrosine kinases: role in cancer progression. Curr Oncol. 2006;13(5):191–3.
44. Veale D, Ashcroft T, Marsh C, Gibson GJ, Harris AL. Epidermal growth factor receptors in non-small cell lung cancer. Br J Cancer. 1987;55(5):513–6.
45. Maruyama IN. Mechanisms of activation of receptor tyrosine kinases: monomers or dimers. Cells. 2014;3(2):304–30.
46. Oakie A, Wang R. Beta-cell receptor tyrosine kinases in controlling insulin secretion and Exocytotic machinery: c-kit and insulin receptor. Endocrinology. 2018;159(11):3813–21.
47. Buck E, Gokhale PC, Koujak S, Brown E, Eyzaguirre A, Tao N, et al. Compensatory insulin receptor (IR) activation on inhibition of insulin-like growth factor-1 receptor (IGF-1R): rationale for cotargeting IGF-1R and IR in cancer. Mol Cancer Ther. 2010;9(10):2652–64.
48. Ferrara N, Gerber HP, LeCouter J. The biology of VEGF and its receptors. Nat Med. 2003;9(6):669–76.
49. Grunewald FS, Prota AE, Giese A, Ballmer-Hofer K. Structure-function analysis of VEGF receptor activation and the role of coreceptors in angiogenic signaling. Biochim Biophys Acta. 2010;1804(3):567–80.
50. Rogers MA, Fantauzzo KA. The emerging complexity of PDGFRs: activation, internalization and signal attenuation. Biochem Soc Trans. 2020;48(3):1167–76.
51. Wahlen E, Olsson F, Raykova D, Soderberg O, Heldin J, Lennartsson J. Activated EGFR and PDGFR internalize in separate vesicles and downstream AKT and ERK1/2 signaling are differentially impacted by cholesterol depletion. Biochem Biophys Res Commun. 2023;665:195–201.
52. Westermark B. Platelet-derived growth factor in glioblastoma-driver or biomarker? Ups J Med Sci. 2014;119(4):298–305.
53. Brahmi M, Lesluyes T, Dufresne A, Toulmonde M, Italiano A, Mir O, et al. Expression and prognostic significance of PDGF ligands and receptors across soft tissue sarcomas. ESMO Open. 2021;6(1):100037.
54. De Coninck S, De Smedt R, Lintermans B, Goossens S, Van Vlierberghe P. PDGFR-β as a new therapeutic target in T-cell acute lymphoblastic leukemia. Blood. 2022;140(Supplement 1):11513–4.
55. Tang Y, Zang H, Wen Q, Fan S. AXL in cancer: a modulator of drug resistance and therapeutic target. J Exp Clin Cancer Res. 2023;42(1):148.
56. Yadav M, Sharma A, Patne K, Tabasum S, Suryavanshi J, Rawat L, et al. AXL signaling in cancer: from molecular insights to targeted therapies. Signal Transduct Target Ther. 2025;10(1):37.
57. Tanaka M, Siemann DW. Therapeutic targeting of the Gas6/Axl signaling pathway in cancer. Int J Mol Sci. 2021;22(18):9953.
58. Aplin AE. Axl of evil? J Invest Dermatol. 2011;131(12):2343–5.
59. AlDoughaim M, AlSuhebany N, AlZahrani M, AlQahtani T, AlGhamdi S, Badreldin H, et al. Cancer biomarkers and precision oncology: a review of recent trends and innovations. Clin Med Insights Oncol. 2024;18:11795549241298541.
60. Dutta PR, Maity A. Cellular responses to EGFR inhibitors and their relevance to cancer therapy. Cancer Lett. 2007;254(2):165–77.
61. Druker BJ, Talpaz M, Resta DJ, Peng B, Buchdunger E, Ford JM, et al. Efficacy and safety of a specific inhibitor of the BCR-ABL tyrosine kinase in chronic myeloid leukemia. N Engl J Med. 2001;344(14):1031–7.
62. Molina MA, Codony-Servat J, Albanell J, Rojo F, Arribas J, Baselga J. Trastuzumab (herceptin), a humanized anti-Her2 receptor monoclonal antibody, inhibits basal and activated Her2 ectodomain cleavage in breast cancer cells. Cancer Res. 2001;61(12):4744–9.
63. Rassy E, Rached L, Pistilli B. Antibody drug conjugates targeting HER2: clinical development in metastatic breast cancer. Breast. 2022;66:217–26.

64. Piperdi B, Perez-Soler R. Role of erlotinib in the treatment of non-small cell lung cancer: clinical outcomes in wild-type epidermal growth factor receptor patients. Drugs. 2012;72(Suppl 0 (0 1)):11–9.
65. Hosomi Y, Morita S, Sugawara S, Kato T, Fukuhara T, Gemma A, et al. Gefitinib alone versus Gefitinib plus chemotherapy for non-small-cell lung cancer with mutated epidermal growth factor receptor: NEJ009 study. J Clin Oncol. 2020;38(2):115–23.
66. Yang JC, Schuler M, Popat S, Miura S, Park K, Passaro A, et al. Afatinib for the treatment of non-small cell lung cancer harboring uncommon EGFR mutations: an updated database of 1023 cases brief report. Front Oncol. 2022;12:834704.
67. Liang SK, Ko JC, Yang JC, Shih JY. Afatinib is effective in the treatment of lung adenocarcinoma with uncommon EGFR p.L747P and p.L747S mutations. Lung Cancer. 2019;133:103–9.
68. Zöchbauer-Müller S, Kaserer B, Prosch H, Cseh A, Solca F, Bauer MJ, et al. Case report: Afatinib treatment in a patient with NSCLC harboring a rare EGFR exon 20 mutation. Front Oncol. 2020;10:593852.
69. Soria JC, Ohe Y, Vansteenkiste J, Reungwetwattana T, Chewaskulyong B, Lee KH, et al. Osimertinib in untreated EGFR-mutated advanced non-small-cell lung cancer. N Engl J Med. 2018;378(2):113–25.
70. Wang X, Wang L, Yu Q, Liu Z, Li C, Wang F, et al. The effectiveness of Lapatinib in HER2-positive metastatic breast cancer patients pretreated with multiline anti-HER2 treatment: a retrospective study in China. Technol Cancer Res Treat. 2021;20:15330338211037812.
71. Planchard D, Smit EF, Groen HJM, Mazieres J, Besse B, Helland Å, et al. Dabrafenib plus trametinib in patients with previously untreated BRAF(V600E)-mutant metastatic non-small-cell lung cancer: an open-label, phase 2 trial. Lancet Oncol. 2017;18(10):1307–16.
72. Atkins MB, Lee SJ, Chmielowski B, Tarhini AA, Cohen GI, Truong TG, et al. Combination Dabrafenib and Trametinib versus combination Nivolumab and Ipilimumab for patients with advanced BRAF-mutant melanoma: the DREAMseq trial-ECOG-ACRIN EA6134. J Clin Oncol. 2023;41(2):186–97.
73. Chapman PB, Hauschild A, Robert C, Haanen JB, Ascierto P, Larkin J, et al. Improved survival with Vemurafenib in melanoma with BRAF V600E mutation. N Engl J Med. 2011;364(26):2507–16.
74. Strumberg D, Richly H, Hilger RA, Schleucher N, Korfee S, Tewes M, et al. Phase I clinical and pharmacokinetic study of the Novel Raf kinase and vascular endothelial growth factor receptor inhibitor BAY 43-9006 in patients with advanced refractory solid tumors. J Clin Oncol. 2005;23(5):965–72.
75. Wilhelm SM, Carter C, Tang L, Wilkie D, McNabola A, Rong H, et al. BAY 43-9006 exhibits broad spectrum oral antitumor activity and targets the RAF/MEK/ERK pathway and receptor tyrosine kinases involved in tumor progression and angiogenesis. Cancer Res. 2004;64(19):7099–109.
76. Schmid TA, Gore ME. Sunitinib in the treatment of metastatic renal cell carcinoma. Ther Adv Urol. 2016;8(6):348–71.
77. Mulet-Margalef N, Garcia-Del-Muro X. Sunitinib in the treatment of gastrointestinal stromal tumor: patient selection and perspectives. Onco Targets Ther. 2016;9:7573–82.
78. Sternberg CN, Davis ID, Mardiak J, Szczylik C, Lee E, Wagstaff J, et al. Pazopanib in locally advanced or metastatic renal cell carcinoma: results of a randomized phase III trial. J Clin Oncol. 2010;28(6):1061–8.
79. Plimack ER, Powles T, Stus V, Gafanov R, Nosov D, Waddell T, et al. Pembrolizumab plus Axitinib versus Sunitinib as first-line treatment of advanced renal cell carcinoma: 43-month follow-up of the phase 3 KEYNOTE-426 study. Eur Urol. 2023;84(5):449–54.
80. Luo GL, Guo BQ, Wu LX, Shen YX, Xie T. Application of Tivantinib for hepatocellular carcinoma: a meta-analysis study. Evid Based Complement Alternat Med. 2022;2022:1976788.
81. Rimassa L, Assenat E, Peck-Radosavljevic M, Pracht M, Zagonel V, Mathurin P, et al. Tivantinib for second-line treatment of MET-high, advanced hepatocellular carcinoma

(METIV-HCC): a final analysis of a phase 3, randomised, placebo-controlled study. Lancet Oncol. 2018;19(5):682–93.
82. Kudo M, Morimoto M, Moriguchi M, Izumi N, Takayama T, Yoshiji H, et al. A randomized, double-blind, placebo-controlled, phase 3 study of tivantinib in Japanese patients with MET-high hepatocellular carcinoma. Cancer Sci. 2020;111(10):3759–69.
83. Drilon A, Laetsch TW, Kummar S, DuBois SG, Lassen UN, Demetri GD, et al. Efficacy of Larotrectinib in TRK fusion-positive cancers in adults and children. N Engl J Med. 2018;378(8):731–9.
84. Malik SM, Maher VE, Bijwaard KE, Becker RL, Zhang L, Tang SW, et al. U.S. Food and Drug Administration approval: crizotinib for treatment of advanced or metastatic non-small cell lung cancer that is anaplastic lymphoma kinase positive. Clin Cancer Res. 2014;20(8):2029–34.
85. Shaw AT, Ou SH, Bang YJ, Camidge DR, Solomon BJ, Salgia R, et al. Crizotinib in ROS1-rearranged non-small-cell lung cancer. N Engl J Med. 2014;371(21):1963–71.
86. Zhou C, Solomon B, Loong HH, Park K, Pérol M, Arriola E, et al. First-line Selpercatinib or chemotherapy and Pembrolizumab in RET fusion-positive NSCLC. N Engl J Med. 2023;389(20):1839–50.
87. Kitamura S, Maeda T, Yanagi T. Vandetanib inhibits cell growth in EGFR-expressing cutaneous squamous cell carcinoma. Biochem Biophys Res Commun. 2020;531(3):396–401.
88. Ottmann OG, Druker BJ, Sawyers CL, Goldman JM, Reiffers J, Silver RT, et al. A phase 2 study of imatinib in patients with relapsed or refractory Philadelphia chromosome-positive acute lymphoid leukemias. Blood. 2002;100(6):1965–71.
89. Krishnan V, Schmidt F, Nawaz Z, Venkatesh PN, Lee KL, Ren X, et al. A single-cell atlas identifies pretreatment features of primary imatinib resistance in chronic myeloid leukemia. Blood. 2023;141(22):2738–55.
90. van Oosterom AT, Judson IR, Verweij J, Stroobants S, Dumez H, Donato di Paola E, et al. Update of phase I study of imatinib (STI571) in advanced soft tissue sarcomas and gastrointestinal stromal tumors: a report of the EORTC Soft Tissue and Bone Sarcoma Group. Eur J Cancer. 2002;38(Suppl 5):S83–7.
91. Voso MT, Larson RA, Jones D, Marcucci G, Prior T, Krauter J, et al. Midostaurin in patients with acute myeloid leukemia and FLT3-TKD mutations: a subanalysis from the RATIFY trial. Blood Adv. 2020;4(19):4945–54.
92. Choueiri TK, Hessel C, Halabi S, Sanford B, Michaelson MD, Hahn O, et al. Cabozantinib versus sunitinib as initial therapy for metastatic renal cell carcinoma of intermediate or poor risk (Alliance A031203 CABOSUN randomised trial): progression-free survival by independent review and overall survival update. Eur J Cancer. 2018;94:115–25.
93. Brose MS, Robinson B, Sherman SI, Krajewska J, Lin C-C, Vaisman F, et al. Cabozantinib for radioiodine-refractory differentiated thyroid cancer (COSMIC-311): a randomised, double-blind, placebo-controlled, phase 3 trial. Lancet Oncol. 2021;22(8):1126–38.
94. Chan A, Delaloge S, Holmes FA, Moy B, Iwata H, Harvey VJ, et al. Neratinib after trastuzumab-based adjuvant therapy in patients with HER2-positive breast cancer (ExteNET): a multicentre, randomised, double-blind, placebo-controlled, phase 3 trial. Lancet Oncol. 2016;17(3):367–77.
95. Pillay V, Allaf L, Wilding AL, Donoghue JF, Court NW, Greenall SA, et al. The plasticity of oncogene addiction: implications for targeted therapies directed to receptor tyrosine kinases. Neoplasia. 2009;11(5):448–58, 2 p following 58
96. Chakravarti A, Loeffler JS, Dyson NJ. Insulin-like growth factor receptor I mediates resistance to anti-epidermal growth factor receptor therapy in primary human glioblastoma cells through continued activation of phosphoinositide 3-kinase signaling. Cancer Res. 2002;62(1):200–7.
97. Curran MA, Montalvo W, Yagita H, Allison JP. PD-1 and CTLA-4 combination blockade expands infiltrating T cells and reduces regulatory T and myeloid cells within B16 melanoma tumors. Proc Natl Acad Sci USA. 2010;107(9):4275–80.

98. Watanabe H, Ichihara E, Kayatani H, Makimoto G, Ninomiya K, Nishii K, et al. VEGFR2 blockade augments the effects of tyrosine kinase inhibitors by inhibiting angiogenesis and oncogenic signaling in oncogene-driven non-small-cell lung cancers. Cancer Sci. 2021;112(5):1853–64.
99. Zhao D, Qin C, Fan X, Li Y, Gu B. Inhibitory effects of quercetin on angiogenesis in larval zebrafish and human umbilical vein endothelial cells. Eur J Pharmacol. 2014;723:360–7.
100. Li X, He X, Lin B, Li L, Deng Q, Wang C, et al. Quercetin limits tumor immune escape through PDK1/CD47 Axis in melanoma. Am J Chin Med. 2024;52(2):541–63.
101. Li G, Fang M, Zhou Y, Liu X, Tian P, Mei F. Afatinib overcoming resistance to icotinib and osimertinib in NSCLC with leptomeningeal metastasis in patients with acquired EGFR L858R/T790M or L858R/S768I mutations: two case reports. Heliyon. 2023;9(10):e20690.
102. Chen Y, Jia L, Li Y, Cui W, Wang J, Zhang C, et al. Efficacy and safety of PD-1 inhibitors plus anti-angiogenesis tyrosine kinase inhibitors with or without transarterial chemo(embolization) for unresectable hepatocellular carcinoma: a meta-analysis. Front Oncol. 2024;14:1364345.
103. Sabarwal A, Agarwal R, Singh RP. Fisetin inhibits cellular proliferation and induces mitochondria-dependent apoptosis in human gastric cancer cells. Mol Carcinog. 2017;56(2):499–514.
104. Sabarwal A, van Rooyen JC, Caburet J, Avgenikos M, Dheeraj A, Ali M, et al. A novel 4′-brominated derivative of fisetin induces cell cycle arrest and apoptosis and inhibits EGFR/ERK1/2/STAT3 pathways in non-small-cell lung cancer without any adverse effects in mice. FASEB J. 2022;36(12):e22654.
105. Goel H, Kumar R, Tanwar P, Upadhyay TK, Khan F, Pandey P, et al. Unraveling the therapeutic potential of natural products in the prevention and treatment of leukemia. Biomed Pharmacother. 2023;160:114351.
106. St-Germain ME, Gagnon V, Mathieu I, Parent S, Asselin E. Akt regulates COX-2 mRNA and protein expression in mutated-PTEN human endometrial cancer cells. Int J Oncol. 2004;24(5):1311–24.
107. Sabarwal A, Wedel J, Liu K, Zurakowski D, Chakraborty S, Flynn E, et al. A combination therapy using an mTOR inhibitor and Honokiol effectively induces autophagy through the modulation of AXL and Rubicon in renal cancer cells and restricts renal tumor growth following organ transplantation. Carcinogenesis. 2022;43(4):360–70.
108. Sabarwal A, Chakraborty S, Mahanta S, Banerjee S, Balan M, Pal S. A novel combination treatment with Honokiol and rapamycin effectively restricts c-met-induced growth of renal cancer cells, and also inhibits the expression of tumor cell PD-L1 involved in immune escape. Cancers (Basel). 2020;12(7):1782.
109. Sabarwal A, Kumar K, Shyanti R, Singh RP. Curcumin in cancer prevention. In: Rani V, Yadav UCS, editors. Functional food and human health. Singapore: Springer Singapore; 2018. p. 329–74.
110. Chakraborty G, Jain S, Kale S, Raja R, Kumar S, Mishra R, et al. Curcumin suppresses breast tumor angiogenesis by abrogating osteopontin-induced VEGF expression. Mol Med Rep. 2008;1(5):641–6.
111. Zhang J, Yu J, Xie R, Chen W, Lv Y. Combinatorial anticancer effects of curcumin and sorafenib towards thyroid cancer cells via PI3K/Akt and ERK pathways. Nat Prod Res. 2016;30(16):1858–61.
112. Cross DA, Ashton SE, Ghiorghiu S, Eberlein C, Nebhan CA, Spitzler PJ, et al. AZD9291, an irreversible EGFR TKI, overcomes T790M-mediated resistance to EGFR inhibitors in lung cancer. Cancer Discov. 2014;4(9):1046–61.
113. Druker BJ, Sawyers CL, Kantarjian H, Resta DJ, Reese SF, Ford JM, et al. Activity of a specific inhibitor of the BCR-ABL tyrosine kinase in the blast crisis of chronic myeloid leukemia and acute lymphoblastic leukemia with the Philadelphia chromosome. N Engl J Med. 2001;344(14):1038–42.
114. Lu Z, Jiang G, Blume-Jensen P, Hunter T. Epidermal growth factor-induced tumor cell invasion and metastasis initiated by dephosphorylation and downregulation of focal adhesion kinase. Mol Cell Biol. 2001;21(12):4016–31.

115. Gao Y, Chen MK, Chu YY, Yang L, Yu D, Liu Y, et al. Nuclear translocation of the receptor tyrosine kinase c-MET reduces the treatment efficacies of olaparib and gemcitabine in pancreatic ductal adenocarcinoma cells. Am J Cancer Res. 2021;11(1):236–50.
116. von Minckwitz G, Jonat W, Fasching P, du Bois A, Kleeberg U, Luck HJ, et al. A multicentre phase II study on gefitinib in taxane- and anthracycline-pretreated metastatic breast cancer. Breast Cancer Res Treat. 2005;89(2):165–72.
117. Esteller M, Silva JM, Dominguez G, Bonilla F, Matias-Guiu X, Lerma E, et al. Promoter hypermethylation and BRCA1 inactivation in sporadic breast and ovarian tumors. J Natl Cancer Inst. 2000;92(7):564–9.
118. Scheel C, Weinberg RA. Phenotypic plasticity and epithelial-mesenchymal transitions in cancer and normal stem cells? Int J Cancer. 2011;129(10):2310–4.
119. Lei ZN, Teng QX, Gupta P, Zhang W, Narayanan S, Yang DH, et al. Cabozantinib reverses Topotecan resistance in human non-small cell lung cancer NCI-H460/TPT10 cell line and tumor xenograft model. Front Cell Dev Biol. 2021;9:640957.
120. Li G, Wang S, Deng D, Xiao Z, Dong Z, Wang Z, et al. Fluorinated chitosan to enhance Transmucosal delivery of Sonosensitizer-conjugated catalase for Sonodynamic bladder cancer treatment post-intravesical instillation. ACS Nano. 2020;14(2):1586–99.
121. Mi S, Lu J, Sun M, Li Z, Zhang H, Neilly MB, et al. MicroRNA expression signatures accurately discriminate acute lymphoblastic leukemia from acute myeloid leukemia. Proc Natl Acad Sci USA. 2007;104(50):19971–6.
122. Khaliq AM, Rajamohan M, Saeed O, Mansouri K, Adil A, Zhang C, et al. Spatial transcriptomic analysis of primary and metastatic pancreatic cancers highlights tumor microenvironmental heterogeneity. Nat Genet. 2024;56(11):2455–65.
123. Memon D, Schoenfeld AJ, Ye D, Fromm G, Rizvi H, Zhang X, et al. Clinical and molecular features of acquired resistance to immunotherapy in non-small cell lung cancer. Cancer Cell. 2024;42(2):209–24.e9.
124. Tang N, Wang L, Esko J, Giordano FJ, Huang Y, Gerber H-P, et al. Loss of HIF-1α in endothelial cells disrupts a hypoxia-driven VEGF autocrine loop necessary for tumorigenesis. Cancer Cell. 2004;6(5):485–95.
125. Goyette MA, Duhamel S, Aubert L, Pelletier A, Savage P, Thibault MP, et al. The receptor tyrosine kinase AXL is required at multiple steps of the metastatic Cascade during HER2-positive breast cancer progression. Cell Rep. 2023;42(12):113604.
126. Rayford A, Gartner F, Ramnefjell MP, Lorens JB, Micklem DR, Aanerud M, et al. AXL expression reflects tumor-immune cell dynamics impacting outcome in non-small cell lung cancer patients treated with immune checkpoint inhibitor monotherapy. Front Immunol. 2024;15:1444007.
127. Baines CP, Zhang J, Wang GW, Zheng YT, Xiu JX, Cardwell EM, et al. Mitochondrial PKCepsilon and MAPK form signaling modules in the murine heart: enhanced mitochondrial PKCepsilon-MAPK interactions and differential MAPK activation in PKCepsilon-induced cardioprotection. Circ Res. 2002;90(4):390–7.
128. Arachiche A, Augereau O, Decossas M, Pertuiset C, Gontier E, Letellier T, et al. Localization of PTP-1B, SHP-2, and Src exclusively in rat brain mitochondria and functional consequences. J Biol Chem. 2008;283(36):24406–11.
129. Vaughn AE, Deshmukh M. Glucose metabolism inhibits apoptosis in neurons and cancer cells by redox inactivation of cytochrome c. Nat Cell Biol. 2008;10(12):1477–83.
130. Tan M, Li P, Klos KS, Lu J, Lan KH, Nagata Y, et al. ErbB2 promotes Src synthesis and stability: novel mechanisms of Src activation that confer breast cancer metastasis. Cancer Res. 2005;65(5):1858–67.
131. Jin Y, Cai Q, Shenoy AK, Lim S, Zhang Y, Charles S, et al. Src drives the Warburg effect and therapy resistance by inactivating pyruvate dehydrogenase through tyrosine-289 phosphorylation. Oncotarget. 2016;7(18):25113–24.
132. Martinez-Outschoorn UE, Pestell RG, Howell A, Tykocinski ML, Nagajyothi F, Machado FS, et al. Energy transfer in "parasitic" cancer metabolism: mitochondria are the powerhouse and Achilles' heel of tumor cells. Cell Cycle. 2011;10(24):4208–16.

133. Miedlich SU, Zalutskaya A, Zhu ED, Demay MB. Phosphate-induced apoptosis of hypertrophic chondrocytes is associated with a decrease in mitochondrial membrane potential and is dependent upon Erk1/2 phosphorylation. J Biol Chem. 2010;285(24):18270–5.
134. Bose S, Zhang C, Le A. Glucose metabolism in cancer: the Warburg effect and beyond. Adv Exp Med Biol. 2021;1311:3–15.
135. Cappuzzo F, Marchetti A, Skokan M, Rossi E, Gajapathy S, Felicioni L, et al. Increased MET gene copy number negatively affects survival of surgically resected non-small-cell lung cancer patients. J Clin Oncol. 2009;27(10):1667–74.
136. Ravichandran M, Maddalo D. Applications of CRISPR-Cas9 for advancing precision medicine in oncology: from target discovery to disease modeling. Front Genet. 2023;14:1273994.
137. Mok TS, Wu YL, Thongprasert S, Yang CH, Chu DT, Saijo N, et al. Gefitinib or carboplatin-paclitaxel in pulmonary adenocarcinoma. N Engl J Med. 2009;361(10):947–57.
138. Huang D, Miller M, Ashok B, Jain S, Peppas NA. CRISPR/Cas systems to overcome challenges in developing the next generation of T cells for cancer therapy. Adv Drug Deliv Rev. 2020;158:17–35.
139. Choueiri TK, Halabi S, Sanford BL, Hahn O, Michaelson MD, Walsh MK, et al. Cabozantinib versus Sunitinib as initial targeted therapy for patients with metastatic renal cell carcinoma of poor or intermediate risk: the Alliance A031203 CABOSUN trial. J Clin Oncol. 2017;35(6):591–7.
140. Sainsbury JR, Farndon JR, Needham GK, Malcolm AJ, Harris AL. Epidermal-growth-factor receptor status as predictor of early recurrence of and death from breast cancer. Lancet. 1987;1(8547):1398–402.
141. Carty M, Bowie AG. Recent insights into the role of toll-like receptors in viral infection. Clin Exp Immunol. 2010;161(3):397–406.

The ERBB Family of Receptor Tyrosine Kinases: Key Players in Cancer

Diana Drago-Garcia, Ilaria Marrocco, Donatella Romaniello, and Mary Luz Uribe

1 Introduction

The ERBB family of receptor tyrosine kinases (RTKs), comprising ERBB1 (EGFR), ERBB2 (HER2), ERBB3 (HER3), and ERBB4 (HER4), plays a crucial role in various cellular processes, including proliferation, differentiation, and survival [1]. Among these, the epidermal growth factor receptor (EGFR) stands out as one of the most extensively studied and clinically significant members of this family. Since its discovery in the 1980s [2], EGFR has been the subject of intense research due to its unique role in normal cellular function and pathological conditions, particularly in cancer development and progression [3].

The importance of EGFR in tumorigenesis has made it a prime target for cancer therapy, leading to the development of numerous targeted therapies. These include small-molecule tyrosine kinase inhibitors (TKIs) and monoclonal antibodies

D. Drago-Garcia
Stem Cell and Cancer Biology Laboratory, The Francis Crick Institute, London, UK
e-mail: diana.dragogarcia@crick.ac.uk

I. Marrocco
Department of Life Sciences and Public Health, Università Cattolica del Sacro Cuore, Rome, Italy
e-mail: ilaria.marrocco@unicatt.it

D. Romaniello
Department of Medical and Surgical Sciences, University of Bologna, Bologna, Italy

IRCCS Azienda Ospedaliero-Universitaria di Bologna, Bologna, Italy
e-mail: donatella.romaniello@unibo.it

M. L. Uribe (✉)
Instituto de Neurociencias, Consejo Superior de Investigaciones Científicas (CSIC), and Universidad Miguel Hernández (UMH), Sant Joan d'Alacant, Spain
e-mail: muribe@umh.es

A. Sabarwal et al. (eds.), *Receptor Tyrosine Kinases in Cancer*, Cancer Drug Discovery and Development, https://doi.org/10.1007/978-3-031-93894-8_2

(mAbs), which have revolutionized the treatment landscape for various malignancies, such as non-small cell lung cancer (NSCLC), colorectal cancer (CRC), and head and neck squamous cell carcinoma (HNSCC) [4]. However, the development of drug resistance mechanisms has created significant challenges for the long-term effectiveness of these therapies, highlighting the need for continuous research and the development of new therapeutic strategies.

This chapter will provide a comprehensive overview of the ERBB family, with a particular focus on EGFR. We will delve into the structural characteristics of ERBB receptors, their signaling pathways, and their role in the pathogenesis of different cancers. Furthermore, we will explore the mechanisms of drug resistance that have emerged in response to ERBB-targeted therapies and review the current landscape of approved small-molecule kinase inhibitors and those in clinical trials.

2 Structure and Function of the ERBB Family of Receptors

2.1 The ERBB Family

The evolution of the ERBB family of receptors is a fascinating example of gene duplication and functional diversification across species. Evolutionary studies have revealed that invertebrate model species typically possess a single EGFR homologue, such as the *Drosophila* Egf receptor (DER) and the *C. elegans* Let-23 receptor [5]. The vertebrate ERBB family expanded through whole genome duplication (WGD) events. The first round (1R) of WGD created two genes, further duplicated during the second round (2R), giving rise to the four ERBB genes found in most tetrapods [5, 6]. Interestingly, teleost fish have seven ERBB genes due to an additional fish-specific genome duplication (FSGD or 3R), with duplicates of *EGFR*, *HER3*, and *HER4*, while *HER2* remains single-copy [7]. This makes the ERBB family one of the most preserved RTK families after the FSGD. A particularly intriguing case is found in platy fishes and swordtails, where a tandem duplication of one fish *EGFR* gene produced a third version called *xmrk* (*Xiphophorus* melanoma receptor kinase) [8]. This gene acquired mutations that allow constitutive dimerization and ligand-independent signaling, a mechanism of oncogenic RTK activation also observed in some human cancers. Recent studies have provided new insights into the functional implications of ERBB family evolution. For instance, the loss of tyrosine kinase activity in HER3, previously thought to be mammal-specific, occurs in the fish homologues as well, suggesting an early evolutionary event that shaped the functional interactions within the family [9].

The ERBB family members exhibit distinct and overlapping tissue expression patterns, contributing to their diverse roles in development and adult physiology. EGFR is widely expressed in epithelial, mesenchymal, and neuronal tissues, playing crucial roles in developing and maintaining various organs [10]. It is particularly abundant in the skin, liver, and gastrointestinal tract. HER2, while expressed in many tissues, is notably present in the heart, nervous system, and breast tissue [11]. Its expression is often amplified in certain cancers, particularly breast cancer [12]. HER3 and HER4 show more restricted expression patterns compared to EGFR and HER2. HER3 is expressed in various tissues, including the nervous system, gastrointestinal tract, and mammary glands [13]. HER4 is predominantly expressed in the heart, central nervous system, and mammary glands [14]. The differential expression of these receptors allows for tissue-specific responses to ERBB ligands and contributes to the fine-tuning of cellular processes in different organs.

The ERBB receptors play critical roles in embryonic development, influencing various processes such as cell proliferation, differentiation, migration, and survival. Knockout studies in mice have demonstrated that all four ERBB receptors are essential for normal development, with each receptor contributing to specific developmental processes. EGFR is crucial for epithelial development and plays a vital role in the formation of the skin, lungs, and gastrointestinal tract [15, 16]. HER2 is essential for cardiac and nervous system development, with knockout mice showing severe defects in these tissues [17]. HER3 is important for the development of the peripheral nervous system and the formation of cardiac cushions [18, 19]. HER4 plays key roles in mammary gland development, central nervous system myelination, and cardiac development [20, 21].

In adult physiology, the ERBB family continues participating in tissue homeostasis, repair, and regeneration. EGFR is particularly important in maintaining the integrity and function of epithelial tissues. In the skin, EGFR signaling regulates keratinocyte proliferation, differentiation, and migration, contributing to wound healing and epidermal barrier function [22]. In the liver, EGFR participates in hepatocyte proliferation and liver regeneration following injury [23–25]. HER2 and HER4 play a role in cardiac function. They are involved in cardiomyocyte survival and adaptations to stress, contributing to the heart's ability to respond to increased workload [26–28]. In the nervous system, ERBB signaling is important for maintaining myelination and supporting neuron-glia interactions [29–31].

2.2 *Structure of ERBB Receptors and Their Ligands*

EGF-family ligands and ERBB receptors are initially synthesized as glycosylated transmembrane proteins. The ligands undergo proteolytic cleavage at the cell surface by specific proteases, which release mature growth factors. These mature ligands are characterized by an EGF-like consensus sequence that contains six conserved cysteine residues [32]. There are 11 known EGF-like peptide growth factors, which can be divided into 3 groups based on their binding specificity [33]. The first group, including EGF, TGFα, amphiregulin, and epigen, binds exclusively to EGFR. The second group, comprising epiregulin, betacellulin, and heparin-binding EGF, can bind both EGFR and HER4. The third group consists of neuregulins, with NRG1 and NRG2 binding to HER3 and HER4, while NRG3 and NRG4 bind only to HER4. EGF-like ligands are synthesized as type I transmembrane precursor proteins, consisting of an extracellular component, a transmembrane segment, and a short intracellular region [34]. The extracellular portion contains one or more EGF-like motifs, characterized by a hexameric cysteine sequence that forms three intramolecular disulfide bonds. This motif is both necessary and sufficient for ERBB receptor binding [35]. The extracellular domain of these ligands undergoes a process called protein ectodomain shedding, facilitated by various members of the ADAM (a disintegrin and metalloproteinase) family [36].

The ERBB receptors share a common structural architecture comprising three main domains (Fig. 1)—an extracellular ligand-binding domain, a single transmembrane domain, and an intracellular tyrosine kinase (TK) domain [37–42]. The extracellular region is composed of four distinct domains (I–IV): two leucine-rich repeat domains that form the ligand-binding pocket (L1 and L2) and two cysteine-rich domains (CR1 and CR2) [40, 43]. Domain II contains the dimerization arm, which is crucial for receptor-receptor interactions [44]. In the absence of a ligand, the extracellular domain adopts a tethered conformation where domain II interacts with domain IV, concealing the dimerization arm [45]. Ligand binding induces a

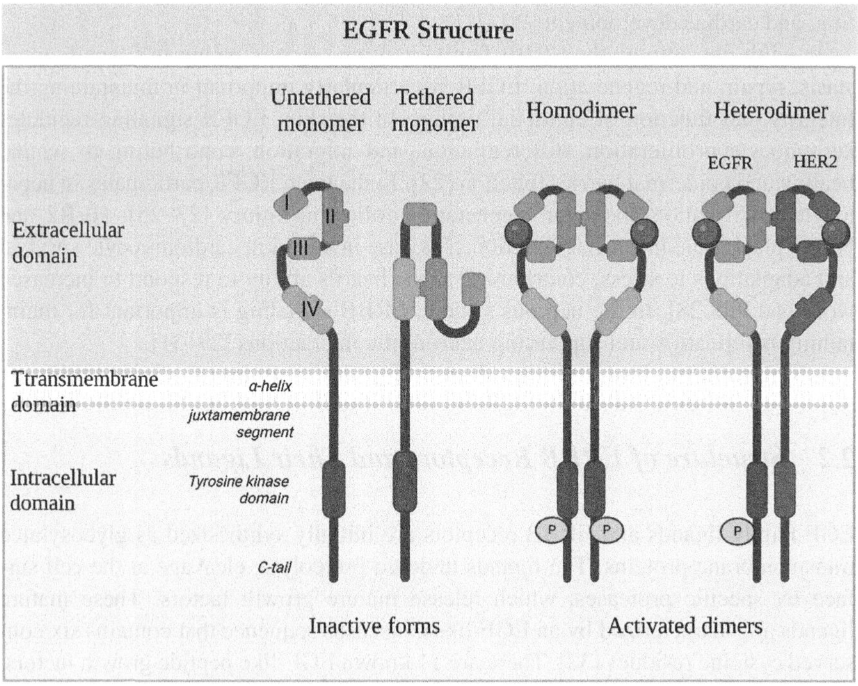

Fig. 1 Diagram of EGFR structure and activation. Domain organization of EGFR, showing the extracellular module (domains I–IV), transmembrane helix, intracellular kinase domain, and C-terminal tail. Ligand-induced conformational changes in EGFR illustrate the transition from the autoinhibited tethered monomer to the active dimeric state (homo- and heterodimeric). The diagram shows the release of the dimerization arm in domain II and the formation of back-to-back dimers upon ligand binding

conformational change, exposing the dimerization arm and facilitating receptor dimerization [46, 47] (Fig. 1). The single α-helical transmembrane domain of ERBB receptors plays a crucial role in connecting extracellular and intracellular regions, actively participating in receptor dimerization and signal transduction (reviewed in [48]). This domain is involved in both homo- and heterodimerization within cellular membranes, contributing to the receptors' signaling capabilities. The intracellular region is essential for their signaling functions and comprises several key domains [49]. The juxtamembrane segment, located immediately after the transmembrane region, is critical for regulating kinase activity and downstream signaling pathways [50, 51]. The tyrosine kinase domain, characterized by its N-lobe and C-lobe structure, is responsible for phosphorylating tyrosine residues, with the C-terminal tail serving as a docking site for signaling proteins upon phosphorylation [52, 53]. ERBB receptors contain varying numbers of tyrosine phosphorylation sites, with specific distributions for each receptor subtype. EGFR possesses 12 such sites, HER2 has 6, and HER3 contains 11. While HER4 also features phosphorylation sites, they have not been thoroughly characterized [54–56].

Each ERBB receptor possesses distinct intracellular domains that contribute to their unique signaling capabilities. EGFR possesses a fully functional structure with all active domains. HER2 has a constitutively active extracellular domain, making it the preferred dimerization partner despite lacking a known ligand. HER3, while considered a pseudokinase due to its impaired kinase domain, can still form active heterodimers with other ERBB receptors. Lastly, HER4 shares structural similarities with EGFR but uniquely undergoes alternative splicing in its juxtamembrane and C-terminal regions, resulting in various isoforms [49].

2.3 Structure and Signaling Pathways Activated by EGFR

Like all members of the ERBB receptor family, EGFR is structured with an extracellular ligand-binding domain, a transmembrane region, and a cytoplasmic tyrosine kinase domain (Fig. 1). The extracellular region consists of four domains: I and III are involved in ligand binding, while II and IV are involved in receptor dimerization [57]. Upon ligand binding, EGFR forms homo- or heterodimers with other ERBB family members, activating its intrinsic tyrosine kinase activity [44, 46, 47]. This activation involves the formation of an asymmetric kinase dimer, where the C-lobe of one kinase domain (the "activator") interacts with the N-lobe of the second kinase domain (the "receiver"), leading to its activation [58]. This activation triggers a cascade of signaling events that regulate various cellular processes (Fig. 2), including proliferation, differentiation, migration, and survival [1, 3, 32, 59].

The two primary downstream signaling pathways activated by EGFR are the PI3K/Akt/PTEN/mTOR and the RAS/RAF/MEK/ERK cascades [60]. In the PI3K/Akt pathway, which is crucial for cell survival and metabolism, phosphorylated EGFR serves as a docking site for PI3K, which generates PIP3 and activates Akt.

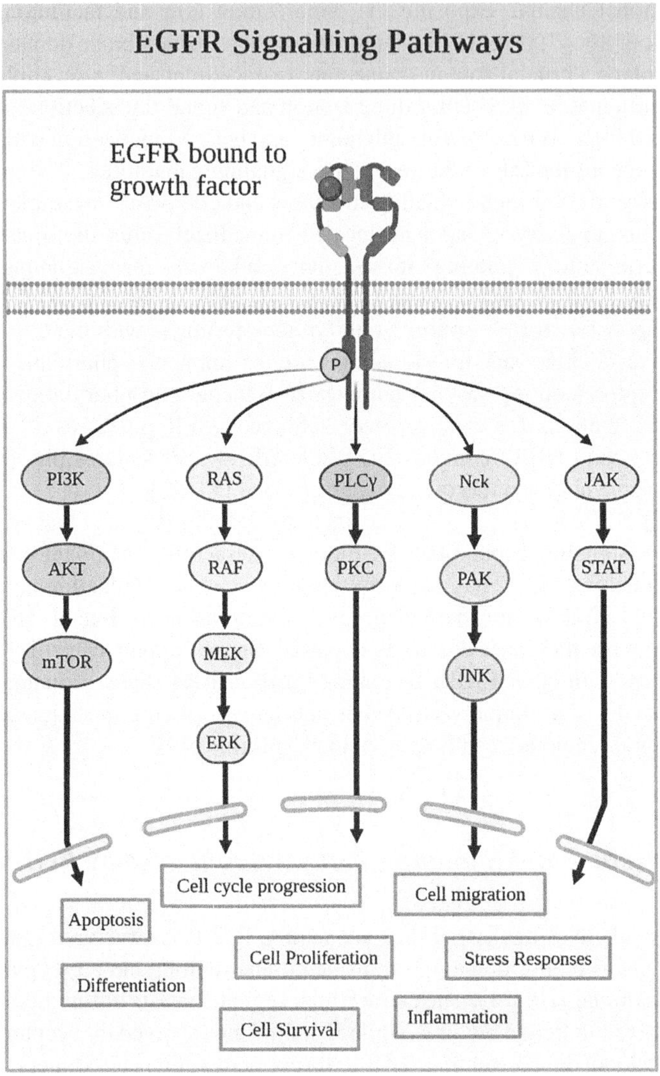

Fig. 2 Simplified schematic of EGFR signaling pathways. The figure illustrates key components and major signaling cascades activated upon EGFR stimulation, including the RAS/RAF/MEK pathway controlling cell proliferation and the PI3K/AKT pathway regulating cell survival. Ligand binding triggers EGFR dimerization, autophosphorylation, and recruitment of adaptor proteins, leading to the activation of downstream effectors

This leads to the activation of mTOR, which promotes cell cycle progression, suppresses apoptosis, and regulates protein synthesis and cell growth [61, 62]. The RAS/RAF/MEK/ERK pathway is initiated when adaptor proteins like Grb2 and Shc bind to phosphorylated EGFR. This interaction activates RAS, leading to a phosphorylation cascade through RAF, MEK, and ERK. Activated ERK then translocates to the nucleus, where it induces the expression of transcription factors that stimulate cell cycle progression [63, 64]. In addition to ERK, ERBB receptors can activate other MAPK pathways, including p38 and JNK [59]. These pathways are involved in stress responses, inflammation, and apoptosis.

Phospholipase C-gamma (PLCγ) is another important effector of EGFR signaling [65]. PLCγ binds directly to phosphorylated ERBB receptors and becomes activated. It then cleaves PIP2 into DAG and IP3, leading to calcium release and protein kinase C (PKC) activation. This pathway is involved in cell migration and differentiation [66, 67]. EGFR signaling also involves other pathways, such as the Nck/PAK cascade, which promotes cell survival and migration [68, 69]. Additionally, EGFR undergoes endocytosis and degradation, regulated by proteins like Cbl, which plays a crucial role in receptor trafficking and signal attenuation [70]. ERBB receptors interact with and activate SRC family kinases, which can further amplify the EGFR signaling by phosphorylating additional tyrosine residues on the receptors [71]. On the other hand, EGFR can also activate the JAK/STAT pathway, particularly in response to certain ligands [72]. This activation can occur either directly or through crosstalk with cytokine receptors. Activated STATs dimerize and translocate to the nucleus, where they regulate expression of genes involved in cell proliferation and survival [72].

The complexity of EGFR signaling is further increased by its crosstalk with other signaling pathways and the involvement of non-coding RNAs in regulating its activity [3]. This intricate network of signaling events allows for fine-tuned control of cellular responses to environmental stimuli. The dysregulation of these signaling pathways, often due to ERBB receptor overexpression or activating mutations, is implicated in various cancers and other diseases, making ERBB receptors important therapeutic targets.

3 ERBB Receptors in Cancer

Improper activation of EGFR results in increased malignant cell survival, proliferation, invasion, and metastasis. Due to its relevance and high expression in cancer tissues, EGFR was one of the first targets to be exploited in cancer treatment.

3.1 ERBB Receptor Overexpression and Mutations in Cancer

EGFR overexpression is associated with resistance to cancer treatment, and mutations in the extracellular and tyrosine kinase domain are often found in tumors. Deletion mutations in the extracellular domain of EGFR are categorized as EGFRvI, EGFRvII (deletion exons 14–15), EGFRvIII (deletion exons 2–7), EGFRvIV (deletion exons 25–27), and EGFRvV (deletion exons 25–28). EGFRvI is characterized by the loss of the whole extracellular domain leaving a sequence that resembles the v-erbB oncoprotein [73]. EGFRvIII is the most common variant in glioblastoma [74, 75]. Mutations in the tyrosine kinase domain are known as activating mutations because they lead to constitutive activation of EGFR. These mutations are present in different cancer types, being the most common in lung cancer (Fig. 3). Other less-known aberrations, such as the EGFR kinase domain duplications (KDDs), are tandem duplications of EGFR exons 18–25. While EGFR KDDs are the most common (55.4%), KDDs have been identified in other members of the family, such as HER2 (27.2%), HER3 (11.5%), and HER4 (5.9%) [76]. Mutations in the juxtamembrane

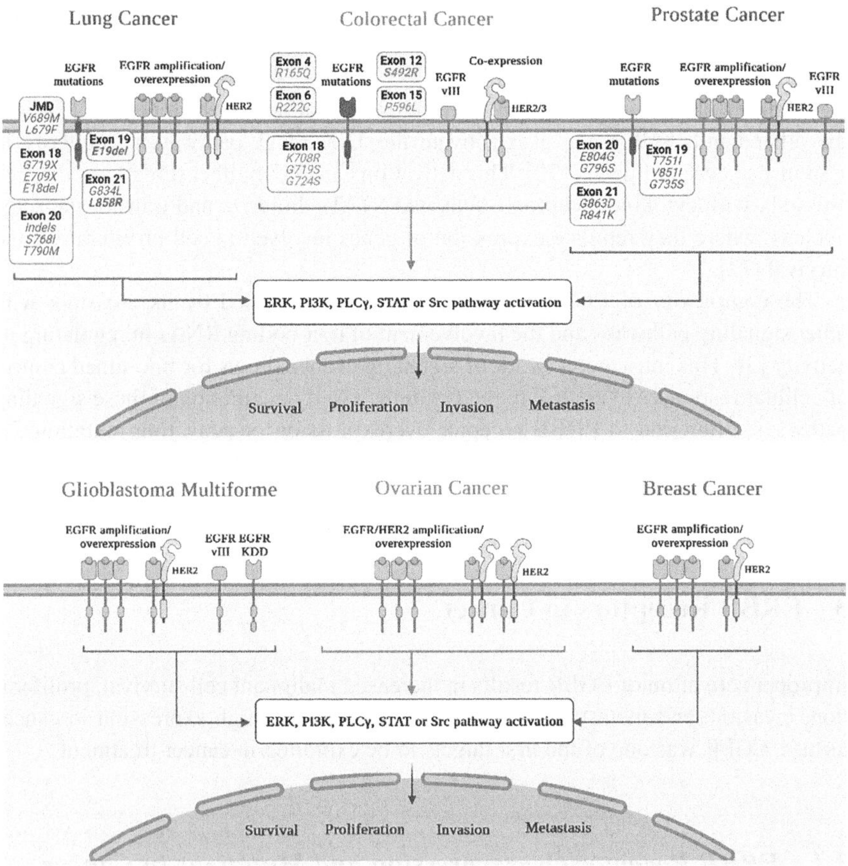

Fig. 3 EGFR alterations across diverse cancer types: a spectrum of mutations and structural changes. This figure illustrates various EGFR modifications, including point mutations and deletions affecting crucial domains. While receptor overexpression and gene amplification are the predominant mechanisms of EGFR activation, internal deletions resulting in truncated variants also play a significant role. Among these, EGFRvIII stands out as the most prevalent truncated form, contributing to aberrant EGFR signaling in multiple cancer types. EGFR kinase domain duplications (KDDs) are novel oncogenic drivers which importance is unraveling in different malignancies

region are rare but have also been described in cancer as a cause of constitutive activation [51]. Fusions involving EGFR, ERBB2, and ERBB4 are also rare genomic alterations occurring in less than 1% of cancer patients, but they represent potentially actionable targets that are most commonly found in specific cancer types such as high-grade gliomas, esophageal carcinomas, and ovarian cancers [77].

3.1.1 Glioblastoma Multiforme

Glioblastoma multiforme (GBM) is the most common primary nervous system cancer in adults. EGFR is amplified in 40–60% of malignant GBM tumors, and these amplifications are accompanied by gene rearrangements that produce N-truncated versions of the EGFR receptor lacking the extracellular domain [74, 75, 78, 79]. The most common deletion variant in gliomas is EGFRvIII, which accompanies up to 60% of the tumors with EGFR amplifications [74, 75]. The EGFRvIII variant can dimerize in a ligand-independent manner, and it is constitutively active, which results in increased growth and malignancy [80, 81]. EGFR KDDs are a newly identified group of oncogenic aberrations, most commonly found in gliomas, appearing with a frequency ranging between 0.58% and 2.4% of the total glioma cases [76]. Translocations that fuse EGFR to several partners can be found most frequently in glioblastoma, with EGFR-SEPT14 being the most frequent functional gene fusion [77]. HER2 is expressed in 41% of primary GBMs, correlating to poor prognosis and immune markers, suggesting a role for HER2 and tumor immune response in gliomas [82, 83].

3.1.2 Lung Cancer

Lung cancer is the leading cause of cancer-associated deaths, being a highly heterogeneous disease; it is classified as small cell lung cancer (SCLC), non-small cell lung cancer (NSCLC), and lung neuroendocrine tumors. NSCLC is the most common type of lung cancer and is subdivided into lung adenocarcinoma (LUAD), lung squamous cell carcinoma (LUSC), and large-cell lung carcinoma (LCLC) [84]. In lung cancers, the most common aberrations of EGFR are mutations in the tyrosine kinase domain. These are found in 31.6% of NSCLCs, corresponding to 36.5% of LUAD cases and 15% of LUSCs [85].

The most common mutations in NSCLCs are the in-frame deletions in exon 19 and the point mutation L858R in exon 21, both accounting for up to 90% of the EGFR-activating mutations [85, 86]. The most common exon 19 deletions in NSCLCs are p.E746_A750del (64.6%), p.L747_P753 > S (8.4%), p.L747_T751del (4.0%), p.L747_A750 > P (3.4%), p.E746-S752 > V(2) (3.2%), p.E746_S752 > V (1.6%), and p.L747_S752del (1.4%) [87]. Other mutations in the tyrosine kinase domain of EGFR are known as "uncommon EGFR mutations," representing the remaining 10–20%. The most prevalent uncommon EGFR mutations include a heterogeneous group of alterations within exons 18 to 21. Exon 18 mutations represent around 4–6% of the total EGFR mutations in NSCLCs, with G719X being the most common [88, 89]. The G719X EGFR mutation refers to a group of point mutations that result in the substitution of the glycine in position 719 to alanine (G719A), cysteine (G719C), serine (G719S), and aspartic acid (G719D). Other less common exon 18 mutations include E709X and exon 18 deletion [88, 90]. EGFR exon 20 insertion accounts for up to 9% of the total EGFR mutations in LUAD, making them the third most common clinically relevant EGFR mutation in NSCLC [91]. EGFR exon 20 insertions are diverse, as 122 different types have been identified in NSCLCs. The most common EGFR exon 20 insertion mutations are Asp770_Asn77ins (28.7%) and Val769_Asp770ins (20.5%) [92]. Other rarer EGFR exon 20 mutations are the S768I and T790M, which are found in less than 1% and 0.5% of lung cancer patients, respectively [93, 94]. However, T790M frequency increases significantly in patients who relapse after treatment, where it accounts for 50–60% of all EGFR mutations [94].

From the EGFR extracellular domain deletion mutations, EGFRvIII is described to be present in 16–40% of NSCLCs, but the real proportion is believed to be much lower [95–98]. The presence of EGFRvIII negatively correlates with mutations in the kinase domain, as it is almost absent in LUADS, and it has been described in 5–42% of LUSCs and 14% of LCLCs [95, 99–101]. Mutations in the juxtamembrane domain of EGFR have been reported in NSCLC, with V689M and L679F being clinically relevant and proven to activate EGFR at a level comparable to well-characterized oncogenic mutations in the tyrosine kinase domain itself [51]. Lastly, EGFR overexpression can be detected in 13% of NSCLCs, although gene amplification of 3–17 copies with no detected structural rearrangements was found in only 6% of the cases [102]. Moreover, gene fusions such as EGFR-RAD51, EGFR-PURB, EGFR-TNS3, EGFR- KIF5B, and EGFR-ZCCHC have been identified in NSCLC as the detection methods for genetic aberrations that continue to improve [103–106].

3.1.3 Ovarian Cancer

Ovarian cancer is the deadliest of the gynecological cancers; the prognosis is often poor as it is usually detected in advanced stages [107]. Epithelial ovarian carcinomas are classified as mucinous, endometrioid, clear cell, low-grade serous (LGSC), or high-grade serous carcinomas (HGSC), the latter being the most frequently diagnosed [108].

EGFR amplification is present in 12% of HGSCs, and low copy number increase was found in 44% of them in the same cohort [109]. HER2 amplification is found in 7% of the serous ovarian carcinoma cases, and low copy number increases are found in 14% of them [110]. An increased copy number of EGFR was associated with an increased copy number of HER2 [109]. Additionally, HER2 gene amplification has been observed in 3–10% of epithelial ovarian carcinomas, with a relatively high incidence of 14.4% in mucinous ovarian carcinoma [111, 112]. No mutations have been identified in EGFR, but HER2 exon 20 insertion mutations, although rare, have been described in serous ovarian carcinomas [109]. Another less common aberration found in ovarian cancer are ERBB4 gene fusions, which are the most common ERBB family gene fusions in ovarian cancer [77].

3.1.4 Gastrointestinal Cancers

The overexpression of ERBB receptors in gastrointestinal cancers, particularly colorectal cancer (CRC), has become an increasingly important area of research due to its implications for cancer progression and treatment. CRC is the third most common cancer worldwide, and it is an extremely heterogeneous disease. Due to its long developing time, it accumulates genetic alterations, and few mutations are common to most colorectal tumors, but each has a unique combination [113].

EGFR is expressed in up to 75% of CRC tumors, and its expression is associated with poor prognosis and increased risk of metastasis [114–116]. The addition of anti-EGFR inhibitors to chemotherapy improved the progression-free survival (PFS) of familial CRCs [117]. The benefits of anti-EGFR therapies are being actively explored in non-KRAS-mutated colorectal cancer tumors [118].

The EGFR mutations G63R, E114K, R165Q, R222C, S492R, and P596L (ectodomain mutations) and K708R, E709K, G719S, G724S, and L858R (TK domain mutations) were identified in CRC and are shown to be oncogenic [119, 120]. From these mutations, S492R is absent in treatment-naïve CRC tumors as it is associated with acquired resistance to EGFR inhibitor treatment [121]. Mutations in exons 19, 20, and 21 have also been identified, appearing to have a higher incidence in Korean and Japanese studies [122, 123]. Aside from mutations, the EGFRvIII ectodomain deletion has been identified in 8.11% of colorectal cancer patients [124]. Although less described, EGFR amplification and the EGFR-SEPT14 fusion in a metastatic colorectal cancer patient have been reported [125].

In CRCs, HER2 is expressed in 89% and HER3 in 83% of the tumors, with 77% showing expression of both receptors [126]. HER2 amplification and somatic mutations are present in 7% of CRCs, and the S310F, L755S, V777L, V842I, and L866M ERBB2-activating mutations have been identified, but their clinical potential is still yet to be uncovered [127–129].

In gastric cancer (GC), EGFR overexpression and HER2 amplification are observed in 27–64% and 6–30%, respectively. However, their prognostic value remains controversial, with conflicting reports on their association with patient outcomes [130]. Some studies suggest HER2 heterogeneity correlates with poor prognosis [131], with higher prevalence in those near the gastroesophageal junction (GEJ). HER3 and HER4 are expressed in 20.7% and 13.3% of GC, respectively, with high HER3 expression correlating to poor survival [130]. HER2 gene fusions are commonly found in GEJ cancers, with a frequency of 3.12% in Chinese patients [132].

In esophageal cancer, EGFR is upregulated compared to the normal mucosa, and its expression correlated with tumor severity and patient survival [133, 134]. EGFR amplification has been identified in esophageal tumors, and it has been significantly associated with lymph node metastasis [133, 135].

3.1.5 Breast Cancer

Breast cancer (BC) is the most commonly diagnosed cancer worldwide, and it is the leading cause of cancer-associated deaths in women [136]. BC is a highly heterogeneous disease, and it is classified based on histological or molecular attributes. The histological subtypes for BC are ductal noninvasive (DCIS), ductal invasive (IDC), lobular noninvasive (LCIS), lobular invasive (ILC), inflammatory breast cancer (IBC), and metastatic breast cancer (MBC). The molecular subtypes are luminal A, luminal B, normal-like, HER2-positive, and triple-negative breast cancer (TNBC). IDC accounts for approximately 80% of BC cases; from the molecular subtypes, luminal A is the most common, accounting for about 70% of the total tumors.

In BC cells, increased levels of EGFR have been associated with increased proliferative and angiogenic activity. Although EGFR overexpression can be found in all BC subtypes, it is more common in the most aggressive subtypes, TNBC and IBC [137, 138]. A report claims that EGFRvIII is found in almost 80% of breast carcinomas [78]. However, breast tumors are negative for EGFRvIII, with its expression being extremely rare, and it is accepted that it does not contribute to the malignant phenotype [139, 140].

HER2 is amplified in 30% of BC tumors, and it is significantly associated with poor prognosis and relapse. Fifteen percent of these tumors are stage I, with no lymph node involvement, highlighting this aberration's clinical relevance [141].

3.1.6 Prostate Cancer

EGFR expression was detected in 18–36% of prostate tumors, showing a strong association with higher grade, advanced stages, and recurrence [142, 143]. EGFR amplification was detected in 3.3% of the studied cases, with only one case showing high-level EGFR amplification of more than 20 gene copies [143]. In prostate cancer, 13% of patients are described to carry a mutation in the tyrosine kinase domain of EGFR [142]. These mutations found include T751I, V851I, and G735S in exon 19; E804G and G796S in exon 20; and G863D and R841K in exon 21 [142, 144]. The EGFRvIII deletion variant is detected in 6% of prostate cancer tumors; the presence of the variant relates to the tumor stage, supporting an oncogenic role for this mutation in prostate tumors [124, 145]. HER2 was detected in 80% of prostate tumors, with evidence of gene amplification [146].

3.1.7 Head and Neck Squamous Cell Carcinoma

Despite the high correlation of HNSCC with HPV infection, numerous studies have revealed significant mutations and aberrations in ERBB family members. For instance, EGFR is overexpressed or mutated in 18–90% of HNSCC cases, establishing its role as a key oncogene in this type of cancer. HER2 has also been identified as amplified in around 7% of HNSCC patients, with gene copy number alteration ranging from 2% to 46%. Furthermore, a mean of 4% of HNSCC cases showed HER2 missense mutations, and 6% of these mutations were predicted to be detrimental to ERBB function. HER3 and HER4 also exhibit aberrant expression in around 46% and 26% of HNSCC, respectively, and their roles in cancer progression potentially depend on subcellular localization [147]. Mutations in the tyrosine kinase domain and the extracellular domain of EGFR have been reported in HNSCC. Extracellular domain mutations have been only identified as associated with cetuximab resistance, and mutations in the tyrosine kinase domain are in exons 18 to 21 [148, 149]. These findings highlight the complex interplay of ERBB family members in head and neck squamous cell carcinoma and underline their importance as potential targets for personalized therapies.

3.2 *Non-genetic Alterations of ERBB Receptors in Cancer*

Aside from the genetic mechanisms, the ERBB family also uses non-genetic adaptations to promote tumor progression and survival. These include DNA methylation, acetylation, and phenotypical transitions. DNA can be methylated at position 5 of the pyrimidine ring of the cytosine residues (CpG sites) to form 5-methylcytosine. The presence of multiple CpG sites in the promoter of a gene can silence its expression. CpG sites in the promoter and the gene body region of EGFR in lung adenocarcinoma have been identified and linked to EGFR expression levels [150]. EGFR acetylation by CREB-binding protein acetyltransferase can enhance its function [151]. Treatment with histone deacetylase inhibitors (HDACi) enhances EGFR acetylation and EGFR phosphorylation, showing the relevance of EGFR acetylation as a resistance mechanism to HDACi [151]. Phenotypic transformations such as the activation of the epithelial to mesenchymal transition and the gain of stem-like features have been observed as a mechanism of acquired resistance to EGFR/HER2 inhibitors in NSCLC [152–154]. As these changes do not involve genetic alterations, they are reversible. Interestingly, it has been observed that EGFR inhibitor-resistant lung cancer cell lines can revert their resistance when implanted in an animal [154]. These studies highlight the complexity of the regulation of EGFR and the other ERBB members and the variety of mechanisms that cancer cells can use to promote their survival.

3.3 *Crosstalk with Other Signaling Pathways in Cancer*

Different malignancies show a strong dependency on the activity of the ERBB receptors, making them strategic therapeutic targets. However, resistance mechanisms arise due to their ability to mediate intricate crosstalk between pathways. The most known examples include the ERK signaling pathway, PI3K/Akt/PTEN/mTOR, PLCγ, STAT, and Src [155]. There is increasing evidence supporting the interplay between ERBB and Wnt signaling. The Min/+ model of colorectal carcinogenesis, characterized by high Wnt activation, shows PGE2 upregulation and transactivation of EGFR [156]. Wnt3a can stimulate cell proliferation and motility through EGFR activation by mediating complex crosstalk between ß-catenin and ERK pathways [157]. The convergence of EGFR/HER2 and Wnt signaling is required for Wnt-mediated tumorigenesis in mammary gland tumors [158].

The members of the ERBB family are strongly associated with cancer promotion, progression, and metastasis. New aberrations and mechanisms that cancer cells use to survive and resist treatment through the ERBB receptors are still being discovered. Advances in technology will provide a more in-depth understanding of the full extent of mutations, alterations, and epigenetic mechanisms governing the ERBB receptors' role in cancer.

4 Targeting ERBB Receptors in Cancer Therapy

ERBB-targeted therapies are a cornerstone in the treatment of various cancers, particularly NSCLC but also breast, colorectal, pancreatic, and head and neck cancers (Table 1). These therapies include small-molecule TKIs such as erlotinib, gefitinib, and afatinib, as well as mAbs like cetuximab [159]. For patients with common EGFR mutations, TKIs have shown impressive response rates and median PFS [160]. However, the efficacy of these therapies varies depending on the specific EGFR mutation, with some uncommon mutations showing variable responses. Advances in genomic profiling have also helped identify patient subgroups more likely to benefit from these therapies.

4.1 EGFR-Targeted Therapies

4.1.1 Tyrosine Kinase Inhibitors (TKIs)

The introduction of these small inhibitors in clinical practice has greatly improved the survival of patients with EGFR-mutated lung cancer, compared to standard chemotherapy regimens. Additionally, in comparison to chemotherapy, EGFR TKIs show a safer profile, with the most common side effects affecting the skin, gastrointestinal tract, and lungs [161].

First-generation EGFR TKIs belong to the class of anilinoquinazolines and act as reversible ATP-competitive inhibitors. Their binding to the ATP-binding pocket of EGFR blocks the kinase activity of the receptor, thus preventing its autophosphorylation and, consequently, the activation of its downstream signaling pathways. Erlotinib and gefitinib, belonging to this category, showed an improved PFS when compared to standard chemotherapy in patients with NSCLC carrying activating mutations in EGFR [162–164]. Despite the initial great response, within

Table 1 ERBB-targeting drugs approved by the FDA

Target receptor	Class	Molecule (trade name)	Year of approval	Company	Indication
EGFR	TKI	*Gefitinib (Iressa)*	2003	AstraZeneca	NSCLC
		Erlotinib (Tarceva)	2004	OSI Pharms	NSCLC, PC
EGFR/ HER2/ HER4		*Afatinib (Gilotrif)*	2013	Boehringer Ingelheim	NSCLC
EGFR		*Osimertinib (Tagrisso)*	2015	AstraZeneca	NSCLC
EGFR/ HER2/ HER4		*Dacomitinib (Vizimpro)*	2018	Pfizer	NSCLC
EGFR		*Mobocertinib (Exkivity)*	2021	Takeda Pharms USA	NSCLC
		Lazertinib (Lazcluze)	2024	Janssen Biotech	NSCLC
	mAb	*Cetuximab (Erbitux)*	2004	ImClone	CRC, HNSCC
		Panitumumab (Vectibix)	2006	Amgen	CRC
		Necitumumab (Portrazza)	2015	Eli Lilly Co	NSCLC
EGFR/MET	BsAb	*Amivantamab (Rybrevant)*	2021	Janssen Biotech	NSCLC
EGFR/ HER2	TKI	*Lapatinib (Tykerb)*	2007	Novartis	BC
EGFR/ HER2/ HER4		*Neratinib (Nerlynx)*	2017	Puma Biotech	BC
HER2		*Tucatinib (Tukysa)*	2020	Seagen	BC, CRC
	mAb	*Trastuzumab (Herceptin)*	1998	Genentech	BC, GC, GEJ cancer
		Pertuzumab (Perjeta)	2012	Genentech	BC
		Margetuximab (Margenza)	2020	MacroGenics Inc.	BC
	ADC	*Trastuzumab emtansine (Kadcyla)*	2013	Genentech	BC
		Trastuzumab deruxtecan (Enhertu)	2019	Daiichi Sankyo	NSCLC, BC, GC, GEJ cancer

TKI tyrosine kinase inhibitor, *mAb* monoclonal antibody, *BsAb* bispecific antibody, *ADC* antibody-drug conjugate, *NSCLC* non-small cell lung cancer, *PC* pancreatic cancer, *CRC* colorectal cancer, *HNSCC* head and neck squamous cell carcinoma, *BC* breast cancer, *GC* gastric cancer, *GEJ* gastroesophageal junction

10–14 months, patients eventually develop resistance. Second-generation drugs were also developed; these anilinoquinazoline-based inhibitors irreversibly bind to the ATP-binding pocket in the kinase domains of not only EGFR but also other members of the ERBB family of receptors. Currently, there are two drugs, approved in clinic, belonging to this category, namely, afatinib and dacomitinib. Afatinib was superior in terms of PFS to chemotherapy in two randomized phase III trials

(LUX-Lung 3 and LUX-Lung 6) [165, 166]. Additionally, afatinib was proved to be superior to gefitinib in terms of PFS, but not overall survival (OS), in the LUX-Lung 7a phase IIb trial [167, 168]. Moreover, when dacomitinib was compared to gefitinib in the ARCHER 1050 phase III trial, the promising results in terms of PFS and OS led to its approval for NSCLC with EGFR-activating mutations [169, 170]. It has to be considered that, in addition to the increased potency, second-generation TKIs, compared to first-generation drugs, come with a higher risk of toxicity [171], which may be attributable to the blockage of wild-type EGFR, as well as other ERBB receptors.

To counteract T790M-mediated resistance to first- and second-generation TKIs, the third-generation inhibitor osimertinib was developed. This small molecule presents a pyrimidine backbone and a covalent warhead that allow the establishment of a strong and irreversible bond with the cysteine 797 of T790M-EGFR [159], while sparing wild-type EGFR, thus limiting potential side effects. The AURA3 clinical trial, comparing osimertinib to chemotherapy in patients with EGFR-mutated lung cancer after progression to first-generation drugs, showed improved PFS and limited side effects for the osimertinib-treated group, leading to its approval in 2015 for T790M+ NSCLC [172]. Later on, the FLAURA3 trial demonstrated that osimertinib is superior to first-generation TKIs, in terms of PFS and OS, in untreated patients expressing EGFR-activating mutations [173, 174]. Based on these results, osimertinib was approved in 2018 as a first-line therapy for patients with EGFR-mutated NSCLC. Unfortunately, as observed with previous generation TKIs, resistance to osimertinib inevitably occurs, in both first- and second-line treatment regimens [175, 176].

4.1.2 Monoclonal Antibodies (mAbs)

mAbs display a different mechanism of action compared to TKIs. Cetuximab is a human-mouse chimeric mAb containing the human IgG1 constant region. It binds to the extracellular domain III of EGFR, preventing its interaction with the ligand, thus inhibiting receptor activation and shutting down downstream signaling pathways [177]. Additionally, cetuximab is able to induce antibody-dependent cellular cytotoxicity (ADCC), to cause internalization and degradation of EGFR [177, 178], as well as to reduce cell proliferation via inducing G1 arrest, caused by increased levels of the cyclin-dependent kinase inhibitor p27kip1 [179, 180]. Moreover, this mAb activates apoptosis by increasing the levels of Bax and decreasing the levels of Bcl-2 [181]. Cetuximab was initially approved by the FDA (Food and Drug Administration), in 2004, for the treatment of metastatic CRC (mCRC), either in combination with chemotherapy as first-line treatment or as monotherapy, after failure of chemotherapy. Retrospective subgroup analysis of data obtained from two clinical trials (CRYSTAL and OPUS trials), testing cetuximab in combination with chemotherapy in mCRC, showed a benefit of the combination for patients expressing wild-type KRAS, while no clinical benefit was observed for KRAS-mutated mCRC patients [182–184]. Consequently, the approval of cetuximab was restricted to patients with EGFR-positive, RAS wild-type mCRC. Additionally, cetuximab is indicated for HNSCC, both in locally advanced disease (in combination with radiotherapy) and in recurrent/metastatic disease (in combination with chemotherapy) [185].

Panitumumab, a fully human anti-EGFR mAb, is also approved for mCRC. Cetuximab and panitumumab bind different, albeit partially overlapping, epitopes in the ligand binding domain of EGFR [186]. However, panitumumab, differently from cetuximab, is an IgG2 mAb. This distinction reflects a difference in the mechanism of action of the two anti-EGFR mAbs. Specifically, cetuximab, being a IgG1, can instigate the ADCC, while the IgG2 backbone of panitumumab is less active in this regard [185]. Another anti-EGFR antibody, necitumumab, demonstrated a benefit, when combined with chemotherapy, in advanced squamous NSCLC [187]. This human mAb is indicated, in combination with chemotherapy, for first-line treatment of patients with metastatic squamous NSCLC.

4.1.3 Bispecific Antibodies

Unlike common mutations, rare mutations are associated with reduced and heterogeneous responses to EGFR kinase inhibitors. Exon20Ins are detected in 4–10% of patients with EGFR-mutated NSCLC, and they are associated with the de novo resistance to EGFR TKIs [188]. In this scenario, the results of the phase I trial CHRYSALIS, testing the bispecific antibody (BsAb) amivantamab, targeting EGFR and the hepatocyte growth factor receptor, MET, in patients expressing Exon20Ins EGFR after failure of platinum-based chemotherapy, were extremely promising. The study showed a response rate of 40% with a median response duration of 11.1 months in 81 patients [189]. This led, in May 2021, to the FDA approval of amivantamab for the treatment of Exon20Ins EGFR-positive NSCLC after failure of chemotherapy [190]. The efficacy of amivantamab combined with chemotherapy in first-line settings for the treatment of Exon20Ins-positive patients was later demonstrated in the phase III trial PAPILLON [191], leading to its approval in March 2024 for this new indication.

Amivantamab is a fully human IgG1 BsAb active against both EGFR and MET. This BsAb is able to block ligand-induced activation, induce downregulation of both receptors [192], and activate ADCC [192–194]. Amivantamab was tested in NSCLC patients carrying EGFR common mutations in first-line settings in combination with the third-generation EGFR TKI lazertinib (MARIPOSA phase III trial) [195] and as second-line therapy, in combination with chemotherapy or with lazertinib and chemotherapy, in patients after progression with osimertinib (MARIPOSA-2 phase III trial) [196]. In both trials, the use of amivantamab improved PFS. Following these promising results, the EGFR-MET BsAb received FDA approval in 2024 for the treatment of EGFR-mutated (Exon19Del or L858R) NSCLC as first-line therapy in combination with lazertinib and as second-line therapy, in combination with chemotherapy after progression with an EGFR TKI.

4.2 HER2-Targeted Therapies

4.2.1 TKIs

Three HER2 TKIs are currently approved by the FDA: lapatinib, neratinib, and tucatinib. Lapatinib is a reversible inhibitor of both HER2 and EGFR. When combined, in a phase III trial, with capecitabine in previously treated patients with HER2-positive BC, lapatinib showed superiority in prolonging time to progression, compared to capecitabine alone [197, 198]. In 2007, based on these results, the FDA granted approval to lapatinib for this indication. Because preclinical data in models of HER2-positive BC showed synergistic effects when lapatinib was combined with trastuzumab, a HER2-targeting mAb [199], this combination was tested in a phase III clinical trial as second line of therapy in patients with HER2-positive MBC, giving positive results [200].

Neratinib is an irreversible inhibitor of EGFR, HER2, and HER4, which causes a decrease in the phosphorylation of ERK and AKT, an increase in p27, and, consequently, cell cycle arrest [201, 202]. Based on the results of the phase III trial ExteNET [203], neratinib was approved for the extended adjuvant treatment of women with early-stage HER2-positive BC, after adjuvant trastuzumab-based therapy. The subsequent phase III trial NALA, which compared neratinib plus capecitabine versus lapatinib plus capecitabine in patients with MBC, following at least two previous anti-HER2 therapies, showed an improved PFS and a CNS (central nervous system) disease intervention for the neratinib containing combination [204]. In light of these results, the FDA approved neratinib in combination with capecitabine for the treatment of women with advanced HER2-positive BC who have already received two or more prior anti-HER2 therapies.

The fact that lapatinib and neratinib inhibit EGFR and HER2 in a similar way might partially explain the high frequency of gastrointestinal and dermatological side effects observed with these two small inhibitors [205, 206]. On the other hand, tucatinib is a small ATP-competitive and selective inhibitor of HER2, able to block HER2 phosphorylation and its downstream signaling pathways, causing cytotoxicity in HER2-amplified BC cell lines [207]. Tucatinib was tested in combination with trastuzumab and capecitabine in the phase II trial HER2CLIMB in previously treated patients with HER2-positive MBC. The addition of tucatinib to trastuzumab and capecitabine demonstrated a clinical benefit, even for patients with brain metastases [208, 209]. Based on these results, in 2020, tucatinib was granted FDA approval. HER2 overexpression/amplification has been additionally described in a relatively small percentage of CRC cases, with higher incidence for RAS wild-type tumors [210–212]. The phase II MOUNTAINEER study, combining tucatinib with trastuzumab in chemotherapy-refractory, HER2+, RAS wild-type unresectable or mCRC, showed clinically relevant anti-tumor efficacy for the combination [213]. Due to the efficacy and manageable safety profile, the combination of tucatinib and trastuzumab received accelerated approval by the FDA in January 2023.

4.2.2 Monoclonal Antibodies

The first HER2-targeting drug to be approved by the FDA is the humanized mAb trastuzumab. This antibody can inhibit the growth of in vitro and in vivo models of HER2-overexpressing BC [214, 215]. Moreover, the antibody acts by activating the ADCC [214]. Of note, preclinical studies showed an additive effect of the combination of trastuzumab and chemotherapy [216]. Currently, trastuzumab is approved by the FDA for the treatment of HER2-overexpressing BC, GC, and GEJ adenocarcinoma. The first approval of trastuzumab in 1998 for metastatic HER2+ BC is due to the results of a phase III trial, which tested the combination of the mAb with chemotherapy in patients with HER2-positive MBC cancer that progressed following prior chemotherapy-based regimens [217]. Later, based on the results of two clinical trials that demonstrated the efficacy of combining adjuvant trastuzumab with chemotherapy for women with HER2-positive breast cancer [218], the FDA approved trastuzumab in adjuvant settings. The efficacy of adjuvant trastuzumab in patients with HER2+ breast cancer was later confirmed by other two clinical trials [217, 219]. Because it has been shown that HER2 is overexpressed in GC, several in vitro and in vivo studies have been conducted, and they proved the efficacy of blocking HER2 with trastuzumab in models of GC [220, 221]. The ToGa (Trastuzumab for Gastric Cancer) trial tested the efficacy of combining trastuzumab with chemotherapy in patients with HER2+ GC and GEJ cancer [222]. The addition of trastuzumab to chemotherapy improved the OS, and the trial showed that the survival benefit for the combination group depended on the levels of expression of HER2 (higher for HER2-overexpressing tumors). Based on these results, the FDA approved this HER2-targeting mAb for patients with HER2-overexpressing metastatic GC or GEJ adenocarcinoma.

Pertuzumab is a humanized mAb targeting HER2, preventing its heterodimerization with other ERBB receptors, shutting down the downstream signaling pathways, and activating ADCC [223–225]. Moreover, this mAb binds to a different site in the extracellular domain of the receptor compared to trastuzumab [224], allowing the combination of two antibodies to better inhibit the receptor. The advantage of combining pertuzumab and trastuzumab has been demonstrated in preclinical and clinical studies in models of HER2-positive BC [225–229]. Pertuzumab was approved by the FDA in 2012, in combination with trastuzumab and docetaxel, for the treatment of patients with HER-positive MBC, based on the results of the CLEOPATRA trial [226, 230]. Final analysis of the CLEOPATRA trial showed an 8-year landmark OS rate of 37% for the group pertuzumab plus trastuzumab plus docetaxel versus 23% in the group trastuzumab plus docetaxel, clearly demonstrating a significant advantage of the dual blocking HER2 for the treatment of women with HER2+ MBC [231].

A third approved antibody targeting HER2 is margetuximab, a chimeric mAb that was designed so that the Fc region has increased affinity for CD16A, the activating subtype of the FcR (Fc receptor), and decreased affinity toward CD32B, the inhibitory subtype of the FcR, causing a more efficient natural killer-mediated ADCC [232]. Margetuximab plus chemotherapy was compared to trastuzumab plus chemotherapy in previously treated women with HER2-positive MBC in the phase III clinical trial SOPHIA. Because PFS was slightly improved in the margetuximab-treated group compared to the trastuzumab-treated group (5.8 versus 4.9 months, respectively), the FDA, in 2020, granted approval to margetuximab. Later updates on the SOPHIA trial showed no improvement in terms of OS when comparing the margetuximab- versus the trastuzumab-treated group [233].

4.2.3 Antibody-Drug Conjugates (ADCs)

ADCs are molecules, composed of three main components (a mAb, a linker, and a cytotoxic payload) used to deliver cytotoxic agents to cancer cells. The binding of the antibody to its target receptor causes internalization of the ADC, followed by lysosomal degradation, release of the cytotoxic agent within the cell, and resulting cell death [234]. Their efficacy has been shown for different types of tumor, including breast, gastric, and colorectal cancers. To date, the FDA has granted approval to two ADCs targeting HER2, both having trastuzumab as mAb.

Trastuzumab emtansine (T-DM1) contains as cytotoxic agent emtansine, a maytansine derivative that acts as an anti-microtubule agent, inducing G2-M cell cycle arrest and consequent cell death [235]. The EMILIA trial showed that T-DM1 as monotherapy improved PFS and OS while reducing toxicity, compared to lapatinib plus capecitabine, in previously treated women with HER2-positive advanced BC [236]. Results from this trial led, in 2013, to the FDA approval of T-DM1 for HER2+ MBC. In 2019, the FDA approved T-DM1 as an adjuvant agent for HER2+ MBC patients with invasive residual disease after neoadjuvant trastuzumab and taxane therapy. This approval followed the results of the phase III clinical trial KATHERINE, which showed that the risk of recurrence of IBC or death was 50% lower in the group that received adjuvant T-DM1, compared to the group that received trastuzumab [237].

Trastuzumab deruxtecan (T-DXd) carries as cytotoxic agent deruxtecan, a topoisomerase I inhibitor, which causes DNA damage and apoptosis. T-DXd inhibits the growth of in vitro and in vivo models of different types of tumors, including BC, GC, and pancreatic cancer (PC), expressing either high or low levels of HER2. In this preclinical study, this ADC was additionally proven to be active against tumors resistant to T-DM1 [238]. The first approval by the FDA of T-DXd dates back to 2019 and referred to women with unresectable or HER2+ MBC after failure of two or more prior anti-HER2-based regimens in the metastatic setting. This approval followed the results of the DESTINY-Breast01 phase II trial where T-DXd displayed durable antitumor activity [239]. The phase III trial DESTINY-Breast03 showed superiority of T-DXd in second-line settings for HER2+ metastatic breast cancer, compared to T-DM1 [240]. By 2022, T-DXd was granted FDA approval for the treatment of patients with low-expressing HER2 MBC, following the results of the DESTINY-Breast04 trial, which showed an improvement in terms of PFS and OS for patients with HER2 low metastatic breast cancer who had received one or two previous lines of chemotherapy, when treated with trastuzumab deruxtecan compared to chemotherapy [241]. Another two indications for T-DXd are HER2+ gastric and gastroesophageal cancer and HER2-mutated metastatic NSCLC, based on the results of the DESTINY-Gastric01 [242] and the DESTINY-Lung01 trials [243], respectively.

5 Drug Resistance Mechanisms: Primary and Acquired Resistance

Targeting members of the ERBB family of receptors through the development and approval of several inhibitors has significantly improved the clinical management of various ERBB-driven tumors [3, 4, 160]. However, despite remarkable initial responses, both monoclonal antibodies (mAbs) and tyrosine kinase inhibitors (TKIs) have shown limited long-term efficacy. Therapeutic failures are primarily driven by the rapid emergence of different drug resistance mutations and activation of compensatory pathways. In this regard, extensive studies have been focused to determining whether resistance is of the primary (intrinsic) or secondary (acquired) type [244].

Primary resistance refers to tumors that exhibit minimal or no response to treatment, preventing patients from achieving stable disease or leading to disease progression within 6 months of an initial clinical response. In contrast, secondary resistance typically develops after extended periods of treatment (Fig. 4). Usually, primary resistance is related to pre-existing and "on-target" mutations, while acquired resistance can be divided into two major categories: target-dependent (on-target) characterized by the acquisition of a secondary mutation and target-independent (off-target) generally associated with the activation of compensatory signaling pathways [245, 246].

5.1 Resistance to EGFR-Targeted Therapies

So far, several anti-EGFR mAbs have been clinically approved as chemo- or radio-sensitizers in EGFR-expressing solid tumors, such as advanced CRC, HNSCC, GEJ adenocarcinoma, NSCLC, and glioma. Although mAb therapies have extended PFS and OS, their efficacy remains modest. Indeed, in mCRC, the response rate is lower than 30% in unselected patients. Mutation analysis has identified a subpopulation of no responders due to intrinsic activating genetic alterations in EGFR downstream signaling molecules, such as KRAS, NRAS, BRAF, and PIK3CA [247–249]. Thus,

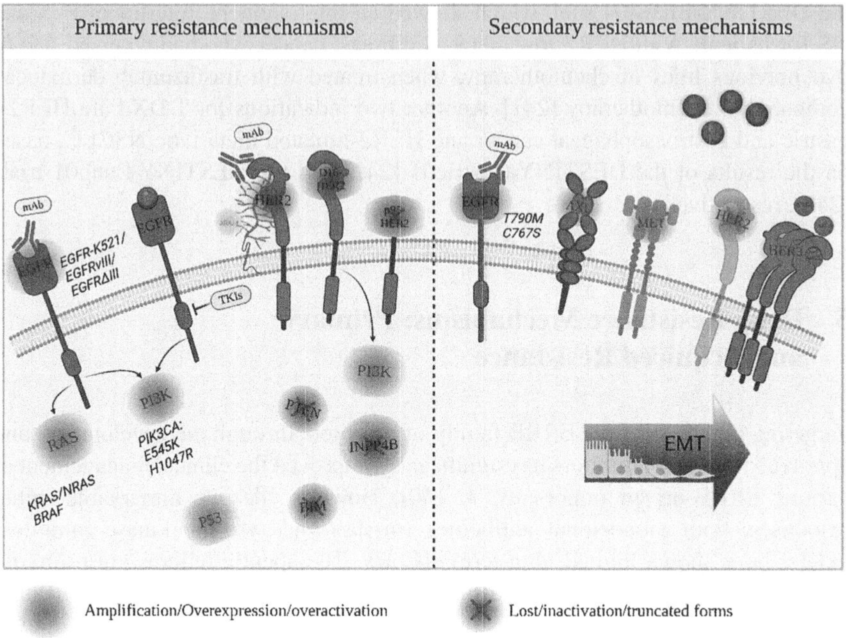

Fig. 4 Common resistance mechanisms to ERBB-targeted therapies. This schematic illustrates the primary (inherent) and secondary (acquired) resistance mechanisms to anti-EGFR and anti-HER2 therapies. (Left) Primary resistance to anti-EGFR mAbs involves activating mutations in RAS, BRAF, and PIK3 pathways, as well as EGFR variants like K521 and EGFRvIII. For EGFR TKIs, resistance can arise from PTEN loss, P53 alterations, and BIM deletion polymorphisms. Inherent resistance to anti-HER2 mAbs involves MUC-4 expression, p95-HER2 truncation, HER2 D16 splice variant, and PI3K/AKT pathway alterations. (Right) Secondary resistance to anti-EGFR therapy includes "on-target" mutations in the EGFR ATP-binding pocket (e.g., T790M, C767S) and "off-target" mechanisms such as overexpression of other ERBB receptors or RTKs. Acquired resistance to anti-HER2 therapy can occur through increased expression of other ERBB family members, loss of HER2 positivity, EMT via HER4-YAP axis, and upregulation of alternative RTKs like MET and its ligand HGF

the TAILOR and PEAK studies have demonstrated a significant improvement in OS for cetuximab and panitumumab only in RAS wild-type patients [250, 251]. Moreover, "on-target" intrinsic mutations and polymorphisms in the EGFR receptor are also responsible for refractory disease. In HNSCC patients, the expression of the EGFR-K521 variant, characterized by a frequent single nucleotide polymorphism, reduces the affinity for cetuximab to EGFR [252]. As well, the type III mutated EGFRvIII, which can be found in prostate, ovarian, and NSCLC, leads to reduced sensitivity to cetuximab [253].

On the other hand, the development of potent small molecule kinase inhibitors in the early 2000s has emerged as highly specific, effective and with less toxicity in tumors harboring somatic mutations within the TK domain. About 12 EGFR TKIs have been approved worldwide for NSCLC with high benefit for patients with a response rate [164, 174, 254]. However, as for mAbs, inherent primary resistances are present in some individuals in which tumor growth and proliferation are guided by different and concurrent gene alterations. Thus, EGFR becomes one, but not the only, target molecule with weak oncogenic potential making EGFR TKIs less effective. In NSCLC, for example, TP53 is one of the frequent alterations (55–65%) followed by loss of PTEN, a tumor suppressor gene inhibitor of the PI3K/AKT pathway, and mutations in PIK3CA (E545K in exon 9 and H1047R in exon 20) [255–260]. The BIM deletion polymorphism (known as BCL2L1), which induces the formation of an inactive protein impairing EGFR TKI-mediated apoptosis, is another primary resistance found in 12–16% of EGFR-mutated lung tumors [261, 262].

Beyond primary resistance, refractory tumors often develop in responders to either mAbs or TKI treatment. In this case, different acquired resistance mechanisms have been described [263]. Acquisition of the new mutation T790M in the ATP-binding pocket of the EGFR typically occurs in 50–60% of NSCLC after responding to first-generation TKIs. Another on-target mutation, namely, C797S, is detected in 7% or 10–20% of patients receiving osimertinib in first-line or second-line treatment, respectively. The co-presence of T790M and C797S in cis (the same allele) induces resistance to all generations of TKIs, while if the two mutations occur in trans (different alleles), tumors are sensitive to the combination of the first and third generations of TKIs. Other EGFR mutations, such as G796R/S, L792H, L718Q, and G724S, may occur with less frequency [264, 265].

Epithelial to mesenchymal transition (EMT), a process where epithelial cells lose polarity and cell-cell contacts, along with cytoskeleton alterations and increased expression of mesenchymal components, has been observed in HNSCC and CRC patients receiving cetuximab [266–268]. In addition, EMT has been reported in a subset of HER2+/PTEN- BC patients following prolonged treatment with trastuzumab [269]. Finally, activation of "bypass track" pathways or changes in ERBB ligand types and/or expression are characteristic in both resistance to TKIs and antibodies. Overexpression of HER2 and HER3 have been widely described as compensatory pathways sustaining survival and proliferation in different tumor types [13, 270–275]. Also, overexpression of other RTKs and their ligands have been described as bypass track. MET amplification and overexpression of hepatocyte growth factor (HGF) or AXL were found in lung cancer, while insulin growth factor (IGF) amplification has been described in BC [276–278].

5.2 Resistance to HER2-Targeted Therapies

Anti-HER2 therapies are currently approved for breast, gastric, and colorectal cancers, where the receptor is highly overexpressed compared to other solid tumors [279, 280]. Usually mAbs, like trastuzumab, are administered in combination with chemotherapy to increase response rate; however, most patients with advanced tumors develop resistance within 1 year. A critical key of therapy failure is related to cellular heterogeneity that exists within the tumor cell population [131, 281]. Recently, researchers have identified a subpopulation of plastic stem-like cells in BC that can reprogram themselves as possible mechanisms of escape from trastuzumab. Moreover, in the tumor microenvironment, several components are present as a biophysical barrier, like MUC-4, a membrane-associated glycoprotein, masking HER2 epitope and the binding site for trastuzumab [282]. Another intrinsic resistance mechanism described by Anido et al. is represented by the expression of the truncated form p95-HER2, lacking the binding site for trastuzumab, which can be bypassed by switching to lapatinib treatment [283]. Moreover, alternative splicing generates a form of HER2 receptor without the exon 16 in the extracellular domain. However, even if the D16-HER2 isoform still contains the antibody's binding site, BC patients do not respond to trastuzumab [284]. Indeed, constitutive activation due to gain-of-function mutations of PI3K/AKT, loss of PTEN or the tumor suppressor INPP4B, and other intrinsic mechanisms of resistance for both trastuzumab and lapatinib treatment are found in BC patients [285–287].

The activation of compensatory signaling pathways is the leading secondary resistance observed with anti-HER2 therapy. Particularly, an increased expression of other members of the ERBB family or their ligands has been observed in BC and GC [288–290]. Loss of HER2 positivity and EMT mediated by HER4-YAP axis activation are also escape mechanisms of trastuzumab-mediated HER2 inhibition investigated in GC [291–293]. Upregulation of IGFR1, MET and its ligand HGF, and EphA2 receptor are other mechanisms implicated in trastuzumab resistance in BC [294–297].

6 Emerging Therapies and Future Directions

6.1 EGFR-Targeted Therapies

6.1.1 Fourth-Generation TKIs

Resistance to osimertinib is associated with the appearance of a tertiary mutation, namely, C797S, in 6–14% of patients who received the third-generation TKI as first- or second-line treatment, respectively [175, 176]. To overcome C797S mutation, fourth-generation EGFR TKIs have been developed. EAI045 is an allosteric inhibitor able to block L858R/T790M and L858R/T790M/C797S EGFR. However, when tested in cell lines and xenograft models expressing mutant forms of EGFR, this inhibitor was active in arresting cell proliferation and tumor growth only when combined with cetuximab [298]. Because of the inefficacy of EAI045 as a single agent, the same group developed JBJ-125–02, another allosteric inhibitor of EGFR, which showed inhibition of in vitro and in vivo models of tumors expressing L858R/T790M and L858R/T790M/C797S EGFR, as a single agent [299]. Moreover, this compound was even more effective when combined with osimertinib [299], paving the way for potential future combinations to be tested in clinical settings. Of note, neither EAI045 nor JBJ-125–02 was active against Exon19Del EGFR [298, 299].

BLU-945 is a potent reversible inhibitor active against several mutants of EGFR. In preclinical studies, this inhibitor blocked the growth of cell line xenografts and patient-derived xenografts, expressing mutant EGFR, as a single agent, as well as combined with osimertinib [300]. The phase I/II clinical trial SYMPHONY evaluated BLU-945 as a monotherapy and in combination with osimertinib in previously treated patients with EGFR-mutated NSCLC. Despite showing a safe profile and anti-tumor activity [300, 301], the clinical development of BLU-945 was discontinued [302]. Another fourth-generation EGFR TKI under clinical development is BDTX-1535. This small molecule is an irreversible EGFR inhibitor active against different mutants of the receptor, including the osimertinib resistance mutation C797S, and able to penetrate the blood-brain barrier. A phase I/II clinical trial is ongoing (NCT05256290). Early data on 19 patients expressing C797S EGFR who received BDTX-1535 showed a preliminary objective response rate of 42% [303, 304]. Several other fourth-generation inhibitors are currently under clinical development (see Table 2).

6.1.2 PROTACs

Proteolysis-targeting chimeras (PROTACs) are bifunctional small molecules composed of a target binder, a linker, and an E3 ligase binder. They act by binding a protein of interest, typically an oncogenic protein, and by inducing its degradation by the ubiquitin-proteasome system [305, 306]. The first EGFR's PROTAC, based on the structure of gefitinib, was able to degrade EGFR in NSCLC cell lines carrying either Exon19Del or L858R EGFR while sparing wild-type receptors [307]. Moreover, another PROTAC built on the structure of afatinib was able to degrade EGFR in cells expressing L858R + T790M, a mutant EGFR resistant to gefitinib [307]. Since then, several EGFR PROTACs have been developed utilizing the first-, second-, third-, and fourth-generation EGFR TKIs as target binders [308]. Another gefitinib-based PROTAC was shown to have favorable pharmacokinetic properties in mice, paving the way for efficacy studies in vivo for this class of molecules [309]. Indeed, a dacomitinib-based degrader showed anti-tumor activity in Exon19Del EGFR-expressing cell line xenografts [310]. Another approach is to employ antibody-based PROTACs. An EGFR degrader, having cetuximab as a target binder, showed a reduction in the viability of EGFR-mutated NSCLC cell lines [311]. While PROTACs represent an innovative and promising therapeutic strategy, issues related to solubility, cell permeability, and metabolic stability should be addressed before translating these preclinical results into clinical practice.

Table 2 Promising ERBB-targeted therapies in clinical development

Target receptor	Class	Drug	Cancer type	Clinical trial identifier
EGFR	mAb	HLX07	CRC	NCT05239650
EGFR, HER2, HER4	TKI (second-generation)	Poziotinib	NSCLC	NCT03066206
EGFR	TKI (third-generation)	Sunvozertinib	NSCLC	NCT06355609 (WUKONG-32-RW), NCT06195189 (WU-KONG36), NCT06348927 (WUKONG-32), NCT06182761 (WUKONG-9), NCT03974022 (WU-KONG1)
		Limertinib	NSCLC	NCT04143607
		Zipalertinib	NSCLC	NCT05967689 (REZILIENT2), NCT05973773 (REZILIENT3)
		Befotertinib	NSCLC	NCT05007938, NCT06517953 (IcomBine), NCT06041776
		Keynatinib	NSCLC	NCT04824079
		Oritinib (SH-1028)	NSCLC	NCT06080776
		Befotertinib	NSCLC	NCT06015568
		Almonertinib	NSCLC	NCT05994339, NCT06043973, NCT05768490, NCT04841811, NCT06417008, NCT04687241, NCT04951635, NCT04923906, NCT06110663
		Furmonertinib	NSCLC	NCT05607550 (FURVENT), NCT05209256, NCT06674343, NCT04853342
	TKI (fourth-generation)	JIN-A02	NSCLC	NCT05394831
		HS-10375	NSCLC	NCT05435248
		WJ13405	NSCLC	NCT05662670
		BLU-945	NSCLC	NCT04862780 (SYMPHONY)
		BPI-361175	NSCLC	NCT05393466
		BDTX-1535	HGG, NSCLC, GBM	NCT06072586, NCT05256290
		QLH11811	NSCLC	NCT05555212
EGFR, HER2	TKI (fourth-generation)	BAY2927088	NSCLC	NCT05099172, NCT06452277

EGFR	ADC	MRG003	Nasopharyngeal carcinoma, HNSCC, advanced solid tumors	NCT05126719, NCT05751512, NCT05688605, NCT06530914, NCT06509997
	PROTAC	HSK40118	NSCLC	NCT06050980
EGFR/MET	BsAb	Bafisontamab (EMB-01)	GI cancers, NSCLC, solid tumors	NCT05176665, NCT05498389, NCT03797391
		MCLA-129	NSCLC	NCT06015568
	Bs-ADC	AZD9592	Advanced solid tumors	NCT05647122
EGFR/HER3	BsAb	Izalontamab (SI-B001)	NSCLC, HNSCC, esophageal squamous cell carcinoma	NCT05949606, NCT06668961, NCT05668858, NCT05022654, NCT05020457, NCT05054439, NCT05020769, NCT05044897, NCT05943795
	Bs-ADC	Izalontamab brengitecan (BL-B01D1)	NSCLC, solid tumors	NCT05983432, NCT06618287

(continued)

Table 2 (continued)

Target receptor	Class	Drug	Cancer type	Clinical trial identifier
HER2	mAb	Inetetamab	BC, NSCLC	NCT05749016, NCT04941885 (Increase), NCT05621434, NCT05823623, NCT05016544, NCT04736589, NCT04681287
	ADC	Trastuzumab rezetecan (SHR-A1811)	Gynecological malignancies, BC, solid tumors, GC, GEJ cancer, CRC, NSCLC, biliary tract cancer	NCT05814354, NCT06057610, NCT06199973, NCT06430437, NCT06126640, NCT06123494, NCT05424835
		ARX788	BC	NCT06663748, NCT04829604 (ACE-Breast-03), NCT06224673, NCT06578286, NCT04983121, NCT05426486
		Disitamab vedotin	BC, GC, GEJ cancer, CRC, urothelial carcinoma, NSCLC, PC, prostate cancer, ovarian cancer	NCT05911295, NCT05904964 (Rosy Trial), NCT06278870, NCT06221748
		MRG002	BC, GC, GEJ cancer, urothelial cancer	NCT04924699, NCT04492488, NCT05754853
		BDC-1001	BC, solid tumors	NCT05954143, NCT04278144
		FS-1502	BC, solid tumors	NCT05755048, NCT03944499
		XMT-2056	Solid tumors	NCT05514717
		BB-1701	BC, solid tumors	NCT04257110, NCT06188559
		DB-1303	BC, solid tumors, uterine cancer	NCT06265428, NCT05150691, NCT06018337 (DYNASTY-Breast02), NCT06340568
		A166	BC, solid tumors	NCT05311397, NCT05346328
	Ab-based PROTAC	ORM-5029	Solid tumors	NCT05511844

HER2/HER2	BsAb	Zanidatamab	BC, solid tumors, GC, GEJ cancer, biliary tract cancer, GI cancers	NCT06043427, NCT06695845, NCT05027139, NCT04224272, NCT05035836, NCT06282575, NCT05152147 (HERIZON-GEA-01), NCT03929666, NCT06435429
		KN026	BC, GC, GEJ cancer, CRC, biliary tract cancer	NCT06023758, NCT05985707, NCT04165993, NCT05427383, NCT05838066
		MBS301	Solid tumors	NCT03842085
	Bs-ADC	Zanidatamab zovodotin	HER2-expressing cancers	NCT03821233
HER2/HER3	BsAb	Zenocutuzumab	NSCLC, prostate cancer, solid tumors	NCT05588609, NCT02912949
HER3	mAb	HMBD-001	NSCLC, solid tumors	NCT05910827, NCT05057013, NCT05919537
		SIBP-03	HNSCC	NCT06194656
	ADC	Patritumab deruxtecan	GI cancers, solid tumors, BC, NSCLC	NCT06596694, NCT06383884, NCT04965766 (ICARUS BREAST), NCT04699630, NCT04619004 (HERTHENA-Lung01), NCT05569811 (VALENTINE), NCT04676477, NCT05338970 (HERTHENA-Lung02), NCT06686394, NCT06172478
		YL202	Solid tumors, breast cancer, NSCLC	NCT06107686, NCT05653752, NCT06439771
		DB-1310	Solid tumors	NCT05785741

The abbreviations used are as follows: *mAb* monoclonal antibody, *TKI* tyrosine kinase inhibitor, *ADC* antibody-drug conjugate, *Ab* antibody, *PROTAC* proteolysis-targeting chimera, *BsAb* bispecific antibody, *Bs* bispecific, *CRC* colorectal cancer, *NSCLC* non-small cell lung cancer, *HGG* high-grade glioma, *GBM* glioblastoma multiforme, *HNSCC* head and neck squamous cell carcinoma, *GI* gastrointestinal, *BC* breast cancer, *GC* gastric cancer, *GEJ* gastroesophageal junction, *PC* pancreatic cancer. The table only shows active studies from clinicaltrials.gov

6.2 HER2-Targeted Therapies

HER2 alterations, either mutation, amplification, or overexpression, have been found in different solid tumors. HER2 overexpression is particularly correlated with poor prognosis for breast, gastric, bladder, and lung cancer. To address the important clinical issue of drug resistance in HER2-driven tumors, several strategies are under clinical investigation. One consists of using BsAbs to recognize and bind two specific antigens. For instance, zanidatamab (ZW25) and anbenitamab target two epitopes of HER2 extracellular domains [312]. Other BsAbs are designed to target HER2 receptors and immune cells: runimotamab (ERBB2/CD3), which increases T-cell-mediated cytotoxicity, and IBI315, both in phase I trial for advanced and metastatic HER2+ tumors [313, 314]. Another strategy is represented by the ADCs. T-DXd has shown promising results in HER2-positive solid tumors, including those with low HER2 expression. This ADC is expected to become a new standard for treating MBC with low HER2 protein levels [315]. Moreover, the FDA granted accelerated approval to T-DXd for adult patients with unresectable or metastatic HER2-positive solid tumors, regardless of the tissue of origin, making it the first tumor-agnostic HER2-directed therapy and antibody-drug conjugate approved for this broad indication [316].

6.3 HER3-Targeted Therapies

Monotherapy with anti-ERBB mAbs has shown limited efficacy in patients, primarily due to the compensatory activation of feedback regulatory loops. Indeed, neutralizing one member often leads to the upregulation of others [270]. Notably, HER3 has garnered great interest as it is considered the preferred heterodimerization partner, within the ERBB family, forming stable dimers with HER2, despite its impaired kinase activity [317]. Several studies have reported that HER3 signaling contributes to resistance in tumors under anti-EGFR/HER2 treatment: for instance, high HER3 expression in HER2+ BC cells is associated with resistance to hormone therapy, HER2-targeted therapy, and chemotherapy [288, 318]. It has been documented that both HER3 and IGF-1R-mediated signaling pathways contribute to trastuzumab resistance, as knocking down HER3 or IGF-1R increases drug sensitivity [295]. Additionally, Engelman et al. demonstrated that in gefitinib-resistant NSCLC cells, amplification of MET induced AKT activation through HER3 signaling [319]. High expression of NRG1, the HER3 ligand, has also been reported as a resistance mechanism in CRC and a subset of ovarian cancers [320, 321]. For this reason, several anti-HER3 mAbs (patritumab, seribantumab, elgemtumab, lumretuzumab) or ADCs like patritumab deruxtecan (HER3-DXd, also known as A3-1402) are under investigation as single drugs or in combination with other mAbs in different solid tumors [275, 322, 323]. Indeed, based on preclinical data and the results of the ARTHENA-Lung01 phase II study in 2023, the administration of HER3-DXd in advanced NSCLC patients refractory to TKIs demonstrated great efficacy and durable response [324].

The BsAb zenocutuzumab, which inhibits the formation of HER2/HER3 heterodimers by targeting domain I of HER2 and domain III of HER3, has shown efficacy and a tolerated safety profile, particularly in MBC [325, 326]. Duligotuzumab is an example of BsAbs targeting EGFR and HER3 that has demonstrated antitumor activity in HNSCC and NSCLC models resistant to cetuximab and erlotinib, respectively [327, 328].

HER3 vaccines are under preclinical and clinical evaluation, and these promote the production of polyclonal antibodies for the targeted receptor and can trigger an immune response against tumor cells. For instance, pING-hHER3FL, a circular DNA fragment, is able to produce HER3 in full length, being tested in NCT03832855 phase I study. Other future strategies described include antisense oligonucleotides, like EZN-3920, or microRNAs, like miR-450b-3p or miR-205, which have been demonstrated to downregulate HER3 in vitro and inhibit tumor growth [329, 330].

7 Conclusion

The ERBB family of receptor tyrosine kinases plays a pivotal role in cancer biology, influencing tumor growth, survival, and metastasis. Targeted therapies, including monoclonal antibodies and tyrosine kinase inhibitors, have significantly advanced the treatment landscape for cancers driven by ERBB receptors, particularly in breast and non-small cell lung cancers. However, challenges such as the emergence of resistance mechanisms, including compensatory signaling pathways, limit the long-term efficacy of these treatments.

Future research directions should focus on developing ERBB-targeted therapies that are tissue-agnostic, allowing for broader application across various cancer types. This includes exploring combination therapies that can effectively target multiple pathways simultaneously, thereby overcoming resistance. Additionally, advancements in genomic profiling can help identify patient subgroups that are more likely to benefit from specific ERBB-targeted therapies, enhancing personalized treatment approaches.

Moreover, innovative strategies such as antisense oligonucleotides and microRNA-based therapies show promise in downregulating HER3 and other resistance pathways, potentially improving treatment outcomes. As our understanding of the ERBB family evolves, integrating evolutionary insights into therapeutic development may reveal new vulnerabilities that can be exploited in cancer treatment.

In summary, while ERBB-targeted therapies have transformed cancer treatment, ongoing research is essential to address the challenges of resistance and to harness the full potential of these therapies across diverse cancer types.

Acknowledgments M.L.U. is supported by the Fundación General CSIC's ComFuturo program, which has received funding from the European Union's Horizon 2020 under Marie Sklodowska-Curie grant agreement no. 101034263. The figures were created using BioRender.com.

Competing Interests The authors have no conflicts of interest to declare that are relevant to the content of this chapter.

References

1. Yarden Y, Sliwkowski MX. Untangling the ErbB signalling network. Nat Rev Mol Cell Biol. 2001;2:127–37.
2. Cohen S, Ushiro H, Stoscheck C, Chinkers M. A native 170,000 epidermal growth factor receptor-kinase complex from shed plasma membrane vesicles. J Biol Chem. 1982;257:1523–31. https://doi.org/10.1016/s0021-9258(19)68224-4.
3. Uribe ML, Marrocco I, Yarden Y. EGFR in cancer: signaling mechanisms, drugs, and acquired resistance. Cancers (Basel). 2021;13:2748. https://doi.org/10.3390/cancers13112748.
4. Marrocco I, Romaniello D, Yarden Y. Cancer immunotherapy: the dawn of antibody cocktails. In: Methods in molecular biology. Humana Press Inc; 2019. p. 11–51.
5. Brunet FG, Volff JN, Schartl M. Whole genome duplications shaped the receptor tyrosine kinase repertoire of jawed vertebrates. Genome Biol Evol. 2016;8:1600–13. https://doi.org/10.1093/GBE/EVW103.
6. Barberán S, Martín-Durán JM, Cebrià F. Evolution of the EGFR pathway in Metazoa and its diversification in the planarian Schmidtea mediterranea. Sci Rep. 2016;6:1–14. https://doi.org/10.1038/srep28071.
7. Volff JN. Genome evolution and biodiversity in teleost fish. Heredity. 2004;94(3):280–94. https://doi.org/10.1038/sj.hdy.6800635.
8. Gómez A, Volff JN, Hornung U, et al. Identification of a second egfr gene in Xiphophorus uncovers an expansion of the epidermal growth factor receptor family in fish. Mol Biol Evol. 2004;21:266–75. https://doi.org/10.1093/molbev/msh017.
9. Littlefield P, Liu L, Mysore V, et al. Structural analysis of the EGFR/HER3 heterodimer reveals the molecular basis for activating HER3 mutations. Sci Signal. 2014;7:ra114. https://doi.org/10.1126/scisignal.2005786.
10. Chen J, Zeng F, Forrester SJ, et al. Expression and function of the epidermal growth factor receptor in physiology and disease. Physiol Rev. 2016;96:1025–69. https://doi.org/10.1152/physrev.00030.2015.
11. Rubin I, Yarden Y. The basic biology of HER2. Ann Oncol. 2001;12(Suppl 1):S3–8. https://doi.org/10.1093/annonc/12.suppl_1.s3.
12. Moutafi M, Robbins CJ, Yaghoobi V, et al. Quantitative measurement of HER2 expression to subclassify ERBB2 unamplified breast cancer. Lab Investig. 2022;102:1101–8. https://doi.org/10.1038/s41374-022-00804-9.
13. Gandullo-Sánchez L, Ocaña A, Pandiella A. HER3 in cancer: from the bench to the bedside. J Exp Clin Cancer Res. 2022;41:310. https://doi.org/10.1186/s13046-022-02515-x.
14. Segers VFM, Dugaucquier L, Feyen E, et al. The role of ErbB4 in cancer. Cell Oncol. 2020;43:335–52.
15. Miettinen PJ, Berger JE, Meneses J, et al. Epithelial immaturity and multiorgan failure in mice lacking epidermal growth factor receptor. Nature. 1995;376(6538):337–41. https://doi.org/10.1038/376337a0.
16. Threadgill DW, Dlugosz AA, Hansen LA, et al. Targeted disruption of mouse EGF receptor: effect of genetic background on mutant phenotype. Science (1979). 1995;269:230–4. https://doi.org/10.1126/science.7618084.

17. Lee KF, Simon H, Chen H, et al. Requirement for neuregulin receptor erbB2 in neural and cardiac development. Nature. 1995;378:394–8. https://doi.org/10.1038/378394A0.
18. Kim K, Lee D. ERBB3-dependent AKT and ERK pathways are essential for atrioventricular cushion development in mouse embryos. PLoS One. 2021;16:e0259426. https://doi.org/10.1371/journal.pone.0259426.
19. Riethmacher D, Sonnenberg-Riethmacher E, Brinkmann V, et al. Severe neuropathies in mice with targeted mutations in the ErbB3 receptor. Nature. 1997;389(6652):725–30. https://doi.org/10.1038/39593.
20. Tidcombe H, Jackson-Fisher A, Mathers K, et al. Neural and mammary gland defects in ErbB4 knockout mice genetically rescued from embryonic lethality. Proc Natl Acad Sci. 2003;100:8281–6. https://doi.org/10.1073/pnas.1436402100.
21. Gassmann M, Casagranda F, Orloli D, et al. Aberrant neural and cardiac development in mice lacking the ErbB4 neuregulin receptor. Nature. 1995;378(6555):390–4. https://doi.org/10.1038/378390a0.
22. Joly-Tonetti N, Ondet T, Monshouwer M, Stamatas GN. EGFR inhibitors switch keratinocytes from a proliferative to a differentiative phenotype affecting epidermal development and barrier function. BMC Cancer. 2021;21:1–10. https://doi.org/10.1186/s12885-020-07685-5.
23. Scheving LA, Zhang X, Garcia OA, et al. Epidermal growth factor receptor plays a role in the regulation of liver and plasma lipid levels in adult male mice. Am J Physiol Gastrointest Liver Physiol. 2014;306:G370. https://doi.org/10.1152/ajpgi.00116.2013.
24. Natarajan A, Wagner B, Sibilia M. The EGF receptor is required for efficient liver regeneration. Proc Natl Acad Sci USA. 2007;104:17081–6. https://doi.org/10.1073/pnas.0704126104.
25. Zhang H, Shi JH, Jiang H, et al. ZBTB20 regulates EGFR expression and hepatocyte proliferation in mouse liver regeneration. Cell Death Dis. 2018;9(5):1–11. https://doi.org/10.1038/s41419-018-0514-0.
26. Perry M-C, Dufour CR, Eichner LJ, et al. ERBB2 deficiency alters an E2F-1-dependent adaptive stress response and leads to cardiac dysfunction. Mol Cell Biol. 2014;34:4232–43. https://doi.org/10.1128/mcb.00895-14.
27. Bersell K, Arab S, Haring B, Kühn B. Neuregulin1/ErbB4 signaling induces cardiomyocyte proliferation and repair of heart injury. Cell. 2009;138:257–70. https://doi.org/10.1016/j.cell.2009.04.060.
28. Crone SA, Zhao YY, Fan L, et al. ErbB2 is essential in the prevention of dilated cardiomyopathy. Nat Med. 2002;8(5):459–65. https://doi.org/10.1038/nm0502-459.
29. Brinkmann BG, Agarwal A, Sereda MW, et al. Neuregulin-1/ErbB signaling serves distinct functions in myelination of the peripheral and central nervous system. Neuron. 2008;59:581–95. https://doi.org/10.1016/j.neuron.2008.06.028.
30. Budi EH, Patterson LB, Parichy DM. Embryonic requirements for ErbB signaling in neural crest development and adult pigment pattern formation. Development. 2008;135:2603–14. https://doi.org/10.1242/dev.019299.
31. Hu X, Zhu Q, Lou T, et al. Pan-ErbB inhibition impairs cognition via disrupting myelination and aerobic glycolysis in oligodendrocytes. Proc Natl Acad Sci. 2024;121:88–100. https://doi.org/10.1073/pnas.2405152121.
32. Citri A, Yarden Y. EGF–ERBB signalling: towards the systems level. Nat Rev Mol Cell Biol. 2006;7(7):505–16. https://doi.org/10.1038/nrm1962.
33. Jones JT, Akita RW, Sliwkowski MX. Binding specificities and affinities of egf domains for ErbB receptors. FEBS Lett. 1999;447:227–31. https://doi.org/10.1016/S0014-5793(99)00283-5.
34. Gullick WJ. The Type 1 growth factor receptors and their ligands considered as a complex system. Endocr Relat Cancer. 2001;8:75–82. https://doi.org/10.1677/erc.0.0080075.
35. Jacobsen NE, Abadi N, Sliwkowski MX, et al. High-resolution solution structure of the EGF-like domain of Heregulin-α. Biochemistry. 1996;35:3402–17. https://doi.org/10.1021/bi952626l.
36. Dang M, Dubbin K, D'Aiello A, et al. Epidermal Growth Factor (EGF) ligand release by substrate-specific A disintegrin and metalloproteases (ADAMs) involves different protein

kinase C (PKC) isoenzymes depending on the stimulus. J Biol Chem. 2011;286:17704–13. https://doi.org/10.1074/jbc.M110.187823.
37. Ogiso H, Ishitani R, Nureki O, et al. Crystal structure of the complex of human epidermal growth factor and receptor extracellular domains. Cell. 2002;110:775–87. https://doi.org/10.1016/S0092-8674(02)00963-7.
38. Cho HS, Mason K, Ramyar KX, et al. Structure of the extracellular region of HER2 alone and in complex with the Herceptin Fab. Nature. 2003;421(6924):756–60. https://doi.org/10.1038/nature01392.
39. Garrett TPJ, McKern NM, Lou M, et al. Crystal structure of a truncated epidermal growth factor receptor extracellular domain bound to transforming growth factor α. Cell. 2002;110:763–73. https://doi.org/10.1016/S0092-8674(02)00940-6.
40. Ward CW, Hoyne PA, Flegg RH. Insulin and epidermal growth factor receptors contain the cysteine repeat motif found in the tumor necrosis factor receptor. Proteins. 1995;22:141–53. https://doi.org/10.1002/prot.340220207.
41. Cho H-S, Leahy DJ. Structure of the extracellular region of HER3 reveals an interdomain tether. Science (1979). 2002;297:1330–3. https://doi.org/10.1126/science.1074611.
42. Jura N, Shan Y, Cao X, et al. Structural analysis of the catalytically inactive kinase domain of the human EGF receptor 3. Proc Natl Acad Sci. 2009;106:21608–13. https://doi.org/10.1073/pnas.0912101106.
43. Lax I, Bellot F, Howk R, et al. Functional analysis of the ligand binding site of EGF-receptor utilizing chimeric chicken/human receptor molecules. EMBO J. 1989;8:421–7. https://doi.org/10.1002/J.1460-2075.1989.TB03393.X.
44. Lemmon MA, Bu Z, Ladbury JE, et al. Two EGF molecules contribute additively to stabilization of the EGFR dimer. EMBO J. 1997;16:281–94. https://doi.org/10.1093/EMBOJ/16.2.281.
45. Bouyain S, Longo PA, Li S, et al. The extracellular region of ErbB4 adopts a tethered conformation in the absence of ligand. Proc Natl Acad Sci USA. 2005;102:15024–9. https://doi.org/10.1073/pnas.0507591102.
46. Dawson JP, Berger MB, Lin C-C, et al. Epidermal growth factor receptor dimerization and activation require ligand-induced conformational changes in the dimer interface. Mol Cell Biol. 2005;25:7734–42. https://doi.org/10.1128/mcb.25.17.7734-7742.2005.
47. Odaka M, Kohda D, Lax I, et al. Ligand-binding enhances the affinity of dimerization of the extracellular domain of the epidermal growth factor receptor. J Biochem. 1997;122:116–21. https://doi.org/10.1093/oxfordjournals.jbchem.a021718.
48. Cymer F, Schneider D. Transmembrane helix-helix interactions involved in ErbB receptor signaling. Cell Adhes Migr. 2010;4:299–312. https://doi.org/10.4161/cam.4.2.11191.
49. Roskoski R. ErbB/HER protein-tyrosine kinases: structures and small molecule inhibitors. Pharmacol Res. 2014;87:42–59. https://doi.org/10.1016/j.phrs.2014.06.001.
50. Thiel KW, Carpenter G. Epidermal growth factor receptor juxtamembrane region regulates allosteric tyrosine kinase activation. Proc Natl Acad Sci USA. 2007;104:19238–43. https://doi.org/10.1073/pnas.0703854104.
51. Red Brewer M, Choi SH, Alvarado D, et al. The Juxtamembrane region of the EGF receptor functions as an activation domain. Mol Cell. 2009;34:641–51. https://doi.org/10.1016/j.molcel.2009.04.034.
52. Stamos J, Sliwkowski MX, Eigenbrot C. Structure of the epidermal growth factor receptor kinase domain alone and in complex with a 4-anilinoquinazoline inhibitor. J Biol Chem. 2002;277:46265–72. https://doi.org/10.1074/jbc.M207135200.
53. Qiu C, Tarrant MK, Boronina T, et al. In vitro enzymatic characterization of near full length EGFR in activated and inhibited states. Biochemistry. 2009;48:6624–32. https://doi.org/10.1021/bi900755n.
54. Pinet L, Wang YH, Deville C, et al. Structural and dynamic characterization of the C-terminal tail of ErbB2: disordered but not random. Biophys J. 2021;120:1869–82. https://doi.org/10.1016/j.bpj.2021.03.005.
55. Schulze WX, Deng L, Mann M. Phosphotyrosine interactome of the ErbB-receptor kinase family. Mol Syst Biol. 2005;1:2005-0008. https://doi.org/10.1038/msb4100012.

56. Gaviraghi M, Rabellino A, Andolfo A, et al. Direct stimulation of ERBB2 highlights a novel cytostatic signaling pathway driven by the receptor Thr701 phosphorylation. Sci Rep. 2020;10(1):1–17. https://doi.org/10.1038/s41598-020-73835-1.
57. Lemmon MA, Schlessinger J, Ferguson KM. The EGFR family: not so prototypical receptor tyrosine kinases. Cold Spring Harb Perspect Biol. 2014;6:a020768. https://doi.org/10.1101/cshperspect.a020768.
58. Kancha RK, Von Bubnoff N, Duyster J. Asymmetric kinase dimer formation is crucial for the activation of oncogenic EGFRvIII but not for ERBB3 phosphorylation. Cell Commun Signal. 2013;11:1–7. https://doi.org/10.1186/1478-811X-11-39.
59. Katz M, Amit I, Yarden Y. Regulation of MAPKs by growth factors and receptor tyrosine kinases. Biochim Biophys Acta. 2007;1773:1161–76. https://doi.org/10.1016/j.bbamcr.2007.01.002.
60. Mendoza MC, Er EE, Blenis J. The Ras-ERK and PI3K-mTOR pathways: cross-talk and compensation. Trends Biochem Sci. 2011;36:320–8. https://doi.org/10.1016/j.tibs.2011.03.006.
61. Fruman DA, Chiu H, Hopkins BD, et al. The PI3K pathway in human disease. Cell. 2017;170:605–35. https://doi.org/10.1016/j.cell.2017.07.029.
62. Hemmings BA, Restuccia DF. The PI3K-PKB/Akt pathway. Cold Spring Harb Perspect Biol. 2015;7:a026609. https://doi.org/10.1101/cshperspect.a026609.
63. Roskoski R. ERK1/2 MAP kinases: structure, function, and regulation. Pharmacol Res. 2012;66:105–43. https://doi.org/10.1016/j.phrs.2012.04.005.
64. Molina JR, Adjei AA. The Ras/Raf/MAPK pathway. J Thorac Oncol. 2006;1:7–9. https://doi.org/10.1016/S1556-0864(15)31506-9.
65. Chen P, Xie H, Wells A. Mitogenic signaling from the egf receptor is attenuated by a phospholipase C-gamma/protein kinase C feedback mechanism. 2017;7:871–81. https://doi.org/10.1091/MBC.7.6.871.
66. Kadamur G, Ross EM. Mammalian phospholipase C. Annu Rev Physiol. 2013;75:127–54. https://doi.org/10.1146/annurev-physiol-030212-183750.
67. Rebecchi MJ, Pentyala SN. Structure, function, and control of phosphoinositide-specific phospholipase C. Physiol Rev. 2000;80:1291–335.
68. Galisteo ML, Chernoff J, Su YC, et al. The adaptor protein Nck links receptor tyrosine kinases with the serine-threonine kinase Pak1. J Biol Chem. 1996;271:20997–1000. https://doi.org/10.1074/JBC.271.35.20997.
69. Poitras L, Jean S, Islam N, Moss T. PAK interacts with NCK and MLK2 to regulate the activation of Jun N-terminal kinase. FEBS Lett. 2003;543:129–35. https://doi.org/10.1016/S0014-5793(03)00424-1.
70. Oved S, Yarden Y. Molecular ticket to enter cells. Nature. 2002;416(6877):133–6. https://doi.org/10.1038/416133a.
71. Olayioye MA, Beuvink I, Horsch K, et al. ErbB receptor-induced activation of Stat transcription factors is mediated by Src tyrosine kinases. J Biol Chem. 1999;274:17209–18. https://doi.org/10.1074/jbc.274.24.17209.
72. Harrison DA. The JAK/STAT pathway. Cold Spring Harb Perspect Biol. 2012;4:a011205. https://doi.org/10.1101/cshperspect.a011205.
73. Wong AJ, Ruppert JM, Bigner SH, et al. Structural alterations of the epidermal growth factor receptor gene in human gliomas. Proc Natl Acad Sci. 1992;89:2965–9. https://doi.org/10.1073/pnas.89.7.2965.
74. Wikstrand CJ, McLendon RE, Friedman AH, Bigner DD. Cell surface localization and density of the tumor-associated variant of the epidermal growth factor receptor, EGFRvIII. Cancer Res. 1997;57:4130–40.
75. Yamazaki H, Fukui Y, Ueyama Y, et al. Amplification of the structurally and functionally altered epidermal growth factor receptor gene (c- erbB) in human brain tumors. Mol Cell Biol. 1988;8:1816–20. https://doi.org/10.1128/mcb.8.4.1816-1820.1988.
76. Du Z, Brown BP, Kim S, et al. Structure–function analysis of oncogenic EGFR kinase domain duplication reveals insights into activation and a potential approach for therapeutic targeting. Nat Commun. 2021;12:1382. https://doi.org/10.1038/s41467-021-21613-6.

77. Schubert L, Elliott A, Le AT, et al. ERBB family fusions are recurrent and actionable oncogenic targets across cancer types. Front Oncol. 2023;13:1115405. https://doi.org/10.3389/fonc.2023.1115405.
78. Moscatello DK, Holgado-Madruga M, Godwin AK, et al. Frequent expression of a mutant epidermal growth factor receptor in multiple human tumors. Cancer Res. 1995;55:5536–9.
79. Wong AJ, Bigner SH, Bigner DD, et al. Increased expression of the epidermal growth factor receptor gene in malignant gliomas is invariably associated with gene amplification. Proc Natl Acad Sci. 1987;84:6899–903. https://doi.org/10.1073/pnas.84.19.6899.
80. Grandal MV, Zandi R, Pedersen MW, et al. EGFRvIII escapes down-regulation due to impaired internalization and sorting to lysosomes. Carcinogenesis. 2007;28:1408–17. https://doi.org/10.1093/carcin/bgm058.
81. Nishikawa R, Ji XD, Harmon RC, et al. A mutant epidermal growth factor receptor common in human glioma confers enhanced tumorigenicity. Proc Natl Acad Sci. 1994;91:7727–31. https://doi.org/10.1073/pnas.91.16.7727.
82. Mei J, Wang T, Xu R, et al. Clinical and molecular immune characterization of ERBB2 in glioma. Int Immunopharmacol. 2021;94:107499. https://doi.org/10.1016/j.intimp.2021.107499.
83. Zhang C, Burger MC, Jennewein L, et al. ErbB2/HER2-specific NK cells for targeted therapy of glioblastoma. JNCI J Natl Cancer Inst. 2016;108:djv375. https://doi.org/10.1093/jnci/djv375.
84. Thai AA, Solomon BJ, Sequist LV, et al. Lung cancer. Lancet. 2021;398:535–54. https://doi.org/10.1016/S0140-6736(21)00312-3.
85. Kumari N, Singh S, Haloi D, et al. Epidermal growth factor receptor mutation frequency in squamous cell carcinoma and its diagnostic performance in cytological samples: a molecular and immunohistochemical study. World J Oncol. 2019;10:142–50. https://doi.org/10.14740/wjon1204.
86. Graham RP, Treece AL, Lindeman NI, et al. Worldwide frequency of commonly detected EGFR mutations. Arch Pathol Lab Med. 2018;142:163–7. https://doi.org/10.5858/arpa.2016-0579-CP.
87. Su J, Zhong W, Zhang X, et al. Molecular characteristics and clinical outcomes of EGFR exon 19 indel subtypes to EGFR TKIs in NSCLC patients. Oncotarget. 2017;8:111246–57. https://doi.org/10.18632/oncotarget.22768.
88. Beau-Faller M, Prim N, Ruppert A-M, et al. Rare EGFR exon 18 and exon 20 mutations in non-small-cell lung cancer on 10 117 patients: a multicentre observational study by the French ERMETIC-IFCT network. Ann Oncol. 2014;25:126–31. https://doi.org/10.1093/annonc/mdt418.
89. Shi Y, Au JS-K, Thongprasert S, et al. A prospective, molecular epidemiology study of EGFR mutations in Asian patients with advanced non–small-cell lung cancer of adenocarcinoma histology (PIONEER). J Thorac Oncol. 2014;9:154–62. https://doi.org/10.1097/JTO.0000000000000033.
90. Kobayashi Y, Mitsudomi T. Not all epidermal growth factor receptor mutations in lung cancer are created equal: perspectives for individualized treatment strategy. Cancer Sci. 2016;107:1179–86. https://doi.org/10.1111/cas.12996.
91. Arcila ME, Nafa K, Chaft JE, et al. EGFR exon 20 insertion mutations in lung adenocarcinomas: prevalence, molecular heterogeneity, and clinicopathologic characteristics. Mol Cancer Ther. 2013;12:220–9. https://doi.org/10.1158/1535-7163.MCT-12-0620.
92. Yasuda H, Kobayashi S, Costa DB. EGFR exon 20 insertion mutations in non-small-cell lung cancer: preclinical data and clinical implications. Lancet Oncol. 2012;13:e23–31. https://doi.org/10.1016/S1470-2045(11)70129-2.
93. Leventakos K, Kipp BR, Rumilla KM, et al. S768I mutation in EGFR in patients with lung cancer. J Thorac Oncol. 2016;11:1798–801. https://doi.org/10.1016/j.jtho.2016.05.007.
94. Li W, Qiu T, Guo L, et al. Primary and acquired EGFR T790M-mutant NSCLC patients identified by routine mutation testing show different characteristics but may both respond to osimertinib treatment. Cancer Lett. 2018;423:9–15. https://doi.org/10.1016/j.canlet.2018.03.005.

95. Ohtsuka K, Ohnishi H, Furuyashiki G, et al. Clinico-pathological and biological significance of tyrosine kinase domain gene mutations and overexpression of epidermal growth factor receptor for lung adenocarcinoma. J Thorac Oncol. 2006;1:787–95. https://doi.org/10.1016/S1556-0864(15)30407-X.
96. Garcia de Palazzo IE, Adams GP, Sundareshan P, et al. Expression of mutated epidermal growth factor receptor by non-small cell lung carcinomas. Cancer Res. 1993;53:3217–20.
97. Okamoto I, Kenyon LC, Emlet DR, et al. Expression of constitutively activated EGFRvlll in non-small cell lung cancer. Cancer Sci. 2003;94:50–6. https://doi.org/10.1111/j.1349-7006.2003.tb01351.x.
98. Sihto H, Puputti M, Pulli L, et al. Epidermal growth factor receptor domain II, IV, and kinase domain mutations in human solid tumors. J Mol Med. 2005;83:976–83. https://doi.org/10.1007/s00109-005-0699-4.
99. Ji H, Zhao X, Yuza Y, et al. Epidermal growth factor receptor variant III mutations in lung tumorigenesis and sensitivity to tyrosine kinase inhibitors. Proc Natl Acad Sci. 2006;103:7817–22. https://doi.org/10.1073/pnas.0510284103.
100. Ohtsuka K, Ohnishi H, Fujiwara M, et al. Abnormalities of epidermal growth factor receptor in lung squamous-cell carcinomas, adenosquamous carcinomas, and large-cell carcinomas. Cancer. 2007;109:741–50. https://doi.org/10.1002/cncr.22476.
101. Sasaki H, Kawano O, Endo K, et al. EGFRvIII mutation in lung cancer correlates with increased EGFR copy number. Oncol Rep. 2007; https://doi.org/10.3892/or.17.2.319.
102. Reissmann PT, Koga H, Figlin RA, et al. Amplification and overexpression of the cyclin D1 and epidermal growth factor receptor genes in non-small-cell lung cancer. J Cancer Res Clin Oncol. 1999;125:61–70. https://doi.org/10.1007/s004320050243.
103. Zhu Y, Wang W, Xu C, et al. EGFR-RAD51 fusion variant in lung adenocarcinoma and response to erlotinib: a case report. Lung Cancer. 2018;115:131–4. https://doi.org/10.1016/j.lungcan.2017.12.001.
104. Konduri K, Gallant J-N, Chae YK, et al. EGFR fusions as novel therapeutic targets in lung cancer. Cancer Discov. 2016;6:601–11. https://doi.org/10.1158/2159-8290.CD-16-0075.
105. Xu H, Shao C. KIF5B-EGFR fusion: a novel EGFR mutation in lung adenocarcinoma. Onco Targets Ther. 2020;13:8317–21. https://doi.org/10.2147/OTT.S263994.
106. Zhang H, Wang J, Li X, et al. Distribution of EGFR fusions in 35,023 Chinese patients with solid tumors-the frequency, fusion partners and clinical outcome. World J Surg Oncol. 2024;22:194. https://doi.org/10.1186/s12957-024-03463-w.
107. Siegel RL, Miller KD, Fuchs HE, Jemal A. Cancer statistics, 2021. CA Cancer J Clin. 2021;71:7–33. https://doi.org/10.3322/caac.21654.
108. Lisio M-A, Fu L, Goyeneche A, et al. High-grade serous ovarian cancer: basic sciences, clinical and therapeutic standpoints. Int J Mol Sci. 2019;20:952. https://doi.org/10.3390/ijms20040952.
109. Lassus H, Sihto H, Leminen A, et al. Gene amplification, mutation, and protein expression of EGFR and mutations of ERBB2 in serous ovarian carcinoma. J Mol Med. 2006;84:671–81. https://doi.org/10.1007/s00109-006-0054-4.
110. Lassus H. ERBB2 amplification is superior to protein expression status in predicting patient outcome in serous ovarian carcinoma. Gynecol Oncol. 2004;92:31–9. https://doi.org/10.1016/j.ygyno.2003.10.010.
111. Nasioudis D, Gysler S, Latif N, et al. Molecular landscape of ERBB2/HER2 gene amplification among patients with gynecologic malignancies; clinical implications and future directions. Gynecol Oncol. 2024;180:1–5. https://doi.org/10.1016/j.ygyno.2023.11.021.
112. Thouvenin L, Charrier M, Clement S, et al. Ovarian cancer with high-level focal ERBB2 amplification responds to trastuzumab and pertuzumab. Gynecol Oncol Rep. 2021;37:100787. https://doi.org/10.1016/j.gore.2021.100787.
113. Wood LD, Parsons DW, Jones S, et al. The genomic landscapes of human breast and colorectal cancers. Science (1979). 2007;318:1108–13. https://doi.org/10.1126/science.1145720.

114. Goldstein NS, Armin M. Epidermal growth factor receptor immunohistochemical reactivity in patients with American Joint Committee on Cancer Stage IV colon adenocarcinoma. Cancer. 2001;92:1331–46. https://doi.org/10.1002/1097-0142(20010901)92:5<1331::AID-CNCR1455>3.0.CO;2-M.
115. Hemming AW, Davis NL, Kluftinger A, et al. Prognostic markers of colorectal cancer: an evaluation of DNA content, epidermal growth factor receptor, and Ki-67. J Surg Oncol. 1992;51:147–52. https://doi.org/10.1002/jso.2930510304.
116. Mayer A, Takimoto M, Fritz E, et al. The prognostic significance of proliferating cell nuclear antigen, epidermal growth factor receptor, and mdr gene expression in colorectal cancer. Cancer. 1993;71:2454–60. https://doi.org/10.1002/1097-0142(19930415)71:8<2454::AID-CNCR2820710805>3.0.CO;2-2.
117. Zaanan A, Henriques J, Cohen R, et al. Efficacy of anti-EGFR in microsatellite instability metastatic colorectal cancer depending on sporadic or familial origin. JNCI J Natl Cancer Inst. 2021;113:496–500. https://doi.org/10.1093/jnci/djaa072.
118. Kasi PM, Afable MG, Herting C, et al. Anti-EGFR antibodies in the management of advanced colorectal cancer. Oncologist. 2023;28:1034–48. https://doi.org/10.1093/oncolo/oyad262.
119. Kim N, Cho D, Kim H, et al. Colorectal adenocarcinoma-derived EGFR mutants are oncogenic and sensitive to EGFR-targeted monoclonal antibodies, cetuximab and panitumumab. Int J Cancer. 2020;146:2194–200. https://doi.org/10.1002/ijc.32499.
120. Cho J, Bass AJ, Lawrence MS, et al. Colon cancer-derived oncogenic EGFR G724S mutant identified by whole genome sequence analysis is dependent on asymmetric dimerization and sensitive to cetuximab. Mol Cancer. 2014;13:141. https://doi.org/10.1186/1476-4598-13-141.
121. Montagut C, Dalmases A, Bellosillo B, et al. Identification of a mutation in the extracellular domain of the epidermal growth factor receptor conferring cetuximab resistance in colorectal cancer. Nat Med. 2012;18:221–3. https://doi.org/10.1038/nm.2609.
122. Nagahara H, Mimori K, Ohta M, et al. Somatic mutations of epidermal growth factor receptor in colorectal carcinoma. Clin Cancer Res. 2005;11:1368–71. https://doi.org/10.1158/1078-0432.CCR-04-1894.
123. Oh B-Y, Lee R-A, Chung S-S, Kim KH. Epidermal growth factor receptor mutations in colorectal cancer patients. J Korean Soc Coloproctol. 2011;27:127. https://doi.org/10.3393/jksc.2011.27.3.127.
124. Peciak J, Stec WJ, Treda C, et al. Low incidence along with low mRNA levels of EGFR vIII in prostate and colorectal cancers compared to glioblastoma. J Cancer. 2017;8:146–51. https://doi.org/10.7150/jca.16108.
125. Li Y, Zhang H-B, Chen X, et al. A rare EGFR – SEPT14 fusion in a patient with colorectal adenocarcinoma responding to Erlotinib. Oncologist. 2020;25:203–7. https://doi.org/10.1634/theoncologist.2019-0405.
126. Maurer CA, Friess H, Kretschmann B, et al. Increased expression of erbB3 in colorectal cancer is associated with concomitant increase in the level of erbB2. Hum Pathol. 1998;29:771–7. https://doi.org/10.1016/S0046-8177(98)90444-0.
127. Muzny DM, Bainbridge MN, Chang K, et al. Comprehensive molecular characterization of human colon and rectal cancer. Nature. 2012;487:330–7. https://doi.org/10.1038/nature11252.
128. Kavuri SM, Jain N, Galimi F, et al. HER2 activating mutations are targets for colorectal cancer treatment. Cancer Discov. 2015;5:832–41. https://doi.org/10.1158/2159-8290.CD-14-1211.
129. Pectasides E, Bass AJ. ERBB2 emerges as a new target for colorectal cancer. Cancer Discov. 2015;5:799–801. https://doi.org/10.1158/2159-8290.CD-15-0730.
130. Arienti C, Pignatta S, Tesei A. Epidermal growth factor receptor family and its role in gastric cancer. Front Oncol. 2019;9:1308. https://doi.org/10.3389/fonc.2019.01308.
131. Kaito A, Kuwata T, Tokunaga M, et al. HER2 heterogeneity is a poor prognosticator for HER2-positive gastric cancer. World J Clin Cases. 2019;7:1964–77. https://doi.org/10.12998/wjcc.v7.i15.1964.
132. Guan Y, Wang Y, Li H, et al. Molecular and clinicopathological characteristics of ERBB2 gene fusions in 32,131 Chinese patients with solid tumors. Front Oncol. 2022;12:986674. https://doi.org/10.3389/fonc.2022.986674.

133. Jiang D, Li X, Wang H, et al. The prognostic value of EGFR overexpression and amplification in esophageal squamous cell carcinoma. BMC Cancer. 2015;15:377. https://doi.org/10.1186/s12885-015-1393-8.
134. Song J, Shi W, Zhang Y, et al. Epidermal growth factor receptor and B7-H3 expression in esophageal squamous tissues correlate to patient prognosis. Onco Targets Ther. 2016;9:6257–63. https://doi.org/10.2147/OTT.S111691.
135. Guo K, Wang W-P, Jiang T, et al. Assessment of epidermal growth factor receptor mutation/copy number and K-ras mutation in esophageal cancer. J Thorac Dis. 2016;8:1753–63. https://doi.org/10.21037/jtd.2016.06.17.
136. Bray F, Laversanne M, Sung H, et al. Global cancer statistics 2022: GLOBOCAN estimates of incidence and mortality worldwide for 36 cancers in 185 countries. CA Cancer J Clin. 2024;74:229–63. https://doi.org/10.3322/caac.21834.
137. Guérin M, Gabillot M, Mathieu M, et al. Structure and expression of c- erb B-2 and EGF receptor genes in inflammatory and non-inflammatory breast cancer: prognostic significance. Int J Cancer. 1989;43:201–8. https://doi.org/10.1002/ijc.2910430205.
138. Rakha EA, El-Sayed ME, Green AR, et al. Prognostic markers in triple-negative breast cancer. Cancer. 2007;109:25–32. https://doi.org/10.1002/cncr.22381.
139. Jungbluth AA, Stockert E, Huang HJS, et al. A monoclonal antibody recognizing human cancers with amplification/overexpression of the human epidermal growth factor receptor. Proc Natl Acad Sci. 2003;100:639–44. https://doi.org/10.1073/pnas.232686499.
140. Rae JM, Scheys JO, Clark KM, et al. EGFR and EGFRvIII expression in primary breast cancer and cell lines. Breast Cancer Res Treat. 2004;87:87–95. https://doi.org/10.1023/B:BREA.0000041585.26734.f9.
141. Slamon DJ, Clark GM, Wong SG, et al. Human breast cancer: correlation of relapse and survival with amplification of the HER-2/neu oncogene. Science (1979). 1987;235:177–82. https://doi.org/10.1126/science.3798106.
142. Peraldo-Neia C, Migliardi G, Mello-Grand M, et al. Epidermal growth factor receptor (EGFR) mutation analysis, gene expression profiling and EGFR protein expression in primary prostate cancer. BMC Cancer. 2011;11:31. https://doi.org/10.1186/1471-2407-11-31.
143. Schlomm T, Kirstein P, Iwers L, et al. Clinical significance of epidermal growth factor receptor protein overexpression and gene copy number gains in prostate cancer. Clin Cancer Res. 2007;13:6579–84. https://doi.org/10.1158/1078-0432.CCR-07-1257.
144. Cai CQ, Peng Y, Buckley MT, et al. Epidermal growth factor receptor activation in prostate cancer by three novel missense mutations. Oncogene. 2008;27:3201–10. https://doi.org/10.1038/sj.onc.1210983.
145. Olapade-Olaopa EO, Moscatello DK, MacKay EH, et al. Evidence for the differential expression of a variant EGF receptor protein in human prostate cancer. Br J Cancer. 2000;82:186–94. https://doi.org/10.1054/bjoc.1999.0898.
146. Zhau HE, Sen WD, Zhou J, et al. Expression of c- erb B-2/neu proto-oncogene in human prostatic cancer tissues and cell lines. Mol Carcinog. 1992;5:320–7. https://doi.org/10.1002/mc.2940050413.
147. Palumbo C, Benvenuto M, Focaccetti C, et al. Recent findings on the impact of ErbB receptors status on prognosis and therapy of head and neck squamous cell carcinoma. Front Med (Lausanne). 2023;10:1066021. https://doi.org/10.3389/FMED.2023.1066021/PDF.
148. Vatte C, Al Amri AM, Cyrus C, et al. Tyrosine kinase domain mutations of *EGFR* gene in head and neck squamous cell carcinoma. Onco Targets Ther. 2017;10:1527–33. https://doi.org/10.2147/OTT.S132187.
149. Nair S, Trummell HQ, Rajbhandari R, et al. Novel EGFR ectodomain mutations associated with ligand-independent activation and cetuximab resistance in head and neck cancer. PLoS One. 2020;15:e0229077. https://doi.org/10.1371/journal.pone.0229077.
150. Xu Z, Qin F, Yuan L, et al. EGFR DNA methylation correlates with EGFR expression, immune cell infiltration, and overall survival in lung adenocarcinoma. Front Oncol. 2021;11:691915. https://doi.org/10.3389/fonc.2021.691915.

151. Song H, Li C-W, Labaff AM, et al. Acetylation of EGF receptor contributes to tumor cell resistance to histone deacetylase inhibitors. Biochem Biophys Res Commun. 2011;404:68–73. https://doi.org/10.1016/j.bbrc.2010.11.064.
152. Cheng L, Gou L, Wei T, Zhang J. GBP1 promotes erlotinib resistance via PGK1-activated EMT signaling in non-small cell lung cancer. Int J Oncol. 2020;57:858–70. https://doi.org/10.3892/ijo.2020.5086.
153. Wu L, Yu Y, Xu L, et al. TROY modulates cancer stem-like cell properties and gefitinib resistance through EMT signaling in non–small cell lung cancer. Front Genet. 2022;13:881875. https://doi.org/10.3389/fgene.2022.881875.
154. Haga Y, Marrocco I, Noronha A, et al. Host-dependent phenotypic resistance to EGFR tyrosine kinase inhibitors. Cancer Res. 2021;81:3862. https://doi.org/10.1158/0008-5472.CAN-20-3555.
155. Yarden Y, Pines G. The ERBB network: at last, cancer therapy meets systems biology. Nat Rev Cancer. 2012;12:553–63. https://doi.org/10.1038/nrc3309.
156. Moran AE, Hunt DH, Javid SH, et al. Apc deficiency is associated with increased Egfr activity in the intestinal enterocytes and adenomas of C57BL/6J-Min/+ mice. J Biol Chem. 2004;279:43261–72. https://doi.org/10.1074/jbc.M404276200.
157. Kim S-E, Choi K-Y. EGF receptor is involved in WNT3a-mediated proliferation and motility of NIH3T3 cells via ERK pathway activation. Cell Signal. 2007;19:1554–64. https://doi.org/10.1016/j.cellsig.2007.02.003.
158. Schroeder JA, Adriance MC, McConnell EJ, et al. ErbB-β-catenin complexes are associated with human infiltrating ductal breast and murine mammary tumor virus (MMTV)-Wnt-1 and MMTV-c-Neu transgenic carcinomas. J Biol Chem. 2002;277:22692–8. https://doi.org/10.1074/jbc.M201975200.
159. Abourehab MAS, Alqahtani AM, Youssif BGM, Gouda AM. Globally approved EGFR inhibitors: insights into their syntheses, target kinases, biological activities, receptor interactions, and metabolism. Molecules. 2021;26:6677. https://doi.org/10.3390/molecules26216677.
160. Levantini E, Maroni G, Del Re M, Tenen DG. EGFR signaling pathway as therapeutic target in human cancers. Semin Cancer Biol. 2022;85:253–75. https://doi.org/10.1016/j.semcancer.2022.04.002.
161. Huang J, Meng L, Yang B, et al. Safety profile of epidermal growth factor receptor tyrosine kinase inhibitors: a disproportionality analysis of FDA adverse event reporting system. Sci Rep. 2020;10:4803. https://doi.org/10.1038/s41598-020-61571-5.
162. Zhou C, Wu Y-L, Chen G, et al. Erlotinib versus chemotherapy as first-line treatment for patients with advanced EGFR mutation-positive non-small-cell lung cancer (OPTIMAL, CTONG-0802): a multicentre, open-label, randomised, phase 3 study. Lancet Oncol. 2011;12:735–42. https://doi.org/10.1016/S1470-2045(11)70184-X.
163. Mitsudomi T, Morita S, Yatabe Y, et al. Gefitinib versus cisplatin plus docetaxel in patients with non-small-cell lung cancer harbouring mutations of the epidermal growth factor receptor (WJTOG3405): an open label, randomised phase 3 trial. Lancet Oncol. 2010;11:121–8. https://doi.org/10.1016/S1470-2045(09)70364-X.
164. Rosell R, Carcereny E, Gervais R, et al. Erlotinib versus standard chemotherapy as first-line treatment for European patients with advanced EGFR mutation-positive non-small-cell lung cancer (EURTAC): a multicentre, open-label, randomised phase 3 trial. Lancet Oncol. 2012;13:239–46. https://doi.org/10.1016/S1470-2045(11)70393-X.
165. Sequist LV, Yang JC-H, Yamamoto N, et al. Phase III study of Afatinib or cisplatin plus pemetrexed in patients with metastatic lung adenocarcinoma with EGFR mutations. J Clin Oncol. 2013;31:3327–34. https://doi.org/10.1200/JCO.2012.44.2806.
166. Wu Y-L, Zhou C, Hu C-P, et al. Afatinib versus cisplatin plus gemcitabine for first-line treatment of Asian patients with advanced non-small-cell lung cancer harbouring EGFR mutations (LUX-Lung 6): an open-label, randomised phase 3 trial. Lancet Oncol. 2014;15:213–22. https://doi.org/10.1016/S1470-2045(13)70604-1.

167. Park K, Tan E-H, O'Byrne K, et al. Afatinib versus gefitinib as first-line treatment of patients with EGFR mutation-positive non-small-cell lung cancer (LUX-Lung 7): a phase 2B, open-label, randomised controlled trial. Lancet Oncol. 2016;17:577–89. https://doi.org/10.1016/S1470-2045(16)30033-X.
168. Paz-Ares L, Tan E-H, O'Byrne K, et al. Afatinib versus gefitinib in patients with EGFR mutation-positive advanced non-small-cell lung cancer: overall survival data from the phase IIb LUX-Lung 7 trial. Ann Oncol. 2017;28:270–7. https://doi.org/10.1093/annonc/mdw611.
169. Mok TS, Cheng Y, Zhou X, et al. Improvement in overall survival in a randomized study that compared dacomitinib with gefitinib in patients with advanced non–small-cell lung cancer and EGFR-activating mutations. J Clin Oncol. 2018;36:2244–50. https://doi.org/10.1200/JCO.2018.78.7994.
170. Wu Y-L, Cheng Y, Zhou X, et al. Dacomitinib versus gefitinib as first-line treatment for patients with EGFR-mutation-positive non-small-cell lung cancer (ARCHER 1050): a randomised, open-label, phase 3 trial. Lancet Oncol. 2017;18:1454–66. https://doi.org/10.1016/S1470-2045(17)30608-3.
171. Zhao Y, Cheng B, Chen Z, et al. Toxicity profile of epidermal growth factor receptor tyrosine kinase inhibitors for patients with lung cancer: a systematic review and network meta-analysis. Crit Rev Oncol Hematol. 2021;160:103305. https://doi.org/10.1016/j.critrevonc.2021.103305.
172. Mok TS, Wu Y-L, Ahn M-J, et al. Osimertinib or platinum–pemetrexed in EGFR T790M–positive lung cancer. N Engl J Med. 2017;376:629–40. https://doi.org/10.1056/NEJMoa1612674.
173. Ramalingam SS, Vansteenkiste J, Planchard D, et al. Overall survival with osimertinib in untreated, EGFR-mutated advanced NSCLC. N Engl J Med. 2020;382:41–50. https://doi.org/10.1056/NEJMoa1913662.
174. Soria J-C, Ohe Y, Vansteenkiste J, et al. Osimertinib in untreated EGFR-mutated advanced non–small-cell lung cancer. N Engl J Med. 2018;378:113–25. https://doi.org/10.1056/NEJMoa1713137.
175. Chmielecki J, Gray JE, Cheng Y, et al. Candidate mechanisms of acquired resistance to first-line osimertinib in EGFR-mutated advanced non-small cell lung cancer. Nat Commun. 2023;14:1070. https://doi.org/10.1038/s41467-023-35961-y.
176. Chmielecki J, Mok T, Wu Y-L, et al. Analysis of acquired resistance mechanisms to osimertinib in patients with EGFR-mutated advanced non-small cell lung cancer from the AURA3 trial. Nat Commun. 2023;14:1071. https://doi.org/10.1038/s41467-023-35962-x.
177. Mendelsohn J, Baselga J. Status of epidermal growth factor receptor antagonists in the biology and treatment of cancer. J Clin Oncol. 2003;21:2787–99. https://doi.org/10.1200/JCO.2003.01.504.
178. Baysal H, De Pauw I, Zaryouh H, et al. The right partner in crime: unlocking the potential of the anti-EGFR antibody cetuximab via combination with natural killer cell chartering immunotherapeutic strategies. Front Immunol. 2021;12:737311. https://doi.org/10.3389/fimmu.2021.737311.
179. Peng D, Fan Z, Lu Y, et al. Anti-epidermal growth factor receptor monoclonal antibody 225 up-regulates p27KIP1 and induces G1 arrest in prostatic cancer cell line DU145. Cancer Res. 1996;56:3666–9.
180. Wu X, Rubin M, Fan Z, et al. Involvement of p27KIP1 in G1 arrest mediated by an anti-epidermal growth factor receptor monoclonal antibody. Oncogene. 1996;12:1397–403.
181. Huang SM, Bock JM, Harari PM. Epidermal growth factor receptor blockade with C225 modulates proliferation, apoptosis, and radiosensitivity in squamous cell carcinomas of the head and neck. Cancer Res. 1999;59:1935–40.
182. Van Cutsem E, Köhne C-H, Hitre E, et al. Cetuximab and chemotherapy as initial treatment for metastatic colorectal cancer. N Engl J Med. 2009;360:1408–17. https://doi.org/10.1056/NEJMoa0805019.
183. Au H-J, Karapetis CS, O'Callaghan CJ, et al. Health-related quality of life in patients with advanced colorectal cancer treated with cetuximab: overall and KRAS-specific results of the

NCIC CTG and AGITG CO.17 trial. J Clin Oncol. 2009;27:1822–8. https://doi.org/10.1200/JCO.2008.19.6048.
184. Bokemeyer C, Köhne C-H, Ciardiello F, et al. FOLFOX4 plus cetuximab treatment and RAS mutations in colorectal cancer. Eur J Cancer. 2015;51:1243–52. https://doi.org/10.1016/j.ejca.2015.04.007.
185. García-Foncillas J, Sunakawa Y, Aderka D, et al. Distinguishing features of cetuximab and panitumumab in colorectal cancer and other solid tumors. Front Oncol. 2019;9:849. https://doi.org/10.3389/fonc.2019.00849.
186. Voigt M, Braig F, Göthel M, et al. Functional dissection of the epidermal growth factor receptor epitopes targeted by panitumumab and cetuximab. Neoplasia. 2012;14:1023-IN3. https://doi.org/10.1593/neo.121242.
187. Park K, Cho EK, Bello M, et al. Efficacy and safety of first-line necitumumab plus gemcitabine and cisplatin versus gemcitabine and cisplatin in east Asian patients with stage IV squamous non-small cell lung cancer: a subgroup analysis of the phase 3, open-label, randomized SQUIRE study. Cancer Res Treat. 2017;49:937–46. https://doi.org/10.4143/crt.2016.423.
188. Vyse S, Huang PH. Targeting EGFR exon 20 insertion mutations in non-small cell lung cancer. Signal Transduct Target Ther. 2019;4:5. https://doi.org/10.1038/s41392-019-0038-9.
189. Park K, Haura EB, Leighl NB, et al. Amivantamab in EGFR exon 20 insertion–mutated non–small-cell lung cancer progressing on platinum chemotherapy: initial results from the CHRYSALIS phase I study. J Clin Oncol. 2021;39:3391–402. https://doi.org/10.1200/JCO.21.00662.
190. Chon K, Larkins E, Chatterjee S, et al. FDA approval summary: amivantamab for the treatment of patients with non–small cell lung cancer with EGFR exon 20 insertion mutations. Clin Cancer Res. 2023;29:3262–6. https://doi.org/10.1158/1078-0432.CCR-22-3713.
191. Zhou C, Tang K-J, Cho BC, et al. Amivantamab plus chemotherapy in NSCLC with EGFR exon 20 insertions. N Engl J Med. 2023;389:2039–51. https://doi.org/10.1056/NEJMoa2306441.
192. Moores SL, Chiu ML, Bushey BS, et al. A novel bispecific antibody targeting EGFR and cMet is effective against EGFR inhibitor–resistant lung tumors. Cancer Res. 2016;76:3942–53. https://doi.org/10.1158/0008-5472.CAN-15-2833.
193. Grugan KD, Dorn K, Jarantow SW, et al. Fc-mediated activity of EGFR x c-Met bispecific antibody JNJ-61186372 enhanced killing of lung cancer cells. MAbs. 2017;9:114–26. https://doi.org/10.1080/19420862.2016.1249079.
194. Vijayaraghavan S, Lipfert L, Chevalier K, et al. Amivantamab (JNJ-61186372), an Fc enhanced EGFR/cMet bispecific antibody, induces receptor downmodulation and antitumor activity by monocyte/macrophage trogocytosis. Mol Cancer Ther. 2020;19:2044–56. https://doi.org/10.1158/1535-7163.MCT-20-0071.
195. Cho BC, Lu S, Felip E, et al. Amivantamab plus lazertinib in previously untreated EGFR-mutated advanced NSCLC. N Engl J Med. 2024;391:1486–98. https://doi.org/10.1056/NEJMoa2403614.
196. Passaro A, Wang J, Wang Y, et al. Amivantamab plus chemotherapy with and without lazertinib in EGFR-mutant advanced NSCLC after disease progression on osimertinib: primary results from the phase III MARIPOSA-2 study. Ann Oncol. 2024;35:77–90. https://doi.org/10.1016/j.annonc.2023.10.117.
197. Cameron D, Casey M, Press M, et al. A phase III randomized comparison of lapatinib plus capecitabine versus capecitabine alone in women with advanced breast cancer that has progressed on trastuzumab: updated efficacy and biomarker analyses. Breast Cancer Res Treat. 2008;112:533–43. https://doi.org/10.1007/s10549-007-9885-0.
198. Geyer CE, Forster J, Lindquist D, et al. Lapatinib plus capecitabine for HER2-positive advanced breast cancer. N Engl J Med. 2006;355:2733–43. https://doi.org/10.1056/NEJMoa064320.

199. Konecny GE, Pegram MD, Venkatesan N, et al. Activity of the dual kinase inhibitor lapatinib (GW572016) against HER-2-overexpressing and trastuzumab-treated breast cancer cells. Cancer Res. 2006;66:1630–9. https://doi.org/10.1158/0008-5472.CAN-05-1182.
200. Blackwell KL, Burstein HJ, Storniolo AM, et al. Randomized study of lapatinib alone or in combination with trastuzumab in women with ErbB2-positive, trastuzumab-refractory metastatic breast cancer. J Clin Oncol. 2010;28:1124–30. https://doi.org/10.1200/JCO.2008.21.4437.
201. Rabindran SK, Discafani CM, Rosfjord EC, et al. Antitumor activity of HKI-272, an orally active, irreversible inhibitor of the HER-2 tyrosine kinase. Cancer Res. 2004;64:3958–65. https://doi.org/10.1158/0008-5472.CAN-03-2868.
202. Wissner A, Mansour TS. The development of HKI-272 and related compounds for the treatment of cancer. Arch Pharm (Weinheim). 2008;341:465–77. https://doi.org/10.1002/ardp.200800009.
203. Chan A, Delaloge S, Holmes FA, et al. Neratinib after trastuzumab-based adjuvant therapy in patients with HER2-positive breast cancer (ExteNET): a multicentre, randomised, double-blind, placebo-controlled, phase 3 trial. Lancet Oncol. 2016;17:367–77. https://doi.org/10.1016/S1470-2045(15)00551-3.
204. Saura C, Oliveira M, Feng Y-H, et al. Neratinib plus capecitabine versus lapatinib plus capecitabine in HER2-positive metastatic breast cancer previously treated with ≥ 2 HER2-directed regimens: phase III NALA trial. J Clin Oncol. 2020;38:3138–49. https://doi.org/10.1200/JCO.20.00147.
205. Frankel C, Palmieri FM. Lapatinib side-effect management. Clin J Oncol Nurs. 2010;14:223–33. https://doi.org/10.1188/10.CJON.223-233.
206. Sodergren SC, Copson E, White A, et al. Systematic review of the side effects associated with anti-HER2-targeted therapies used in the treatment of breast cancer, on behalf of the EORTC quality of life group. Target Oncol. 2016;11:277–92. https://doi.org/10.1007/s11523-015-0409-2.
207. Kulukian A, Lee P, Taylor J, et al. Preclinical activity of HER2-selective tyrosine kinase inhibitor tucatinib as a single agent or in combination with trastuzumab or docetaxel in solid tumor models. Mol Cancer Ther. 2020;19:976–87. https://doi.org/10.1158/1535-7163.MCT-19-0873.
208. Curigliano G, Mueller V, Borges V, et al. Tucatinib versus placebo added to trastuzumab and capecitabine for patients with pretreated HER2+ metastatic breast cancer with and without brain metastases (HER2CLIMB): final overall survival analysis. Ann Oncol. 2022;33:321–9. https://doi.org/10.1016/j.annonc.2021.12.005.
209. Murthy RK, Loi S, Okines A, et al. Tucatinib, trastuzumab, and capecitabine for HER2-positive metastatic breast cancer. N Engl J Med. 2020;382:597–609. https://doi.org/10.1056/NEJMoa1914609.
210. Benson AB, Venook AP, Al-Hawary MM, et al. Colon cancer, version 2.2021, NCCN clinical practice guidelines in oncology. J Natl Compr Cancer Netw. 2021;19:329–59. https://doi.org/10.6004/jnccn.2021.0012.
211. Sartore-Bianchi A, Amatu A, Porcu L, et al. HER2 positivity predicts unresponsiveness to EGFR-targeted treatment in metastatic colorectal cancer. Oncologist. 2019;24:1395–402. https://doi.org/10.1634/theoncologist.2018-0785.
212. Sartore-Bianchi A, Trusolino L, Martino C, et al. Dual-targeted therapy with trastuzumab and lapatinib in treatment-refractory, KRAS codon 12/13 wild-type, HER2-positive metastatic colorectal cancer (HERACLES): a proof-of-concept, multicentre, open-label, phase 2 trial. Lancet Oncol. 2016;17:738–46. https://doi.org/10.1016/S1470-2045(16)00150-9.
213. Strickler JH, Cercek A, Siena S, et al. Tucatinib plus trastuzumab for chemotherapy-refractory, HER2-positive, RAS wild-type unresectable or metastatic colorectal cancer (MOUNTAINEER): a multicentre, open-label, phase 2 study. Lancet Oncol. 2023;24:496–508. https://doi.org/10.1016/S1470-2045(23)00150-X.

214. Carter P, Presta L, Gorman CM, et al. Humanization of an anti-p185HER2 antibody for human cancer therapy. Proc Natl Acad Sci. 1992;89:4285–9. https://doi.org/10.1073/pnas.89.10.4285.
215. Pietras RJ, Pegram MD, Finn RS, et al. Remission of human breast cancer xenografts on therapy with humanized monoclonal antibody to HER-2 receptor and DNA-reactive drugs. Oncogene. 1998;17:2235–49. https://doi.org/10.1038/sj.onc.1202132.
216. Baselga J, Norton L, Albanell J, et al. Recombinant humanized anti-HER2 antibody (Herceptin) enhances the antitumor activity of paclitaxel and doxorubicin against HER2/neu overexpressing human breast cancer xenografts. Cancer Res. 1998;58:2825–31.
217. Slamon DJ, Leyland-Jones B, Shak S, et al. Use of chemotherapy plus a monoclonal antibody against HER2 for metastatic breast cancer that overexpresses HER2. N Engl J Med. 2001;344:783–92. https://doi.org/10.1056/NEJM200103153441101.
218. Romond EH, Perez EA, Bryant J, et al. Trastuzumab plus adjuvant chemotherapy for operable HER2-positive breast cancer. N Engl J Med. 2005;353:1673–84. https://doi.org/10.1056/NEJMoa052122.
219. Piccart-Gebhart MJ, Procter M, Leyland-Jones B, et al. Trastuzumab after adjuvant chemotherapy in HER2-positive breast cancer. N Engl J Med. 2005;353:1659–72. https://doi.org/10.1056/NEJMoa052306.
220. Matsui Y, Inomata M, Tojigamori M, et al. Suppression of tumor growth in human gastric cancer with HER2 overexpression by an anti-HER2 antibody in a murine model. Int J Oncol. 2005;27:681–5.
221. Tanner M, Hollmén M, Junttila TT, et al. Amplification of HER-2 in gastric carcinoma: association with Topoisomerase IIα gene amplification, intestinal type, poor prognosis and sensitivity to trastuzumab. Ann Oncol. 2005;16:273–8. https://doi.org/10.1093/annonc/mdi064.
222. Bang Y-J, Van Cutsem E, Feyereislova A, et al. Trastuzumab in combination with chemotherapy versus chemotherapy alone for treatment of HER2-positive advanced gastric or gastro-oesophageal junction cancer (ToGA): a phase 3, open-label, randomised controlled trial. Lancet. 2010;376:687–97. https://doi.org/10.1016/S0140-6736(10)61121-X.
223. Baselga J. A new anti-ErbB2 strategy in the treatment of cancer: prevention of ligand-dependent ErbB2 receptor heterodimerization. Cancer Cell. 2002;2:93–5. https://doi.org/10.1016/S1535-6108(02)00098-3.
224. Franklin MC, Carey KD, Vajdos FF, et al. Insights into ErbB signaling from the structure of the ErbB2-pertuzumab complex. Cancer Cell. 2004;5:317–28. https://doi.org/10.1016/S1535-6108(04)00083-2.
225. Scheuer W, Friess T, Burtscher H, et al. Strongly enhanced antitumor activity of trastuzumab and pertuzumab combination treatment on HER2-positive human xenograft tumor models. Cancer Res. 2009;69:9330–6. https://doi.org/10.1158/0008-5472.CAN-08-4597.
226. Baselga J, Cortés J, Kim S-B, et al. Pertuzumab plus trastuzumab plus docetaxel for metastatic breast cancer. N Engl J Med. 2012;366:109–19. https://doi.org/10.1056/NEJMoa1113216.
227. Gianni L, Pienkowski T, Im Y-H, et al. Efficacy and safety of neoadjuvant pertuzumab and trastuzumab in women with locally advanced, inflammatory, or early HER2-positive breast cancer (NeoSphere): a randomised multicentre, open-label, phase 2 trial. Lancet Oncol. 2012;13:25–32. https://doi.org/10.1016/S1470-2045(11)70336-9.
228. Nahta R, Hung M-C, Esteva FJ. The HER-2-targeting antibodies trastuzumab and pertuzumab synergistically inhibit the survival of breast cancer cells. Cancer Res. 2004;64:2343–6. https://doi.org/10.1158/0008-5472.CAN-03-3856.
229. von Minckwitz G, Procter M, de Azambuja E, et al. Adjuvant pertuzumab and trastuzumab in early HER2-positive breast cancer. N Engl J Med. 2017;377:122–31. https://doi.org/10.1056/NEJMoa1703643.
230. Swain SM, Kim S-B, Cortés J, et al. Pertuzumab, trastuzumab, and docetaxel for HER2-positive metastatic breast cancer (CLEOPATRA study): overall survival results from a randomised, double-blind, placebo-controlled, phase 3 study. Lancet Oncol. 2013;14:461–71. https://doi.org/10.1016/S1470-2045(13)70130-X.

231. Swain SM, Miles D, Kim S-B, et al. Pertuzumab, trastuzumab, and docetaxel for HER2-positive metastatic breast cancer (CLEOPATRA): end-of-study results from a double-blind, randomised, placebo-controlled, phase 3 study. Lancet Oncol. 2020;21:519–30. https://doi.org/10.1016/S1470-2045(19)30863-0.
232. Nordstrom JL, Gorlatov S, Zhang W, et al. Anti-tumor activity and toxicokinetics analysis of MGAH22, an anti-HER2 monoclonal antibody with enhanced Fcγ receptor binding properties. Breast Cancer Res. 2011;13:R123. https://doi.org/10.1186/bcr3069.
233. Rugo HS, Im S-A, Cardoso F, et al. Margetuximab versus trastuzumab in patients with previously treated HER2-positive advanced breast cancer (SOPHIA): final overall survival results from a randomized phase 3 trial. J Clin Oncol. 2023;41:198–205. https://doi.org/10.1200/JCO.21.02937.
234. Burris HA. Developments in the use of antibody-drug conjugates. American Society of Clinical Oncology Educational Book; 2013. p. e99–e102. https://doi.org/10.14694/EdBook_AM.2013.33.e99.
235. Barok M, Joensuu H, Isola J. Trastuzumab emtansine: mechanisms of action and drug resistance. Breast Cancer Res. 2014;16:209. https://doi.org/10.1186/bcr3621.
236. Verma S, Miles D, Gianni L, et al. Trastuzumab Emtansine for HER2-positive advanced breast cancer. N Engl J Med. 2012;367:1783–91. https://doi.org/10.1056/NEJMoa1209124.
237. von Minckwitz G, Huang C-S, Mano MS, et al. Trastuzumab Emtansine for residual invasive HER2-positive breast cancer. N Engl J Med. 2019;380:617–28. https://doi.org/10.1056/NEJMoa1814017.
238. Ogitani Y, Aida T, Hagihara K, et al. DS-8201a, a novel HER2-targeting ADC with a novel DNA topoisomerase I inhibitor, demonstrates a promising antitumor efficacy with differentiation from T-DM1. Clin Cancer Res. 2016;22:5097–108. https://doi.org/10.1158/1078-0432.CCR-15-2822.
239. Modi S, Saura C, Yamashita T, et al. Trastuzumab deruxtecan in previously treated HER2-positive breast cancer. N Engl J Med. 2020;382:610–21. https://doi.org/10.1056/NEJMoa1914510.
240. Cortés J, Kim S-B, Chung W-P, et al. Trastuzumab deruxtecan versus trastuzumab emtansine for breast cancer. N Engl J Med. 2022;386:1143–54. https://doi.org/10.1056/NEJMoa2115022.
241. Modi S, Jacot W, Yamashita T, et al. Trastuzumab deruxtecan in previously treated HER2-low advanced breast cancer. N Engl J Med. 2022;387:9–20. https://doi.org/10.1056/NEJMoa2203690.
242. Shitara K, Bang Y-J, Iwasa S, et al. Trastuzumab deruxtecan in previously treated HER2-positive gastric cancer. N Engl J Med. 2020;382:2419–30. https://doi.org/10.1056/NEJMoa2004413.
243. Li BT, Smit EF, Goto Y, et al. Trastuzumab deruxtecan in HER2-mutant non–small-cell lung cancer. N Engl J Med. 2022;386:241–51. https://doi.org/10.1056/NEJMoa2112431.
244. Morgillo F, Bareschino MA, Bianco R, et al. Primary and acquired resistance to anti-EGFR targeted drugs in cancer therapy. Differentiation. 2007;75:788–99. https://doi.org/10.1111/j.1432-0436.2007.00200.x.
245. Benavente S, Huang S, Armstrong EA, et al. Establishment and characterization of a model of acquired resistance to epidermal growth factor receptor targeting agents in human cancer cells. Clin Cancer Res. 2009;15:1585–92. https://doi.org/10.1158/1078-0432.CCR-08-2068.
246. Cai W-Q, Zeng L-S, Wang L-F, et al. The latest battles between EGFR monoclonal antibodies and resistant tumor cells. Front Oncol. 2020;10:1249. https://doi.org/10.3389/fonc.2020.01249.
247. Afrăsânie V-A, Marinca MV, Alexa-Stratulat T, et al. KRAS, NRAS, BRAF, HER2 and microsatellite instability in metastatic colorectal cancer – practical implications for the clinician. Radiol Oncol. 2019;53:265–74. https://doi.org/10.2478/raon-2019-0033.
248. Bokemeyer C, Bondarenko I, Hartmann JT, et al. Efficacy according to biomarker status of cetuximab plus FOLFOX-4 as first-line treatment for metastatic colorectal cancer: the OPUS study. Ann Oncol. 2011;22:1535–46. https://doi.org/10.1093/annonc/mdq632.

249. Pentheroudakis G, Kotoula V, De Roock W, et al. Biomarkers of benefit from cetuximab-based therapy in metastatic colorectal cancer: interaction of EGFR ligand expression with RAS/RAF, PIK3CA genotypes. BMC Cancer. 2013;13:49. https://doi.org/10.1186/1471-2407-13-49.
250. Qin S, Li J, Wang L, et al. Efficacy and tolerability of first-line cetuximab plus leucovorin, fluorouracil, and oxaliplatin (FOLFOX-4) versus FOLFOX-4 in patients with RAS wild-type metastatic colorectal cancer: the open-label, randomized, phase III TAILOR trial. J Clin Oncol. 2018;36:3031–9. https://doi.org/10.1200/JCO.2018.78.3183.
251. Schwartzberg LS, Rivera F, Karthaus M, et al. PEAK: a randomized, multicenter phase II study of panitumumab plus modified fluorouracil, leucovorin, and oxaliplatin (mFOLFOX6) or bevacizumab plus mFOLFOX6 in patients with previously untreated, unresectable, wild-type KRAS exon 2 metastatic colorectal. J Clin Oncol. 2014;32:2240–7. https://doi.org/10.1200/JCO.2013.53.2473.
252. Braig F, Kriegs M, Voigtlaender M, et al. Cetuximab resistance in head and neck cancer is mediated by EGFR-K521 polymorphism. Cancer Res. 2017;77:1188–99. https://doi.org/10.1158/0008-5472.CAN-16-0754.
253. Patel D, Lahiji A, Patel S, et al. Monoclonal antibody cetuximab binds to and down-regulates constitutively activated epidermal growth factor receptor vIII on the cell surface. Anticancer Res. 2007;27:3355–66.
254. Maemondo M, Inoue A, Kobayashi K, et al. Gefitinib or chemotherapy for non–small-cell lung cancer with mutated EGFR. N Engl J Med. 2010;362:2380–8. https://doi.org/10.1056/NEJMoa0909530.
255. Chaft JE, Arcila ME, Paik PK, et al. Coexistence of PIK3CA and other oncogene mutations in lung adenocarcinoma–rationale for comprehensive mutation profiling. Mol Cancer Ther. 2012;11:485–91. https://doi.org/10.1158/1535-7163.MCT-11-0692.
256. Kim Y, Lee B, Shim JH, et al. Concurrent genetic alterations predict the progression to target therapy in EGFR-mutated advanced NSCLC. J Thorac Oncol. 2019;14:193–202. https://doi.org/10.1016/j.jtho.2018.10.150.
257. Li J, Yen C, Liaw D, et al. PTEN, a putative protein tyrosine phosphatase gene mutated in human brain, breast, and prostate cancer. Science (1979). 1997;275:1943–7. https://doi.org/10.1126/science.275.5308.1943.
258. Spoerke JM, O'Brien C, Huw L, et al. Phosphoinositide 3-kinase (PI3K) pathway alterations are associated with histologic subtypes and are predictive of sensitivity to PI3K inhibitors in lung cancer preclinical models. Clin Cancer Res. 2012;18:6771–83. https://doi.org/10.1158/1078-0432.CCR-12-2347.
259. Wang F, Diao X, Zhang X, et al. Identification of genetic alterations associated with primary resistance to EGFR-TKIs in advanced non-small-cell lung cancer patients with EGFR sensitive mutations. Cancer Commun. 2019;39:1–15. https://doi.org/10.1186/s40880-019-0354-z.
260. Yamamoto C, Basaki Y, Kawahara A, et al. Loss of PTEN expression by blocking nuclear translocation of EGR1 in gefitinib-resistant lung cancer cells harboring epidermal growth factor receptor–activating mutations. Cancer Res. 2010;70:8715–25. https://doi.org/10.1158/0008-5472.CAN-10-0043.
261. Cho EN, Kim EY, Jung JY, et al. BCL2-like 11 intron 2 deletion polymorphism is not associated with non-small cell lung cancer risk and prognosis. Lung Cancer. 2015;90:106–10. https://doi.org/10.1016/j.lungcan.2015.07.017.
262. Ebi H, Oze I, Nakagawa T, et al. Lack of association between the BIM deletion polymorphism and the risk of lung cancer with and without EGFR mutations. J Thorac Oncol. 2015;10:59–66. https://doi.org/10.1097/JTO.0000000000000371.
263. Zahavi D, Weiner L. Monoclonal antibodies in cancer therapy. Antibodies. 2020;9:34. https://doi.org/10.3390/antib9030034.
264. Brindel A, Althakfi W, Barritault M, et al. Uncommon EGFR mutations in lung adenocarcinoma: features and response to tyrosine kinase inhibitors. J Thorac Dis. 2020;12:4643–50. https://doi.org/10.21037/jtd-19-3790.

265. Zhang Q, Zhang X-C, Yang J-J, et al. EGFR L792H and G796R: two novel mutations mediating resistance to the third-generation EGFR tyrosine kinase inhibitor osimertinib. J Thorac Oncol. 2018;13:1415–21. https://doi.org/10.1016/j.jtho.2018.05.024.
266. Cheng H, Fertig EJ, Ozawa H, et al. Decreased SMAD4 expression is associated with induction of epithelial-to-mesenchymal transition and cetuximab resistance in head and neck squamous cell carcinoma. Cancer Biol Ther. 2015;16:1252–8. https://doi.org/10.1080/15384047.2015.1056418.
267. Kimura I, Kitahara H, Ooi K, et al. Loss of epidermal growth factor receptor expression in oral squamous cell carcinoma is associated with invasiveness and epithelial-mesenchymal transition. Oncol Lett. 2016;11:201–7. https://doi.org/10.3892/ol.2015.3833.
268. Oliveras-Ferraros C, Vazquez-Martin A, Cufí S, et al. Stem cell property epithelial-to-mesenchymal transition is a core transcriptional network for predicting cetuximab (Erbitux™) efficacy in KRAS wild-type tumor cells. J Cell Biochem. 2011;112:10–29. https://doi.org/10.1002/jcb.22952.
269. Oliveras-Ferraros C, Corominas-Faja B, Cufí S, et al. Epithelial-to-mesenchymal transition (EMT) confers primary resistance to trastuzumab (Herceptin). Cell Cycle. 2012;11:4020–32. https://doi.org/10.4161/cc.22225.
270. Mancini M, Gal H, Gaborit N, et al. An oligoclonal antibody durably overcomes resistance of lung cancer to third-generation EGFR inhibitors. EMBO Mol Med. 2018;10:294–308. https://doi.org/10.15252/emmm.201708076.
271. Maron R, Schechter B, Nataraj NB, et al. Inhibition of a pancreatic cancer model by cooperative pairs of clinically approved and experimental antibodies. Biochem Biophys Res Commun. 2019;513:219–25. https://doi.org/10.1016/j.bbrc.2019.03.204.
272. Marrocco I, Romaniello D, Vaknin I, et al. Upfront admixing antibodies and EGFR inhibitors preempts sequential treatments in lung cancer models. EMBO Mol Med. 2021;13:e13144. https://doi.org/10.15252/emmm.202013144.
273. Miano C, Morselli A, Pontis F, et al. NRG1/ERBB3/ERBB2 axis triggers anchorage-independent growth of basal-like/triple-negative breast cancer cells. Cancers (Basel). 2022;14:1603. https://doi.org/10.3390/cancers14071603.
274. Romaniello D, Mazzeo L, Mancini M, et al. A combination of approved antibodies overcomes resistance of lung cancer to osimertinib by blocking bypass pathways. Clin Cancer Res. 2018;24:5610–21. https://doi.org/10.3390/cancers12092394.
275. Romaniello D, Marrocco I, Belugali Nataraj N, et al. Targeting HER3, a catalytically defective receptor tyrosine kinase, prevents resistance of lung cancer to a third-generation EGFR kinase inhibitor. Cancers (Basel). 2020;12:2394. https://doi.org/10.3390/cancers12092394.
276. Jiao D, Chen Y, Liu X, et al. Targeting MET endocytosis or degradation to overcome HGF-induced gefitinib resistance in EGFR-sensitive mutant lung adenocarcinoma. Biochem Biophys Res Commun. 2023;682:371–80. https://doi.org/10.1016/j.bbrc.2023.10.037.
277. Lei T, Xu T, Zhang N, et al. Anlotinib combined with osimertinib reverses acquired osimertinib resistance in NSCLC by targeting the c-MET/MYC/AXL axis. Pharmacol Res. 2023;188:106668. https://doi.org/10.1016/j.phrs.2023.106668.
278. Peng Y, Tan J. The relationship between IGF pathway and acquired resistance to tyrosine kinase inhibitors in cancer therapy. Front Biosci (Landmark Ed). 2023;28:163. https://doi.org/10.31083/j.fbl2808163.
279. Ma C, Wang X, Guo J, et al. Challenges and future of HER2-positive gastric cancer therapy. Front Oncol. 2023;13:1080990. https://doi.org/10.3389/fonc.2023.1080990.
280. Zhu K, Yang X, Tai H, et al. HER2-targeted therapies in cancer: a systematic review. Biomark Res. 2024;12:16. https://doi.org/10.1186/s40364-024-00565-1.
281. Bedard PL, Cardoso F, Piccart-Gebhart MJ. Stemming resistance to HER-2 targeted therapy. J Mammary Gland Biol Neoplasia. 2009;14:55–66. https://doi.org/10.1007/s10911-009-9116-x.
282. Price-Schiavi SA, Jepson S, Li P, et al. Rat Muc4 (sialomucin complex) reduces binding of anti-ErbB2 antibodies to tumor cell surfaces, a potential mechanism for herceptin resistance. Int J Cancer. 2002;99:783–91. https://doi.org/10.1002/ijc.10410.

283. Anido J, Scaltriti M, Bech Serra JJ, et al. Biosynthesis of tumorigenic HER2 C-terminal fragments by alternative initiation of translation. EMBO J. 2006;25:3234–44. https://doi.org/10.1038/sj.emboj.7601191.
284. Castiglioni F, Tagliabue E, Campiglio M, et al. Role of exon-16-deleted HER2 in breast carcinomas. Endocr Relat Cancer. 2006;13:221–32. https://doi.org/10.1677/erc.1.01047.
285. Berns K, Horlings HM, Hennessy BT, et al. A functional genetic approach identifies the PI3K pathway as a major determinant of trastuzumab resistance in breast cancer. Cancer Cell. 2007;12:395–402. https://doi.org/10.1016/j.ccr.2007.08.030.
286. Gewinner C, Wang ZC, Richardson A, et al. Evidence that inositol polyphosphate 4-phosphatase type II is a tumor suppressor that inhibits PI3K signaling. Cancer Cell. 2009;16:115–25. https://doi.org/10.1016/j.ccr.2009.06.006.
287. Nagata Y, Lan K-H, Zhou X, et al. PTEN activation contributes to tumor inhibition by trastuzumab, and loss of PTEN predicts trastuzumab resistance in patients. Cancer Cell. 2004;6:117–27. https://doi.org/10.1016/j.ccr.2004.06.022.
288. Garrett JT, Olivares MG, Rinehart C, et al. Transcriptional and posttranslational up-regulation of HER3 (ErbB3) compensates for inhibition of the HER2 tyrosine kinase. Proc Natl Acad Sci. 2011;108:5021–6. https://doi.org/10.1073/pnas.1016140108.
289. Ritter CA, Perez-Torres M, Rinehart C, et al. Human breast cancer cells selected for resistance to trastuzumab in vivo overexpress epidermal growth factor receptor and ErbB ligands and remain dependent on the ErbB receptor network. Clin Cancer Res. 2007;13:4909–19. https://doi.org/10.1158/1078-0432.CCR-07-0701.
290. Sampera A, Sánchez-Martín FJ, Arpí O, et al. HER-family ligands promote acquired resistance to trastuzumab in gastric cancer. Mol Cancer Ther. 2019;18:2135–45. https://doi.org/10.1158/1535-7163.MCT-19-0455.
291. Pietrantonio F, Caporale M, Morano F, et al. HER2 loss in HER2-positive gastric or gastroesophageal cancer after trastuzumab therapy: implication for further clinical research. Int J Cancer. 2016;139:2859–64. https://doi.org/10.1002/ijc.30408.
292. Seo S, Ryu M-H, Park YS, et al. Loss of HER2 positivity after anti-HER2 chemotherapy in HER2-positive gastric cancer patients: results of the GASTric cancer HER2 reassessment study 3 (GASTHER3). Gastric Cancer. 2019;22:527–35. https://doi.org/10.1007/s10120-018-0891-1.
293. Shi J, Li F, Yao X, et al. The HER4-YAP1 axis promotes trastuzumab resistance in HER2-positive gastric cancer by inducing epithelial and mesenchymal transition. Oncogene. 2018;37:3022–38. https://doi.org/10.1038/s41388-018-0204-5.
294. Browne BC, Crown J, Venkatesan N, et al. Inhibition of IGF1R activity enhances response to trastuzumab in HER-2-positive breast cancer cells. Ann Oncol. 2011;22:68–73. https://doi.org/10.1093/annonc/mdq349.
295. Huang X, Gao L, Wang S, et al. Heterotrimerization of the growth factor receptors erbB2, erbB3, and insulin-like growth factor-I receptor in breast cancer cells resistant to herceptin. Cancer Res. 2010;70:1204–14. https://doi.org/10.1158/0008-5472.CAN-09-3321.
296. Shattuck DL, Miller JK, Carraway KL, Sweeney C. Met receptor contributes to trastuzumab resistance of Her2-overexpressing breast cancer cells. Cancer Res. 2008;68:1471–7. https://doi.org/10.1158/0008-5472.CAN-07-5962.
297. Zhuang G, Brantley-Sieders DM, Vaught D, et al. Elevation of receptor tyrosine kinase EphA2 mediates resistance to trastuzumab therapy. Cancer Res. 2010;70:299–308. https://doi.org/10.1158/0008-5472.CAN-09-1845.
298. Jia Y, Yun C-H, Park E, et al. Overcoming EGFR(T790M) and EGFR(C797S) resistance with mutant-selective allosteric inhibitors. Nature. 2016;534:129–32. https://doi.org/10.1038/nature17960.
299. To C, Jang J, Chen T, et al. Single and dual targeting of mutant EGFR with an allosteric inhibitor. Cancer Discov. 2019;9:926–43. https://doi.org/10.1158/2159-8290.CD-18-0903.
300. Lim SM, Schalm SS, Lee EJ, et al. BLU-945, a potent and selective next-generation EGFR TKI, has antitumor activity in models of osimertinib-resistant non-small-cell lung cancer. Ther Adv Med Oncol. 2024;16. https://doi.org/10.1177/17588359241280689.

301. Elamin YY, Nagasaka M, Shum E, et al. BLU-945 monotherapy and in combination with osimertinib (OSI) in previously treated patients with advanced EGFR-mutant (EGFRm) NSCLC in the phase 1/2 SYMPHONY study. J Clin Oncol. 2023;41:9011. https://doi.org/10.1200/JCO.2023.41.16_suppl.9011.
302. Medicines B. Blueprint medicines highlights 2024 corporate strategy and business priorities at 42nd annual J.P. In: Morgan healthcare conference. 2024. https://ir.blueprintmedicines.com/news-releases/news-release-details/blueprint-medicines-highlights-2024-corporate-strategy-and. Accessed 10 Dec 2024.
303. Johnson ML, Henry JT, Spira AI, et al. A phase 1 study to assess BDTX-1535, an oral EGFR inhibitor, in patients with glioblastoma or non–small-cell lung cancer. J Clin Oncol. 2023;41:TPS9156. https://doi.org/10.1200/JCO.2023.41.16_suppl.TPS9156.
304. Therapeutcics BD. Black diamond therapeutics announces initial phase 2 data demonstrating robust anti-tumor activity of BDTX-1535 in patients with recurrent EGFRm NSCLC who present with a broad spectrum of classical, non-classical, and C797S resistance mutations. 2024. www.blackdiamondtherapeutics.com. Accessed 10 Dec 2024.
305. Burslem GM, Crews CM. Proteolysis-targeting chimeras as therapeutics and tools for biological discovery. Cell. 2020;181:102–14. https://doi.org/10.1016/j.cell.2019.11.031.
306. Sakamoto KM, Kim KB, Kumagai A, et al. Protacs: chimeric molecules that target proteins to the Skp1–Cullin–F box complex for ubiquitination and degradation. Proc Natl Acad Sci. 2001;98:8554–9. https://doi.org/10.1073/pnas.141230798.
307. Burslem GM, Smith BE, Lai AC, et al. The advantages of targeted protein degradation over inhibition: an RTK case study. Cell Chem Biol. 2018;25:67–77.e3. https://doi.org/10.1016/j.chembiol.2017.09.009.
308. Shen J, Chen L, Liu J, et al. EGFR degraders in non-small-cell lung cancer: breakthrough and unresolved issue. Chem Biol Drug Des. 2024;103:14517. https://doi.org/10.1111/cbdd.14517.
309. Cheng M, Yu X, Lu K, et al. Discovery of potent and selective epidermal growth factor receptor (EGFR) bifunctional small-molecule degraders. J Med Chem. 2020;63:1216–32. https://doi.org/10.1021/acs.jmedchem.9b01566.
310. Shi S, Du Y, Huang L, et al. Discovery of novel potent covalent inhibitor-based EGFR degrader with excellent in vivo efficacy. Bioorg Chem. 2022;120:105605. https://doi.org/10.1016/j.bioorg.2022.105605.
311. Vartak R, Deore B, Sanhueza CA, Patel K. Cetuximab-based PROteolysis targeting chimera for effectual downregulation of NSCLC with varied EGFR mutations. Int J Biol Macromol. 2023;252:126413. https://doi.org/10.1016/j.ijbiomac.2023.126413.
312. Klein C, Brinkmann U, Reichert JM, Kontermann RE. The present and future of bispecific antibodies for cancer therapy. Nat Rev Drug Discov. 2024;23:301–19. https://doi.org/10.1038/s41573-024-00896-6.
313. Swain SM, Shastry M, Hamilton E. Targeting HER2-positive breast cancer: advances and future directions. Nat Rev Drug Discov. 2023;22:101–26. https://doi.org/10.1038/s41573-022-00579-0.
314. Lin W, Zhang Y, Yang Y, et al. Anti-PD-1/Her2 bispecific antibody IBI315 enhances the treatment effect of Her2-positive gastric cancer through gasdermin B-cleavage induced pyroptosis. Adv Sci. 2023;10. https://doi.org/10.1002/advs.202303908.
315. NCI. Enhertu approved for any HER2-positive solid cancer – NCI. In: National Cancer Institute. 2024. https://www.cancer.gov/news-events/cancer-currents-blog/2024/fda-enhertu-her2-positive-solid-tumors. Accessed 15 Dec 2024.
316. O'Leary K. First tumor-agnostic approval for an ADC. Nat Med. 2024;30:3392. https://doi.org/10.1038/s41591-024-03389-6.
317. Holbro T, Beerli RR, Maurer F, et al. The ErbB2/ErbB3 heterodimer functions as an oncogenic unit: ErbB2 requires ErbB3 to drive breast tumor cell proliferation. Proc Natl Acad Sci. 2003;100:8933–8. https://doi.org/10.1073/pnas.1537685100.

318. Liu B, Ordonez-Ercan D, Fan Z, et al. Downregulation of erbB3 abrogates erbB2-mediated tamoxifen resistance in breast cancer cells. Int J Cancer. 2007;120:1874–82. https://doi.org/10.1002/ijc.22423.
319. Engelman JA, Zejnullahu K, Mitsudomi T, et al. MET amplification leads to gefitinib resistance in lung cancer by activating ERBB3 signaling. Science (1979). 2007;316:1039–43. https://doi.org/10.1126/science.1141478.
320. Sheng Q, Liu X, Fleming E, et al. An activated ErbB3/NRG1 autocrine loop supports in vivo proliferation in ovarian cancer cells. Cancer Cell. 2010;17:298–310. https://doi.org/10.1016/j.ccr.2009.12.047.
321. Yonesaka K, Zejnullahu K, Okamoto I, et al. Activation of ERBB2 signaling causes resistance to the EGFR-directed therapeutic antibody cetuximab. Sci Transl Med. 2011;3:99ra86. https://doi.org/10.1126/scitranslmed.3002442.
322. Majumder A. HER3: toward the prognostic significance, therapeutic potential, current challenges, and future therapeutics in different types of cancer. Cells. 2023;12:2517. https://doi.org/10.3390/cells12212517.
323. Gaborit N, Abdul-Hai A, Mancini M, et al. Examination of HER3 targeting in cancer using monoclonal antibodies. Proc Natl Acad Sci. 2015;112:839–44. https://doi.org/10.1073/pnas.1423645112.
324. Yu HA, Goto Y, Hayashi H, et al. HERTHENA-Lung01, a phase II trial of patritumab deruxtecan (HER3-DXd) in epidermal growth factor receptor–mutated non–small-cell lung cancer after epidermal growth factor receptor tyrosine kinase inhibitor therapy and platinum-based chemotherapy. J Clin Oncol. 2023;41:5363–75. https://doi.org/10.1200/JCO.23.01476.
325. Hamilton EP, Petit T, Pistilli B, et al. Clinical activity of MCLA-128 (zenocutuzumab), trastuzumab, and vinorelbine in HER2 amplified metastatic breast cancer (MBC) patients (pts) who had progressed on anti-HER2 ADCs. J Clin Oncol. 2020;38:3093. https://doi.org/10.1200/JCO.2020.38.15_suppl.3093.
326. Alsina M, Boni V, Schellens JHM, et al. First-in-human phase 1/2 study of MCLA-128, a full length IgG1 bispecific antibody targeting HER2 and HER3: final phase 1 data and preliminary activity in HER2+ metastatic breast cancer (MBC). J Clin Oncol. 2017;35:2522. https://doi.org/10.1200/JCO.2017.35.15_suppl.2522.
327. De Pauw I, Wouters A, Van den Bossche J, et al. Dual targeting of epidermal growth factor receptor and HER3 by MEHD7945A as monotherapy or in combination with cisplatin partially overcomes cetuximab resistance in head and neck squamous cell carcinoma cell lines. Cancer Biother Radiopharm. 2017;32:229–38. https://doi.org/10.1089/cbr.2017.2216.
328. Huang S, Li C, Armstrong EA, et al. Dual targeting of EGFR and HER3 with MEHD7945A overcomes acquired resistance to EGFR inhibitors and radiation. Cancer Res. 2013;73:824–33. https://doi.org/10.1158/0008-5472.CAN-12-1611.
329. Zhao Z, Li R, Sha S, et al. Targeting HER3 with miR-450b-3p suppresses breast cancer cells proliferation. Cancer Biol Ther. 2014;15:1404–12. https://doi.org/10.4161/cbt.29923.
330. Iorio MV, Casalini P, Piovan C, et al. microRNA-205 regulates HER3 in human breast cancer. Cancer Res. 2009;69:2195–200. https://doi.org/10.1158/0008-5472.CAN-08-2920.

Epidermal Growth Factor Receptor in COPD and Lung Cancer: Is Inflammation the Culprit?

Hina Agraval, Jiten Sharma, and Umesh C. S. Yadav

Abbreviations

EGFR	epidermal growth factor receptor
COPD	chronic obstructive pulmonary diseases
CSE	cigarette smoke extract
ROS	reactive oxygen species
LC	lung cancer
NSCLC	non-small cell lung cancer
SCLC	small cell lung cancer
MMPs	matrix metalloproteinases
DEGs	differentially expressed genes
TLR	toll-like receptor
TKIs	tyrosine kinase inhibitors
EMT	epithelial-mesenchymal transition

Authors Hina Agraval and Jiten Sharma have equally contributed to this chapter.

H. Agraval
Department of Medicine, National Jewish Health, Denver, CO, USA
e-mail: agravalh@njhealth.org

J. Sharma
Division of Molecular Medicine and Gene Therapy, Lund Stem Cell Center, Lund University, Lund, Sweden
e-mail: jiten.sharma@med.lu.se

U. C. S. Yadav (✉)
Special Center for Molecular Medicine, Jawaharlal Nehru University, New Delhi, India
e-mail: ucsyadav@mail.jnu.ac.in

A. Sabarwal et al. (eds.), *Receptor Tyrosine Kinases in Cancer*, Cancer Drug Discovery and Development, https://doi.org/10.1007/978-3-031-93894-8_3

1 Introduction

Chronic obstructive pulmonary disease (COPD) and lung cancer (LC) represent two highly pervasive public health challenges with devastating impacts on individuals and healthcare systems worldwide [1, 2]. Despite their distinct clinical manifestations, these diseases share significant epidemiological and biological intersections, primarily driven by common risk factors like cigarette smoking (CS), exposure to pollutants, xenobiotics, chemical fumes, genetic factors, and underlying inflammatory processes. Inflammation drives cancer growth and metastasis through several interconnected mechanisms that create a supportive tumor microenvironment. Chronic inflammation resulting from the continuous production of pro-inflammatory cytokines, reactive oxygen species (ROS), and growth factors during COPD collectively fuel cellular survival, hyperproliferation, and airway remodeling including mucus hypersecretion. Thus, in both COPD and LC, the role of growth factor receptors becomes a critical point of convergence. Epidermal growth factor receptor (EGFR), an important tyrosine kinase receptor, plays a crucial role in chronic lung diseases like COPD and LC where its dysregulation is linked with disease severity and complex pathophysiology. In this chapter, we provide the evidence of dysregulated EGFR pathway in the intricate relationship between COPD and LC, highlighting the potential role of inflammation in modulating EGFR pathways, which act as a pivotal factor in the pathogenesis and progression of these diseases.

COPD, a complex respiratory condition, is characterized by persistent airflow limitation, often resulting in chronic bronchitis and emphysema. LC, predominantly categorized into non-small cell lung cancer (NSCLC) and small cell lung cancer (SCLC), arises from uncontrolled proliferation of airway epithelial cells. Tobacco smoke exposure, the most significant risk factor for both conditions, is responsible for up to 90% of the cases [3]. The insidious nature of tobacco smoke lies in its ability to induce oxidative stress, DNA damage, and chronic inflammatory responses within the pulmonary system [4]. Such processes not only set the stage for the development of COPD but also foster an environment conducive to oncogenic transformations, thereby linking COPD and LC from across pathophysiological domains. Research reveals a staggering correlation between COPD and increased risk for LC development, independent of other factors such as smoking history, age, and sex [5]. Individuals with COPD face a four- to six-fold higher risk of developing LC than those without COPD, indicating that airflow obstruction itself may serve as a potent risk factor for LC, beyond the risks of shared carcinogenic exposure [6]. Approximately 70–80% of LC cases occur in individuals with pre-existing COPD, underscoring a significant dose-response relationship between the severity of airflow obstruction and LC incidence [7, 8]. Moreover, the presence of emphysema, a common COPD component, is notably associated with an elevated risk of squamous cell carcinoma and SCLC, further substantiating the clinically intertwined nature of these diseases [9].

Understanding the clinical intricacies of two diseases offers critical insights into shared pathophysiological pathways that may highlight opportunities for targeted therapeutic interventions. An array of diverse molecular players has been identified as contributors to the pathogenesis of both COPD and LC; here only key players have been discussed in detail. Notably, ROS generate oxidative stress leading to chronic inflammation and DNA mutations. Cytokines such as tumor necrosis factor-alpha (TNF-α) and interleukins (IL) like IL-6 sustain inflammatory cycles that not only exacerbate COPD symptoms but also promote a microenvironment that is fertile for tumor growth and progression. Among the several molecular players, matrix metalloproteinases (MMPs) facilitate pathological tissue remodeling, contributing to both COPD-induced structural damage and the invasive potential of LC cells. Genetic and epigenetic modifications, including mutation, DNA methylation, and alterations in microRNAs, further contribute to the dysregulated cellular behavior observed in both diseases, emphasizing the need for integrative therapeutic strategies.

The EGFR signaling pathway is central to both COPD and LC development and progression. EGFR, a transmembrane receptor tyrosine kinase, upon activation, initiates a cascade of downstream signaling pathways, including the PI3K/AKT and RAS/RAF/MEK/ERK pathways [10, 11]. These pathways are crucial for cell survival and growth, rendering EGFR a key player in oncogenesis, particularly in NSCLC. Dysregulated EGFR signaling contributes to uncontrolled epithelial cell growth, the pathological hallmark of LC, and is also implicated in excessive mucus production in COPD [12]. Understanding of the EGFR's role in disease biology began with the foundational discovery of epidermal growth factor (EGF) by Stanley Cohen and continues to this day where targeting EGFR mutations has become an important therapeutic interventions in LC. This knowledge underscores the critical role of EGFR as an oncogenic driver [13] and a therapeutic target in NSCLC, particularly when its propensity for overexpression or mutation across various cancer types has been reported [14–16]. The intertwined pathogenesis of COPD and LC and EGFR-driven shared inflammatory and molecular pathways position EGFR as a central therapeutic target in these conditions. Continued research into the modulation of EGFR signaling holds the promise of not only elucidating common disease mechanisms but also developing innovative and targeted therapies that could address the dual burden of these pervasive respiratory conditions. Understanding these connections can lead to integrated, efficient, and personalized treatment modalities considering the overlapping etiologies and pathophysiology of COPD and LC.

2 Overlapping Pathophysiological Mechanisms in COPD and Lung Cancer

Despite their distinct clinical identities, both COPD and LC are intimately linked through several shared risk factors and various players that contribute to disease development and progression. Understanding these players provides critical insights into shared pathways and potential therapeutic targets. Some of the key molecular components are discussed below.

2.1 Reactive Oxygen Species (ROS)

Both COPD and LC are characterized by increased oxidative stress and elevated ROS levels. These highly reactive molecules can cause DNA damage, protein modification, and lipid peroxidation, which aggravate chronic inflammation and propel tumorigenesis. COPD is marked by increased ROS from sources such as CS, mitochondrial dysfunction, and inflammatory cells like neutrophils and macrophages [17]. This oxidative stress is compounded by depletion of antioxidants, particularly glutathione, which is crucial for cellular defense [17]. In LC, apart from extracellular CS exposure, the metabolic reprogramming in the transformed cells results in elevated ROS production [18], which is essential for tumor growth and progression. The accumulation of ROS activates key transcription factors, including NF-κB, AP-1, and Nrf2, influencing inflammatory and antioxidant responses. NF-κB is essential for regulating inflammation and cellular responses to oxidative stress [19]. In COPD, NF-κB sustains chronic inflammation in lung tissues, while in LC, it promotes anti-apoptotic factors that enhance cancer cell survival and facilitate immune evasion. Given the crucial role of NF-κB in health and disease, its signaling and nuclear translocation are tightly regulated. Inhibitors of nuclear factor-κB (IκBs) sequester NF-κB in the cytoplasm under resting conditions. Upon activation, IκBs undergo inhibitory phosphorylation, leading to their degradation and the subsequent release of NF-κB subunits, allowing it to enter the nucleus and activate target gene transcription [20].

ROS can mediate both direct and indirect translocation of NF-κB subunits and downstream target gene activation. Phosphorylation of NF-κB subunit RelA at serine-276 is necessary for activating certain NF-κB-dependent genes [21]. This modification allows RelA to interact with coactivators CBP/p300 and the positive transcription elongation factor b [21]. The phosphorylation is facilitated by protein kinase A catalytic subunit (PKAc) and seems to be regulated by ROS [22], as antioxidant treatments can inhibit both the phosphorylation and the interaction with these coactivators [22]. ROS can also activate NF-κB activity by modulating IκBα. H_2O_2 exposure of the cells influences IκBα phosphorylation on tyrosine-42 and other tyrosine residues, which is distinct from the typical phosphorylation sites at serine-32 and serine-36 that promote IκBα degradation [23]. This alternative phosphorylation can occur through casein kinase II, seemingly independent of IκB kinase (IKK) activation [23].

ROS and NF-κB signaling can also act at a crossroads, integrating oxidative stress with receptor tyrosine kinases such as EGFR-mediated cellular responses, which contribute to tumor progression and therapeutic resistance. In head and neck squamous cell carcinoma, the IKK-NF-κB signaling pathway frequently intersects with the EGFR-AP1 pathway, promoting malignant behaviors in cancer cells [24]. IKKα and IKKβ can cooperatively activate NF-κB and EGFR/AP1 signaling pathways as silencing of both IKKs decreased the expression and activity of NF-κB and EGFR. Additionally, RelA is essential for the co-expression of components within the EGFR/AP1 pathway, emphasizing the interaction between NF-κB and EGFR signaling [24].

Overall, the interplay between ROS, IκBα, and IKK highlights a regulatory network of post-translational modification that modulates NF-κB activity and its downstream signaling pathways playing an important role in COPD and LC.

2.2 Cytokines and Inflammatory Mediators

Chronic inflammation is a hallmark of COPD and a known promoter of tumor growth in LC. While cigarette smoking, a common factor in both conditions, triggers an inflammatory response in the lungs, the etiology of inflammation in COPD and LC differs slightly. In COPD, smoking causes chronic airway inflammation and tissue destruction through cytokines like TNF-α and IL-6 [25]. In LC, smoking contributes to a cancer-promoting environment, with cytokines like IL-1β and IL-6 supporting tumor growth and immune evasion [26]. Despite these differences, both mechanisms ultimately result in creating and sustaining a microenvironment that is rich in pro-inflammatory molecules including TNF-α and interleukins (e.g., IL-6, IL-1β) [27]. Elevated levels of pro-inflammatory cytokines like TNF-α and IL-6 have been found in COPD, pneumonia, and LC patients, indicating a state of persistent inflammation that actively drives disease progression and exacerbates comorbid conditions [28]. In particular, IL-6 is significantly increased in LC patients compared to those with COPD or pneumonia, highlighting its potential as a biomarker for disease severity and progression. Furthermore, chronic inflammation, sustained by cytokines such as IL-1β, plays a critical role in creating a tumor-promoting microenvironment in LC [28]. The environment rich in IL-1β facilitates tumor growth, angiogenesis, and metastasis through the activation of key signaling pathways like MAPK and NF-κB [29], which are implicated in the recruitment and activation of immunosuppressive cells within tumors. The overlap in inflammatory pathways between COPD and LC suggests a shared pathogenic mechanism where chronic inflammation acts as a common denominator. Thus, targeting specific cytokines such as IL-6 and IL-1β could help mitigate inflammation and potentially decrease the risk of developing LC in COPD patients. Moreover, identifying cytokine profiles and inflammatory mediators may offer predictive insights into disease exacerbation, providing a basis for early intervention. As research continues to elucidate these connections, the role of cytokines in bridging the pathophysiological gap between COPD and LC will remain a focal point of future research for developing integrated treatment approaches and improving patient outcomes.

2.3 Matrix Metalloproteinases (MMPs)

MMPs are a group of zinc-dependent enzymes that degrade extracellular matrix components and facilitate tissue remodeling. MMPs have been shown to play a crucial role in the pathophysiology of both COPD and LC, linking these two conditions through their activities. Among the various MMPs, MMP-2, -7, -8, -9, and -13 have been particularly associated with the progression and severity of COPD. LC -related MMPs, particularly MMP-2 and -9, play vital roles in COPD by contributing to the dysregulation of protease/antiprotease balance, leading to tissue destruction, airway remodeling, and chronic inflammation. Elevated level of MMP-2, specifically in COPD patients, serves as a potential biomarker, distinguishing it from asthma [30]. The involvement of MMPs extends to the epithelial-mesenchymal transition (EMT), a process significant for airway remodeling in COPD and crucial for tumor growth and metastasis in LC [31]. Increased expression of MMP-2 in LC correlates with malignancy characteristics such as tumor size, lymph node metastasis, and reduced patient survival, indicating its role in cancer progression [32]. Moreover, elevated level of MMP-9 is associated with severity and clinical history of COPD, correlating positively with inflammatory markers like COX-2 and PGE-2 [33, 34]. MMP-7, also known as matrilysin, is also implicated in COPD and involved in the pathophysiology of LC. In COPD, MMP-7 (T > C) gene variant (rs10502001) is associated with an increased risk of COPD, with the "CC" genotype showing a notable correlation to higher mRNA expression levels of MMP-7 in patient samples [35]. MMP-7 plays a critical role in tumor progression by promoting spread through air spaces (STAS) of lung adenocarcinoma [36]. Histological analysis of STAS LC samples reveals MMP-7 as an independent predictor of STAS incidence in LC and correlates with aggressive tumor behavior and poor prognosis [36]. MMP-12 also known as macrophage metalloelastase or macrophage elastase expressed in inflamed lung tissues, such as during COPD, can drive alveolar destruction and emphysema. IL-4 secreted by the infiltrating basophils in the lungs can activate the macrophages to upregulate and secrete MMP-12 and aggravate tissue damage [37]. MMP-12 is also a poor prognostic factor for lung adenocarcinoma and can be targeted with penfluridol. MMP-12 was shown to promote EMT, which was ameliorated by penfluridol treatment to the cells [38].

The dual role of MMPs in both COPD and lung cancer underscores the importance of these enzymes in creating a pro-inflammatory microenvironment conducive to disease progression and serves as a critical link between COPD and lung cancer through their roles in tissue remodeling, inflammation, and disease progression (Fig. 1).

2.4 Genetic and Epigenetic Modifications

Smoking induces both genetic and epigenetic modifications in the lung tissue of COPD patients by increasing vulnerabilities and susceptibility to cancer development. Alterations in gene expression, driven by DNA methylation and histone modification, can significantly contribute to both COPD and LC by altering cellular behavior and responses to environmental stimuli. Recent studies have identified smoking-induced genetic and epigenetic modifications as crucial links between these two diseases. Chromatin-modifying enzymes, which mediate DNA

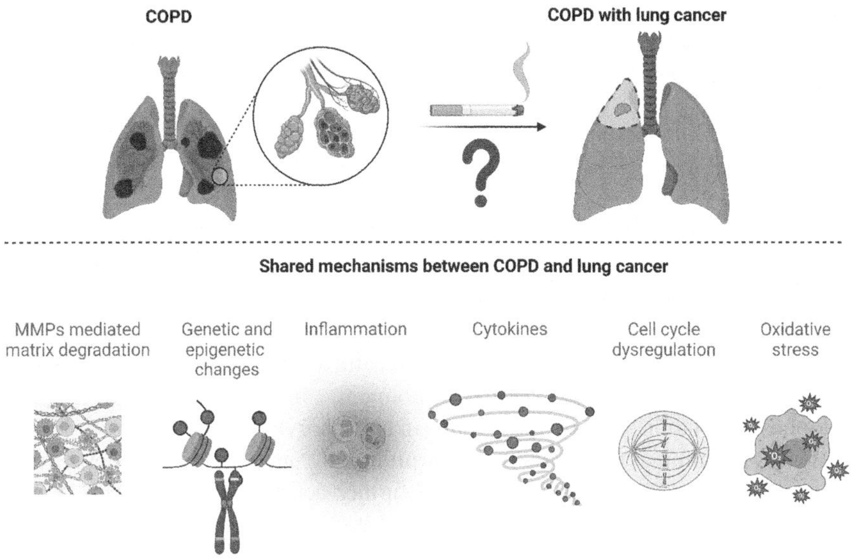

Fig. 1 Overlapping mechanisms between COPD and lung cancer

methylation and histone modifications, play a significant role in response to CS, altering gene expression patterns and impacting downstream pathways [39]. Research utilizing C57BL/6 J mice and human lung tissues has shown that CS exposure leads to increased expression of various chromatin modifiers, while others like DNMT1, DNMT3a, and HDAC families are downregulated [40]. Specifically, targeted histone marks such as H3K56ac and H4K12ac are induced by CS, underscoring their genotoxic stress effects on transcription regulation [40]. These findings suggest the potential for devising epigenetic-based therapies for COPD and LC. Further, a study exploring the molecular links between COPD and LC identified genes downregulated by CS exposure being common in both diseases. Researchers exposed the airway epithelial cells to cigarette smoke extract (CSE) and performed transcriptome analysis and reported that it led to downregulation and promoter hypermethylation of genes like CCNA1, SNCA, and ZNF549 [41]. Promoter hypermethylation of these genes in lung adenocarcinoma, but not in normal or COPD lungs, suggested their involvement in LC onset among COPD patients [41]. Such methylation patterns in sputum DNA from lung adenocarcinoma cases pointed to potential noninvasive biomarkers for early LC detection in high-risk groups like smokers with COPD.

Integrative analysis of gene expression in blood and lung tissues has identified shared differentially expressed genes (DEGs) as potential biomarkers for COPD, leading to the identification of 12 hub gene-network clusters [42]. These clusters, including IRAK2 and MECOM, are linked to pathways involving protein degradation, inflammatory responses, and immune regulation, serving as potential indicators for COPD diagnosis and prognosis. Notably, some COPD hub genes are recognized as prognostic markers in LC also, highlighting the intersection between these two diseases [42]. These findings emphasize the importance of cigarette smoking-driven genetic and epigenetic modifications in elevating LC risk among COPD patients.

3 EGFR Signaling and Inflammation

EGFR belongs to the ErbB family of receptor tyrosine kinases (RTKs) and includes members such as ErbB1/HER1, ErbB2/HER2/Neu, ErbB3/HER3, and ErbB4/HER4. Ligand binding to the receptor's extracellular domain induces dimerization, which brings intracellular kinase domains into proximity for trans-autophosphorylation. This process results in phosphorylation at the tyrosine residue present in C-terminal tail of the receptor, promoting interactions with downstream effectors such as Shc1 and Grb2. EGFR further activates PI3K/Akt and MAPK pathways (Fig. 2), both of which lead to cell proliferation, differentiation, migration, and apoptosis [43]. In chronic lung diseases like COPD and LC, EGFR signaling intricately intersects with chronic inflammation, revealing a complex and interconnected pathophysiology of the two conditions. EGFR, being essential for cellular growth and differentiation, can be transactivated by various factors like G-protein-coupled receptor (GPCR) agonists and cytokines, as has been shown in the studies investigating IL-8 and IL-1β-dependent shedding of EGFR ligands [44]. This process involves a disintegrin and metalloproteinases particularly ADAM10-mediated ligand shedding and receptor transactivation and highlights the complexity of inflammatory signaling pathways [44]. The EGFR-ligand system also plays a crucial role in the defense response of epithelial cells during chronic inflammatory skin disorders and wound healing [45]. Upon compromised epidermal integrity, EGFR signaling is activated in neutrophils, macrophages, and mast cells to promote rapid repair. EGFR interacts with toll-like receptors (TLRs) on keratinocytes, enhancing the expression of antimicrobial peptides such as defensins and cathelicidins. TGF-α induces TLR5 and TLR9 expression, increasing IL-8 production and antimicrobial activity [45]. EGFR signaling, triggered by innate immune mediators like IL-37, facilitates keratinocyte migration, proliferation, and enhanced adaptive immune response [46]. Moreover, in aggressive cancers such as inflammatory breast cancer (IBC), EGFR plays a role in regulating cancer stem-like cells through the COX-2 pathway, a key mediator of inflammation [47]. Here, EGFR signaling influences the expression of COX-2, which in turn modulates tumor cell invasion and proliferation [47], highlighting a critical intersection of oncogenic and inflammatory pathways. In the context of non-cancerous diseases, EGFR's involvement in inflammation-related pathologies is also evident. In obesity-induced kidney diseases, for example, EGFR pathway activation exacerbates renal inflammation and oxidative stress, contributing to disease progression. Rodent models of diabetic

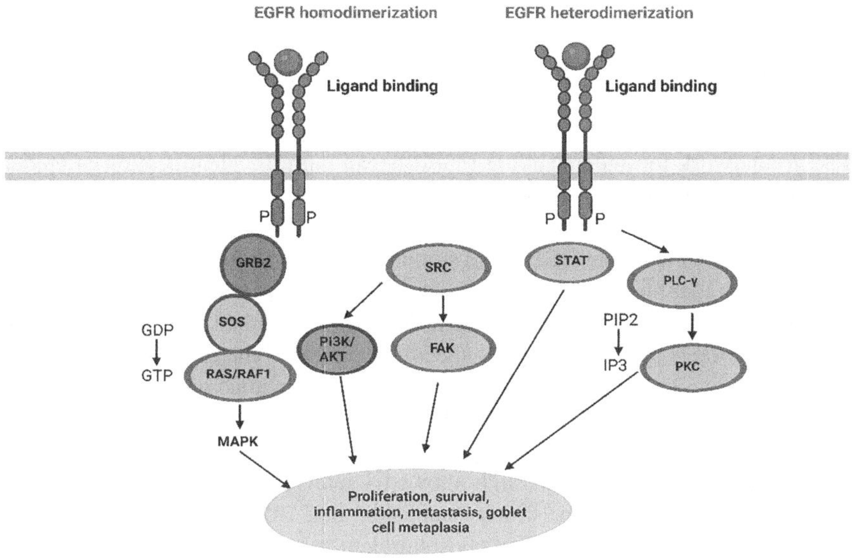

Fig. 2 Schematics showing EGFR signaling

nephropathy treated with EGFR inhibitors demonstrated that EGFR inhibition significantly reduced renal inflammation, oxidative stress, and fibrosis in obesity-related nephropathy [48]. The study further revealed that hyperlipidemia activates EGFR via a c-Src/EGFR complex and TLR4-mediated phosphorylation, underscoring the detrimental role of EGFR in exacerbating kidney disease [48].

In hepatocellular carcinoma, EGFR interacts with inflammatory cytokines, such as CXCL5 and CXCL8, through PI3K/ERK1/2 signaling pathway, regulating tumor proliferation and metastasis [49]. These interactions illustrate the potential of targeting EGF-EGFR axis to manage not only tumor growth but also the inflammatory milieu that supports cancer progression. Additionally, studies on natural compounds such as gramine (also called donaxine; an indole alkaloid) reveal the therapeutic potential of targeting EGFR-mediated pathways. By inhibiting EGFR tyrosine kinase phosphorylation, gramine suppresses downstream signaling cascades like PI3K/Akt/mTOR and JAK/STAT3, critical in inflammation and oncogenic signaling [50]. This disruption leads to reduced expression of pro-inflammatory and proliferative markers, affirming EGFR's role in mediating inflammation-related oncogenic processes [50].

In summary, EGFR serves as a critical nexus in the convergence of growth signaling and inflammation, with its pathways offering valuable targets for therapeutic intervention across various diseases characterized by inflammatory components such as COPD.

4 Role of EGFR in COPD and Lung Cancer

4.1 History

The groundwork for understanding EGFR began with the identification of epidermal growth factor (EGF) by Stanley Cohen in the early 1960s. Cohen, awarded the Nobel Prize in 1986 along with Rita Levi-Montalcini for their discovery of growth factors, demonstrated that EGF could stimulate cellular proliferation. This was followed by the identification of a receptor for EGF, known as EGFR, and its characterization in 1970s. Researchers, including Stanley Cohen, revealed that EGF exerted its effects by binding to this specific receptor on the cell surface, initiating signal transduction pathways that lead to different cellular responses [51]. This important understanding was followed by early molecular and clinical advances around EGFR in the 1980s. The EGFR gene was cloned and characterized via sequencing [52]. This breakthrough allowed scientists to delve into the structure and function of this protein, which established that EGFR is a receptor tyrosine kinase, a category of receptors that phosphorylate target proteins on tyrosine amino acid residues. In the 1980s–1990s, majority of the research was focused on understanding the role of EGFR in cancer. It was discovered that EGFR was overexpressed or mutated in various cancers, including lung, breast, and colorectal cancers [53]. These findings highlighted its role in tumorigenesis and identified EGFR as one of the central molecules for cancer research and a potential target for the development of targeted therapies against cancer [54]. The understanding of EGFR's role as an oncogenic driver propelled the field of targeted therapies designed to specifically inhibit EGFR activity. This era saw the emergence of monoclonal antibodies and small molecule tyrosine kinase inhibitors (TKIs) aimed at blocking EGFR and curbing tumor growth [54]. However, with the modern therapeutic developments in the 2000s, a shift toward personalized medicine and tailoring treatments based on the genetic profile of tumors began. EGFR inhibitors received FDA approval, with drugs such as gefitinib (Iressa) [55] and erlotinib (Tarceva) [56] becoming available for the treatment of NSCLC with specific EGFR mutations. However, soon the resistance of the cancer cells against these treatments became challenging [57]. Ongoing research is geared to address this issue including the development of next-generation inhibitors focused on targeting resistance mechanisms. While immunotherapies have transformed the treatment landscape for many cancers by leveraging the body's immune system to fight tumors, not all patients respond favorably to these treatments [58]. Moreover, resistance to immunotherapy can also develop, necessitating diverse therapeutic strategies. Next-generation EGFR inhibitors provide an essential alternative or complementary approach, addressing cases where immunotherapy might be less effective or unsuitable [58]. The new generation TKIs should be designed to overcome specific resistance pathways and provide more effective treatment options for patients.

4.2 EGFR in Lung Cancer

Dysregulated activation of EGFR pathway is a key characteristic of LC pathology and involves altered ligand production, EGFR gene amplification, receptor mutations or deletions, and persistent activation. In head and neck cancers, EGFR signaling is frequently dysregulated, with mutations in the HER1 gene—especially in exons 18–21 of HER1 kinase domain—identified as critical drivers of the disease [59]. Additionally, alterations in EGFR tyrosine kinase domain have been reported in about a quarter of NSCLC cases, where these mutations are often linked to increased receptor expression in 75% of cases [60]. Mouse model studies have demonstrated that deletion of LRIG1, a membrane-associated HER1 suppressor, or intracellular regulator Mig6 heightens susceptibility to LC. This observation is consistent with human data showing downregulation of LRIG1 and MIG6 in squamous cell carcinoma and adenocarcinomas and activated EGFR pathway [61, 62]. These mutations lead to increased cell proliferation and reduced apoptosis, irrespective of the presence of extracellular ligands. Moreover, beyond gene mutations, there are reports of elevated EGFR gene copy numbers, which are linked with a better response to TKIs [63]. In adenocarcinoma patients, there may be instances of both EGFR mutations and elevated gene copies, while others could exhibit either one of these alterations [64]. Sholl et al. found that among 65 never-smoking Asian women with adenocarcinoma, 80% had EGFR mutations, including 9 cases with gene amplification. All patients demonstrated exon 19 mutations and solid histological growth patterns. Significantly, these patients showed the worst overall survival compared to those with tumors lacking EGFR amplification [65]. EGFR oncogenic mutants and associated signaling proteins are frequently found in exosomes derived from LC patient samples. These exosomes can activate translational signaling pathways and regulate the expression of target genes such as VEGF, anti-apoptotic Bcl-xL, and the cell cycle inhibitor p27. EGFR can be transferred to endothelial cells via exosomes, triggering the MAPK and AKT pathways and boosting VEGF expression, which in turn enhances anti-apoptotic gene expression and promotes unanchored growth [66]. Beyond angiogenesis, EGFR-containing exosomes can also enter host macrophages, impairing antiviral immunity and facilitating immune evasion [67].

Inflammation-induced EGFR activation is now recognized as a major driver of LC. A recent study showed that IL-17R and EGFR interact through adaptor proteins, specifically TNF receptor-associated factor 4 (TRAF4) and Act1, which then recruits c-Src to facilitate IL-17A-mediated EGFR transactivation [68]. Another study showed that in NSCLC, a rapid increase in TNF-α levels could be an indicator of universal adaptive response to EGFR inhibition. EGFR activation has been shown to suppress TNF-α mRNA level via miR-21. Conversely, EGFR inhibition leads to loss of miR-21, leaving stable TNF-α mRNA. Additionally, TNF-α-induced NF-κB activation results in a positive feedback loop that enhances TNF-α transcription. Inhibition of TNF-α signaling significantly increases the sensitivity of NSCLC cells expressing wild-type EGFR or EGFR wild-type patient-derived xenograft (PDX) model to EGFR inhibition [69]. Blocking of TNF-α also improves the efficacy of EGFR inhibition in cells having EGFR oncogenes [69]. Recent findings suggest that acute influenza infection can promote LC growth by reprogramming the tumor microenvironment, which is associated with poorer overall survival rates. Dutta et al. recently reported that activation of IL-17-EGFR-TRAF4 axis results in heightened inflammation during severe influenza A virus (IAV) infection. Gefitinib, an EGFR inhibitor, reduced the severity of IAV infections. [70, 71]

EGFR mutations leading to EGFR TKI resistance present a significant challenge for NSCLC patients. A recent research by Chen et al. identified that samples resistant to EGFR TKIs exhibited higher levels of cytotoxic factors, such as granzyme A and PD-L1. Notably, around 33.3% (11 out of 33) of the treated samples were categorized as "hot" tumors, particularly among patients with the EGFR L858R mutation, where the rate was even higher at 46.7% (7 out of 15). Significant elevation in the effector cell populations was observed in these "hot" tumors with TKI resistance. Additionally, increased expression of CD8A, CD8B, granzyme B (GZMB), and GZMK was observed in "hot" tumors after TKI resistance. Further, a four-gene effector cell signature demonstrated a strong association with improved survival outcomes in external immunotherapy databases [72].

IL-6 is associated with the proliferation and differentiation of lung tumor cells and is also associated with acquired resistance to anti-EGFR therapies [73]. Gao et al. were the first to demonstrate that mutations in the EGFR kinase domain led to increased IL-6 levels, suggesting that targeting IL-6 could be beneficial in LC treatment. IL-6 binds to its receptor IL-6R and transmits signals through gp130 and activates downstream pathways such as JAK/STAT, ERK, and PI3K leading to transcription of target genes [74]. This pathway is associated with various processes, including cell viability and growth, differentiation, EMT, metastasis and invasion, and angiogenesis. Studies also indicate that T790M replacement in exon 20 of kinase domain of EGFR confers resistance to therapy with EGFR inhibitor afatinib in NSCLC cell lines by activating STAT3 [75–77]. Additionally, mutant EGFR also activates TGF-β, which subsequently triggers IL-6/gp130/STAT pathway. The activation of this pathway induces resistance against anti-EGFR drugs such as erlotinib [78].

Elevated levels of lysosomal proteases and dysregulated lysosomal metabolism are linked to LC progression [79]. Notably, ligand-induced lysosomal polyubiquitination and degradation of EGFR is one of the critical mechanisms of EGFR regulation [80, 81]. This process aids in the removal of EGFR receptors from the cell surface, thereby desensitizing cells to mitogenic signals. The internalization of active receptors begins with their recruitment to clathrin-coated pits, leading to fusion with early endosomes. Here, free EGFR can recycle back to the cell surface, while EGF-bound receptors are sorted into early endosomes through a receptor kinase-dependent mechanism and subsequently transported to late endosomes. During this sorting, the sequestration of the cytoplasmic domain prevents the receptor from interacting with cytoplasmic target molecules, resulting in inactivation of EGFR signaling. Late endosomes then fuse with lysosomes, initiating the degradation of ligand-bound receptors via lysosomal proteases [82, 83]. Notably, the pH of the endocytic compartments decreases along the endocytic route, which is mainly regulated by vacuolar ATPase activity. The EGF-EGFR complex remains stable in the mildly acidic conditions of early and late endosomes but ultimately degraded in the lysosome [84]. Targeting this degradation pathway of EGFR with antibodies or TKIs could be a promising therapeutic strategy for lung cancer (Fig. 3) [85].

4.3 EGFR in COPD

Excessive mucin production is a hallmark of COPD, and dysregulated EGFR signaling is now believed to contribute to increased mucus production and secretion. Studies have shown that increased immunoreactivity of EGFR correlates with excessive mucus cell metaplasia in the airways of asthma patients and bronchial biopsies from smokers [86, 87]. Increased expression of HER2 and HER3 has been linked to high mucus production in these patients. A comparative study by Saied et al. reported that EGFR expression was significantly higher in healthy smokers and COPD patients compared to non-smoking control. COPD patients exhibited notably higher EGFR expression than healthy smokers, and a subgroup of COPD ex-smokers had significantly higher EGFR expression than current smoker COPD patients. EGFR levels showed a significant negative correlation with FEV1 in healthy smoker and smokers with COPD and positive correlation with mucin production [88]. Additionally, significant positive correlation was observed between

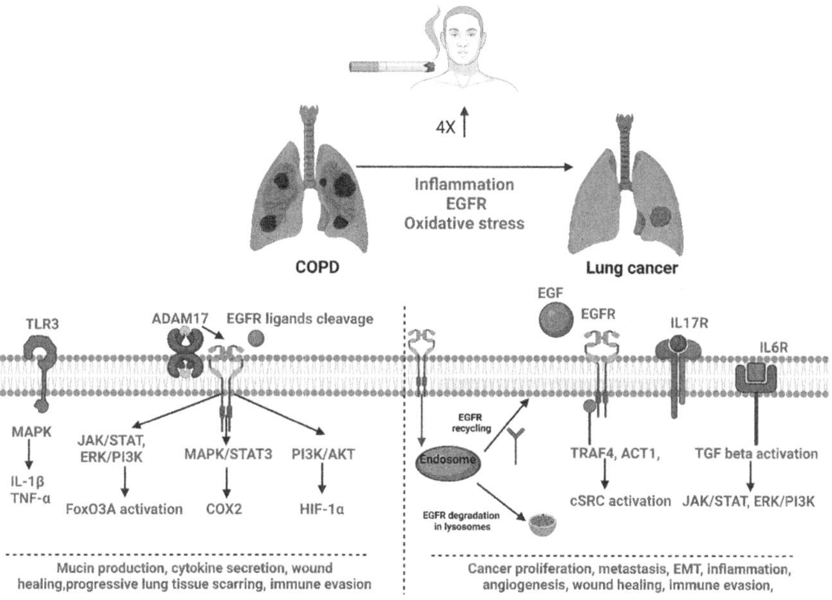

Fig. 3 EGFR dysregulation during COPD and LC

EGFR expression and smoking index in healthy smokers suggesting that EGFR regulates mucus production in COPD patients [88].

ADAM17, formerly known as TACE, is a widely expressed sheddase that releases various soluble bioactive ligands from airway epithelial and myeloid cells. It plays a crucial role in airway epithelial cis- and trans-signaling [89]. ADAM17 facilitates ectodomain shedding and can cleave nearly all EGFR ligands, thereby mediating both autocrine and paracrine EGFR signaling in airway epithelium [89]. Both EGFR and ADAM17 contribute to the progression of COPD through processes such as mucin production, cytokine secretion, fibrotic wound healing in airway epithelium, abnormal airway proliferation, maintenance of barrier integrity, and progressive lung tissue scarring [90]. Recently, it was shown that HB-EGF expression is significantly higher in COPD patients compared to non-smokers and smokers without COPD. This increase in HB-EGF correlates positively with increased expression of N-cadherin, vimentin, and collagen deposition. Mechanistically, HB-EGF contributes to airway remodeling by influencing airway fibrosis and EMT in human bronchial epithelial (HBE) cells and lung fibroblasts [91].

Another study found that aberrant EGFR activation in the airways of COPD patients leads to increased PI3K/Akt-mediated phosphorylation of FoxO3A. This resulted in reduced levels of nuclear FoxO3A and elevated chemokine expression. In COPD, decreased nuclear FoxO3A was linked to increased phosphorylation of EGFR and Akt. Treatment with quercetin, LY294002, or erlotinib was shown to restore nuclear FoxO3A levels and reduce both IL-8 expression and the phosphorylation of Akt, EGFR, and FoxO3A [92]. In an early-stage COPD mouse model, treatment with Chuan Bei Pi Pa dropping pills (CBPP) resulted in improved lung function and reduced pulmonary fibrosis. This compound was found to downregulate key pathways, including the EGFR and MLC2 signaling pathways. Recent studies have highlighted that EGFR also plays a role in modulating the wound healing response to SARS-CoV infection. Inhibition of EGFR signaling may block pathogenic fibrotic responses to viral infections including SARS-CoV, suggesting potential avenues for future research [93–95].

A study examined novel characteristics of NSCLC associated with COPD by evaluating methylation profiles and changes in EGFR signaling. Given the increased sensitivity to smoking in COPD patients, coexistence of NSCLC with COPD could represent a subset of smoking-associated NSCLC. This study found that IL-12Rβ2 and Wif-1 methylation and HER2 upregulation were common in COPD. Conversely, EGFR mutation was more prevalent in non-COPD group, and EGFR expression showed an inverse correlation with %FEV1 [96].

During COPD, the imbalance between oxidants and antioxidants is a key factor in disease progression. In a CS-induced rat COPD model, vitamin E treatment reduced inflammation, apoptosis, and ROS by blocking EGFR/MAPK pathway, which further blocked p-STAT3 translocation to the nucleus. However, COX2 upregulation reduced the protective effects of vitamin E in COPD rats [97].

HIF-1α is a crucial transcription factor that manages hypoxic responses and regulates oxygen balance, hypoxia adaptation, inflammatory responses, and tumor development. Zhang et al. recently found that both protein and mRNA levels of EGFR and HIF-1α were elevated in BALF samples from COPD patients compared to healthy individuals. In COPD, HIF-1α was overexpressed, leading to increased expression of inflammatory factors through activation of EGFR/PI3K/AKT pathway. This pathway, activated by pulmonary inflammation, further enhanced HIF-1α expression in a feedback loop, thereby exacerbating the pathological changes associated with COPD [98].

Long-term activation of TLR3 leads to inflammation and impairs lung function in COPD. Src and EGFR are essential for the phosphorylation of Tyr-759 and Tyr-858 during TLR3 activation. Recent studies have shown that both TLR3 and tyrosine kinase inhibitors can reduce cigarette smoke- and polyIC-induced airway inflammation (marked by elevated levels of IL-1β and TNF-α), airway hyperreactivity, and remodeling through inhibition of EGFR/TLR3/MAPK signaling pathway. Therefore, targeting this pathway may be a potent strategy for treating airway inflammation and remodeling in COPD [99, 100].

VEGF signaling is essential for angiogenesis and is upregulated by EGFR expression. In contrast, VEGF can also be upregulated independently of EGFR signaling [101, 102], which may contribute to resistance against EGFR inhibitors. Both VEGF and IL-8, along with other pro-angiogenic factors, can affect the inflammatory response in ways similar to EGFR signaling. In EGFR-mutant NSCLCs, increased EGFR signaling enhanced VEGF level through mechanisms that are not associated with hypoxia. This increase in VEGF can promote resistance to EGFR TKIs. Clinical trials have demonstrated that adding anti-VEGF therapy to EGFR TKIs significantly improves patient outcomes [103–105].

5 EGFR as a Therapeutic Target for COPD and Lung Cancer

5.1 Lung Cancer

Most therapeutics targeting EGFR have primarily focused on HER1 and HER2. The EGFR pathway can be inhibited through two main approaches: the use of small molecule TKIs (Table 1) that interfere with intracellular EGFR signaling and monoclonal antibodies that prevent receptor-ligand interactions. Small molecule inhibitors approved for LC patients with HER1 mutations have demonstrated 71.2% response rate, compared to 1.1% response rate with carboplatin and paclitaxel [106, 107]. In contrast, monoclonal antibodies such as cetuximab and trastuzumab, tested in LC patients with HER1 mutations, did not show positive results in phase II clinical trials [108]. Studies have demonstrated that first-generation TKIs, such as gefitinib [107, 109] and erlotinib [110, 111], offer survival benefits over standard platinum-based chemotherapy, with improved response rates and significantly enhanced progression-free survival. Subsequently, second-generation TKIs like

Table 1 EGFR TKIs in lung cancer clinical trials

Clinical trial	EGFR TKI	Population	EGFR mutation type
NCT01929200	Icotinib x 2 years	Stages II and III	Exon 19 or 21
NCT04410796	Osimertinib	Stage II	Exon 19/EGFR L858R or others
NCT05493501	Aumolertinib Osimertinib	Stage III	T790M resistance mutation
NCT05281406 (PACE-LUNG)	Osimertinib + CBDCA/ CDDP + PEM	Stage II	Exon19del or L858R EGFR mutation
NCT01996098	Chemotherapy plus icotinib x 6 or 12 months	Stages II and III	Exon 19 or 21
NCT01746251	Afatinib x 2 years	Stages I–III	EGFR mutation, prior chemotherapy allowed
NCT02971501 NCT04181060 NCT04988607	Bevacizumab	Stage II Stage III Stage II	EGFR-activating mutations in exon 19 EGFR exon 19 and 20 mutations EGFR exon 21 L858R mutation
NCT03909334	Ramucirumab (osimertinib + ramucirumab)	Stage II	Exon 19 deletion
NCT04487080 (MARIPOSA)	Amivantamab (amivantamab + lazertinib, osimertinib + placebo, lazertinib + placebo)	Stage III	Exon 19 dell and exon 21 L858R substitution EGFR mutations and the EGFR T790M+ resistance mutation
NCT0140214	Erlotinib x 2 years	Stage III	Exon 19 or 21
NCT01405079	Erlotinib x 2 years	Stages II and III	Exon 19 deletion or L858R
NCT01683175	Erlotinib x 2 years	Stage III	Exon 19 deletion or L858R
WJOG6410L	Gefitinib x 2 years	Stages II and III	Exon 19 deletion or L858R

afatinib and dacomitinib showed better progression-free survival than their first-generation counterparts [112, 113]. However, they were associated with increased cytotoxicity [112, 114].

5.1.1 Combination Therapies

As established by FLAURA study, osimertinib is currently the standard treatment for previously untreated EGFR-mutated NSCLC (NCT02296125). It demonstrated better outcomes compared to platinum-based therapy in T790M-positive NSCLC patients. However, patients eventually developed resistance to osimertinib. It is now recognized that TKI therapy alone cannot effectively target innate TKI-resistant cancer cells, prompting studies into combination strategies with platinum-based chemotherapy. A phase III trial compared gefitinib alone to a combination of gefitinib with platinum-based chemotherapy. It revealed a significant improvement in progression-free survival (PFS) (median, 16 months vs. 8 months; HR, 0.51; 95%

CI, 0.39–0.66). Similarly, a phase III NEJ009 trial evaluated gefitinib alone against gefitinib combined with carboplatin and pemetrexed (CGP) that demonstrated a notable increase in PFS in the combination group.

Following the approval of osimertinib, it has been increasingly used in combination therapies in place of gefitinib. The phase III FLAURA2 trial (NCT04035486) is currently evaluating first-line osimertinib with or without platinum-pemetrexed in 587 patients with EGFR-mutated locally advanced or metastatic NSCLC. The combination of osimertinib and chemotherapy has shown a manageable safety and tolerability profile in this patient population. These observations are based on initial data from the primary endpoint of progression-free survival, which supports further assessment in the FLAURA2 randomized phase. Overall, combination therapies are emerging as better options, demonstrating survival benefits in first-line treatment settings and phase III trials. More clinical trials are necessary to establish combination therapy as a viable option for LC patients.

5.1.2 Combining EGFR TKI with VEGF Inhibitor

VEGF and EGFR signaling are closely linked in LC progression, leading to exploration of combination therapies that involve VEGF inhibition with anti-VEGF antibody bevacizumab alongside EGFR TKI erlotinib. This combination has demonstrated increased PFS in patients, although no overall survival benefits have been observed. Further, NEJ026 and ARTemis/CTONG1509 trials showed significant improvements in PFS with erlotinib alone and in combination with bevacizumab: NEJ026 trial reported a median PFS of 16.9 months versus 13.3 months (HR, 0.605; 95% CI, 0.417–0.877), while ARTemis trial reported 17.9 months versus 11.2 months' PFS (HR, 0.55; 95% CI, 0.41–0.73). These outcomes suggest that combination therapies have better clinical outcomes than the single therapy treatment options.

5.1.3 MET Targeting Options

MET signaling is responsible for osimertinib resistance by evading EGFR signaling, and MET dysregulation is reported in 6–24% of patients [115]. MET-targeting drugs such as savolitinib, tepotinib, and capmatinib were given to patients (resistant to osimertinib) as single agents or combined with osimertinib in a clinical study. The overall response rate (32%) and median of progression-free survival (5.3 months) have been recorded in EGFR-mutated NSCLC patient having MET-driven resistance receiving savolitinib and osimertinib together [116]. Amivantamab is emerging as an alternative option for addressing resistant EGFR mutations including del19, L858R, T790M, C797S, and exon 20 insertions [117, 118]. Telisotuzumab vedotin (Teliso-V), a potent microtubule inhibitor, is an anti-c-Met monoclonal antibody that is combined with monomethyl auristatin E [119]. Interim analysis of the safety and efficacy profile indicates that Teliso-V and osimertinib given together achieved an objective response rate of 58% in patients who had been earlier administered first-line osimertinib [120].

5.1.4 C797S-Targeting Therapy

The C797S EGFR mutation contributes to drug resistance by disrupting the interaction with osimertinib [121]. Fourth-generation allosteric EGFR inhibitors are currently being studied, as they bind to EGFR at a site distant from the ATP-binding pocket. This unique binding mechanism selectively alters EGFR configuration and permits evasion of C797S-mediated resistance. Both in vivo and in vitro studies have shown that these inhibitors exhibit promising anticancer effects [76]. One such inhibitor, JBJ-04-125-02, specifically targets L858R mutation and inhibits the signaling of triple mutant (L858R/T790M/C797S) EGFR. When combined with osimertinib, JBJ-04-125-02 led to greater overall efficacy compared to either of the drugs alone [122]. Another drug, BLU-701, a selective and potent fourth-generation EGFR inhibitor, is being assessed in patients with EGFR-mutated NSCLC previously treated with EGFR TKIs, including osimertinib. Additionally, the efficacy of BLU-701 in C797X mutations containing patients is to be assessed in a phase II expansion cohort [123]. Its combination with osimertinib has shown better outcomes in PDX model. Furthermore, monoclonal antibodies such as brigatinib, when paired with cetuximab or panitumumab, have also shown antitumor efficacy in preclinical studies targeting triple mutation with C797S in cis and T790M, activating the EGFR mutation [124–126].

5.2 COPD

COPD shares common pathogenic pathways with LC, and numerous studies have indicated that COPD increases the risk of developing NSCLC, independent of patient age or the extent of cigarette smoke exposure. Recently, Wu et al. assessed the effect of COPD on survival among NSCLC patients receiving EGFR TKI therapy. This study utilized data from Taiwan's National Health Insurance Research Database, covering the period from 1995 to 2013. Patients treated with EGFR TKIs were categorized into two groups COPD and non-COPD, and age, gender, comorbidities, and cancer therapies were adjusted. The survival analysis revealed that patients with COPD experienced significantly worse outcomes: the median overall survival was 2.04 years among those having COPD compared to 2.28 years for those without COPD ($p < 0.001$), and the median PFS was 0.62 years versus 0.69 years, respectively ($p < 0.001$). These findings indicate that LC patients receiving TKI had poorer survival outcomes when they had pre-existing COPD [127].

A population-based cohort study conducted in the United States from September 2018 to 2023 examined the relationship between FDA-approved EGFR inhibitors—gefitinib, erlotinib, afatinib, and osimertinib—and the risk of developing

COPD. Among 975,797 patients, 12,214 were users of EGFR inhibitors, while 963,583 were non-users. The study found that the use of EGFR inhibitors was related with lower risk of new-onset COPD, with osimertinib showing the lowest risk, followed by afatinib, gefitinib, and erlotinib. Notably, the risk reduction was more significant for emphysema compared to chronic bronchitis. Additionally, EGFR inhibitors were found to lower COPD risk in both smokers and non-smokers [128].

Furthermore, a clinical investigation used BIBW2948 among COPD patients suffering from mucus hypersecretion. This multicenter, double-blind, placebo-controlled study aimed to evaluate safety and mucin-related outcomes. Unfortunately, among the 48 participants, there was no significant reduction in mucin stores or gene expression in the treatment group. Furthermore, an increase in adverse effects was observed with higher doses, including elevated liver enzymes and a decline in forced expiratory volume in 1 s (FEV1) [129]. Ex vivo analyses indicate that higher doses may be more effective in both inhibiting EGFR and reducing mucin levels; however, potential adverse effects should be anticipated.

6 Summary and Future Perspectives

COPD is a risk factor for LC, which extends beyond their common etiology. Both conditions are determined by oxidative imbalance in the lung tissue and are augmented by associated inflammation, cellular proliferation, EMT, and senescence. Additionally, genetic predisposition and altered epigenetic regulation of gene expression are observed in both diseases [130]. Alternatively, COPD could also act as LC driver by enhancing oxidative stress and chronic exposure to pro-inflammatory cytokines, increased DNA damage and concurrent suppression of DNA repair processes, and enhanced cell growth. Inflammation is a major source of reactive nitrogen and oxygen species, which remain persistently elevated in COPD [131]. Given that nearly all cancers emerge from inflammation, and various inflammatory diseases can predispose individuals to cancer, it is plausible that chronic inflammation in COPD could drive LC development [132].

Increased EGFR activation and inflammation are common components in both COPD and LC. COPD can adversely affect the survival of LC patients, particularly those undergoing treatment with EGFR TKIs. Although there is dysregulation in EGFR signaling associated with COPD, the inhibition of EGFR activation has shown clinical benefits primarily in LC treatment [127]. However, the complexities of EGFR signaling including different mutations that play critical role in cancer development, aggravating the cancer growth, and developing therapy resistance need to be unraveled. Future research should also focus on understanding the pharmacokinetics, side effects, and potential clinical efficacy of both existing drug candidates and those in pipeline. Strategies for targeting EGFR in COPD and LC may involve inhibiting ligand binding, manipulating ligand release, and affecting receptor dimerization and activation by metalloproteinases. Other key approaches could include restoring the separation between EGFR ligands and receptors, enhancing the binding affinity of antagonists, and limiting downstream EGFR signaling pathways [133]. Additionally, a combination approach using inhibitors and monoclonal antibodies should be clinically implemented, along with a deeper understanding of resistance mechanisms to develop new therapeutic strategies. Another crucial aspect that needs serious effort is the active molecular profiling of patient biopsies to tailor personalized therapies. This approach could lead to improved survival outcomes for both COPD and LC patients [134].

Acknowledgments The authors acknowledge BioRender for providing some of the figure templates. Dr. Umesh C S Yadav acknowledges the grant support from the Department of Biotechnology (DBT) (project no. BT/PR42671/MED/30/2347/2021).

Informed Consent Statement Not applicable.

Data Availability Statement Not applicable.

Conflicts of Interest The authors declare no conflict of interest.

Declaration of Generative AI and AI-Assisted Technologies in the Writing Process During the preparation of this work, the authors used ChatGPT to improve language and readability. After using this tool/service, the authors reviewed and edited the content as needed and take full responsibility for the content of the publication.

References

1. Bray F, Laversanne M, Sung H, et al. Global cancer statistics 2022: GLOBOCAN estimates of incidence and mortality worldwide for 36 cancers in 185 countries. CA Cancer J Clin. 2024;74(3):229–63.
2. Aaron SD, Vandemheen KL, Whitmore GA, et al. Early diagnosis and treatment of COPD and asthma—a randomized, controlled trial. N Engl J Med. 2024;390(22):2061–73.
3. Corrales L, Rosell R, Cardona AF, et al. Lung cancer in never smokers: the role of different risk factors other than tobacco smoking. Crit Rev Oncol Hematol. 2020;148:102895.
4. Caliri AW, Tommasi S, Besaratinia A. Relationships among smoking, oxidative stress, inflammation, macromolecular damage, and cancer. Mutat Res/Rev Mutat Res. 2021;787:108365.
5. Husebø GR, Nielsen R, Hardie J, et al. Risk factors for lung cancer in COPD–results from the Bergen COPD cohort study. Respir Med. 2019;152:81–8.
6. Wang DC, Shi L, Zhu Z, Gao D, Zhang Y. Genomic mechanisms of transformation from chronic obstructive pulmonary disease to lung. Semin Cancer Biol. 2017;42:52–9. https://doi.org/10.1016/j.semcancer.2016.11.001.
7. Young RP, Hopkins RJ. Chronic obstructive pulmonary disease (COPD) and lung cancer screening. Trans Lung Cancer Res. 2018;7(3):347.
8. Ytterstad E, Moe PC, Hjalmarsen A. COPD in primary lung cancer patients: prevalence and mortality. Int J Chron Obstruct Pulmon Dis. 2016;11:625–36.
9. Kim YS, Kim EY, Ahn HK, et al. Prognostic significance of CT-emphysema score in patients with advanced squamous cell lung cancer. J Thorac Dis. 2016;8(8):1966.
10. Zhang B, Zhang Y, Jiang X, et al. JMJD8 promotes malignant progression of lung cancer by maintaining EGFR stability and EGFR/PI3K/AKT pathway activation. J Cancer. 2021;12(4):976.

11. Liang Y, Zhang T, Jing S, et al. 20(S)-Ginsenoside Rg3 inhibits lung cancer cell proliferation by targeting EGFR-mediated Ras/Raf/MEK/ERK pathway. Am J Chin Med. 2021;49(03):753–65.
12. Kanai K, Koarai A, Shishikura Y, et al. Cigarette smoke augments MUC5AC production via the TLR3-EGFR pathway in airway epithelial cells. Respir Investig. 2015;53(4):137–48.
13. Gallant J-N, Sheehan JH, Shaver TM, et al. EGFR kinase domain duplication (EGFR-KDD) is a novel oncogenic driver in lung cancer that is clinically responsive to afatinib. Cancer Discov. 2015;5(11):1155–63.
14. Li X, Zhao L, Chen C, et al. Can EGFR be a therapeutic target in breast cancer? Biochim Biophys Acta (BBA)-Rev Cancer. 2022;1877(5):188789.
15. Troiani T, Martinelli E, Capasso A, et al. Targeting EGFR in pancreatic cancer treatment. Curr Drug Targets. 2012;13(6):802–10.
16. Nagalakshmi K, Jamil K, Pingali U, et al. Epidermal growth factor receptor (EGFR) mutations as biomarker for head and neck squamous cell carcinomas (HNSCC). Biomarkers. 2014;19(3):198–206.
17. McGuinness AJA, Sapey E. Oxidative stress in COPD: sources, markers, and potential mechanisms. J Clin Med. 2017;6(2):21.
18. Martínez-Reyes I, Chandel NS. Cancer metabolism: looking forward. Nat Rev Cancer. 2021;21(10):669–80.
19. Choudhury S, Ghosh S, Gupta P, et al. Inflammation-induced ROS generation causes pancreatic cell death through modulation of Nrf2/NF-κB and SAPK/JNK pathway. Free Radic Res. 2015;49(11):1371–83.
20. Zinatizadeh MR, Schock B, Chalbatani GM, et al. The nuclear factor kappa B (NF-kB) signaling in cancer development and immune diseases. Genes Dis. 2021;8(3):287–97.
21. Nowak DE, Tian B, Jamaluddin M, et al. RelA Ser276 phosphorylation is required for activation of a subset of NF-κB-dependent genes by recruiting cyclin-dependent kinase 9/cyclin T1 complexes. Mol Cell Biol. 2008;28:3623.
22. Jamaluddin M, Wang S, Boldogh I, et al. TNF-α-induced NF-κB/RelA Ser276 phosphorylation and enhanceosome formation is mediated by an ROS-dependent PKAc pathway. Cell Signal. 2007;19(7):1419–33.
23. Schoonbroodt S, Ferreira VR, Best-Belpomme M, et al. Crucial role of the amino-terminal tyrosine residue 42 and the carboxyl-terminal PEST domain of IκBα in NF-κB activation by an oxidative stress. J Immunol. 2000;164(8):4292–300.
24. Nottingham LK, Yan CH, Yang X, et al. Aberrant IKKα and IKKβ cooperatively activate NF-κB and induce EGFR/AP1 signaling to promote survival and migration of head and neck cancer. Oncogene. 2014;33(9):1135–47.
25. Xiong Z, Leme AS, Ray P, et al. CX3CR1+ lung mononuclear phagocytes spatially confined to the interstitium produce TNF-α and IL-6 and promote cigarette smoke-induced emphysema. J Immunol. 2011;186(5):3206–14.
26. Lee JM, Tsuboi M, Kim ES, et al. Overcoming immunosuppression and pro-tumor inflammation in lung cancer with combined IL-1β and PD-1 inhibition. Future Oncol. 2022;18(27):3085–100.
27. Hogea P, Tudorache E, Fira-Mladinescu O, et al. Serum and Bronchoalveolar lavage fluid levels of cytokines in patients with lung cancer and chronic lung disease: a prospective comparative study. J Pers Med. 2023;13(6):998.
28. Chen J, Li X, Huang C, et al. Change of serum inflammatory cytokines levels in patients with chronic obstructive pulmonary disease, pneumonia and lung cancer. Technol Cancer Res Treat. 2020;19:1533033820951807.
29. Alharbi KS, Fuloria NK, Fuloria S, et al. Nuclear factor-kappa B and its role in inflammatory lung disease. Chem Biol Interact. 2021;345:109568.
30. Mahor D, Kumari V, Vashisht K, et al. Elevated serum matrix metalloprotease (MMP-2) as a candidate biomarker for stable COPD. BMC Pulm Med. 2020;20:1–9.

31. Agraval H, Kandhari K, Yadav UCS. MMPs as potential molecular target in epithelial-to-mesenchymal transition driven COPD progression. Life Sci. 2024;352:122874.
32. Han L, Sheng B, Zeng Q, et al. Correlation between MMP2 expression in lung cancer tissues and clinical parameters: a retrospective clinical analysis. BMC Pulm Med. 2020;20:1–9.
33. Verma AK, Pandey AK, Singh A, et al. Increased serum levels of matrix-metalloproteinase-9, cyclo-oxygenase-2 and prostaglandin E-2 in patients with chronic obstructive pulmonary disease (COPD). Indian J Clin Biochem. 2022;37:1–9.
34. Agraval H, Sharma JR, Prakash N, et al. Fisetin suppresses cigarette smoke extract-induced epithelial to mesenchymal transition of airway epithelial cells through regulating COX-2/MMPs/β-catenin pathway. Chem Biol Interact. 2022;351:109771.
35. Kumar S, Swaroop S, Sahu A, et al. Association of MMP7 T> C gene variant (rs10502001) and expression in chronic obstructive pulmonary disease. DNA Cell Biol. 2023;42(9):548–53.
36. Yamada K, Kadota K, Fujimoto S, et al. MMP-7 expression is associated with a higher rate of tumor spread through air spaces in resected lung adenocarcinomas. Lung Cancer. 2023;175:125–30.
37. Gharib SA, Manicone AM, Parks WC. Matrix metalloproteinases in emphysema. Matrix Biol. 2018;73:34–51.
38. Hung W-Y, Lee W-J, Cheng G-Z, et al. Blocking MMP-12-modulated epithelial-mesenchymal transition by repurposing penfluridol restrains lung adenocarcinoma metastasis via uPA/uPAR/TGF-β/Akt pathway. Cell Oncol. 2021;44:1087–103.
39. Seiler CL, Song JM, Kotandeniya D, et al. Inhalation exposure to cigarette smoke and inflammatory agents induces epigenetic changes in the lung. Sci Rep. 2020;10(1):11290.
40. Sundar IK, Rahman I. Gene expression profiling of epigenetic chromatin modification enzymes and histone marks by cigarette smoke: implications for COPD and lung cancer. Am J Phys Lung Cell Mol Phys. 2016;311(6):L1245–L58.
41. Tessema M, Tassew DD, Yingling CM, et al. Identification of novel epigenetic abnormalities as sputum biomarkers for lung cancer risk among smokers and COPD patients. Lung Cancer. 2020;146:189–96.
42. Banaganapalli B, Mallah B, Alghamdi KS, et al. Integrative weighted molecular network construction from transcriptomics and genome wide association data to identify shared genetic biomarkers for COPD and lung cancer. PLoS One. 2022;17(10):e0274629.
43. Purba ER, Saita EI, Maruyama IN. Activation of the EGF receptor by ligand binding and oncogenic mutations: the "rotation model". Cells. 2017;6(2) https://doi.org/10.3390/cells6020013. [published Online First: 2017/06/03].
44. Tabei Y, Nakajima Y. IL-1β-activated PI3K/AKT and MEK/ERK pathways coordinately promote induction of partial epithelial–mesenchymal transition. Cell Commun Signal. 2024;22(1):392.
45. Pastore S, Mascia F, Mariani V, et al. The epidermal growth factor receptor system in skin repair and inflammation. J Invest Dermatol. 2008;128(6):1365–74.
46. Tjabringa GS, Aarbiou J, Ninaber DK, et al. The antimicrobial peptide LL-37 activates innate immunity at the airway epithelial surface by transactivation of the epidermal growth factor receptor. J Immunol. 2003;171(12):6690–6.
47. Wang X, Reyes ME, Zhang D, et al. EGFR signaling promotes inflammation and cancer stem-like activity in inflammatory breast cancer. Oncotarget. 2017;8(40):67904.
48. Fang Q, Zou C, Zhong P, et al. EGFR mediates hyperlipidemia-induced renal injury via regulating inflammation and oxidative stress: the detrimental role and mechanism of EGFR activation. Oncotarget. 2016;7(17):24361.
49. Chen W, Tan M, Zhang H, et al. Signaling molecules in the microenvironment of hepatocellular carcinoma. Funct Integr Genomics. 2024;24(5):146.
50. Ramu A, Kathiresan S, Ramadoss H, et al. Gramine attenuates EGFR-mediated inflammation and cell proliferation in oral carcinogenesis via regulation of NF-κB and STAT3 signaling. Biomed Pharmacother. 2018;98:523–30.

51. Cohen S, Carpenter G. Human epidermal growth factor: isolation and chemical and biological properties. Proc Natl Acad Sci. 1975;72(4):1317–21.
52. Ishii S, Xu Y-H, Stratton RH, et al. Characterization and sequence of the promoter region of the human epidermal growth factor receptor gene. Proc Natl Acad Sci. 1985;82(15):4920–4.
53. Xu Y-H, Richert N, Ito S, et al. Characterization of epidermal growth factor receptor gene expression in malignant and normal human cell lines. Proc Natl Acad Sci. 1984;81(23):7308–12.
54. Modjtahedi H, Styles JM, Dean CJ. The human EGF receptor as a target for cancer therapy: six new rat mAbs against the receptor on the breast carcinoma MDA-MB 468. Br J Cancer. 1993;67(2):247–53.
55. Cohen MH, Williams GA, Sridhara R, et al. United States Food and Drug Administration drug approval summary: gefitinib (ZD1839; Iressa) tablets. Clin Cancer Res. 2004;10(4):1212–8.
56. Cohen MH, Johnson JR, Chen Y-F, et al. FDA drug approval summary: erlotinib (Tarceva®) tablets. Oncologist. 2005;10(7):461–6.
57. Tian X, Gu T, Lee M-H, et al. Challenge and countermeasures for EGFR targeted therapy in non-small cell lung cancer. Biochim Biophys Acta (BBA)-Rev Cancer. 2022;1877(1):188645.
58. Ma L-R, Li J-X, Tang L, et al. Immune checkpoints and immunotherapy in non-small cell lung cancer: novel study progression, challenges and solutions. Oncol Lett. 2021;22(5):787.
59. Sharma SV, Bell DW, Settleman J, et al. Epidermal growth factor receptor mutations in lung cancer. Nat Rev Cancer. 2007;7(3):169–81.
60. Shigematsu H, Gazdar AF. Somatic mutations of epidermal growth factor receptor signaling pathway in lung cancers. Int J Cancer. 2006;118(2):257–62.
61. Regales L, Gong Y, Shen R, et al. Dual targeting of EGFR can overcome a major drug resistance mutation in mouse models of EGFR mutant lung cancer. J Clin Invest. 2009;119(10):3000–10.
62. Lu L, Teixeira VH, Yuan Z, et al. LRIG1 regulates cadherin-dependent contact inhibition directing epithelial homeostasis and pre-invasive squamous cell carcinoma development. J Pathol. 2013;229(4):608–20. https://doi.org/10.1002/path.4148. [published Online First: 2012/12/05].
63. Mukohara T, Engelman JA, Hanna NH, et al. Differential effects of gefitinib and cetuximab on non–small-cell lung cancers bearing epidermal growth factor receptor mutations. J Natl Cancer Inst. 2005;97(16):1185–94.
64. Li AR, Chitale D, Riely GJ, et al. EGFR mutations in lung adenocarcinomas: clinical testing experience and relationship to EGFR gene copy number and immunohistochemical expression. J Mol Diagn. 2008;10(3):242–8.
65. Bethune G, Bethune D, Ridgway N, et al. Epidermal growth factor receptor (EGFR) in lung cancer: an overview and update. J Thorac Dis. 2010;2(1):48–51. [published Online First: 2010/03/01]
66. Al-Nedawi K, Meehan B, Kerbel RS, et al. Endothelial expression of autocrine VEGF upon the uptake of tumor-derived microvesicles containing oncogenic EGFR. Proc Natl Acad Sci. 2009;106(10):3794–9.
67. Gao L, Wang L, Dai T, et al. Tumor-derived exosomes antagonize innate antiviral immunity. Nat Immunol. 2018;19(3):233–45.
68. Li X, Bechara R, Zhao J, et al. IL-17 receptor-based signaling and implications for disease. Nat Immunol. 2019;20(12):1594–602. https://doi.org/10.1038/s41590-019-0514-y. [published Online First: 2019/11/21].
69. Gong K, Guo G, Gerber DE, et al. TNF-driven adaptive response mediates resistance to EGFR inhibition in lung cancer. J Clin Invest. 2018;128(6):2500–18. https://doi.org/10.1172/JCI96148.
70. Dutta A, Hung C-Y, Chen T-C, et al. An IL-17-EGFR-TRAF4 axis contributes to the alleviation of lung inflammation in severe influenza. Commun Biol. 2023;6(1):600. https://doi.org/10.1038/s42003-023-04982-0.

71. Garmendia I, Varthaman A, Marmier S, et al. Acute influenza infection promotes lung tumor growth by reprogramming the tumor microenvironment. Cancer Immunol Res. 2023;11(4):530–45. https://doi.org/10.1158/2326-6066.Cir-22-0534. [published Online First: 2023/03/09].
72. Chen Q, Xia L, Wang J, et al. EGFR-mutant NSCLC may remodel TME from non-inflamed to inflamed through acquiring resistance to EGFR-TKI treatment. Lung Cancer. 2024;192:107815. https://doi.org/10.1016/j.lungcan.2024.107815.
73. Ray K, Ujvari B, Ramana V, et al. Cross-talk between EGFR and IL-6 drives oncogenic signaling and offers therapeutic opportunities in cancer. Cytokine Growth Factor Rev. 2018;41:18–27. https://doi.org/10.1016/j.cytogfr.2018.04.002.
74. Gao SP, Mark KG, Leslie K, et al. Mutations in the EGFR kinase domain mediate STAT3 activation via IL-6 production in human lung adenocarcinomas. J Clin Invest. 2007;117(12):3846–56. https://doi.org/10.1172/jci31871. [published Online First: 2007/12/07].
75. Pao W, Miller VA, Politi KA, et al. Acquired resistance of lung adenocarcinomas to gefitinib or erlotinib is associated with a second mutation in the EGFR kinase domain. PLoS Med. 2005;2(3):e73. https://doi.org/10.1371/journal.pmed.0020073. [published Online First: 2005/03/02].
76. Jia Y, Yun CH, Park E, et al. Overcoming EGFR(T790M) and EGFR(C797S) resistance with mutant-selective allosteric inhibitors. Nature. 2016;534(7605):129–32. https://doi.org/10.1038/nature17960. [published Online First: 2016/06/03].
77. Wang S, Tsui ST, Liu C, et al. EGFR C797S mutation mediates resistance to third-generation inhibitors in T790M-positive non-small cell lung cancer. J Hematol Oncol. 2016;9(1):59. https://doi.org/10.1186/s13045-016-0290-1.
78. Yao Z, Fenoglio S, Gao DC, et al. TGF-beta IL-6 axis mediates selective and adaptive mechanisms of resistance to molecular targeted therapy in lung cancer. Proc Natl Acad Sci USA. 2010;107(35):15535–40. https://doi.org/10.1073/pnas.1009472107. [published Online First: 2010/08/18].
79. Wiley HS, Burke PM. Regulation of receptor tyrosine kinase signaling by endocytic trafficking. Traffic. 2001;2(1):12–8.
80. Steffan JJ, Williams BC, Welbourne T, et al. HGF-induced invasion by prostate tumor cells requires anterograde lysosome trafficking and activity of Na+-H+ exchangers. J Cell Sci. 2010;123(7):1151–9.
81. Dykes SS, Gray AL, Coleman DT, et al. The Arf-like GTPase Arl8b is essential for three-dimensional invasive growth of prostate cancer in vitro and xenograft formation and growth in vivo. Oncotarget. 2016;7(21):31037.
82. Futter CE, Pearse A, Hewlett LJ, et al. Multivesicular endosomes containing internalized EGF-EGF receptor complexes mature and then fuse directly with lysosomes. J Cell Biol. 1996;132(6):1011–23.
83. Authier F, Métioui M, Bell AW, et al. Negative regulation of epidermal growth factor signaling by selective proteolytic mechanisms in the endosome mediated by cathepsin B. J Biol Chem. 1999;274(47):33723–31.
84. Alwan HAJ, van Zoelen EJJ, van Leeuwen JEM. Ligand-induced lysosomal epidermal growth factor receptor (EGFR) degradation is preceded by proteasome-dependent EGFR De-ubiquitination *. J Biol Chem. 2003;278(37):35781–90. https://doi.org/10.1074/jbc.M301326200.
85. Sung WJ, Kim D, Zhu A, et al. The lysosome as a novel therapeutic target of EGFR-mediated tumor inflammation. Front Pharmacol. 2022;13 https://doi.org/10.3389/fphar.2022.1050758.
86. O'Donnell R, Richter A, Ward J, et al. Expression of ErbB receptors and mucins in the airways of long term current smokers. Thorax. 2004;59(12):1032–40.
87. Inoue H, Akimoto K, Homma T, et al. Airway epithelial dysfunction in asthma: relevant to epidermal growth factor receptors and airway epithelial cells. J Clin Med. 2020;9(11) https://doi.org/10.3390/jcm9113698. [published Online First: 2020/11/22]

88. Saied EM, Bediwy AS. Expression of epidermal growth factor receptor (EGFR) in the bronchial epithelium of patients with chronic obstructive pulmonary disease (COPD). Eur Respir J. 2011;38(Suppl 55):3823.
89. Scheller J, Chalaris A, Garbers C, et al. ADAM17: a molecular switch to control inflammation and tissue regeneration. Trends Immunol. 2011;32(8):380–7.
90. Deshmukh HS, Case LM, Wesselkamper SC, et al. Metalloproteinases mediate mucin 5AC expression by epidermal growth factor receptor activation. Am J Respir Crit Care Med. 2005;171(4):305–14. https://doi.org/10.1164/rccm.200408-1003OC. [published Online First: 2004/11/09].
91. Eapen MS, Sharma P, Thompson IE, et al. Heparin-binding epidermal growth factor (HB-EGF) drives EMT in patients with COPD: implications for disease pathogenesis and novel therapies. Lab Investig. 2019;99(2):150–7. https://doi.org/10.1038/s41374-018-0146-0. [published Online First: 2018/11/20].
92. Ganesan S, Unger BL, Comstock AT, et al. Aberrantly activated EGFR contributes to enhanced IL-8 expression in COPD airways epithelial cells via regulation of nuclear FoxO3A. Thorax. 2013;68(2):131–41. https://doi.org/10.1136/thoraxjnl-2012-201719. [published Online First: 2012/10/27].
93. Venkataraman T, Frieman MB. The role of epidermal growth factor receptor (EGFR) signaling in SARS coronavirus-induced pulmonary fibrosis. Antivir Res. 2017;143:142–50. https://doi.org/10.1016/j.antiviral.2017.03.022.
94. Venkataraman T, Coleman Christopher M, Frieman MB. Overactive epidermal growth factor receptor signaling leads to increased fibrosis after severe acute respiratory syndrome coronavirus infection. J Virol. 2017;91(12) https://doi.org/10.1128/jvi.00182-17.
95. Kalinowski A, Ueki I, Min-Oo G, et al. EGFR activation suppresses respiratory virus-induced IRF1-dependent CXCL10 production. Am J Phys Lung Cell Mol Phys. 2014;307(2):L186–L96. https://doi.org/10.1152/ajplung.00368.2013.
96. Suzuki M, Wada H, Yoshino M, et al. Molecular characterization of chronic obstructive pulmonary disease-related non-small cell lung cancer through aberrant methylation and alterations of EGFR signaling. Ann Surg Oncol. 2010;17(3):878–88. https://doi.org/10.1245/s10434-009-0739-3.
97. Zhao H, Gong J, Li L, et al. Vitamin E relieves chronic obstructive pulmonary disease by inhibiting COX2-mediated p-STAT3 nuclear translocation through the EGFR/MAPK signaling pathway. Lab Investig. 2022;102(3):272–80. https://doi.org/10.1038/s41374-021-00652-z.
98. Zhang HX, Yang JJ, Zhang SA, et al. HIF-1α promotes inflammatory response of chronic obstructive pulmonary disease by activating EGFR/PI3K/AKT pathway. Eur Rev Med Pharmacol Sci. 2018;22(18):6077–84. https://doi.org/10.26355/eurrev_201809_15946. [published Online First: 2018/10/04].
99. Jiang J-j, Chen S-m, Li H-y, et al. TLR3 inhibitor and tyrosine kinase inhibitor attenuate cigarette smoke/poly I:C-induced airway inflammation and remodeling by the EGFR/TLR3/MAPK signaling pathway. Eur J Pharmacol. 2021;890:173654. https://doi.org/10.1016/j.ejphar.2020.173654.
100. Stowell N, Seideman J, Raymond H, et al. Long-term activation of TLR3 by Poly(I:C) induces inflammation and impairs lung function in mice. Respir Res. 2009;10:43. https://doi.org/10.1186/1465-9921-10-43.
101. Wang L, Xu Z, Chen B, et al. The role of vascular endothelial growth factor in small-airway Remodelling in a rat model of chronic obstructive pulmonary disease. Sci Rep. 2017;7(1):41202. https://doi.org/10.1038/srep41202.
102. Mura M, Binnie M, Han B, et al. Functions of Type II pneumocyte-derived vascular endothelial growth factor in alveolar structure, acute inflammation, and vascular permeability. Am J Pathol. 2010;176(4):1725–34. https://doi.org/10.2353/ajpath.2010.090209.
103. Tabernero J. The role of VEGF and EGFR inhibition: implications for combining anti–VEGF and anti–EGFR agents. Mol Cancer Res. 2007;5(3):203–20. https://doi.org/10.1158/1541-7786.Mcr-06-0404.

104. Le X, Nilsson M, Goldman J, et al. Dual EGFR-VEGF pathway inhibition: a promising strategy for patients with EGFR-mutant NSCLC. J Thorac Oncol. 2021;16(2):205–15. https://doi.org/10.1016/j.jtho.2020.10.006.
105. Kanazawa H. Role of vascular endothelial growth factor in the pathogenesis of chronic obstructive pulmonary disease. Med Sci Monit. 2007;13(11):Ra189-95. [published Online First: 2007/10/31]
106. Mok TS, Wu Y-L, Thongprasert S, et al. Gefitinib or carboplatin–paclitaxel in pulmonary adenocarcinoma. N Engl J Med. 2009;361(10):947–57.
107. Maemondo M, Inoue A, Kobayashi K, et al. Gefitinib or chemotherapy for non–small-cell lung cancer with mutated EGFR. N Engl J Med. 2010;362(25):2380–8.
108. Hanna N, Lilenbaum R, Ansari R, et al. Phase II trial of cetuximab in patients with previously treated non–small-cell lung cancer. J Clin Oncol. 2006;24(33):5253–8.
109. Mitsudomi T, Morita S, Yatabe Y, et al. Gefitinib versus cisplatin plus docetaxel in patients with non-small-cell lung cancer harbouring mutations of the epidermal growth factor receptor (WJTOG3405): an open label, randomised phase 3 trial. Lancet Oncol. 2010;11(2):121–8.
110. Zhou C, Wu Y-L, Chen G, et al. Erlotinib versus chemotherapy as first-line treatment for patients with advanced EGFR mutation-positive non-small-cell lung cancer (OPTIMAL, CTONG-0802): a multicentre, open-label, randomised, phase 3 study. Lancet Oncol. 2011;12(8):735–42. https://doi.org/10.1016/S1470-2045(11)70184-X.
111. Rosell R, Carcereny E, Gervais R, et al. Erlotinib versus standard chemotherapy as first-line treatment for European patients with advanced EGFR mutation-positive non-small-cell lung cancer (EURTAC): a multicentre, open-label, randomised phase 3 trial. Lancet Oncol. 2012;13(3):239–46.
112. Park K, Tan E-H, O'Byrne K, et al. Afatinib versus gefitinib as first-line treatment of patients with EGFR mutation-positive non-small-cell lung cancer (LUX-Lung 7): a phase 2B, open-label, randomised controlled trial. Lancet Oncol. 2016;17(5):577–89.
113. Wu Y-L, Cheng Y, Zhou X, et al. Dacomitinib versus gefitinib as first-line treatment for patients with EGFR-mutation-positive non-small-cell lung cancer (ARCHER 1050): a randomised, open-label, phase 3 trial. Lancet Oncol. 2017;18(11):1454–66.
114. Mok TS, Cheng Y, Zhou X, et al. Improvement in overall survival in a randomized study that compared dacomitinib with gefitinib in patients with advanced non–small-cell lung cancer and EGFR-activating mutations. J Clin Oncol. 2018;36(22):2244–50.
115. Rosário M, Birchmeier W. How to make tubes: signaling by the Met receptor tyrosine kinase. Trends Cell Biol. 2003;13(6):328–35.
116. Ahn M, De Marinis F, Bonanno L, et al. EP08. 02-140 MET biomarker-based preliminary efficacy analysis in SAVANNAH: savolitinib+ osimertinib in EGFRm NSCLC post-osimertinib. J Thorac Oncol. 2022;17(9):S469–S70.
117. Cho BC, Simi A, Sabari J, et al. Amivantamab, an epidermal growth factor receptor (EGFR) and mesenchymal-epithelial transition factor (MET) bispecific antibody, designed to enable multiple mechanisms of action and broad clinical applications. Clin Lung Cancer. 2023;24(2):89–97.
118. Yun J, Lee S-H, Kim S-Y, et al. Antitumor activity of amivantamab (JNJ-61186372), an EGFR–MET bispecific antibody, in diverse models of EGFR exon 20 insertion–driven NSCLC. Cancer Discov. 2020;10(8):1194–209.
119. Wang J, Anderson MG, Oleksijew A, et al. ABBV-399, a c-Met antibody–drug conjugate that targets both MET–amplified and c-Met–overexpressing tumors, irrespective of MET pathway dependence. Clin Cancer Res. 2017;23(4):992–1000.
120. Goldman JW, Horinouchi H, Cho BC, et al. Phase 1/1b study of telisotuzumab vedotin (Teliso-V)+ osimertinib (Osi), after failure on prior Osi, in patients with advanced, c-Met overexpressing, EGFR-mutated non-small cell lung cancer (NSCLC). American Society of Clinical Oncology; 2022.
121. Tsai CM, Yen GC, Sun FM, et al. Assessment of the anti-invasion potential and mechanism of select cinnamic acid derivatives on human lung adenocarcinoma cells. Mol Pharm.

2013;10(5):1890–900. https://doi.org/10.1021/mp3006648. [published Online First: 2013/04/09].
122. To C, Jang J, Chen T, et al. Single and dual targeting of mutant EGFR with an allosteric inhibitor. Cancer Discov. 2019;9(7):926–43.
123. Wang S, Song Y, Liu D. EAI045: the fourth-generation EGFR inhibitor overcoming T790M and C797S resistance. Cancer Lett. 2017;385:51–4.
124. Zhou Z, Zhao Y, Shen S, et al. Durable clinical response of lung adenocarcinoma harboring EGFR 19Del/T790M/in trans-C797S to combination therapy of first-and third-generation EGFR tyrosine kinase inhibitors. J Thorac Oncol. 2019;14(8):e157–e59.
125. Uchibori K, Inase N, Araki M, et al. Brigatinib combined with anti-EGFR antibody overcomes osimertinib resistance in EGFR-mutated non-small-cell lung cancer. Nat Commun. 2017;8(1):14768.
126. Wang X, Zhou L, Yin JC, et al. Lung adenocarcinoma harboring EGFR 19del/C797S/T790M triple mutations responds to brigatinib and anti-EGFR antibody combination therapy. J Thorac Oncol. 2019;14(5):e85–8.
127. Wu CC, Rau KM, Lee WC, et al. Presence of Chronic Obstructive Pulmonary Disease (COPD) impair survival in lung cancer patients receiving epidermal growth factor receptor-tyrosine kinase inhibitor (EGFR-TKI): a Nationwide, population-based cohort study. J Clin Med. 2019;8(7) https://doi.org/10.3390/jcm8071024. [published Online First: 2019/07/25]
128. Huang P-C, Ma KS-K. Reduced risk of incident chronic obstructive pulmonary disease following EGFR-targeted Therapies in Patients With Lung Cancer. In: C46 clinical and physiologic metrics of respiratory disease and COPD: from diagnosis to prognosis; 2024. p. A5737-A37.
129. Woodruff PG, Wolff M, Hohlfeld JM, et al. Safety and efficacy of an inhaled epidermal growth factor receptor inhibitor (BIBW 2948 BS) in chronic obstructive pulmonary disease. Am J Respir Crit Care Med. 2010;181(5):438–45.
130. Durham AL, Adcock IM. The relationship between COPD and lung cancer. Lung Cancer. 2015;90(2):121–7. https://doi.org/10.1016/j.lungcan.2015.08.017. [published Online First: 2015/09/14].
131. Hussain SP, Hofseth LJ, Harris CC. Radical causes of cancer. Nat Rev Cancer. 2003;3(4):276–85.
132. Schetter AJ, Heegaard NH, Harris CC. Inflammation and cancer: interweaving microRNA, free radical, cytokine and p53 pathways. Carcinogenesis. 2010;31(1):37–49. https://doi.org/10.1093/carcin/bgp272. [published Online First: 2009/12/04].
133. Vallath S, Hynds RE, Succony L, et al. Targeting EGFR signalling in chronic lung disease: therapeutic challenges and opportunities. Eur Respir J. 2014;44(2):513–22. https://doi.org/10.1183/09031936.00146413. [published Online First: 2014/01/18].
134. Araki T, Kanda S, Horinouchi H, et al. Current treatment strategies for EGFR-mutated non-small cell lung cancer: from first line to beyond osimertinib resistance. Jpn J Clin Oncol. 2023;53(7):547–61. https://doi.org/10.1093/jjco/hyad052. [published Online First: 2023/06/06].

Role of Receptor Tyrosine Kinase AXL in Cancer

Rahul Kumar, Pranay Tanwar, Dhiraj Bhatia, and Raghu Solanki

1 Introduction

Receptor tyrosine kinases (RTKs) are the transmembrane receptors that play a pivotal role in cancer progression and have gained attention due to their therapeutic potential oncology [1]. RTKs have a distinctive structure comprising four regions: extracellular domain (ECD), transmembrane domain (TM), juxtamembrane domain (JM), and tyrosine kinase domain (TKD) [2, 3]. Genetic alterations such as mutations, amplifications, and overexpression of specific RTKs are critical in establishing conditions that promote tumor growth. Aberrant RTK signaling has been linked to cancer cell proliferation, survival, and progression, suggesting the importance of RTK dysregulation in oncogenesis [4–6]. Targeted therapies against RTKs aim to disrupt key cellular processes involved in cancer development.

AXL is the member of the TAM (Tyro-3, AXL, and Mer) family of RTKs, overexpressed in various cancers including ovarian, lung, and breast cancer [7, 8]. The discovery of an unrecognized transforming gene, which was later named as AXL, was first reported by Bryan and colleagues in 1991 in two individuals with chronic myeloid leukemia (CML) [9]. In 1994, the structural and functional similarities between Tyro-3 and MERTK were discovered [10, 11]. AXL plays a role in promoting tumor growth and metastasis by activating signaling pathways such as PI3K/Akt and MAPK/Erk, and it has been implicated in the development of treatment

R. Kumar (✉) · P. Tanwar
Dr B. R. A. Institute Rotary Cancer Hospital, All India Institute of Medical Sciences, New Delhi, India
e-mail: rahul.kumar@aiims.edu; pranaytanwar@gmail.com

D. Bhatia · R. Solanki (✉)
Department of Biological Sciences and Engineering, Indian Institute of Technology Gandhinagar, Palaj-382355, Gandhinagar, Gujarat, India
e-mail: dhiraj.bhatia@iitgn.ac.in; raghu.solanki@iitgn.ac.in

A. Sabarwal et al. (eds.), *Receptor Tyrosine Kinases in Cancer*, Cancer Drug Discovery and Development, https://doi.org/10.1007/978-3-031-93894-8_4

resistance. As a result, AXL has emerged as a promising therapeutic target. Inhibition of AXL using small-molecule inhibitor or silencing with short hairpin RNA can suppress tumor progression and re-sensitize treatment-resistant cancer cells. For example, the US Food and Drug Administration (FDA) has granted fast-track approval to bemcentinib (R428/BGB324), a first AXL inhibitor for advanced metastatic non-small cell lung cancer (NSCLC) with STK11 mutation [12]. Ongoing clinical trials are exploring combination therapies that pair AXL inhibition with immunotherapies or chemotherapeutic agents. Developing novel combination strategies involving AXL remains a critical area of research to enhance cancer treatment outcomes.

Growth arrest-specific factor 6 (GAS6) and protein S (PROS1) are two well-characterized vitamin K-dependent ligands that can activate TAM receptors produced by various cell types [13, 14]. Although Tyro-3, AXL, and Mer share some similar functions, they exhibit distinct expression patterns and unique activation mechanism. GAS6 binds to all three TAM receptors, showing the highest affinity for AXL, while PROS1 preferentially binds to Mer and Tyro-3 [15]. AXL dysregulation is often associated with aggressive cancer phenotypes and has been observed in various cancers, including breast cancer, lung cancer, pancreatic ductal adenocarcinoma, and clear cell renal cell carcinoma [16–21]. Overexpression or hyperactivation of AXL correlates with poor prognosis and is a hallmark of treatment resistance, metastasis, and tumor progression [16, 22].

AXL-mediated therapeutic resistance poses a significant challenge in therapeutic management. AXL enables tumor cells to evade the effects of chemotherapy, targeted therapies, and immunotherapies by promoting mechanisms such as epithelial-to-mesenchymal transition (EMT), which facilitates cancer cell invasiveness and metastasis [23, 24]. Upregulation of anti-apoptotic proteins (such as survivin, BCL-XL, and BCL-2) enhances cell survival under treatment pressure, and immune suppression leads to impairing immune system-mediated tumor clearance [25–27]. In the last decade, AXL has emerged as a prominent therapeutic target. Targeting AXL can suppress tumor growth by reducing metastasis and angiogenesis [28, 29]. Various strategies to target AXL include small-molecule tyrosine inhibitors that block AXL's kinase activity, monoclonal antibodies (mAb) that specifically bind to AXL to inhibit its function, and ligand-blocking agents that prevent ligand (e.g., GAS6) binding and subsequent activation of AXL [21, 30, 31].

In this chapter, we discuss the structural and functional features of AXL. We also examine AXL's activation, regulatory mechanisms, and prominent role in cancer development, particularly proliferation, survival, immune evasion, EMT, and angiogenesis. Finally, we also address the translational relevance by exploring AXL as an emerging therapeutic target that enhances patient prognosis, promotes tumor regression, and reduces tumor burden. In general, this chapter aims to improve our understanding of AXL biology and its implications in therapeutic management.

2 Structure and Function of AXL

The AXL receptor is encoded by the AXL gene positioned at locus 19q13.2 on chromosome 19. Its genomic region encompasses 20 exons and translates into an 80–140 kDa single-pass transmembrane receptor. Like other RTKs, AXL comprises three primary domains—an ECD, TD, and KD. The ECD comprises two immunoglobulin-like (Ig-like) repeats and two fibronectin type III (FNIII) repeats [8, 32, 33] (Fig. 1a).

The Ig-like repeats facilitate the binding of ligand such as growth arrest-specific protein 6 (GAS6) and protein S (Pros1) and are regulated by the FNIII repeat [34, 35]. The TD span over the plasma membrane provides structural support and ensures signal transmission integrity [36]. Following the TD, intracellular KD contains autophosphorylation sites and phosphorylate-specific Tyr residues, including Tyr779, Tyr821, and Tyr866, upon dimerization induced by GAS6 binding with AXL's ECD [8, 9]. These phosphorylated Tyr residues act as docking sites for various downstream signaling pathways, such as PI3K-AKT, MAPK-ERK, TGF-β, and STAT, and are involved in cell survival, proliferation, and immune modulation [37–39].

In pathological conditions, AXL signaling was found dysregulated in various malignancies (Fig. 1b) and significantly promotes cancer progression by facilitating tumor cell proliferation and survival, promoting metastasis through epithelial-to-mesenchymal transition (EMT), and conferring resistance to chemotherapy and targeted therapies [23, 40, 41] (Fig. 2).

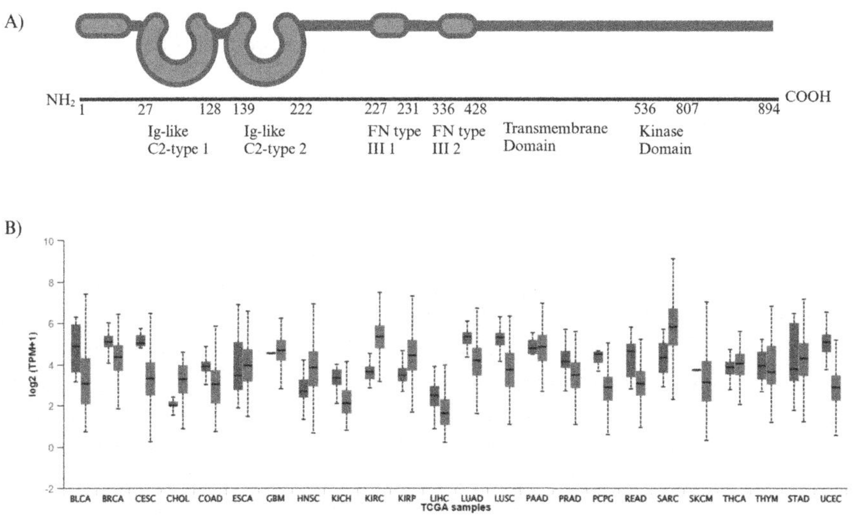

Fig. 1 Structure and expression of AXL. (**a**) The structure of the human AXL receptor aligned with its corresponding amino acid residue. (**b**) AXL gene expression across various cancers in the TGCA database described by UALCAN. Blue and red show normal and tumor samples, respectively

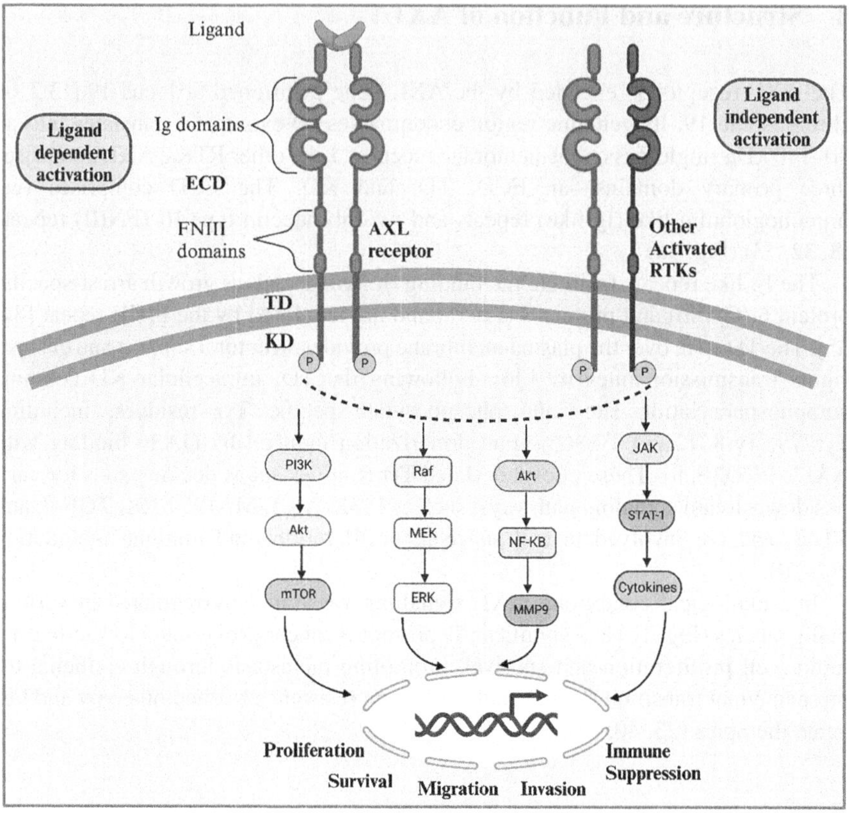

Fig. 2 Schematic representation of AXL activation and its downstream signaling pathway. Upon activation by interaction with classical ligands or other activated RTKs, AXL signaling mediates downstream signaling pathways and leads to proliferation, survival, migration, invasion, and immune suppression

Moreover, it also supports angiogenesis and tumor immune evasion by suppressing anti-tumor immune responses [23, 28]. Beyond cancer, AXL has been implicated in various other diseases. In infectious diseases, AXL facilitates viral entry and replication, as seen with the Zika virus, where it acts as a cofactor for infection in neural progenitor cells, contributing to viral pathogenesis [42, 43]. In fibrotic conditions, such as pulmonary, intestinal, and liver fibrosis, AXL signaling stimulates the fibrogenic pathways and the accumulation of extracellular matrix (ECM), aggravating tissue scarring and organ dysfunction [44–46]. Furthermore, reduced AXL activity may minimize tissue damage and excessive immune activation in autoimmune disorders such as encephalomyelitis (EAE) by inhibiting inflammatory reactions [47, 48]. These functions highlight the significance of AXL in non-cancerous disorders and its potential as a target for treatment in diseases ranging from chronic infections to immune-mediated illness.

3 Activation of AXL and Its Regulation

AXL signaling is a multifaceted process, involving changes in structural conformation. In its inactive state, AXL exits as a monomer embedded within the plasma membrane in a dormant conformation. Ligands such as Gas6, Tubby, Tubby-like protein 1 (Tulp-1), galectin-3, or Pros1 interact with the Ig-like enrich in ECD and are stabilized by the FNIII repeats [8, 49, 50]. Upon binding or interaction with ligand, a conformational change occurs, which induces receptor dimerization by bringing two AXL receptors into close proximity. The dimerization process aligns their intracellular KD and cross-phosphorylates key Tyr residues within each other. It will create docking sites for further downstream effector molecules and initiate various intracellular pathways, such as PI3K/AKT for cell survival, MAPK/ERK for proliferation, and STAT pathways for transcriptional regulation [37, 51–53].

Beyond its primary activation by the ligand, AXL can also be activated through mechanisms independent of the canonical AXL-Gas6 interaction. These alternative pathways involve transactivation by other receptor tyrosine kinases (RTKs) like EGFR, HER2, or MET and interactions with integrins and extracellular matrix components [54–56]. Additionally, intracellular signaling molecules, including kinases, can directly phosphorylate AXL and trigger downstream signaling cascades. This non-canonical activation enables AXL to function in diverse cellular contexts, particularly where ligand availability is limited [57].

The study also revealed differences in the binding preferences of TAM receptors to GAS6 and Pros1. Tyro-3 binds preferentially to Pros1, while AXL exhibits a stronger affinity for GAS6 [58]. Pros1 interacts exclusively with Tyro-3 and Mer, whereas GAS6 binds to all three receptors, AXL, Tyro-e, and Mer. Among these, AXL demonstrates the highest overall binding affinity for GAS6. Mer, although showing weak affinity for both ligands, exhibits enhanced binding strength in the presence of increased local concentrations of phosphatidylserine. Additionally, the crystal structures of TAM receptors indicate that the binding and activation of Tyro-3 and AXL by Pros1 and GAS6 require Zn^{2+}, Ca^{2+}, and Ni^{2+} ions highlighting the importance of ion interactions for proper structural rearrangements of TAM receptors [59, 60].

Promoter methylation also plays a significant role in modulating AXL expression, where hypermethylation suppresses its transcription and hypomethylation facilitates its upregulation [61, 62]. Several transcription factors (TFs) orchestrate AXL expression at the transcriptional level. These include HIF-α, which promotes AXL under hypoxic conditions; AP-1, which integrates stress and growth signals; and YAP1/TEAD, which links AXL expression to the Hippo signaling pathway. Other regulators like CREB and SP-1 are involved in response to cyclic AMP and growth-related stimuli, respectively, while MZF1 contributes to developmental and cellular stress contexts [63–66].

Additionally, interactions with microRNAs (such as miR-33a-5p, miR-34a, and miR-93-5p) significantly regulate AXL expression [67–69]. PTMs such as glycosylation at key sites, including Asn43, Asn157, Asn198, Asn339, Asn345, and

Asn401, play a crucial role in protein stability and ligand-binding affinity [70]. The activation and regulation of AXL underpin its role in normal physiological conditions for maintaining cellular homeostasis.

4 Role of AXL in Cancer

AXL is an RTK that plays a significant role in cancer by promoting cell proliferation, survival, migration, invasion, angiogenesis, stemness, and immunosuppression (Fig. 3) [40]. Its dysregulation and overexpression are strongly associated with aggressive tumor behavior, therapy resistance, and poor clinical outcomes across various cancer types [71, 72].

AXL enables cancer cells to adapt to a harsh surrounding microenvironment, evade immune surveillance, and metastasize to distant sites [73]. Additionally, AXL signaling regulates the remodeling of the extracellular matrix and the release of cytokines, which fosters the tumor-aggressive phenotype [70, 74]. The role of AXL in driving cancer development and progression is detailed below:

4.1 AXL Promotes Cell Proliferation and Survival

AXL significantly promotes tumor proliferation and survival by inducing downstream signaling pathways such as PI3K/Akt, RAS/RAF/MEK/ERK, and NF-κβ [11]. The PI3K/AKT pathway mainly enhances cell survival by inhibiting pro-apoptotic factors, while the RAS/RAF/MEK/ERK pathway promotes mitogenic signaling, fostering rapid tumor growth [75, 76]. Furthermore, AXL promotes cell survival by modulating NF-κβ nuclear trafficking and raising the expression of anti-apoptotic markers like BCL-XL, BCL-2, and survivin while concurrently lowering the expression of pro-apoptotic markers such as BAD and caspase 3 [22, 77]. The proliferation of tumor cells can be suppressed by inhibiting the AXL signaling pathway, significantly extending survival.

4.2 AXL-Mediated Cell Migration and Invasion

AXL signaling plays a key role in enhancing cell invasion by promoting mesenchymal characteristics and activating additional pathways that increase cell motility. For example, AXL can stimulate downstream effectors of receptor tyrosine kinase such as MAPK, PI3K/AKT, and JAK/STAT pathways, to drive actin remodeling, cell migration, and cancer cell survival [78]. Furthermore, AXL signaling converges on the integrin pathway through NCK2/ILK that regulates cell adhesion and further facilitates metastatic progression. Ryu Kanzaki et al. demonstrated that

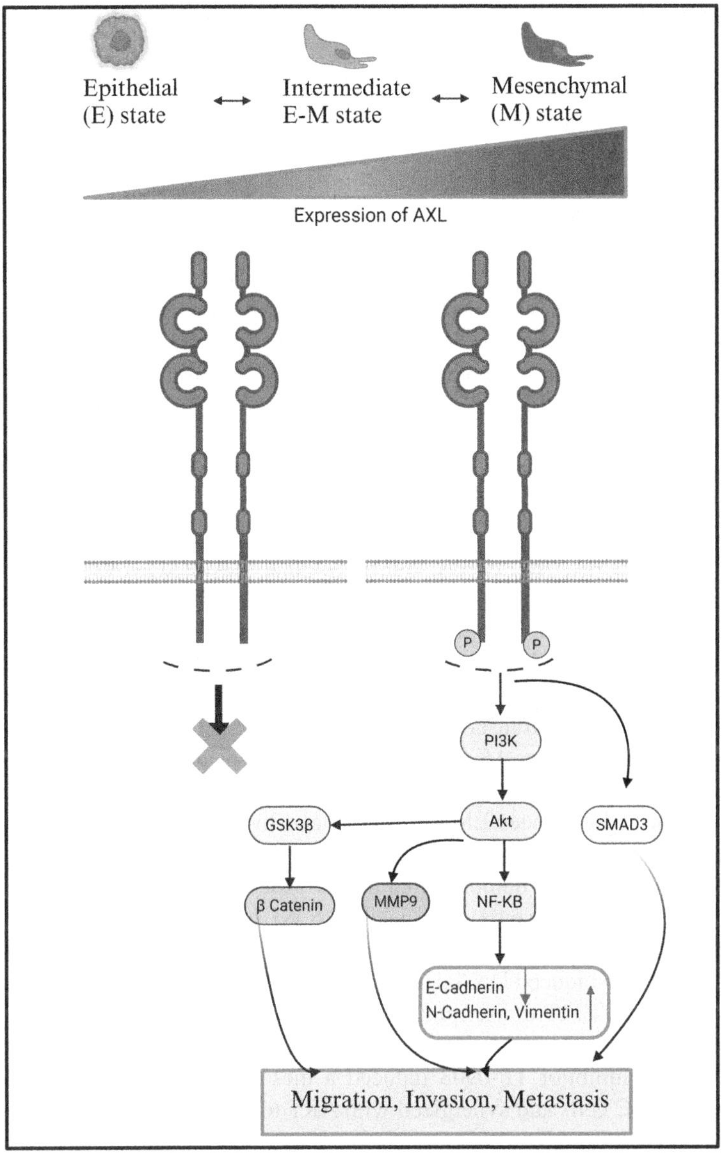

Fig. 3 The change in AXL expression pattern between epithelial and mesenchymal states during the EMT. As long as the EMT process continues, high AXL expression triggers the intracellular pathway responsible for migration, invasion, and metastasis

recombinant Gas6 enhances H1299 cell migration, while conditioned media derived from LCAFhTERT cells activated AXL in H1299 cells and promoted their migration [79]. Silencing GAS6 significantly reduced both AXL activation and H1299 cell migration induced by conditioned media from LCAFhTERT cells.

4.3 AXL-Driven Epithelial-to-Mesenchymal Transition (EMT)

Increasing evidence suggests that the EMT, a fundamental developmental process, plays a significant role in tumor metastasis. During EMT, polarized epithelial cells lose their junctional architecture and apical-basal polarity, transitioning into motile mesenchymal cells. The role of AXL in EMT has been extensively reported in various cancers including breast, ovarian, NSCLC, and pancreatic cancers. In breast cancer, AXL overexpression is associated with poorer prognosis with metastases exhibiting higher levels of AXL expression. Furthermore, AXL has been identified as a downstream effector of EMT, promoting tumor growth and invasiveness [80]. Notably, AXL knockdown has been shown to improve survival by preventing the spread of highly metastatic breast cancer cell lines from the mammary glands to lymph nodes and other organs. This finding provided the first in vivo study directly linking AXL to metastasis [81].

The mesenchymal "Mes" subtype of ovarian cancer, associated with poor prognosis, exhibits a pathway addiction that can be therapeutically targeted. This subtype is characterized by amplified signaling of the AXL; extensive crosstalk with other RTKs such as cMET, EGFR, and HER2; and sustained activation of extracellular signal-regulated kinase (ERK) [82]. This prolonged ERK activation drives the transcription factor Slug, promoting mesenchymal traits and tumor progression. Emily N. Arner and colleagues discovered that AXL activation leads to the phosphorylation of TANK-binding kinase 1 (TBK1), which subsequently activates AKT3 and Snail, a key transcription factor driving EMT [83]. They demonstrated that TBK1 phosphorylates AKT3 and directly interacts with it relying on the multiprotein complex mTORC1. Activated AKT3 binds to Snail and promotes its nuclear accumulation and enhances EMT, as evidenced by increased abundance of EMT markers.

Mirko Terragno et al. demonstrated that artesunate (ART), an antimalarial drug known to generate reactive oxygen species (ROS), exhibits greater lethality against epithelial breast cancer cell lines compared to mesenchymal ones [84]. The study revealed that ART-induced DNA damage, apoptotic cell death, and ROS levels were modulated by the EMT transcription factor ZEB1. In triple-negative breast cancer (TNBC), AXL is overexpressed and maintains a mesenchymal phenotype. Treatment with the AXL inhibitor TP-0903 induced a mesenchymal-to-epithelial transition (MET) in TNBC cells and synergized with ART to amplify ROS production, DNA damage, and cell death.

4.4 AXL-Induced Angiogenesis

AXL is significant in promoting angiogenesis or forming new blood vessels required for metastasis. Activation of AXL can be achieved through ligand-mediated or independent of ligand by dimerization with other activated RTKs. Further, AXL triggers downstream signaling pathways such as PI3K/Akt/ERK and STAT3 that regulate the expression of pro-angiogenic proteins such as fibroblast growth factor (FGF), vascular endothelial growth factor (VEGF), and platelet-derived growth factor (PDGF) [85–88]. These factors aid in forming a tumor's vasculature and promote endothelial cell migration, proliferation, and survival. As a result of VEGF-A-mediated PI3K/Akt pathway activation, AXL has been linked to promoting proliferation and tube development in human umbilical vein endothelial cells (HUVEC), required for angiogenesis [89–91]. AXL and GAS6 knockdown studies have shown decreased endothelial tube formation, by inhibiting the production of pro-angiogenic proteins including angiopoietin-2 (Ang-2) [28, 29].

AXL-mediated signaling also enhances the response of endothelial cells to hypoxic conditions by stabilizing hypoxia-inducible factor-1 alpha (HIF-1α), a key transcription factor involved in the upregulation of angiogenesis-related genes [92]. Furthermore, AXL-induced angiogenesis is associated with increased vascular permeability, which enables efficient nutrient delivery and waste removal in tumors. AXL expression is associated with increased microvessel density, aggressive tumor activity, and poor prognosis [93, 94]. By driving angiogenesis, AXL creates a tumor microenvironment that supports metastatic dissemination.

4.5 AXL-Associated Stemness Regulation

In the last few years, AXL emerged as a key driver of cancer stem cell (CSC) properties in cancer, contributing to self-renewal and tumor maintenance [95]. AXL initiates signaling pathways that promote the expression of stem cell markers such as bone gamma-carboxyglutamate protein 1 (Bglap1), aldehyde dehydrogenase 1 (ALDH1), cell division cycle 2a (Cdc2a), cluster of differentiation 44 (CD44), CD133, and enhancer of zeste homolog 2 (EZH2) [23, 40, 96, 97]. Increased expression of these markers may enhance chemotherapy resistance in many cancers including squamous cell carcinoma [98]. The AXL-PYK2 (proline-rich tyrosine kinase 2)-PKCα (protein kinase C alpha) axis is known to have a significant contribution in maintaining stemness signature in triple-negative breast cancer (TNBC) by activating STAT3, SMAD3, TAZ, and FRA1 as well as the pluripotent transcription factors such as Nanog and Oct4 [99].

Furthermore, AXL-induced stemness is closely linked with EMT, as acquiring mesenchymal traits often accompanies the emergence of stem-like properties [98, 100]. AXL contributes to tumor heterogeneity by maintaining stemness properties and helps tumor cells adapt and survive under unfavorable conditions. Therapeutic strategies inhibiting AXL could reduce CSC-mediated tumor initiation, progression, and relapses, improving long-term clinical outcomes.

4.6 AXL Regulates Immunosuppression

In the last decade, immunotherapy revolutionized cancer treatment by activating the host immune system to destroy or eliminate tumor cells. Despite recent considerable advancements, drug resistance remains to be a major challenge in clinical implication. AXL acts as a negative inflammatory mediator, which inhibits immune-related signaling that activates immune cells such as dendritic cells, macrophages, and natural killer cells [101, 102]. Studies have shown that AXL could downregulate the production of pro-inflammatory cytokines and chemokines along with the toll-like receptor (TLR)-dependent inflammatory response and thus prevent the autoimmune response of immune cells [103–105]. AXL activation also has a role in immune invasion by upregulating the co-expression of BCL-2 and Twist protein [106, 107]. In head and neck cancer, AXL provides therapeutic resistance through the expression of programmed cell death ligand 1 (PD-L1) mediated by PI3K/Akt pathway [108]. It was noted that patients with elevated AXL and PD-L1 expression have lower overall survival rates and a decreased antitumor immune response [109].

5 Exploring AXL as a Therapeutic Intervention

Given its importance in cancer growth, metastasis, and drug resistance, AXL emerged as a promising therapeutic opportunity. Targeting AXL has the potential to increase chemotherapy effectiveness by sensitizing cancer cells to treatment. A diverse array of AXL inhibitors is currently under development and encompasses several therapeutic modalities. These include small-molecule tyrosine kinase inhibitors (TKIs) that block AXL kinase activity, anti-AXL monoclonal antibodies targeting the receptor's extracellular domain, and soluble decoy receptors that neutralize the ligand GAS6. Additionally, nucleic acid aptamers are short, single-stranded RNA or DNA molecules with a high selective binding affinity for their target molecules, making them promising therapeutic agents. For example, Ji An Hwang et al. developed novel and modified DNA aptamers specifically designed to bind to AXL receptors [110]. They evaluated the impact of these aptamers on human lung cancer cells that had developed resistance to EGFR TKIs through AXL signaling activation.

5.1 Small-Molecule Inhibitors of AXL

Small-molecule inhibitors (SMIs) targeting AXL have emerged as a significant therapeutic approach to treating AXL-mediated tumor progression, metastasis, and therapy resistance. These inhibitors primarily target the intracellular kinase domain of AXL, preventing its autophosphorylation and subsequent activation of downstream signaling cascades, which are required for cell survival, proliferation, and migration.

Several AXL inhibitors are currently under investigation, each contributing uniquely to cancer therapy. For example, bemcentinib (by Rigel Pharmaceuticals and BerGenBio) is a selective orally bioavailable SMI that has shown efficacy in clinical trials for hematologic malignancies and solid tumors [111]. BMS-777607 (by Aslan Pharmaceuticals) is a multi-kinase inhibitor targeting AXL, MET, and RON, with significant potential to inhibit metastasis and tumor progression [112, 113]. TP-0903 (by Tolero Pharmaceuticals) is another next-generation AXL inhibitor, demonstrating promising results in preclinical studies in reversing drug resistance and enhancing chemotherapy efficacy [114].

NA80xl, DP3975, and SGI-7079 (NCT00409968) are other small-molecule kinase type I inhibitors that have shown strong preclinical activity in cancers with high AXL expression [115]. These SMIs competes with the ATP binding site and further inhibits the catalytic activity of the kinase. Other inhibitors, such as bosutinib (NCT02228382, NCT03023319), crizotinib (NCT02737501, NCT02435108), cabozantinib (NCT01908426, NCT01865747), gilteritinib (NCT02927262, NCT02456883), S49076 (ISRCTN00759419), amuvatinib (NCT01357395, NCT00894894), sunitinib (NCT00706706, NCT02691793), and sitravatinib (NCT03680521), are known to target AXL alongside other RTKs like PDGFR, KIT, MET, and VEGFR, making them versatile targets for multi-pathway inhibition. NPS-1034 is a type II dual inhibitor of AXL, TNFRSF1A, and MET that has shown the potential to act on activated or altered MET or AXL and suppress cancer progression [116].

5.2 *Monoclonal Antibodies and Antibody-Drug Conjugates Targeting AXL*

SMIs often exhibit cytotoxicity, and off-target effects result from genetic changes or lack of specificity against the target. To overcome these limitations, several monoclonal antibodies (mAbs) or antibody-drug conjugates (ADCs) that target AXL are being developed to increase the specificity and efficacy against tumor cells. These antibodies inhibit AXL-mediated processes like proliferation, migration, invasion, etc., thus enhancing immune-mediated tumor cell destruction.

mAbs offer a promising approach by explicitly blocking the extracellular region of the receptor, preventing its activation, and inhibiting subsequent downstream signaling pathways. Tilvestamab (ARGX-112) is a humanized Ig mAb that shows a strong affinity with Ig-rich AXL and prevents its interaction with the ligand GAS6 [117]. Thus, tilvestamab inhibits ligand-mediated AXL receptor activation and promotes receptor internalization followed by degradation. Likewise, YW327.6S2 (phage-derived mAb) and CDX-0168 are also designed to block the ligand-mediated activation of the AXL pathway and downregulate further signaling [118, 119].

Preclinical studies in the human non-small cell lung cancer cell (NSCLC) model showed that YW327.6S2 enhanced the inhibitory effect of erlotinib, while CDX-0168 inhibited AXL phosphorylation and induced the release of cytokines and chemokines, leading to tumor cell death via antibody-dependent cellular cytotoxicity (ADCC). In Kaposi sarcoma, MAb173 targets AXL and induces receptor degradation, substantially reducing tumor growth by increasing tumor cell apoptosis [120]. Similarly, 20G7-D9 reduced tumor growth, migration, and invasion in TNBC-derived xenografts [121].

Antibody-drug conjugates (ADCs) have further enhanced the utility of AXL-targeting strategies by combining specific AXL inhibition with cytotoxic payload delivery. Enapotamab vedotin, DAXL-88-MMAE, mipasetamab uzoptirine, and mecbotamab vedotin represent ADCs that deliver potent anti-tumor agents directly to AXL-expressing cells, increasing efficacy while reducing off-target toxicity [122–125]. These advancements highlight the therapeutic promise of mAb and ADCs in targeting AXL, particularly in aggressive and treatment-resistant cancers.

5.3 Natural Compounds as AXL Inhibitors

Natural compounds are substances derived from natural products that offer anticancer properties [126, 127]. It could be classified into different classes based on their origin and structural features. One such naturally occurring terpenoid is dihydroartemisinic acid (DHA) [128]. It was isolated from the sweet wormwood plant (*Artemisia annua*) and initially identified as an anti-malarial drug. The active derivative artemisinin is known to target the miR-34a/miR-7/JARID2 pathway, downregulating the AXL expression and thus inhibiting cancer progression in prostate cancer. Furthermore, mistletoe is an extract of European mistletoe (*Viscum album*) contains bioactive compounds known to inhibit AXL-mediated disease progression and overcome cisplatin and erlotinib resistance in NSCLC cells [129]. Ongoing research intends to use natural products as an alternative to traditional AXL targeting SMIs and limit its side effects.

5.4 Combination Therapies Targeting AXL

Over the past few years, a combined therapeutic approach presents as an intriguing strategy to enhance treatment effectiveness by sensitizing tumor cells and limiting resistance. It targets AXL and other RTKs simultaneously or during therapy and can significantly enhance antitumor efficacy by addressing multiple cancer-driven mechanisms, including drug resistance [11]. One promising strategy is combining an AXL and ATR inhibitor, which upregulates the expression of Bim, further increasing apoptosis, suppressing cell migration, and improving patient outcomes in pleural mesothelioma [130].

Another strategy combines AXL inhibitors with immune checkpoint inhibitors (ICIs), such as anti-PD-1. For example, in NSCLC cell lines, treatment with an AXL inhibitor alongside erlotinib restored sensitivity to the drug. BGB324, either alone or in combination with nivolumab (an anti-PD-1 antibody), improved survival in mice with mesenchymal glioblastoma tumors [131]. Another clinical trial evaluated BGB324 in combination with anti-PD1 immunotherapy in patients with advanced NSCLC (NCT0318457) [132]. Seven of the 29 patients showed partial responses, and 40% of those with AXL-positive biopsies had objective responses.

AXL-driven immune evasion mechanisms, including suppression of antigen presentation and promotion of an immunosuppressive tumor microenvironment, can hinder the effectiveness of ICIs. Blocking AXL restores immune surveillance and enhances the antitumor activity of ICIs [133, 134]. The activation of AXL through interaction with other RTKs suggests that combination therapy targeting both AXL and its interacting partners may be necessary in clinical trials. Overexpression of AXL in esophageal cancer has been shown to increase cisplatin resistance by modulating c-ABL/p73 signaling [135]. AXL has also been reported to be activated by various chemotherapeutic agents in acute myeloid leukemia (AML), indicating that combining AXL inhibition with chemotherapy could be a potential treatment strategy for AML [136]. Marrocco et al. demonstrated that polytherapeutic and synergistic combinations of monoclonal antibodies and EGFR-specific TKIs may prevent resistance formation in EGFR mutant lung cancer models by inhibiting the MAPK and PI3K/AKT pathways [137].

These results suggest that future research should focus on AXL inhibitors in combination with other targeted therapies or chemotherapeutics to maximize clinical benefits. The role of AXL in chemoresistance indicates that reducing AXL in conjugation with inhibitors of other molecular targets may provide a promising therapeutic strategy for cancer treatment.

6 Conclusion and Future Perspective

AXL plays a crucial role in both cancer progression and immune response. In normal tissues, AXL is typically expressed at lower levels compared to diseased tissues, underscoring its potential oncogenic role. Although AXL mutations are rare in human cancers, the receptors are frequently overexpressed in most malignancies during tumor progression or in response to therapeutic treatments promote tumor growth, metastasis, and drug resistance. Genetic and pharmacological inhibition of GAS6/AXL signaling in tumor models has provided valuable insights into how this pathway drives tumor progression and therapy resistance. This chapter begins by providing an overview of AXL's structure, function, regulation mechanism, and signaling pathways.

Numerous therapeutic strategies have been developed to target AXL including small-molecule inhibitors, monoclonal antibodies, natural compounds, nucleotide aptamers, and soluble receptors. We explore emerging AXL-targeted approaches in cancer therapy. These AXL-targeted treatments, whether used alone or in combination with other small-molecule inhibitors or conventional chemotherapy, are expected to improve survival outcomes for cancer patients. However, despite the promising potential of AXL-targeted therapies, several challenges remain. While AXL small-molecule inhibitors demonstrate strong therapeutic efficacy, their off-target effects can lead to the inhibition of other kinases and unanticipated toxicities, thereby limiting their clinical uses. Additionally, molecularly targeted therapies, particularly multi-targeted AXL tyrosine kinase inhibitors (TKIs), have been associated with significant adverse effects. Moving forward, further research is needed to unravel AXL molecular signaling networks and identify reliable predictive biomarkers. This will help determine which patients may derive clinical benefits while minimizing potential side effects. Further research is needed to thoroughly investigate the cellular and molecular mechanisms through which AXL signaling promotes tumor growth, in order to develop optimal anti-AXL combination therapies.

Acknowledgments The authors would like to thank All India Institute of Medical Sciences, New Delhi, India, and IIT Gandhinagar for providing necessary facilities. R.S. acknowledges the Science and Engineering Research Board (SERB), now known as Anusandhan National Research Foundation (ANRF), Government of India, for financial support through the National Post Doctoral Fellowship (NPDF). The figures are created with BioRender.com.

References

1. Tomuleasa C, Tigu AB, Munteanu R, Moldovan CS, Kegyes D, Onaciu A, Gulei D, Ghiaur G, Einsele H, Croce CM. Therapeutic advances of targeting receptor tyrosine kinases in cancer. Signal Transduct Target Ther. 2024;9(1):201.
2. Lemmon MA, Schlessinger J. Cell signaling by receptor tyrosine kinases. Cell. 2010;141(7):1117–34.
3. Zhang N, Li Y. Receptor tyrosine kinases: biological functions and anticancer targeted therapy. MedComm. 2023;4(6):e446.
4. Saraon P, Pathmanathan S, Snider J, Lyakisheva A, Wong V, Stagljar I. Receptor tyrosine kinases and cancer: oncogenic mechanisms and therapeutic approaches. Oncogene. 2021;40(24):4079–93.
5. Kumar R, Goel H, Solanki R, Rawat L, Tabasum S, Tanwar P, Pal S, Sabarwal A. Recent developments in receptor tyrosine kinase inhibitors: A promising mainstay in targeted cancer therapy. Med Drug Discov. 2024;23:100195.
6. Du Z, Lovly CM. Mechanisms of receptor tyrosine kinase activation in cancer. Mol Cancer. 2018;17:1–3.
7. Axelrod H, Pienta KJ. Axl as a mediator of cellular growth and survival. Oncotarget. 2014;5(19):8818.
8. Linger RM, Keating AK, Earp HS, Graham DK. TAM receptor tyrosine kinases: biologic functions, signaling, and potential therapeutic targeting in human cancer. Adv Cancer Res. 2008;100:35–83.
9. O'Bryan JP, Frye RA, Cogswell PC, Neubauer A, Kitch B, Prokop C, Espinosa R, Le Beau MM, Earp HS, Liu ET. Axl, a transforming gene isolated from primary human myeloid leukemia cells, encodes a novel receptor tyrosine kinase. Mol Cell Biol. 1991;11(10):5016–31.

10. Ohashi K, Mizuno K, Kuma KI, Miyata T, Nakamura T. Cloning of the cDNA for a novel receptor tyrosine kinase, Sky, predominantly expressed in brain. Oncogene. 1994;9(3):699–705.
11. Graham DK, Dawson TL, Mullaney DL, Snodgrass HR, Earp HS. Cloning and mRNA expression analysis of a novel human protooncogene, c-mer. Cell Growth Differ. 1994;5(6):647–57.
12. Li H, Liu Z, Liu L, Zhang H, Han C, Girard L, Park H, Zhang A, Dong C, Ye J, Rayford A. AXL targeting restores PD-1 blockade sensitivity of STK11/LKB1 mutant NSCLC through expansion of TCF1+ CD8 T cells. Cell Rep Med. 2022;3(3):100554.
13. Stitt TN, Conn G, Goret M, Lai C, Bruno J, Radzlejewski C, Mattsson K, Fisher J, Gies DR, Jones PF, Masiakowski P. The anticoagulation factor protein S and its relative, Gas6, are ligands for the Tyro 3/Axl family of receptor tyrosine kinases. Cell. 1995;80(4):661–70.
14. Varnum BC, Young C, Elliott G, Garcia A, Bartley TD, Fridell YW, Hunt RW, Trail G, Clogston C, Toso RJ, Yanagihara D. Axl receptor tyrosine kinase stimulated by the vitamin K-dependent protein encoded by growth-arrest-specific gene 6. Nature. 1995;373(6515):623–6.
15. Verma A, Warner SL, Vankayalapati H, Bearss DJ, Sharma S. Targeting Axl and Mer kinases in cancer. Mol Cancer Ther. 2011;10(10):1763–73.
16. Gay CM, Balaji K, Byers LA. Giving AXL the axe: targeting AXL in human malignancy. Br J Cancer. 2017;116(4):415–23.
17. Elkabets M, Pazarentzos E, Juric D, Sheng Q, Pelossof RA, Brook S, Benzaken AO, Rodon J, Morse N, Yan JJ, Liu M. AXL mediates resistance to PI3Kα inhibition by activating the EGFR/PKC/mTOR axis in head and neck and esophageal squamous cell carcinomas. Cancer Cell. 2015;27(4):533–46.
18. Colavito SA. AXL as a target in breast cancer therapy. J Oncol. 2020;2020(1):5291952.
19. Zhang G, Wang M, Zhao H, Cui W. Function of Axl receptor tyrosine kinase in non-small cell lung cancer. Oncol Lett. 2018;15(3):2726–34.
20. Song X, Wang H, Logsdon CD, Rashid A, Fleming JB, Abbruzzese JL, Gomez HF, Evans DB, Wang H. Overexpression of receptor tyrosine kinase Axl promotes tumor cell invasion and survival in pancreatic ductal adenocarcinoma. Cancer. 2011;117(4):734–43.
21. Yu H, Liu R, Ma B, Li X, Yen HY, Zhou Y, Krasnoperov V, Xia Z, Zhang X, Bovc AM, Buscarini M. Axl receptor tyrosine kinase is a potential therapeutic target in renal cell carcinoma. Br J Cancer. 2015;113(4):616–25.
22. Wium M, Ajayi-Smith AF, Paccez JD, Zerbini LF. The role of the receptor tyrosine kinase Axl in carcinogenesis and development of therapeutic resistance: an overview of molecular mechanisms and future applications. Cancers. 2021;13(7):1521.
23. Asiedu MK, Beauchamp-Perez FD, Ingle JN, Behrens MD, Radisky DC, Knutson KL. AXL induces epithelial-to-mesenchymal transition and regulates the function of breast cancer stem cells. Oncogene. 2014;33(10):1316–24.
24. Arner EN, Alzhanova D, Westcott JM, Hinz S, Tiron CE, Blø M, Mai A, Virtakoivu R, Phinney N, Nord S, Aguilera KY. AXL-TBK1 driven AKT3 activation promotes metastasis. Sci Signal. 2024;17(867):eado6057.
25. Linger RM, Cohen RA, Cummings CT, Sather S, Migdall-Wilson J, Middleton DH, Lu X, Barón AE, Franklin WA, Merrick DT, Jedlicka P. Mer or Axl receptor tyrosine kinase inhibition promotes apoptosis, blocks growth and enhances chemosensitivity of human non-small cell lung cancer. Oncogene. 2013;32(29):3420–31.
26. Nguyen KQ, Tsou WI, Kotenko S, Birge RB. TAM receptors in apoptotic cell clearance, autoimmunity, and cancer. Autoimmunity. 2013;46(5):294–7.
27. Han JU, Tian R, Yong B, Luo C, Tan P, Shen J, Peng T. Gas6/Axl mediates tumor cell apoptosis, migration and invasion and predicts the clinical outcome of osteosarcoma patients. Biochem Biophys Res Commun. 2013;435(3):493–500.
28. Li Y, Ye X, Tan C, Hongo JA, Zha J, Liu J, Kallop D, Ludlam MJ, Pei L. Axl as a potential therapeutic target in cancer: role of Axl in tumor growth, metastasis and angiogenesis. Oncogene. 2009;28(39):3442–55.
29. Tanaka M, Siemann DW. Axl signaling is an important mediator of tumor angiogenesis. Oncotarget. 2019;10(30):2887.

30. Goyette MA, Côté JF. AXL receptor tyrosine kinase as a promising therapeutic target directing multiple aspects of cancer progression and metastasis. Cancers. 2022;14(3):466.
31. Liu Z, Chen L, Zhang J, Yang J, Xiao X, Shan L, Mao W. Recent discovery and development of AXL inhibitors as antitumor agents. Eur J Med Chem. 2024;272:116475.
32. Tang Y, Zang H, Wen Q, Fan S. AXL in cancer: a modulator of drug resistance and therapeutic target. J Exp Clin Cancer Res. 2023;42(1):148.
33. Yamagata M, Sanes JR, Weiner JA. Synaptic adhesion molecules. Curr Opin Cell Biol. 2003;15(5):621–32.
34. Sasaki T, Knyazev PG, Clout NJ, Cheburkin Y, Göhring W, Ullrich A, Timpl R, Hohenester E. Structural basis for Gas6–Axl signalling. EMBO J. 2006;25(1):80–7.
35. van der Meer JH, van der Poll T, van 't Veer C. TAM receptors, Gas6, and protein S: roles in inflammation and hemostasis. Blood, the journal of the American society of. Hematology. 2014;123(16):2460–9.
36. Li E, Hristova K. Role of receptor tyrosine kinase transmembrane domains in cell signaling and human pathologies. Biochemistry. 2006;45(20):6241–51.
37. Brandão L, Migdall-Wilson J, Eisenman K, Graham DK. TAM receptors in leukemia: expression, signaling, and therapeutic implications. Crit Rev Oncog. 2011;16(1–2):47–63.
38. DeRyckere D, Huelse JM, Earp HS, Graham DK. TAM family kinases as therapeutic targets at the interface of cancer and immunity. Nat Rev Clin Oncol. 2023;20(11):755–79.
39. Bauer T, Zagórska A, Jurkin J, Yasmin N, Köffel R, Richter S, Gesslbauer B, Lemke G, Strobl H. Identification of Axl as a downstream effector of TGF-β1 during Langerhans cell differentiation and epidermal homeostasis. J Exp Med. 2012;209(11):2033–47.
40. Zhu C, Wei Y, Wei X. AXL receptor tyrosine kinase as a promising anti-cancer approach: functions, molecular mechanisms and clinical applications. Mol Cancer. 2019;18(1):153.
41. Chandrashekar DS, Karthikeyan SK, Korla PK, Patel H, Shovon AR, Athar M, Netto GJ, Qin ZS, Kumar S, Manne U, Creighton CJ. UALCAN: an update to the integrated cancer data analysis platform. Neoplasia. 2022;25:18–27.
42. Meertens L, Labeau A, Dejarnac O, Cipriani S, Sinigaglia L, Bonnet-Madin L, Le Charpentier T, Hafirassou ML, Zamborlini A, Cao-Lormeau VM, Coulpier M. Axl mediates ZIKA virus entry in human glial cells and modulates innate immune responses. Cell Rep. 2017;18(2):324–33.
43. Zwernik SD, Adams BH, Raymond DA, Warner CM, Kassam AB, Rovin RA, Akhtar P. AXL receptor is required for Zika virus strain MR-766 infection in human glioblastoma cell lines. Mol Ther Oncol. 2021;23:447–57.
44. Yang DC, Gu S, Li JM, Hsu SW, Chen SJ, Chang WH, Chen CH. Targeting the AXL receptor in combating smoking-related pulmonary fibrosis. Am J Respir Cell Mol Biol. 2021;64(6):734–46.
45. Steiner CA, Rodansky ES, Johnson LA, Berinstein JA, Cushing KC, Huang S, Spence JR, Higgins PD. AXL is a potential target for the treatment of intestinal fibrosis. Inflamm Bowel Dis. 2021;27(3):303–16.
46. Grøndal SM, Tutusaus A, Boix L, Reig M, Blø M, Hodneland L, Gausdal G, Jackson A, Garcia de Frutos P, Lorens JB, Morales A. Dynamic changes in immune cell populations by AXL kinase targeting diminish liver inflammation and fibrosis in experimental MASH. Front Immunol. 2024;15:1400553.
47. Weinger JG, Brosnan CF, Loudig O, Goldberg MF, Macian F, Arnett HA, Prieto AL, Tsiperson V, Shafit-Zagardo B. Loss of the receptor tyrosine kinase Axl leads to enhanced inflammation in the CNS and delayed removal of myelin debris during experimental autoimmune encephalomyelitis. J Neuroinflammation. 2011;8:1–8.
48. DuBois JC, Ray AK, Davies P, Shafit-Zagardo B. Anti-Axl antibody treatment reduces the severity of experimental autoimmune encephalomyelitis. J Neuroinflammation. 2020;17:1–4.
49. Caberoy NB, Zhou Y, Li W. Tubby and tubby-like protein 1 are new MerTK ligands for phagocytosis. EMBO J. 2010;29(23):3898–910.

50. Al Kafri N, Hafizi S. Galectin-3 stimulates tyro3 receptor tyrosine kinase and erk signalling, cell survival and migration in human cancer cells. Biomol Ther. 2020;10(7):1035.
51. Yeh CY, Shin SM, Yeh HH, Wu TJ, Shin JW, Chang TY, Raghavaraju G, Lee CT, Chiang JH, Tseng VS, Lee YC. Transcriptional activation of the Axl and PDGFR-α by c-Met through a ras-and Src-independent mechanism in human bladder cancer. BMC Cancer. 2011;11:1–2.
52. Hung CN, Chen M, DeArmond DT, Chiu CH, Limboy CA, Tan X, Kusi M, Chou CW, Lin LL, Zhang Z, Wang CM. AXL-initiated paracrine activation of pSTAT3 enhances mesenchymal and vasculogenic supportive features of tumor-associated macrophages. Cell Rep. 2023;42(9):113067.
53. Scaltriti M, Elkabets M, Baselga J. Molecular pathways: AXL, a membrane receptor mediator of resistance to therapy. Clin Cancer Res. 2016;22(6):1313–7.
54. Meyer AS, Miller MA, Gertler FB, Lauffenburger DA. The receptor AXL diversifies EGFR signaling and limits the response to EGFR-targeted inhibitors in triple-negative breast cancer cells. Sci Signal. 2013;6(287):ra66.
55. Goyette MA, Duhamel S, Aubert L, Pelletier A, Savage P, Thibault MP, Johnson RM, Carmeliet P, Basik M, Gaboury L, Muller WJ. The receptor tyrosine kinase AXL is required at multiple steps of the metastatic cascade during HER2-positive breast cancer progression. Cell Rep. 2018;23(5):1476–90.
56. Salian-Mehta S, Xu M, Wierman ME. AXL and MET crosstalk to promote gonadotropin releasing hormone (GnRH) neuronal cell migration and survival. Mol Cell Endocrinol. 2013;374(1–2):92–100.
57. Fridell YW, Jin Y, Quilliam LA, Burchert A, McCloskey P, Spizz G, Varnum B, Der C, Liu ET. Differential activation of the Ras/extracellular-signal-regulated protein kinase pathway is responsible for the biological consequences induced by the Axl receptor tyrosine kinase. Mol Cell Biol. 1996;16(1):135–45.
58. Tsou WI, Nguyen KQ, Calarese DA, Garforth SJ, Antes AL, Smirnov SV, Almo SC, Birge RB, Kotenko SV. Receptor tyrosine kinases, TYRO3, AXL, and MER, demonstrate distinct patterns and complex regulation of ligand-induced activation. J Biol Chem. 2014;289(37):25750–63.
59. Prouse T, Majumder S, Majumder R. Functions of TAM receptors and ligands protein S and Gas6 in atherosclerosis and cardiovascular disease. Int J Mol Sci. 2024;25(23):12736.
60. Hasanbasic I, Rajotte I, Blostein M. The role of γ-carboxylation in the anti-apoptotic function of gas6. J Thromb Haemost. 2005;3(12):2790–7.
61. Levin PA, Brekken RA, Byers LA, Heymach JV, Gerber DE. Axl receptor axis: a new therapeutic target in lung cancer. J Thorac Oncol. 2016;11(8):1357–62.
62. Safaric Tepes P, Pal D, Lindsted T, Ibarra I, Lujambio A, Jimenez Sabinina V, Senturk S, Miller M, Korimerla N, Huang J, Glassman L. An epigenetic switch regulates the ontogeny of AXL-positive/EGFR-TKi-resistant cells by modulating miR-335 expression. elife. 2021;10:e66109.
63. Goyette MA, Elkholi IE, Apcher C, Kuasne H, Rothlin CV, Muller WJ, Richard DE, Park M, Gratton JP, Côté JF. Targeting Axl favors an antitumorigenic microenvironment that enhances immunotherapy responses by decreasing Hif-1α levels. Proc Natl Acad Sci. 2021;118(29):e2023868118.
64. Badarni M, Prasad M, Balaban N, Zorea J, Yegodayev KM, Joshua BZ, Dinur AB, Grénman R, Rotblat B, Cohen L, Elkabets M. Repression of AXL expression by AP-1/JNK blockage overcomes resistance to PI3Ka therapy. JCI Insight. 2019;4(8):e125341.
65. Saab S, Chang OS, Nagaoka K, Hung MC, Yamaguchi H. The potential role of YAP in Axl-mediated resistance to EGFR tyrosine kinase inhibitors. Am J Cancer Res. 2019;9(12):2719.
66. Adams DE, Zhen Y, Qi X, Shao WH. Axl expression in renal mesangial cells is regulated by Sp1, Ap1, MZF1, and Ep300, and the IL-6/miR-34a pathway. Cells. 2022;11(12):1869.
67. Jiang N, Wang X, Xie X, Liao Y, Liu N, Liu J, Miao N, Shen J, Peng T. lncRNA DANCR promotes tumor progression and cancer stemness features in osteosarcoma by upregulating AXL via miR-33a-5p inhibition. Cancer Lett. 2017;405:46–55.

68. Cho CY, Huang JS, Shiah SG, Chung SY, Lay JD, Yang YY, Lai GM, Cheng AL, Chen LT, Chuang SE. Negative feedback regulation of AXL by miR-34a modulates apoptosis in lung cancer cells. RNA. 2016;22(2):303–15.
69. Yang LG, Cao MZ, Zhang J, Li XY, Sun QL. LncRNA XIST modulates HIF-1A/AXL signaling pathway by inhibiting miR-93-5p in colorectal cancer. Mol Genet Genomic Med. 2020;8(4):e1112.
70. Tanaka M, Siemann DW. Gas6/Axl signaling pathway in the tumor immune microenvironment. Cancers. 2020;12(7):1850.
71. Cardone C, Blauensteiner B, Moreno-Viedma V, Martini G, Simeon V, Vitiello PP, Ciardiello D, Belli V, Matrone N, Troiani T, Morgillo F. AXL is a predictor of poor survival and of resistance to anti-EGFR therapy in RAS wild-type metastatic colorectal cancer. Eur J Cancer. 2020;138:1–10.
72. Bottai G, Raschioni C, Szekely B, Di Tommaso L, Szasz AM, Losurdo A, Gyorffy B, Acs B, Torrisi R, Karachaliou N, Tokes T. AXL-associated tumor inflammation as a poor prognostic signature in chemotherapy-treated triple-negative breast cancer patients. NPJ Breast Cancer. 2016;2:16033.
73. Engelsen AS, Lotsberg ML, Abou Khouzam R, Thiery JP, Lorens JB, Chouaib S, Terry S. Dissecting the role of AXL in cancer immune escape and resistance to immune checkpoint inhibition. Front Immunol. 2022;13:869676.
74. Divine LM, Nguyen MR, Meller E, Desai RA, Arif B, Rankin EB, Bligard KH, Meyerson C, Hagemann IS, Massad M, Thaker PH. AXL modulates extracellular matrix protein expression and is essential for invasion and metastasis in endometrial cancer. Oncotarget. 2016;7(47):77291.
75. Brunet A, Bonni A, Zigmond MJ, Lin MZ, Juo P, Hu LS, Anderson MJ, Arden KC, Blenis J, Greenberg ME. Akt promotes cell survival by phosphorylating and inhibiting a Forkhead transcription factor. Cell. 1999;96(6):857–68.
76. Roberts PJ, Der CJ. Targeting the Raf-MEK-ERK mitogen-activated protein kinase cascade for the treatment of cancer. Oncogene. 2007;26(22):3291–310.
77. Repici A, Ardizzone A, De Luca F, Colarossi L, Prestifilippo A, Pizzino G, Paterniti I, Esposito E, Capra AP. Signaling pathways of AXL receptor tyrosine kinase contribute to the Pathogenetic mechanisms of glioblastoma. Cells. 2024;13(4):361.
78. Son HY, Jeong HK. Immune evasion mechanism and AXL. Front Oncol. 2021;11:756225.
79. Kanzaki R, Naito H, Kise K, Takara K, Eino D, Minami M, Shintani Y, Funaki S, Kawamura T, Kimura T, Okumura M. Gas6 derived from cancer-associated fibroblasts promotes migration of Axl-expressing lung cancer cells during chemotherapy. Sci Rep. 2017;7(1):10613.
80. Antony J, Huang RY. AXL-driven EMT state as a targetable conduit in cancer. Cancer Res. 2017;77(14):3725–32.
81. Gjerdrum C, Tiron C, Høiby T, Stefansson I, Haugen H, Sandal T, Collett K, Li S, McCormack E, Gjertsen BT, Micklem DR. Axl is an essential epithelial-to-mesenchymal transition-induced regulator of breast cancer metastasis and patient survival. Proc Natl Acad Sci. 2010;107(3):1124–9.
82. Antony J, Thiery JP, Huang RY. Epithelial-to-mesenchymal transition: lessons from development, insights into cancer and the potential of EMT-subtype based therapeutic intervention. Phys Biol. 2019;16(4):041004.
83. Guney E, Arruda AP, Parlakgul G, Cagampan E, Min N, Lee GY, Greene L, Tsaousidou E, Inouye K, Han MS, Davis RJ. Aberrant Ca2+ signaling by IP3Rs in adipocytes links inflammation to metabolic dysregulation in obesity. Sci Signal. 2021;14(713):eabf2059.
84. Terragno M, Vetrova A, Semenov O, Sayan AE, Kriajevska M, Tulchinsky E. Mesenchymal–epithelial transition and AXL inhibitor TP-0903 sensitise triple-negative breast cancer cells to the antimalarial compound, artesunate. Sci Rep. 2024;14(1):425.
85. Chen S, Huang CS, Li K, Cheng M, Zhang C, Xiong J, Tian G, Zhou R, Ling R, Wang X, Xiong G. Tumor initiating cells escape tumor immunity via CCL8 from tumor-associated macrophages in mice. J Clin Invest. 2025; https://doi.org/10.1172/JCI180893.

86. Jiang X, Wang J, Deng X, Xiong F, Zhang S, Gong Z, Li X, Cao K, Deng H, He Y, Liao Q. The role of microenvironment in tumor angiogenesis. J Exp Clin Cancer Res. 2020;39:1–9.
87. Patel SA, Nilsson MB, Le X, Cascone T, Jain RK, Heymach JV. Molecular mechanisms and future implications of VEGF/VEGFR in cancer therapy. Clin Cancer Res. 2023;29(1):30–9.
88. Li W, Xiong X, Abdalla A, Alejo S, Zhu L, Lu F, Sun H. HGF-induced formation of the MET–AXL–ELMO2–DOCK180 complex promotes RAC1 activation, receptor clustering, and cancer cell migration and invasion. J Biol Chem. 2018;293(40):15397–418.
89. Cudmore M, Ahmad S, Al-Ani B, Hewett P, Ahmed S, Ahmed A. VEGF-E activates endothelial nitric oxide synthase to induce angiogenesis via cGMP and PKG-independent pathways. Biochem Biophys Res Commun. 2006;345(4):1275–82.
90. Ruan GX, Kazlauskas A. Axl is essential for VEGF-A-dependent activation of PI3K/Akt. EMBO J. 2012;31(7):1692–703.
91. Ruan GX, Kazlauskas A. Lactate engages receptor tyrosine kinases Axl, Tie2, and vascular endothelial growth factor receptor 2 to activate phosphoinositide 3-kinase/Akt and promote angiogenesis. J Biol Chem. 2013;288(29):21161–72.
92. Jiang BH, Liu LZ. PI3K/PTEN signaling in angiogenesis and tumorigenesis. Adv Cancer Res. 2009;102:19–65.
93. Du W, Phinney NZ, Huang H, Wang Z, Westcott J, Toombs JE, Zhang Y, Beg MS, Wilkie TM, Lorens JB, Brekken RA. AXL is a key factor for cell plasticity and promotes metastasis in pancreatic cancer. Mol Cancer Res. 2021;19(8):1412–21.
94. Chai ZT, Zhang XP, Ao JY, Zhu XD, Wu MC, Lau WY, Sun HC, Cheng SQ. AXL overexpression in tumor-derived endothelial cells promotes vessel metastasis in patients with hepatocellular carcinoma. Front Oncol. 2021;11:650963.
95. Engelsen AS, Wnuk-Lipinska K, Bougnaud S, Vatter FA, Tiron C, Villadsen R, Miyano M, Lotsberg ML, Madeleine N, Panahandeh P, Dhakal S. AXL is a driver of stemness in normal mammary gland and breast cancer. iScience. 2020;23(11):101649.
96. Ott M, Litzenburger UM, Sahm F, Rauschenbach KJ, Tudoran R, Hartmann C, Marquez VE, von Deimling A, Wick W, Platten M. Promotion of glioblastoma cell motility by enhancer of zeste homolog 2 (EZH2) is mediated by AXL receptor kinase. PLoS One. 2012;7(10):e47663. https://doi.org/10.1371/journal.pone.0047663.
97. Rankin EB, Giaccia AJ. The receptor tyrosine kinase AXL in cancer progression. Cancers. 2016;8(11):103.
98. Cichoń MA, Szentpetery Z, Caley M, Papadakis E, Mackenzie I, Brennan C, O'toole EA. The receptor tyrosine kinase Axl regulates cell–cell adhesion and stemness in cutaneous squamous cell carcinoma. Oncogene. 2014;33(32):4185–92.
99. Khera L, Vinik Y, Maina F, Lev S. The AXL-PYK2-PKCα axis as a nexus of stemness circuits in TNBC. Life Sci Alliance. 2021;4(6):e202000985.
100. Davidsen KT, Haaland GS, Lie MK, Lorens JB, Engelsen AS. The role of Axl receptor tyrosine kinase in tumor cell plasticity and therapy resistance. In: Biomarkers of the tumor microenvironment: basic studies and practical applications; 2017. p. 351–76.
101. Rothlin CV, Lemke G. TAM receptor signaling and autoimmune disease. Curr Opin Immunol. 2010;22(6):740–6.
102. Lemke G, Rothlin CV. Immunobiology of the TAM receptors. Nat Rev Immunol. 2008;8(5):327–36.
103. Cv R. TAM receptors and pleiotropic inhibitors of the innate immune response. Cell. 2007;131:1124–36.
104. Rothlin CV, Ghosh S, Zuniga EI, Oldstone MB, Lemke G. TAM receptors are pleiotropic inhibitors of the innate immune response. Cell. 2007;131(6):1124–36.
105. Wang R, Chen S, Alhusayan M, Xing B, Liu Y, Band H, Asangani I, Shi J, Tong W. CBL loss or mutations upregulates AXL to promote cytokine-independent growth in myeloid malignancies: AXL upregulation promotes cytokine-independent growth in CBL-mutant CMML. Blood. 2024;144(Supplement 1):4574.

106. Hong CC, Lay JD, Huang JS, Cheng AL, Tang JL, Lin MT, Lai GM, Chuang SE. Receptor tyrosine kinase AXL is induced by chemotherapy drugs and overexpression of AXL confers drug resistance in acute myeloid leukemia. Cancer Lett. 2008;268(2):314–24.
107. Aguilera TA, Giaccia AJ. Molecular pathways: oncologic pathways and their role in T-cell exclusion and immune evasion—a new role for the AXL receptor tyrosine kinase. Clin Cancer Res. 2017;23(12):2928–33.
108. Skinner HD, Giri U, Yang LP, Kumar M, Liu Y, Story MD, Pickering CR, Byers LA, Williams MD, Wang J, Shen L. Integrative analysis identifies a novel AXL–PI3 kinase–PD-L1 signaling axis associated with radiation resistance in head and neck cancer. Clin Cancer Res. 2017;23(11):2713–22.
109. Terry S, Dalban C, Rioux-Leclercq N, Adam J, Meylan M, Buart S, Bougoüin A, Lespagnol A, Dugay F, Moreno IC, Lacroix G. Association of AXL and PD-L1 expression with clinical outcomes in patients with advanced renal cell carcinoma treated with PD-1 blockade. Clin Cancer Res. 2021;27(24):6749–60.
110. Hwang JA, Hur JY, Kim Y, Im JH, Jin SH, Ryu SH, Choi CM. Efficacy of newly discovered DNA aptamers targeting AXL in a lung cancer cell with acquired resistance to Erlotinib. Transl Cancer Res. 2021;10(2):1025.
111. Bhalla S, Fattah FJ, Ahn C, Williams J, Macchiaroli A, Padro J, Pogue M, Dowell JE, Putnam WC, McCracken N, Micklem D. Phase 1 trial of bemcentinib (BGB324), a first-in-class, selective AXL inhibitor, with docetaxel in patients with previously treated advanced non-small cell lung cancer. Lung Cancer. 2023;182:107291.
112. Kasikara C, Davra V, Calianese D, Geng K, Spires TE, Quigley M, Wichroski M, Sriram G, Suarez-Lopez L, Yaffe MB, Kotenko SV. Pan-TAM tyrosine kinase inhibitor BMS-777607 enhances anti–PD-1 mAb efficacy in a murine model of triple-negative breast cancer. Cancer Res. 2019;79(10):2669–83.
113. Zeng JY, Sharma S, Zhou YQ, Yao HP, Hu X, Zhang R, Wang MH. Synergistic activities of MET/RON inhibitor BMS-777607 and mTOR inhibitor AZD8055 to polyploid cells derived from pancreatic cancer and cancer stem cells. Mol Cancer Ther. 2014;13(1):37–48.
114. Zhang Y, Arner EN, Rizvi A, Toombs JE, Huang H, Warner SL, Foulks JM, Brekken RA. AXL inhibitor TP-0903 reduces metastasis and therapy resistance in pancreatic cancer. Mol Cancer Ther. 2022;21(1):38–47.
115. Wu X, Liu X, Koul S, Lee CY, Zhang Z, Halmos B. AXL kinase as a novel target for cancer therapy. Oncotarget. 2014;5(20):9546.
116. Chang YC, Liu CT, Yu CY, Sung WW. NPS-1034 exerts therapeutic efficacy in renal cell carcinoma through multiple targets of MET, AXL, and TNFRSF1A signaling in a metastatic model. Cells. 2024;13(20):1713.
117. Blø M, Nilsson LH, Jackson A, Boniecka A, Toombs JE, Ahmed L, Milde E, Mydel PM, Marti HP, Brekken RA, Gabra H. Tilvestamab, a novel clinical stage humanized anti-AXL function blocking antibody. Eur J Cancer. 2020;138:S44.
118. Ye X, Li Y, Stawicki S, Couto S, Eastham-Anderson J, Kallop D, Weimer R, Wu Y, Pei L. An anti-Axl monoclonal antibody attenuates xenograft tumor growth and enhances the effect of multiple anticancer therapies. Oncogene. 2010;29(38):5254–64.
119. Alvarado D, Vitale L, Murphy M, O'Neill T, Natoli E, Lillquist J, Crew L, Wasiuk A, Weidlick J, Sisson C, Widger J. 550 An Axl-targeting monoclonal antibody that inhibits Axl activity and potently stimulates the innate immune response. J ImmunoTher Cancer. 2020; https://doi.org/10.1136/jitc-2020-SITC2020.0550.
120. Liu R, Gong M, Li X, Zhou Y, Gao W, Tulpule A, Chaudhary PM, Jung J, Gill PS. Induction, regulation, and biologic function of Axl receptor tyrosine kinase in Kaposi sarcoma. Blood. 2010;116(2):297–305.
121. Leconet W, Chentouf M, du Manoir S, Chevalier C, Sirvent A, Aït-Arsa I, Busson M, Jarlier M, Radosevic-Robin N, Theillet C, Chalbos D. Therapeutic activity of anti-AXL antibody against triple-negative breast cancer patient-derived xenografts and metastasis. Clin Cancer Res. 2017;23(11):2806–16.

122. Van Renterghem B, Wozniak A, Castro PG, Franken P, Pencheva N, Sciot R, Schöffski P. Enapotamab vedotin, an AXL-specific antibody-drug conjugate, demonstrates anti-tumor efficacy in patient-derived xenograft models of soft tissue sarcoma. Int J Mol Sci. 2022;23(14):7493.
123. Liu Y, Wei J, Liu J, Ma W, Duan Y, Liu D. Novel AXL-targeted agents overcome FLT3 inhibitor resistance in FLT3-ITD+ acute myeloid leukemia cells. Oncol Lett. 2021;21(5):397.
124. Zammarchi F, Havenith KE, Chivers S, Hogg P, Bertelli F, Tyrer P, Janghra N, Reinert HW, Hartley JA, van Berkel PH. Preclinical development of ADCT-601, a novel pyrrolobenzodiazepine dimer-based antibody–drug conjugate targeting AXL-expressing cancers. Mol Cancer Ther. 2022;21(4):582–93.
125. Shen GG, Briggs E, Crass R, Karumanchi S, Boyle WJ, Baker MG, Martin P. Population pharmacokinetic and exposure-response safety analyses of mecbotamab vedotin (BA3011) in patients with advanced solid tumors. J Clin Oncol. 2023;41(16_suppl):e15010.
126. Goel H, Kumar R, Tanwar P, Upadhyay TK, Khan F, Pandey P, Kang S, Moon M, Choi J, Choi M, Park MN. Unraveling the therapeutic potential of natural products in the prevention and treatment of leukemia. Biomed Pharmacother. 2023;160:114351.
127. Nan Y, Su H, Zhou B, Liu S. The function of natural compounds in important anticancer mechanisms. Front Oncol. 2023;12:1049888.
128. Paccez JD, Duncan K, Sekar D, Correa RG, Wang Y, Gu X, Bashin M, Chibale K, Libermann TA, Zerbini LF. Dihydroartemisinin inhibits prostate cancer via JARID2/miR-7/miR-34a-dependent downregulation of Axl. Oncogenesis. 2019;8(3):14.
129. Kim S, Kim KC, Lee C. Mistletoe (Viscum album) extract targets Axl to suppress cell proliferation and overcome cisplatin-and erlotinib-resistance in non-small cell lung cancer cells. Phytomedicine. 2017;36:183–93.
130. Hirai S, Yamada T, Katayama Y, Ishida M, Kawachi H, Matsui Y, Nakamura R, Morimoto K, Horinaka M, Sakai T, Sekido Y. Effects of combined therapeutic targeting of AXL and ATR on pleural mesothelioma cells. Mol Cancer Ther. 2024;23(2):212–22.
131. Sadahiro H, Kang KD, Gibson JT, Minata M, Yu H, Shi J, Chhipa R, Chen Z, Lu S, Simoni Y, Furuta T. Activation of the receptor tyrosine kinase AXL regulates the immune microenvironment in glioblastoma. Cancer Res. 2018;78(11):3002–13.
132. Krebs M, Brunsvig P, Helland Å, Viñolas N, Aix S, Carcereny E, Gomez MD, Perez JT, Arriola E, Campelo RG, Spicer J. P1. 01-72 a phase II study of selective AXL inhibitor bemcentinib and pembrolizumab in patients with NSCLC refractory to anti-PD (L) 1. J Thorac Oncol. 2019;14(10):S388.
133. Lee W, Kim DK, Synn CB, Lee HK, Park S, Jung DS, Choi Y, Kim JH, Byeon Y, Kim YS, Lee S. Incorporation of SKI-G-801, a novel AXL inhibitor, with anti-PD-1 plus chemotherapy improves anti-tumor activity and survival by enhancing T cell immunity. Front Oncol. 2022;12:821391.
134. Im K, Choi YJ, Kim DH, Kim DS, Ban K, Ji W, Baek IJ, Choi CM, Lee JC, Rho JK. AXL receptor tyrosine kinase inhibition improves the anti-tumor effects of CD8+ T cells by inducing CD103+ dendritic cell-mediated T cell priming. Biochem Biophys Res Commun. 2023;680:7–14.
135. Hong J, Peng D, Chen Z, Sehdev V, Belkhiri A. ABL regulation by AXL promotes cisplatin resistance in esophageal cancer. Cancer Res. 2013;73(1):331–40.
136. Park IK, Mishra A, Chandler J, Whitman SP, Marcucci G, Caligiuri MA. Inhibition of the receptor tyrosine kinase Axl impedes activation of the FLT3 internal tandem duplication in human acute myeloid leukemia: implications for Axl as a potential therapeutic target. Blood. 2013;121(11):2064–73.
137. Marrocco I, Romaniello D, Vaknin I, Drago-Garcia D, Oren R, Uribe ML, Belugali Nataraj N, Ghosh S, Eilam R, Salame TM, Lindzen M. Upfront admixing antibodies and EGFR inhibitors preempts sequential treatments in lung cancer models. EMBO Mol Med. 2021;13(4):e13144.

Regulation of Angiogenesis Via VEGF/ Thrombospondin-1 Signaling

Randolph S. Watnick

1 Introduction

In the earliest stages of cancer, carcinomas (tumors derived from epithelial cells) are physically confined within the epithelial compartment of the tissue from whence they arise. These early lesions (carcinomas in situ) are separated from the tissue parenchyma by the basement membrane [1]. Opposite the basement membrane are a myriad of cells consisting of fibroblasts, myofibroblasts, immune/inflammatory cells, and endothelial cells [2]. In addition to these cell types are extracellular matrix proteins, secreted largely by fibroblasts, which serve as a substrate for tumor cells to attach [2].

In order for tumors to progress to a clinically relevant and potentially lethal disease, they must, in a sequential manner, acquire the capacity to escape the epithelial compartment, invade the local parenchyma, and disseminate systemically. To enable this process, tumor cells must degrade, or induce the degradation of, the basement membrane that separates the epithelial compartment from the parenchyma. Invasion of the tissue parenchyma by the tumor, or, conversely, invasion of stromal cells into the tumor, initiates a phase of tumor progression in which tumor growth becomes dependent on non-cell autonomous processes regulated by paracrine and juxtacrine signaling interactions between the tumor and its microenvironment [3–5].

In other words, in order to expand in size beyond the diffusion limit of oxygen in tissue, a new vasculature must form in and around the tumor. The ingrowth of this vasculature is driven by both soluble and matrix-bound growth factors and enzymes secreted by tumor cells and the stromal cells that comprise the microenvironment [6]. Strong evidence exists, indicating that stromal cells play as central a role in

R. S. Watnick (✉)
Vascular Biology Program, Boston Children's Hospital, Harvard Medical School, Boston, MA, USA
e-mail: randy.watnick@childrens.harvard.edu

A. Sabarwal et al. (eds.), *Receptor Tyrosine Kinases in Cancer*, Cancer Drug Discovery and Development, https://doi.org/10.1007/978-3-031-93894-8_5

matrix remodeling, invasion, and metastasis as the tumor cells themselves [7–9]. Critical to this process is the observation that carcinoma cells are able to reprogram and coopt the surrounding stromal cells to enhance tumor growth [10]. Specifically, tumor-stromal paracrine signaling pathways have been demonstrated to play a major role in the tumorigenesis and subsequent outgrowth of tumors in multiple sites [11–13]. For example, stromal fibroblasts from prostate tumors are able to stimulate tumor formation of immortal but non-transformed prostate epithelial cells when the mixture is injected orthotopically into nude mice [10].

As stated above, tumor angiogenesis is intricately linked to signaling between the tumor and microenvironment. In normal tissue architecture, the epithelial compartment is not vascularized as it is generally only 1–2 cell layers in thickness. The minimal thickness allows oxygen to diffuse readily across the basement membrane and nourish both epithelial cell layers. However, when tumors form and the lumen of epithelial ducts fills with tumor cells, the cells in the center become hypoxic due to their increased distance from existing blood vessels. Thus, in order for tumors to gain access to the vasculature, the basement membrane must be degraded, allowing the blood vessels to grow into the epithelial compartment. Degradation of the basement membrane and ingrowth of blood vessels require tumor cells to secrete proangiogenic growth factors and turn off production of antiangiogenic factors. Further dissemination and metastasis require a recapitulation of that process in the distant metastatic microenvironment. Though a myriad of pro- and antiangiogenic factors have been discovered and studied, the initial attempts to delineate their regulation was examined in a cell autonomous fashion, with most of the attention paid to vascular endothelial growth factor (VEGF) [14–18] and thrombospondin-1 (Tsp-1) [19, 20], two of the major positive and negative regulators of angiogenesis. However, as the importance of the tumor microenvironment became more apparent the study of the regulation of angiogenic factors in stromal cells also increased.

Not only is the regulation of angiogenesis in the tumor microenvironment critical to primary tumor growth but also for metastatic dissemination and growth in distant organs. It is well established that tumors arising in different sites preferentially metastasize to specific organs [21]. For example, prostate cancer metastasizes preferentially to bone and liver, while breast cancer metastasizes to brain, bone, and lung [22]. The ability of a tumor cell to survive and proliferate in a metastatic environment ultimately relies on its ability to augment the angiogenic output of its microenvironment.

2 VEGF

One of the most well-studied and potent inducers of angiogenesis is vascular endothelial growth factor (VEGF). VEGF was initially identified by its ability to induce vessel permeability, as such, it was first called vascular permeability factor or VPF [23, 24]. The regulation of VEGF in tumor cells has been exhaustively studied. Signal transduction pathways leading from receptor tyrosine kinases or oncogenic

Ras and PI3 kinase via the MAPK or Akt pathway lead to increased transcription of VEGF and its subsequent secretion into the extracellular matrix [16]. However, carcinoma cells also secrete proteins into the extracellular space, which do not act directly on endothelial cells but rather modulate VEGF production and secretion by stromal cells in the microenvironment, such as TGF-β, PDGF, and bFGF [25, 26].

Stromal VEGF expression was first demonstrated to be regulated by carcinoma cells in a transgenic mouse model in which GFP, driven by the VEGF promoter, was inserted into the mouse genome [27]. In this model, activation of the VEGF promoter results in the expression of GFP. Examination of tumor xenografts in these VEGF-GFP mice revealed, via fluorescence microscopy, that the stromal fibroblasts that had infiltrated the tumor fluoresced green, indicating that the VEGF promoter had been activated. Strikingly, in normal tissues, there were no fluorescent cells, indicating that the VEGF expression is not required for normal tissue homeostasis. These results indicated that tumors secrete factors that act on cells in the microenvironment to stimulate VEGF expression.

However, while it was clear from these experiments that VEGF expression was being stimulated, it was not evident whether this stimulation was required for tumor growth, supportive of tumor growth, or merely a physiological reaction to local tumor growth. The evidence that stromally produced VEGF was critical for tumor growth was obtained from studies designed to test the efficacy of a human-specific anti-VEGF antibody, bevacizumab (Avastin). In these experiments, human tumor cells were injected into immunocompromised mice, which were subsequently treated with the human-specific VEGF antibody [28]. While the antibody was able to dramatically inhibit tumor growth, the tumors still grew. The authors of the study hypothesized that the continued growth of the treated tumors was due to a residual angiogenic stimulus driven by VEGF produced and secreted from the murine tumor microenvironment, which could not be inhibited by the human-specific antibody. To test their hypothesis, human tumor xenografts were treated with human-specific VEGF antibodies as well as a soluble version of murine VEGFR1 (mFlt) fused to IgG, which acts as a decoy receptor for VEGF [29]. The combination of these two treatment modalities resulted in the complete blockage of tumor growth, demonstrating the importance of the contribution of stromal-produced VEGF.

VEGF (also referred to as VEGF-A) is one member of a family of proteins under the VEGF umbrella, the others of which are as follows: VEGF-B, VEGF-C, VEGF-D, VEGF-E (virally encoded), and placental growth factor (PlGF) [30]. VEGF-C and VEGF-D are primarily implicated in the regulation of lymphangiogenesis [31]. VEGFA, heretofore referred to solely as VEGF, has multiple isoforms generated by alternative splicing, which include $VEGF_{121}$, $VEGF_{165}$, $VEGF_{189}$, and $VEGF_{206}$. $VEGF_{165}$ ($VEGF_{164}$ in mice) is the most commonly expressed isoform. $VEGF_{165}$ is also the most physiologically relevant isoform as it interacts with the extracellular matrix (ECM) more strongly than the highly diffusible $VEGF_{121}$ and less tightly than the extracellular matrix-bound $VEGF_{189}$. There are two additional, though less commonly expressed isoforms: $VEGF_{145}$ and $VEGF_{183}$. The major component of the ECM that mediates the properties of VEGF isoforms is their interaction with heparin [32–34]. For example, $VEGF_{121}$ has a weak affinity for heparin, while $VEGF_{189}$ and $VEGF_{206}$ have very high affinities for heparin, each containing two heparin-binding domains (encoded by exons 6 and 7). $VEGF_{165}$ possesses an intermediate affinity for heparin, containing a single heparin-binding domain, making it partially diffusible and partially ECM-bound.

3 VEGFR1

The first identified receptor of VEGF was VEGFR1 (alternatively denoted Fms-like tyrosine kinase 1, Flt1, in the mouse). VEGF binding to VEGFR1 is mediated by extracellular Immunoglobulin (Ig)-like loops 2 and 3 of VEGFR1 [35]. VEGFR1 signaling and biology appear more complex than those of other VEGFRs. Though VEGFR1 is widely expressed, it possesses weak kinase activity and appears not to be required for endothelial cell function [36]. Studies using mutants of VEGF that have selective receptor binding properties demonstrated that VEGFR1 cooperates with VEGFR2 in inducing gene expression in human umbilical vein endothelial cells (HUVECS); however, VEGFR1 stimulation alone did not lead to unique expression patterns [37]. However, alternate studies have discovered that VEGFR1 plays a role in the tissue-specific release of growth factors, such as HGF from liver sinusoidal endothelial cells, which serves to protect hepatocytes from hepatotoxin-induced damage [38]. In monocytes and macrophages, VGFR1 has been reported to mediate migration in response to VEGF or PlGF [39]. Moreover, VEGFR1 can promote proliferation and migration in some tumor cell lines in response to VEGF or PlGF [40].

VEGFR1 exists as a full-length form and an alternatively spliced, soluble form (sFlt1) [41]. Despite having weaker kinase activity, full-length VEGFR1, as well as the soluble version, both possess a higher binding affinity for VEGF than that of VEGFR2. Binding of VEGF to VEGFR1 has thus been postulated to attenuate activation of VEGFR2. As such, sFlt1 has been demonstrated to act as a negative regulator of angiogenesis [42]. In a pathological setting, overexpression of sFlt1 in the placenta has been causally linked to the pathogenesis of preeclampsia [43, 44]. Specifically, an sFlt1 variant denoted sFlt-14 (containing exon 14) has been found to accumulate in the circulation of women suffering from preeclampsia [45].

Strikingly, deletion of the VEGFR1 tyrosine kinase domain in mice does not disrupt vascular development [46]. Accordingly, VEGFR1 appears not to be required for angiogenic signaling in endothelial cells. Rather, VEGFR1 may function as a decoy receptor to capture VEGF and locally regulate VEGFR2 signaling and the development of angiogenic sprouts [47].

The comparatively weak kinase activity of VEGFR1 has been attributed to the lack of positive regulatory phosphorylation sites in the kinase activation loop [48]. On the other hand, overexpression studies of VEGFR1 in various tissue culture models demonstrated that the majority of the intracellular tyrosine residues were phosphorylated, such as those in positions 794, 1169, 1213, 1242, 1327, and 1333

[48, 49]. However, it cannot be ruled out that these results were artifacts of non-physiological expression levels.

Intriguingly, VEGFR1-specific ligands VEGFB and PlGF have entirely different functions. VEGFB acts on cardiac endothelial cells, where it mediates revascularization following ischemic events [50]. Another unique property of VEGFB is its ability to stimulate fatty acid uptake in endothelial cells, specifically in the heart [51]. VEGFB-dependent fatty acid transport requires VEGFR1 as well as neuropilin 1 (NRP1). Of note, PlGF cannot compensate for the loss of VEGFB in fatty acid transport [51].

4 VEGFR2

The primary receptor on endothelial cells for VEGF mediating its angiogenic activity is VEGFR2. VEGFR2 or KDR (kinase insert domain receptor) in the human and Flk1 (fetal liver kinase-1) in the mouse. VEGFR2 binds VEGFA via its extracellular Ig-like domains 2 and 3, but with a tenfold lower affinity than VEGFR1 [52, 53]. Unlike VEGFR1, VEGFR2 is required for endothelial cell biology during development and in the adult, both physiologically and pathologically. Additionally, because of this status, more is known about VEGFR2 signaling than the other VEGF receptors. In addition to its role in neovascularization, another essential role for VEGFR2 signaling is vascular permeability. Because of this, multiple small-molecular-weight compounds are clinically approved as inhibitors of VEGFR2 kinase activity to abrogate pathological angiogenesis in cancer.

In contrast to VEGFR1, VEGFR2 also binds proteolytically processed VEGFC and VEGFD [54]. Similar to VEGFR1, VEGFR2 is alternatively spliced to a soluble VEGFR2 (sVEGFR2), which is present in various tissues, such as the skin, heart, spleen, kidney, and ovary, and plasma. sVEGFR2 binds VEGFC and prevents binding to VEGFR3, consequently inhibiting lymphatic endothelial cell proliferation [55].

The essential role of VEGFR2 was validated genetically in vivo via the generation of *Vegfr2*$^{-/-}$ KO mice, which die at E8.5 from impaired development of hematopoietic and endothelial cells [56], a phenotype similar to that of the *vegfa*$^{-/-}$ mouse [57, 58]. VEGFR2 is expressed most prominently in endothelial cells and their embryonic progenitor/stem cells, with the highest expression levels observed during embryonic vasculogenesis and angiogenesis [59–61]. VEGFR2 is also upregulated during pathologies associated with neovascularization, such as tumor angiogenesis [59, 62]. VEGFR2 is thus considered to be the main transducer of VEGF activity on endothelial cell differentiation, proliferation, migration, and formation of the vascular tube.

Binding of VEGF to VEGFR2 stimulates receptor dimerization, which is partially mediated by low-affinity homotypic interactions between Ig-like domains on adjacent receptors [63, 64]. Dimerization of VEGFR2 enables trans/autophosphorylation of intracellular tyrosine residues. There are at least 11 tyrosine residues in the intracellular domain of VEGFR2 that are phosphorylated as a result of VEGF binding. The primary phosphorylation sites are tyrosine Y951 in the kinase insert domain, Y1054, Y1059 within the kinase domain, as well as Y1175 and Y1214 in the carboxy-terminal domain [65, 66]. Additionally, tyrosine 1305, 1309, and 1319 have been identified as being phosphorylated subsequent to VEGF treatment [65], though their biological roles in VEGF signaling are not yet clear. Moreover, the juxtamembrane tyrosine Y801 is phosphorylated in vitro using biochemically isolated intracellular domains [67]; it is not known whether Y801 is phosphorylated in the intact VEGFR2. It has been postulated that phosphorylation of Y801 may precede and be a prerequisite for phosphorylation of Y1054 and 1059, which are critical for kinase activity [67, 68],

The Y1175 in human is crucial for endothelial and hematopoietic cell development. Mice with a Y1173F, the corresponding murine residue, knock-in die between E8.5 and E9.5 due to ablation of endothelial and hematopoietic progenitors [69], reminiscent of the phenotype of *vegfr2*$^{-/-}$ mice [56]. In contrast, mice with a Y1212F substitution (corresponding to 1214 in humans) are viable and fertile [69].

VEGFR2 is also expressed on T cells, where the phosphorylation of Y951 functions as a binding site for the T cell-specific adapter molecule (TSAd) [65, 70], also denoted VRAP (VEGF receptor associated protein). On endothelial cells Y951 is phosphorylated following VEGF binding in lls during development as well as in tumor neovascularization [65], but not in quiescent endothelial cells. Finally, phosphorylation of Y1175 generates a site recognized by various signaling mediators like PLCγ [49, 66], as well as the adapter proteins SCK and SHB [71, 72]. Y1175 is also bound by SHCA and GRB2, which subsequently recruit the nucleotide-exchange factor son-of-sevenless (SOS) to VEGFR2 [72, 73]. A myriad of additional signal transduction mediators have been associated with VEGFR2 signaling based on in vitro experiments; however, the in vivo confirmation of their biological significance remains to be established.

VEGF signaling through VEGFR2 also promotes endothelial proliferation via downstream activation of the RAS/RAF/MAPK pathway [74, 75]. The mechanism of VEGFR2-induced RAS activation is postulated to be mediated by protein kinase C (PKC) activation of sphingosine kinase (SPK) [76]. In addition to regulating endothelial proliferation, activation of the RAS signaling pathway stimulates the production of pleiotropic lipid mediators, the prostaglandins (PGI_2) via phospholipase A_2 ($cPLA_2$) [75]. Unlike many other RTKs, VEGFR2 stimulates ERK not via GRB2-SOS-RAS but rather through pY1175-dependent phosphorylation of PLCγ and subsequent activation of PKCs [66]. Phosphorylated PLCγ hydrolyzes phosphatidylinositol (4,5)-bisphosphate (PIP_2), stimulating the accumulation of diacylglycerol (DAG) and inositol 1,4,5-trisphosphate (IP_3). Production of IP_3 subsequently leads to the release of intracellular calcium $[Ca^{2+}]_i$, activating DAG, which activates PKC family members. The PKC isoforms PKCβ, PKCα, and PKCζ also regulate VEGF-mediated proliferation [66, 77, 78]. Specifically, PKC phosphorylation results in activation of protein kinase D (PKD) [79], which induces the nuclear translocation of histone deacetylases (HDAC) 5 and 7 [80, 81], followed by phosphorylation of cAMP-response-element binding protein (CREB) as well as heat shock protein 27 (HSP27) [82].

Angiogenesis is a multistep biological process involving migration and invasion of endothelial cells. Endothelial cells invade and migrate via a protease-mediated degradation of the basement membrane along a concentration gradient of VEGF and other growth factors. Downstream of VEGFR2 activation, multiple signal transducers are recruited to the phosphorylated tyrosine Y951, Y1175, and Y1214, which have been implicated in VEGF-induced migration of endothelial cells. For example, pY951 binds TSAd [65, 70], as validated by knock-in of the Y951F mutation as well a knock-down of TSAd, which blocks VEGF-dependent actin-reorganization and migration, but not proliferation of endothelial cells. In this process, TSAd forms a complex with SRC in a VEGFA-dependent manner [65].

Additionally, phosphorylation of Y1175 recruits SHB, which is subsequently phosphorylated downstream of SRC activation [71]. SHB then binds to FAK [83], which is phosphorylated in VEGF-treated cells [84], thereby regulating cell attachment and migration [85]. Phosphorylation of Y1214 actuates the recruitment of NCK and FYN to VEGFR2, where the NCK/FYN complex mediates phosphorylation of p21-activated protein kinase-2 (PAK-2) and subsequent activation of Cdc42 and p38MAPK [86].

Interestingly, shear stress also stimulates VEGFR2 activation and GAB1 tyrosine phosphorylation in a VEGF-independent manner [87]. GAB1 contains a binding site for the p85 subunit of PI3K [88]. Thus, VEGF induces the production of PIP_3, which in turn activates the GTP-binding Rac, resulting in membrane ruffle formation and cell motility [89]. VEGFR2-induced migration has also been shown to be mediated by heterotrimeric G-proteins and activation of PLCβ3 [90].

Hypoxia is a major regulator of angiogenesis via stimulation of VEGF expression by hypoxia inducible factor 1 and 2 (HIF1 and HIF2). HIFs, along with other hypoxia-regulated genes, mediate the production of other signaling molecules, including epidermal growth factor (EGF), platelet-derived growth factor (PDGF), which collaborate with oncogenic mutations (*vhl, ras, wnt-kras signaling pathway genes*) to augment VEGF expression and VEGF-mediated signaling [91, 92]. Canonical VEGF signaling via VEGFR2 regulates the activities of several kinases and ultimately mediates cell proliferation, migration, survival, and vascular permeability during vasculogenesis and angiogenesis. Elongating blood vessels are comprised of tip and stalk endothelial cells that make up the leading edge the growing vessel. Blood vessel growth mediated by the tip cells occurs along a VEGF gradient, which promotes the formation of filopodia [93–99]. The molecular regulation of these events is mediated by activation of the canonical notch signaling pathway on endothelial cells, mediated mainly, but not exclusively, by the Notch ligand delta-like 4 (DLL4) [100]. Increased notch signaling on adjacent cells functions as a negative feedback loop, reducing VEGF-R2 expression. This negative feedback loop is critical to physiological angiogenesis and prevents abnormal vessel growth associated with pathological conditions such as cancer and wet AMD. In 2014, a noncanonical pathway for VEGFR2 signaling was characterized in neurons [101]. VEGFR2 was found to be highly expressed in retinal neurons, though at significantly lower levels than in endothelial cells. Deletion of VEGFR2 in neurons resulted in abnormal angiogenesis toward neurons due to increased extracellular VEGF. These findings underscored a critical role for neurons in titrating the level of VEGF available for neuronal and tissue vascularization.

In 1993, the findings that anti-VEGF antibodies decreased the growth of tumor cells implanted in immune-deficient mice opened up translational possibilities for targeting VEGF-VEGFR signaling [28]. In addition, it was also demonstrated that inactivation of a single allele of the *vegfa* gene in mice resulted in defective vascular development and early embryonic lethality [57, 58], highlighting the importance of VEGF during embryonic development. Inactivation of both copies of *vegfr2* largely phenocopied *vegfa* single allele deletion [56]. The ability to delete VEGF in target tissues with the advent of *cre-lox* systems created the possibility of assessing the role of VEGF in individual tissues/cells [102]. Numerous studies employing this approach have documented the important role of VEGF in angiogenesis and homeostasis in a variety of pathophysiological circumstances [103].

5 Thrombospondin-1

While much of the attention in the field of angiogenesis has been given to the identification and characterization of pro-angiogenic factors, the studies detailing the role of one of the most potent antiangiogenic proteins, Tsp-1, should not be overlooked. Thrombospondin-1 (Tsp-1) is an endogenous antiangiogenic protein that functions via a multimodal approach. Tsp-1 binds to cell surface receptors CD36 and CD47 on the endothelial cell surface and renders the cells insensitive to both VEGF and bFGF. Tsp-1 also induces caspase-dependent apoptosis mediated by Src-family kinase Fyn signaling downstream from CD36 [104–107]. Tsp-1 also binds to MMP-9 and functionally inactivates it [108, 109]. In tumor cells, Tsp-1 expression is repressed via a signal transduction cascade emanating from PI3-kinase via Rho GTPase to ROCK to Myc, which represses Tsp-1 in a phosphorylation-dependent manner [110]. The Rho-Rock pathway has been shown to be active in several human breast cancer cell lines in which Tsp-1 expression was virtually silenced [110]. Accordingly, this pathway represents the first biochemical elucidation of a cell-autonomous "angiogenic switch."

While the expression of VEGF in the tumor-associated stroma is widely accepted to have a positive correlation with tumor progression [27, 111, 112], the role of thrombospondin-1 (Tsp-1) expression in the tumor-associated stroma is unclear. Tsp-1 expression by epithelial tumor cells is observed infrequently, and ectopic expression of Tsp-1 is inhibitory to tumor growth [19, 110, 113]. Stromal Tsp-1, meanwhile, has been correlated with a desmoplastic response and increased invasiveness in a subset of breast cancers [111, 114, 115], while it has been demonstrated to be inhibitory to early-stage breast cancers [116]. Expression of Tsp-1 by stromal fibroblasts has been shown to be inhibitory to tumor formation and growth [117]. Intriguingly, the same report demonstrated that tumors that arose in an environment high in Tsp-1 eventually overcame the inhibitory effects of this protein by increasing their production of VEGF. Thus, the complex interrelationship between these two proteins and their relative expression levels in the tumor-associated stroma plays a critical role in the induction and maintenance of angiogenesis in human tumors.

The work described above demonstrated that VEGF expression in the stroma is a critical component in tumor-mediated angiogenesis. Conversely, Tsp-1 expression in the tumor-associated stroma can be a potent inhibitor of tumor angiogenesis and growth. The question that arises then is how do tumors stimulate the expression of VEGF in the stroma while concomitantly repressing the expression of Tsp-1?

6 CD36

Thrombospondin-1 exerts its antiangiogenic activity primarily via binding to two of its cognate receptors on the cell surface of endothelial cells: CD36 and CD47. CD36 is a transmembrane glycoprotein that is a member of the class B scavenger receptor family. In addition to functioning as a receptor for Tsp-1, another physiological role of CD36 is as a translocase for long-chain fatty acid translocase [118]. It is via these functions that CD36 mediates a diverse range of physiological processes, including fatty acid metabolism, atherosclerosis, and angiogenesis. In angiogenesis, CD36 mediates the inhibition of endothelial cell migration and proliferation by TSP-1. Significant quantities of CD36 are also found in platelets, macrophages, adipocytes, and endothelial cells [118]. In general, CD36 is expressed on small, rather than large, vessel endothelial cells [119]. CD36 is a well-characterized receptor for TSP-1 in endothelial cells [120], and binding of CD36 by TSP-1 trimers induces clustering and the subsequent activation of Fyn, JNK, and p38MAPK to induce apoptosis through activation of caspase-8- and caspase-9-dependent pathways [120].

CD36 localizes and is enriched in lipid rafts and tetraspanin-enriched microdomains [121, 122]. The tetraspanins are a family of proteins that contain four transmembrane domains [123, 124], which undergo homo- and heterotypic interactions to generate biologically unique regions in the plasma membrane that are enriched in integrins and other membrane proteins. CD36 is a loop-type extracellular protein with both the N- and C-terminal domains containing short intracellular sequences that do not contain any consensus sequences for the docking of signal transduction proteins. Thus, CD36 is thought to induce signal transduction indirectly through complex formation with lipids, integrins, and itself. In fact, it has been demonstrated in platelets that the association of CD36 with the Src family kinase Lyn is mediated by lipids [122]. However, the specific Src family kinase that interacts with CD36 is dependent upon the presence of TSP-1 [125].

In 1997, it was reported that CD36 exists as both monomers and dimers in the plasma membrane [126]. Of note, while the addition of TSP-1 increased the formation of CD36 clusters in the membrane, TSP-1 did not induce dimerization of a soluble form of CD36. It has also been shown that overexpression of CD36 in an immortalized endothelial cell line resulted in 40% of the CD36 molecules localizing to nanoclusters that contained as many as 75 CD36 molecules [127]. These nanoclusters are enriched in Fyn and F-actin. The size of the clusters and the phosphorylation of Fyn increase after addition of TSP-1 or a multivalent anti-CD36 IgM molecule [127]. The formation of nanoclusters is likely facilitated by the fact that CD36 mobility in the plasma membrane is restricted by the cytoskeleton [128].

The clustering of CD36 promotes signal transduction through the recruitment of other membrane proteins, such as integrins, and kinases, such as Fyn and Syk [129]. Since TSP-1 can engage integrins in the absence of CD36, it may form multivalent complexes with these proteins via the integrin-associated protein (IAP) also known as CD47.

7 CD47

CD47 is a 50 kDa transmembrane receptor, also known as integrin-associated protein (IAP), that was initially identified as a protein absent in red blood cells in patients with Rh-null hemolytic anemia [130]. CD47 consists of an extracellular N-terminal IgV domain, five transmembrane domains, and a short C-terminal intracellular tail [131]. Four alternatively spliced isoforms of CD47 exist, differing in the length of their cytoplasmic tails [132]. CD47 has two main roles: as a ligand for the signal-regulatory protein alpha (SIRPα) and as a receptor for TSP-1. The CD47-SIRPα axis delivers inhibitory signals for phagocytosis and conveys a “don’t eat me” signal that has important functions in hematopoiesis and innate immune surveillance [133–135]. Initial studies reported that CD47 is bound by TSP-1 at two peptide motifs on the carboxy-terminal domain [136]. While partial conservation of some of these residues is found in TSP-2 and -4, CD47 only binds to TSP-1 with high affinity [137]. Some molecules of CD47 contain proteoglycan side chains that are important for high-affinity TSP-1 binding and signaling [138]. The interaction with the proteoglycan moieties may be specific to TSP-1.

Thrombospondin 1 binding to CD47 has an inhibitory influence on VEGF signaling. CD47 constitutively associates with VEGFR2 on endothelial [138] and tumor [139] cells. Binding of CD47 by TSP-1 inhibits the interaction of CD47 with VEGFR-2 and disrupts VEGFR2 phosphorylation [138]. Binding of CD47 by TSP-1 thus prevents VEGFR2 autophosphorylation, which, in turn, decreases activation of endothelial nitric oxide synthase (eNOS) required for Akt phosphorylation [125, 140]. Consistent with the receptor clustering seen with CD36, recent evidence suggests that clustering of CD47 is also required for the high-affinity binding of TSP-1 [141]. These results suggest that, in addition to the amount of CD47 available, the distribution pattern of the receptor is important in regulating TSP-1 signaling.

By disrupting TSP-1/CD47 signaling, angiogenesis can be enhanced, which can have therapeutic benefits. Both TSP-1 and CD47 null mice exhibit decreased necrosis and improved wound healing in the cutaneous flap model and in full-thickness skin grafts. Skin grafts implanted on TSP-1 or CD47 null mice survive, while those implanted on wild-type mice fail due to a lack of angiogenesis and tissue perfusion [137]. Similarly, pancreatic islet grafts have improved survival in TSP-1 null mice, due to increased angiogenesis [142]. In cancer, CD47 null mice display augmented VEGF and VEGFR2 expression with a concomitant increase in angiogenesis and tumor progression in a syngeneic murine model of prostate cancer [136].

8 Conclusion

To date, several antiangiogenics have been designed to target VEGF and its receptors for cancer treatment. The prevailing theory was that such drugs would not lead to the development of resistance since the target, namely, the endothelial cell, was genetically stable. However, the therapeutic benefits of these drugs have been modest and transient, and toxicity and refractory disease remain significant drawbacks. Historically, the stimulation of alternate VEGF-independent proangiogenic pathways was thought to promote the resumption of angiogenesis. In recent years, the existence of VEGF isoforms that are expressed differentially in various tissues and cancers has been reported. Furthermore, some of the variants trigger intussusceptive angiogenesis, a form of vascularization used by tumors to escape antiangiogenic therapy. Given the distinct roles of VEGF variants in both angiogenesis modulation and neoplastic progression, it is plausible that they may act as contributors to the poor clinical outcome observed with current VEGF-targeting drugs. Moreover, these drugs target the variants indiscriminately, regardless of whether they contribute to angiogenesis or diminish the angiogenic response. It is thus important to determine the profile of the VEGF transcripts in various cancers. Additionally, the receptors through which these ligands transduce signals also undergo splicing. While data are still sparse on the clinical impact of splice products of canonical VEGF receptors in cancer, their clinical significance mandates further exploration to improve treatment approaches. Concerning VEGF isoforms, it has been shown unequivocally that $VEGF_{165a}$ promotes both sprouting and intussusceptive angiogenesis. Given the importance of SRPK in the formation of the $VEGF_{165a}$ variant, the compound TG0003, which inhibits SRPK, could be employed in order to lower $VEGF_{xxxa}$ and restore the $VEGF_{165a}$/$VEGF_{165b}$ levels. On the other hand, $VEGF_{165b}$ appears to have a dual effect, namely, the stimulation of angiogenesis, albeit weakly, and the competitive inhibition of $VEGF_{165a}$ binding to VEGFR-2. Furthermore, drug carriers and antibodies could improve precision targeting and thus alleviate undesirable off-target effects observed with VEGFR inhibitors. Moreover, metallic nanoparticles, including gold nanoparticles, have been shown to be effective carriers in targeting tumor vessels in preclinical models.

Angiogenesis is a complex process that is stimulated by a myriad of growth factors and cytokines and inhibited by an equally diverse cohort of proteins. Underscoring this complexity is the fact that signaling molecules secreted by tumors that act on stromal cells can often have different, and even opposite, activities with respect to the production of pro- and antiangiogenic factors. Therefore, the composition of the tumor microenvironment as well as the stage of the tumor have profound effects on whether the tumor microenvironment is pro-angiogenic or antiangiogenic. The complex signaling mechanisms described in this chapter constitute multiple potential, and largely untapped, targets for therapeutic intervention to inhibit tumor growth in patients. Ultimately, the strategy of targeting molecules that mediate processes, such as angiogenesis, via tumor-stromal interactions may prove to be hugely successful as the genomic instability and high mutational burden in tumor cells is exceedingly rare in the non-tumorous cells that make up the TME. The hope, then is that antiangiogenic therapy targeting the tumor microenvironment will result in lower rates and incidences of acquired resistance than traditional therapeutic strategies.

References

1. Hanahan D, Weinberg RA. The hallmarks of cancer. Cell. 2000;100(1):57–70.
2. Ronnov-Jessen L, Petersen OW, Bissell MJ. Cellular changes involved in conversion of normal to malignant breast: importance of the stromal reaction. Physiol Rev. 1996;76(1):69–125.
3. Chung LW, Davies R. Prostate epithelial differentiation is dictated by its surrounding stroma. Mol Biol Rep. 1996;23(1):13–9.
4. Tuxhorn JA, Ayala GE, Rowley DR. Reactive stroma in prostate cancer progression. J Urol. 2001;166(6):2472–83.
5. Henshall SM, et al. Altered expression of androgen receptor in the malignant epithelium and adjacent stroma is associated with early relapse in prostate cancer. Cancer Res. 2001;61(2):423–7.
6. Hanahan D, Folkman J. Patterns and emerging mechanisms of the angiogenic switch during tumorigenesis. Cell. 1996;86(3):353–64.
7. Camps JL, et al. Fibroblast-mediated acceleration of human epithelial tumor growth in vivo. Proc Natl Acad Sci USA. 1990;87(1):75–9.
8. Grey AM, et al. Purification of the migration stimulating factor produced by fetal and breast cancer patient fibroblasts. Proc Natl Acad Sci USA. 1989;86(7):2438–42.
9. Picard O, Rolland Y, Poupon MF. Fibroblast-dependent tumorigenicity of cells in nude mice: implication for implantation of metastases. Cancer Res. 1986;46(7):3290–4.
10. Olumi AF, et al. Carcinoma-associated fibroblasts direct tumor progression of initiated human prostatic epithelium. Cancer Res. 1999;59(19):5002–11.
11. Hom YK, et al. Uterine and vaginal organ growth requires epidermal growth factor receptor signaling from stroma. Endocrinology. 1998;139(3):913–21.
12. Donjacour AA, Cunha GR. Stromal regulation of epithelial function. Cancer Treat Res. 1991;53:335–64.
13. Cunha GR, et al. Stromal-epithelial interactions in adult organs. Cell Differ. 1985;17(3):137–48.
14. Akiyama H, et al. Induction of VEGF gene expression by retinoic acid through Sp1-binding sites in retinoblastoma Y79 cells. Invest Ophthalmol Vis Sci. 2002;43(5):1367–74.
15. Damert A, Ikeda E, Risau W. Activator-protein-1 binding potentiates the hypoxia-induciblefactor-1-mediated hypoxia-induced transcriptional activation of vascular-endothelial growth factor expression in C6 glioma cells. Biochem J. 1997;327(Pt 2):419–23.
16. Rak J, et al. Mutant ras oncogenes upregulate VEGF/VPF expression: implications for induction and inhibition of tumor angiogenesis. Cancer Res. 1995;55(20):4575–80.

17. Wojta J, et al. Hepatocyte growth factor increases expression of vascular endothelial growth factor and plasminogen activator inhibitor-1 in human keratinocytes and the vascular endothelial growth factor receptor flk-1 in human endothelial cells. Lab Investig. 1999;79(4):427–38.
18. Xiong S, et al. Up-regulation of vascular endothelial growth factor in breast cancer cells by the heregulin-beta1-activated p38 signaling pathway enhances endothelial cell migration. Cancer Res. 2001;61(4):1727–32.
19. Watnick R, et al. Ras modulates Myc activity to repress thrombospondin-1 expression and increase tumor angiogenesis. Cancer Cell. 2003;3(3):219.
20. Rak J, et al. Oncogenes and tumor angiogenesis: differential modes of vascular endothelial growth factor up-regulation in ras-transformed epithelial cells and fibroblasts. Cancer Res. 2000;60(2):490–8.
21. Chambers AF, Groom AC, MacDonald IC. Dissemination and growth of cancer cells in metastatic sites. Nat Rev Cancer. 2002;2(8):563–72.
22. Pettaway CA, et al. Selection of highly metastatic variants of different human prostatic carcinomas using orthotopic implantation in nude mice. Clin Cancer Res. 1996;2(9):1627–36.
23. Leung DW, et al. Vascular endothelial growth factor is a secreted angiogenic mitogen. Science. 1989;246(4935):1306–9.
24. Senger DR, et al. Tumor cells secrete a vascular permeability factor that promotes accumulation of ascites fluid. Science. 1983;219(4587):983–5.
25. Brogi E, et al. Indirect angiogenic cytokines upregulate VEGF and bFGF gene expression in vascular smooth muscle cells, whereas hypoxia upregulates VEGF expression only. Circulation. 1994;90(2):649–52.
26. Tsai JC, Goldman CK, Gillespie GY. Vascular endothelial growth factor in human glioma cell lines: induced secretion by EGF, PDGF-BB, and bFGF. J Neurosurg. 1995;82(5):864–73.
27. Fukumura D, et al. Tumor induction of VEGF promoter activity in stromal cells. Cell. 1998;94(6):715–25.
28. Kim KJ, et al. Inhibition of vascular endothelial growth factor-induced angiogenesis suppresses tumour growth in vivo. Nature. 1993;362(6423):841–4.
29. Gerber HP, et al. Complete inhibition of rhabdomyosarcoma xenograft growth and neovascularization requires blockade of both tumor and host vascular endothelial growth factor. Cancer Res. 2000;60(22):6253–8.
30. Ferrara N, Adamis AP. Ten years of anti-vascular endothelial growth factor therapy. Nat Rev Drug Discov. 2016;15(6):385–403.
31. Alitalo K, Tammela T, Petrova TV. Lymphangiogenesis in development and human disease. Nature. 2005;438(7070):946–53.
32. Ferrara N. Vascular endothelial growth factor and age-related macular degeneration: from basic science to therapy. Nat Med. 2010;16(10):1107–11.
33. Houck KA, et al. Dual regulation of vascular endothelial growth factor bioavailability by genetic and proteolytic mechanisms. J Biol Chem. 1992;267(36):26031–7.
34. Poltorak Z, et al. VEGF145, a secreted vascular endothelial growth factor isoform that binds to extracellular matrix. J Biol Chem. 1997;272(11):7151–8.
35. Wiesmann C, et al. Crystal structure at 1.7 A resolution of VEGF in complex with domain 2 of the Flt-1 receptor. Cell. 1997;91(5):695–704.
36. Park JE, et al. Placenta growth factor. Potentiation of vascular endothelial growth factor bioactivity, in vitro and in vivo, and high affinity binding to Flt-1 but not to Flk-1/KDR. J Biol Chem. 1994;269(41):25646–54.
37. Yang S, et al. Vascular endothelial growth factor-induced genes in human umbilical vein endothelial cells: relative roles of KDR and Flt-1 receptors. Arterioscler Thromb Vasc Biol. 2002;22(11):1797–803.
38. LeCouter J, et al. Angiogenesis-independent endothelial protection of liver: role of VEGFR-1. Science. 2003;299(5608):890–3.
39. Barleon B, et al. Migration of human monocytes in response to vascular endothelial growth factor (VEGF) is mediated via the VEGF receptor flt-1. Blood. 1996;87(8):3336–43.

40. Yao J, et al. Expression of a functional VEGFR-1 in tumor cells is a major determinant of anti-PlGF antibodies efficacy. Proc Natl Acad Sci USA. 2011;108(28):11590–5.
41. Kendall RL, Thomas KA. Inhibition of vascular endothelial cell growth factor activity by an endogenously encoded soluble receptor. Proc Natl Acad Sci USA. 1993;90(22):10705–9.
42. Ambati BK, et al. Corneal avascularity is due to soluble VEGF receptor-1. Nature. 2006;443(7114):993–7.
43. Clark DE, et al. A vascular endothelial growth factor antagonist is produced by the human placenta and released into the maternal circulation. Biol Reprod. 1998;59(6):1540–8.
44. Maynard SE, et al. Excess placental soluble fms-like tyrosine kinase 1 (sFlt1) may contribute to endothelial dysfunction, hypertension, and proteinuria in preeclampsia. J Clin Invest. 2003;111(5):649–58.
45. Sela S, et al. A novel human-specific soluble vascular endothelial growth factor receptor 1: cell-type-specific splicing and implications to vascular endothelial growth factor homeostasis and preeclampsia. Circ Res. 2008;102(12):1566–74.
46. Hiratsuka S, et al. Flt-1 lacking the tyrosine kinase domain is sufficient for normal development and angiogenesis in mice. Proc Natl Acad Sci USA. 1998;95(16):9349–54.
47. Kappas NC, et al. The VEGF receptor Flt-1 spatially modulates Flk-1 signaling and blood vessel branching. J Cell Biol. 2008;181(5):847–58.
48. Ito N, et al. Identification of vascular endothelial growth factor receptor-1 tyrosine phosphorylation sites and binding of SH2 domain-containing molecules. J Biol Chem. 1998;273(36):23410–8.
49. Cunningham SA, et al. Interactions of FLT-1 and KDR with phospholipase C gamma: identification of the phosphotyrosine binding sites. Biochem Biophys Res Commun. 1997;240(3):635–9.
50. Li X, et al. Reevaluation of the role of VEGF-B suggests a restricted role in the revascularization of the ischemic myocardium. Arterioscler Thromb Vasc Biol. 2008;28(9):1614–20.
51. Hagberg CE, et al. Vascular endothelial growth factor B controls endothelial fatty acid uptake. Nature. 2010;464(7290):917–21.
52. Fuh G, et al. Requirements for binding and signaling of the kinase domain receptor for vascular endothelial growth factor. J Biol Chem. 1998;273(18):11197–204.
53. Shinkai A, et al. Mapping of the sites involved in ligand association and dissociation at the extracellular domain of the kinase insert domain-containing receptor for vascular endothelial growth factor. J Biol Chem. 1998;273(47):31283–8.
54. McColl BK, et al. Plasmin activates the lymphangiogenic growth factors VEGF-C and VEGF-D. J Exp Med. 2003;198(6):863–8.
55. Albuquerque RJ, et al. Alternatively spliced vascular endothelial growth factor receptor-2 is an essential endogenous inhibitor of lymphatic vessel growth. Nat Med. 2009;15(9):1023–30.
56. Shalaby F, et al. Failure of blood-Island formation and vasculogenesis in Flk-1-deficient mice. Nature. 1995;376(6535):62–6.
57. Carmeliet P, et al. Abnormal blood vessel development and lethality in embryos lacking a single VEGF allele. Nature. 1996;380(6573):435–9.
58. Ferrara N, et al. Heterozygous embryonic lethality induced by targeted inactivation of the VEGF gene. Nature. 1996;380(6573):439–42.
59. Millauer B, et al. Glioblastoma growth inhibited in vivo by a dominant-negative Flk-1 mutant. Nature. 1994;367(6463):576–9.
60. Oelrichs RB, et al. NYK/FLK-1: a putative receptor protein tyrosine kinase isolated from E10 embryonic neuroepithelium is expressed in endothelial cells of the developing embryo. Oncogene. 1993;8(1):11–8.
61. Quinn TP, et al. Fetal liver kinase 1 is a receptor for vascular endothelial growth factor and is selectively expressed in vascular endothelium. Proc Natl Acad Sci USA. 1993;90(16):7533–7.
62. Plate KH, et al. Up-regulation of vascular endothelial growth factor and its cognate receptors in a rat glioma model of tumor angiogenesis. Cancer Res. 1993;53(23):5822–7.

63. Ruch C, et al. Structure of a VEGF-VEGF receptor complex determined by electron microscopy. Nat Struct Mol Biol. 2007;14(3):249–50.
64. Yang Y, et al. Direct contacts between extracellular membrane-proximal domains are required for VEGF receptor activation and cell signaling. Proc Natl Acad Sci USA. 2010;107(5):1906–11.
65. Matsumoto T, et al. VEGF receptor-2 Y951 signaling and a role for the adapter molecule TSAd in tumor angiogenesis. EMBO J. 2005;24(13):2342–53.
66. Takahashi T, Ueno H, Shibuya M. VEGF activates protein kinase C-dependent, but Ras-independent Raf-MEK-MAP kinase pathway for DNA synthesis in primary endothelial cells. Oncogene. 1999;18(13):2221–30.
67. Solowiej J, et al. Characterizing the effects of the juxtamembrane domain on vascular endothelial growth factor receptor-2 enzymatic activity, autophosphorylation, and inhibition by axitinib. Biochemistry. 2009;48(29):7019–31.
68. Kendall RL, et al. Vascular endothelial growth factor receptor KDR tyrosine kinase activity is increased by autophosphorylation of two activation loop tyrosine residues. J Biol Chem. 1999;274(10):6453–60.
69. Sakurai Y, et al. Essential role of Flk-1 (VEGF receptor 2) tyrosine residue 1173 in vasculogenesis in mice. Proc Natl Acad Sci USA. 2005;102(4):1076–81.
70. Wu LW, et al. VRAP is an adaptor protein that binds KDR, a receptor for vascular endothelial cell growth factor. J Biol Chem. 2000;275(9):6059–62.
71. Holmqvist K, et al. The adaptor protein shb binds to tyrosine 1175 in vascular endothelial growth factor (VEGF) receptor-2 and regulates VEGF-dependent cellular migration. J Biol Chem. 2004;279(21):22267–75.
72. Warner AJ, et al. The Shc-related adaptor protein, Sck, forms a complex with the vascular-endothelial-growth-factor receptor KDR in transfected cells. Biochem J. 2000;347(Pt 2):501–9.
73. Kroll J, Waltenberger J. The vascular endothelial growth factor receptor KDR activates multiple signal transduction pathways in porcine aortic endothelial cells. J Biol Chem. 1997;272(51):32521–7.
74. Meadows KN, Bryant P, Pumiglia K. Vascular endothelial growth factor induction of the angiogenic phenotype requires Ras activation. J Biol Chem. 2001;276(52):49289–98.
75. Zachary I. Signaling mechanisms mediating vascular protective actions of vascular endothelial growth factor. Am J Physiol Cell Physiol. 2001;280(6):C1375–86.
76. Shu X, et al. Sphingosine kinase mediates vascular endothelial growth factor-induced activation of ras and mitogen-activated protein kinases. Mol Cell Biol. 2002;22(22):7758–68.
77. Wellner M, et al. The proliferative effect of vascular endothelial growth factor requires protein kinase C-alpha and protein kinase C-zeta. Arterioscler Thromb Vasc Biol. 1999;19(1):178–85.
78. Xia P, et al. Characterization of vascular endothelial growth factor's effect on the activation of protein kinase C, its isoforms, and endothelial cell growth. J Clin Invest. 1996;98(9):2018–26.
79. Wong C, Jin ZG. Protein kinase C-dependent protein kinase D activation modulates ERK signal pathway and endothelial cell proliferation by vascular endothelial growth factor. J Biol Chem. 2005;280(39):33262–9.
80. Ha CH, et al. Protein kinase D-dependent phosphorylation and nuclear export of histone deacetylase 5 mediates vascular endothelial growth factor-induced gene expression and angiogenesis. J Biol Chem. 2008;283(21):14590–9.
81. Wang S, et al. Control of endothelial cell proliferation and migration by VEGF signaling to histone deacetylase 7. Proc Natl Acad Sci USA. 2008;105(22):7738–43.
82. Evans IM, Britton G, Zachary IC. Vascular endothelial growth factor induces heat shock protein (HSP) 27 serine 82 phosphorylation and endothelial tubulogenesis via protein kinase D and independent of p38 kinase. Cell Signal. 2008;20(7):1375–84.
83. Holmqvist K, et al. The Shb adaptor protein causes Src-dependent cell spreading and activation of focal adhesion kinase in murine brain endothelial cells. Cell Signal. 2003;15(2):171–9.

84. Abu-Ghazaleh R, et al. Src mediates stimulation by vascular endothelial growth factor of the phosphorylation of focal adhesion kinase at tyrosine 861, and migration and anti-apoptosis in endothelial cells. Biochem J. 2001;360(Pt 1):255–64.
85. Parsons JT. Focal adhesion kinase: the first ten years. J Cell Sci. 2003;116(Pt 8):1409–16.
86. Lamalice L, Houle F, Huot J. Phosphorylation of Tyr1214 within VEGFR-2 triggers the recruitment of Nck and activation of Fyn leading to SAPK2/p38 activation and endothelial cell migration in response to VEGF. J Biol Chem. 2006;281(45):34009–20.
87. Jin ZG, et al. Flow shear stress stimulates Gab1 tyrosine phosphorylation to mediate protein kinase B and endothelial nitric-oxide synthase activation in endothelial cells. J Biol Chem. 2005;280(13):12305–9.
88. Laramée M, et al. The scaffolding adapter Gab1 mediates vascular endothelial growth factor signaling and is required for endothelial cell migration and capillary formation. J Biol Chem. 2007;282(11):7758–69.
89. Cain RJ, Ridley AJ. Phosphoinositide 3-kinases in cell migration. Biol Cell. 2009;101(1):13–29.
90. Bhattacharya R, et al. Distinct role of PLCbeta3 in VEGF-mediated directional migration and vascular sprouting. J Cell Sci. 2009;122(Pt 7):1025–34.
91. Semenza GL. HIF-1: using two hands to flip the angiogenic switch. Cancer Metastasis Rev. 2000;19(1–2):59–65.
92. Semenza GL. HIF-1: mediator of physiological and pathophysiological responses to hypoxia. J Appl Physiol (1985). 2000;88(4):1474–80.
93. De Bock K, Georgiadou M, Carmeliet P. Role of endothelial cell metabolism in vessel sprouting. Cell Metab. 2013;18(5):634–47.
94. Eilken HM, Adams RH. Dynamics of endothelial cell behavior in sprouting angiogenesis. Curr Opin Cell Biol. 2010;22(5):617–25.
95. Gerhardt H. VEGF and endothelial guidance in angiogenic sprouting. Organogenesis. 2008;4(4):241–6.
96. Gerhardt H, et al. VEGF guides angiogenic sprouting utilizing endothelial tip cell filopodia. J Cell Biol. 2003;161(6):1163–77.
97. Hellström M, Phng LK, Gerhardt H. VEGF and Notch signaling: the yin and yang of angiogenic sprouting. Cell Adhes Migr. 2007;1(3):133–6.
98. Potente M, Gerhardt H, Carmeliet P. Basic and therapeutic aspects of angiogenesis. Cell. 2011;146(6):873–87.
99. Ruhrberg C, et al. Spatially restricted patterning cues provided by heparin-binding VEGF-A control blood vessel branching morphogenesis. Genes Dev. 2002;16(20):2684–98.
100. Wacker A, Gerhardt H. Endothelial development taking shape. Curr Opin Cell Biol. 2011;23(6):676–85.
101. Okabe K, et al. Neurons limit angiogenesis by titrating VEGF in retina. Cell. 2014;159(3):584–96.
102. Gerber HP, et al. VEGF is required for growth and survival in neonatal mice. Development. 1999;126(6):1149–59.
103. Chung AS, Ferrara N. Developmental and pathological angiogenesis. Annu Rev Cell Dev Biol. 2011;27:563–84.
104. Dawson DW, et al. CD36 mediates the in vitro inhibitory effects of thrombospondin-1 on endothelial cells. J Cell Biol. 1997;138(3):707–17.
105. Saumet A, et al. Type 3 repeat/C-terminal domain of thrombospondin-1 triggers caspase-independent cell death through CD47/alphavbeta3 in promyelocytic leukemia NB4 cells. Blood. 2005;106(2):658–67.
106. Lamy L, et al. Interactions between CD47 and thrombospondin reduce inflammation. J Immunol. 2007;178(9):5930–9.
107. Henkin J, Volpert OV. Therapies using anti-angiogenic peptide mimetics of thrombospondin-1. Expert Opin Ther Targets. 2011;15(12):1369–86.

108. Bergers G, et al. Matrix metalloproteinase-9 triggers the angiogenic switch during carcinogenesis. Nat Cell Biol. 2000;2(10):737–44.
109. Rodriguez-Manzaneque JC, et al. Thrombospondin-1 suppresses spontaneous tumor growth and inhibits activation of matrix metalloproteinase-9 and mobilization of vascular endothelial growth factor. Proceedings of the National Academy of Sciences of the United States of America. 2001;98(22):12485–90.
110. Watnick RS, et al. Thrombospondin-1 repression is mediated via distinct mechanisms in fibroblasts and epithelial cells. Oncogene. 2014;
111. Brown LF, et al. Vascular stroma formation in carcinoma in situ, invasive carcinoma, and metastatic carcinoma of the breast. Clin Cancer Res. 1999;5(5):1041–56.
112. Mueller MM, Fusenig NE. Tumor-stroma interactions directing phenotype and progression of epithelial skin tumor cells. Differentiation. 2002;70(9–10):486–97.
113. Streit M, et al. Overexpression of thrombospondin-1 decreases angiogenesis and inhibits the growth of human cutaneous squamous cell carcinomas. Am J Pathol. 1999;155(2):441–52.
114. Wong SY, Purdie AT, Han P. Thrombospondin and other possible related matrix proteins in malignant and benign breast disease. An immunohistochemical study. Am J Pathol. 1992;140(6):1473–82.
115. Bertin N, et al. Thrombospondin-1 and -2 messenger RNA expression in normal, benign, and neoplastic human breast tissues: correlation with prognostic factors, tumor angiogenesis, and fibroblastic desmoplasia. Cancer Res. 1997;57(3):396–9.
116. Clezardin P, et al. Expression of thrombospondin (TSP1) and its receptors (CD36 and CD51) in normal, hyperplastic, and neoplastic human breast. Cancer Res. 1993;53(6):1421–30.
117. Filleur S, et al. In vivo mechanisms by which tumors producing thrombospondin 1 bypass its inhibitory effects. Genes Dev. 2001;15(11):1373–82.
118. Silverstein RL, Febbraio M. CD36, a scavenger receptor involved in immunity, metabolism, angiogenesis, and behavior. Sci Signal. 2009;2(72):re3.
119. Luscinskas FW, Lawler J. Integrins as dynamic regulators of vascular function. FASEB J. 1994;8(12):929–38.
120. Jiménez B, et al. Signals leading to apoptosis-dependent inhibition of neovascularization by thrombospondin-1. Nat Med. 2000;6(1):41–8.
121. Miao WM, et al. CD36 associates with CD9 and integrins on human blood platelets. Blood. 2001;97(6):1689–96.
122. Thorne RF, et al. The association between CD36 and Lyn protein tyrosine kinase is mediated by lipid. Biochem Biophys Res Commun. 2006;351(1):51–6.
123. Hemler ME. Tetraspanin proteins promote multiple cancer stages. Nat Rev Cancer. 2014;14(1):49–60.
124. van Deventer S, Arp AB, van Spriel AB. Dynamic plasma membrane organization: a complex symphony. Trends Cell Biol. 2021;31(2):119–29.
125. Sun J, et al. Thrombospondin-1 modulates VEGF-A-mediated Akt signaling and capillary survival in the developing retina. Am J Physiol Heart Circ Physiol. 2009;296(5):H1344–51.
126. Daviet L, et al. Thrombospondin induces dimerization of membrane-bound, but not soluble CD36. Thromb Haemost. 1997;78(2):897–901.
127. Githaka JM, et al. Ligand-induced growth and compaction of CD36 nanoclusters enriched in Fyn induces Fyn signaling. J Cell Sci. 2016;129(22):4175–89.
128. Jaqaman K, et al. Cytoskeletal control of CD36 diffusion promotes its receptor and signaling function. Cell. 2011;146(4):593–606.
129. Kazerounian S, et al. Priming of the vascular endothelial growth factor signaling pathway by thrombospondin-1, CD36, and spleen tyrosine kinase. Blood. 2011;117(17):4658–66.
130. Miller YE, et al. Identification of a cell-surface antigen produced by a gene on human chromosome 3 (cen-q22) and not expressed by Rhnull cells. Am J Hum Genet. 1987;41(6):1061–70.

131. Mawby WJ, et al. Isolation and characterization of CD47 glycoprotein: a multispanning membrane protein which is the same as integrin-associated protein (IAP) and the ovarian tumour marker OA3. Biochem J. 1994;304(Pt 2):525–30.
132. Reinhold MI, et al. In vivo expression of alternatively spliced forms of integrin-associated protein (CD47). J Cell Sci. 1995;108(Pt 11):3419–25.
133. Tsai RK, Rodriguez PL, Discher DE. Self inhibition of phagocytosis: the affinity of 'marker of self' CD47 for SIRPalpha dictates potency of inhibition but only at low expression levels. Blood Cells Mol Dis. 2010;45(1):67–74.
134. McCracken MN, Cha AC, Weissman IL. Molecular pathways: activating T cells after cancer cell phagocytosis from blockade of CD47 "Don't Eat Me" signals. Clin Cancer Res. 2015;21(16):3597–601.
135. Horrigan SK. Replication study: the CD47-signal regulatory protein alpha (SIRPa) interaction is a therapeutic target for human solid tumors. elife. 2017:6.
136. Gao L, et al. CD47 deficiency in tumor stroma promotes tumor progression by enhancing angiogenesis. Oncotarget. 2017;8(14):22406–13.
137. Isenberg JS, et al. Gene silencing of CD47 and antibody ligation of thrombospondin-1 enhance ischemic tissue survival in a porcine model: implications for human disease. Ann Surg. 2008;247(5):860–8.
138. Kaur S, et al. Thrombospondin-1 inhibits VEGF receptor-2 signaling by disrupting its association with CD47. J Biol Chem. 2010;285(50):38923–32.
139. Russell S, et al. Combined therapy with thrombospondin-1 type I repeats (3TSR) and chemotherapy induces regression and significantly improves survival in a preclinical model of advanced stage epithelial ovarian cancer. FASEB J. 2015;29(2):576–88.
140. Greenaway J, et al. Thrombospondin-1 inhibits VEGF levels in the ovary directly by binding and internalization via the low density lipoprotein receptor-related protein-1 (LRP-1). J Cell Physiol. 2007;210(3):807–18.
141. Wang F, et al. Aging-associated changes in CD47 arrangement and interaction with thrombospondin-1 on red blood cells visualized by super-resolution imaging. Aging Cell. 2020;19(10):e13224.
142. Olerud J, et al. Improved vascular engraftment and graft function after inhibition of the angiostatic factor thrombospondin-1 in mouse pancreatic islets. Diabetes. 2008;57(7):1870–7.

Receptor Tyrosine Kinase c-MET: A Biomarker and Therapeutic Target in Cancer

Jee Soo Park, Ji Eun Lee, Myung Eun Lee, Yeong Ji Park, and Won Sik Ham

1 Introduction

Receptor tyrosine kinases (RTKs) are a subclass of tyrosine kinases (TKs) that regulate numerous essential cellular processes, including cell growth, motility, differentiation, and metabolism [1]. Although RTKs are crucial for normal physiology, dysregulation of RTK signaling leads to the development of many human diseases, especially in the development and progression of various cancers [2]. With the advent of the genomic era and the adoption of next-generation sequencing in cancer research and routine clinical practice, genomic studies have identified various alterations in genes encoding RTKs such as *EGFR*, *HER2*, *Met*, and others [3].

In particular, overexpression and activation of the RTK c-MET (mesenchymal–epithelial transition factor) and its ligand, hepatocyte growth factor (HGF), have been observed in various human malignancies, including lung, liver, and gastric cancers, as well as many other solid tumors [4–8]. The *MET* proto-oncogene, located on chromosome 7q31, encodes c-MET. It is a cell surface receptor primarily expressed in epithelial and endothelial cells, where it plays critical roles in embryonic development, tissue repair, and wound healing under physiological conditions [9–13]. HGF, also known as scatter factor, is the only known ligand for c-MET and is synthesized primarily by mesenchymal cells, fibroblasts, and smooth muscle cells, and acts through a paracrine mechanism [14–17]. Under normal conditions, the binding of HGF to c-MET induces receptor dimerization and autophosphorylation, triggering a cascade of downstream signaling pathways such as mitogen-activated protein kinase (MAPK), phosphatidylinositol 3-kinase (PI3K)–AKT, SRC, and signal transducer and activator of transcription (STAT) pathways [18–21].

J. S. Park · J. E. Lee · M. E. Lee · Y. J. Park · W. S. Ham (✉)
Department of Urology and Urological Science Institute, Yonsei University College of Medicine, Seoul, South Korea
e-mail: leejy9611@yuhs.ac; lme0228@yuhs.ac; lightruth17@yuhs.ac; uroham@yuhs.ac

A. Sabarwal et al. (eds.), *Receptor Tyrosine Kinases in Cancer*, Cancer Drug Discovery and Development, https://doi.org/10.1007/978-3-031-93894-8_6

These signaling events regulate various cellular processes, including proliferation, survival, migration, invasion, and angiogenesis [22–25], thereby playing a crucial role in normal human biological functions, mediating embryogenesis, organ development, wound healing, and tissue regeneration [26–29].

However, abnormal activation of the HGF/c-MET axis can lead to uncontrolled proliferation and increase invasive/metastatic abilities of cancer cells [30, 31], which promote the development and progression of multiple cancers [32]. Considering the central role of the c-MET pathway in cancer progression, targeting c-MET has emerged as a promising therapeutic strategy, and several c-MET inhibitors have shown encouraging results in clinical trials for the treatment of cancers with c-MET dysregulation. Several clinical trials have evaluated the therapeutic benefits of c-MET targeted therapies involving various agents, such as small-molecule inhibitors and monoclonal antibodies [33].

Furthermore, the HGF/c-MET axis plays a role in immunomodulation, suggesting a potential immunomodulatory function for this pathway [34]. Thus, the potential of combining c-MET inhibitors with immune checkpoint inhibitors, including programmed cell death protein 1 (PD-1) and cytotoxic T-lymphocyte-associated antigen 4 (CTLA-4) inhibitors, may offer a valuable approach to enhancing antitumor immune responses by reducing the tumor cells' capacity to evade immune detection. In addition to small-molecule inhibitors combined with targeted therapy or immune checkpoint inhibitors, antibody-drug conjugates targeting c-MET have been developed and are being evaluated in clinical trials [35].

This study provides a comprehensive overview of the role of c-MET in cancer development and as a therapeutic target in cancer. We aimed to explore the molecular mechanisms underlying c-MET dysregulation, its impact on cancer progression, and the current and emerging therapeutic strategies for targeting c-MET in cancer. As our understanding of c-MET biology continues to evolve, we anticipate that c-MET-targeted therapies will play an increasingly important role in the personalized treatment of cancer.

2 Structure of c-MET

2.1 Domain Architecture

Initially, c-MET is composed of an alpha chain and a beta chain linked by disulfide bonds. The c-MET precursor undergoes glycosylation and proteolytic cleavage between R307 and S308 by furin protease. After modification, the mature c-MET contains a smaller α-subunit (32 kDa) and a β-subunit (120 kDa) [36, 37]. The extracellular α-subunit includes amino acids 1–307 of the semaphorin (SEMA) domain located at the N-terminus. The β-subunit has eight domains, including a SEMA domain, a plexin-semaphorin-integrin (PSI) domain, four consecutive immunoglobulin-plexin-transcription factor (IPT1–4) domains, a transmembrane helix, and an intracellular kinase domain [37, 38]. The SEMA domain is crucial not only for c-MET dimerization but also for HGF ligand binding. The PSI domain, which is located below the SEMA domain, regulates HGF binding affinity. The IPT domains also contribute to the modulation of the interaction between c-MET and HGF [38, 39]. The intracellular domain of c-MET contains phosphorylation sites and comprises juxtamembrane (JM) domain, TK domain, and multifunctional docking site (MFDS). Similar to c-MET, mature HGF is structurally composed of α- and β-subunits following proteolytic cleavage [40]. The α-subunit contains four consecutive kringle (K1–K4) domains and an N-terminal (N) domain, whereas the β-subunit includes a serine protease homology (SPH) domain (Fig. 1a) [37, 40].

2.2 Ligand Binding Site

c-MET, an RTK family member, has two distinct binding sites for its ligand HGF. One is located in the SEMA domain, which interacts with the SPH domain of HGF, and the other is in the IPT3/4 domain, which contacts the K1 and the N-terminal domains of the HGF α-subunit (Fig. 1a) [37]. Some studies have suggested that the SPH domain interacts with the bottom surface of the SEMA domain, exhibiting a lower affinity than the IPT3/4-K1/N interaction [37, 39]. However, the SPH–SEMA interaction is functionally crucial for HGF-induced c-MET activation

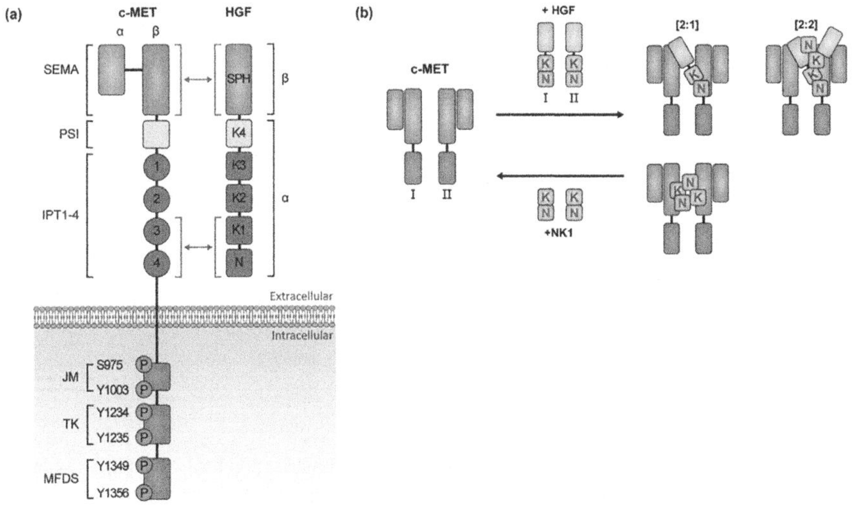

Fig. 1 Structures of c-MET and HGF. (**a**) The c-MET is composed of α- and β-chains. Their intracellular domains interact with a ligand called HGF. The extracellular domains contain several phosphosites, which are involved in activating downstream cascades. (**b**) Active c-MET/HGF complex: HGF binds to c-MET, promoting its dimerization in either [2:1] or [2:2] stoichiometry. NK1, a natural splicing variant of HGF, can also induce active c-MET dimerization. HGF, hepatocyte growth factor; SEMA, semaphorin; TK, tyrosine kinase domain; JM, juxtamembrane domain; MFDS, multifunctional docking site

by stabilizing this connection [37]. Because HGF binds to dimeric c-MET, the TK domain is phosphorylated at Y1234 and Y1235. This leads to autophosphorylation at Y1349 and Y1356 within the MFDS domain, which activates downstream signaling pathways [19, 41].

c-MET can also be activated by NK1, a natural splicing variant of HGF. NK1, the N-terminal domain of HGF, forms an HGF dimer through the head-to-tail dimerization of the NK1 region (Fig. 1b) [37, 42]. A study has demonstrated that mutations in the dimer formation sites of the NK1 fragment (N127, V140, and K144) caused a loss of c-MET activity, whereas mutations in full-length HGF did not affect c-MET activation. Thus, c-MET activation may depend on the structural stability of HGF [42]. Furthermore, HGF has a heparin-binding site in its N domain, which can regulate c-MET activation by enhancing its binding affinity [37].

2.3 Structural Variations and Isoforms

The association between c-MET and HGF leads to structural changes in c-MET. The intracellular TK domain of c-MET was activated, resulting in exposure to the MFDS [38]. The C-terminal domain contains binding sites for effector molecules, such as growth factor receptor-bound protein 2 (GRB2) and STAT3, in the context of the HGF/c-MET signaling pathway (Fig. 2) [19, 38, 43]. The structure of c-MET can also vary depending on the HGF isoform that binds to it. As mentioned earlier, HGF has a natural splicing variant referred to as NK1. Both the c-MET/HGF and c-MET/NK1 complexes induce activation of downstream cascades, although they exhibit

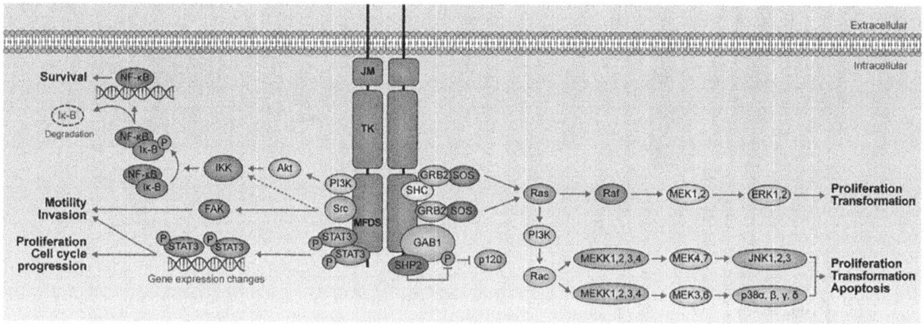

Fig. 2 c-MET signaling pathways. HGF-induced c-MET dimerization leads to autophosphorylation of the multifunctional docking domain (MFDS) and recruits downstream signaling effectors. Major effectors include growth factor receptor-bound protein 2 (GRB2), GRB2-associated adaptor protein 1 (GAB1), phosphatidylinositol 3-kinase (PI3K), son of sevenless (SOS), rat sarcoma oncogene homolog (RAS), signal transducer and activator of transcription 3 (STAT3), SRC, SRC homology protein tyrosine phosphatase 2 (SHP2), SRC homology domain c-terminal adaptor homolog (SHC), focal adhesion kinase, AKT, and mTOR. These molecules regulate processes such as cell proliferation, motility, migration, apoptosis, cell cycle progression, and transformation. In various cancers, c-MET signaling pathways also play an important role in modulating the tumor microenvironment

structural differences. The c-MET dimer interacts with monomeric HGF components [c-MET I-SEMA and HGF-N, -K2, -K3, -SPH; c-MET II-SEMA and HGF-K1] as well as dimeric HGF [c-MET I-SEMA and HGF I-N, -K2, -K3, -SPH; c-MET II-SEMA and HGF I-K1 and HGF II-SPH]. By contrast, NK1 recruits two c-MET molecules symmetrically and forms an active complex only in the dimeric state (Fig. 1b) [37, 44]. Additionally, c-MET produces variants with different structures through alternative mRNA splicing. Splicing site mutations in c-MET accelerate tumorigenesis in several tissues, including the kidney, as c-MET is involved in various mechanisms such as cell proliferation, survival, epithelial–mesenchymal transition, and tumor angiogenesis [19, 36].

3 Mechanism of Activation

3.1 *Ligand Binding and Dimerization*

HGF binding to c-MET may trigger the dimerization of c-MET into an active complex. As mentioned earlier, c-MET has two extracellular HGF-binding sites (the SEMA and IPT domains) and interacts with HGF in either 1:2 or 2:2 stoichiometry. When the two ligands bind to c-MET, HGF II acts as a scaffold for HGF I and promotes c-MET dimerization (Fig. 1b). HGF also regulates the stability of the c-MET/HGF complex through heparin binding within the N domain [37]. However, receptor dimerization is possible even without HGF ligand-binding. c-MET overexpression induces ligand-independent dimerization, although it does not form a stable complex [1]. Hence, the HGF/c-MET signaling pathway becomes hyperactivated, further promoting tumor progression, metastasis, and drug resistance [1, 45]. Some studies have shown that c-MET is upregulated in many patients with renal cell carcinoma (RCC), which may be associated with bone metastasis [41]. Therefore, HGF-induced c-MET dimerization is a crucial starting point for regulating the tumor environment.

3.2 *Autophosphorylation Process*

HGF binding induces c-MET dimerization and autophosphorylation at the catalytic sites Y1234 and Y1235. The activated TK domain leads to the exposure of the C-terminus, resulting in autophosphorylation of the MFDS domain (Y1349 and Y1356) (Fig. 1a) [19]. This facilitates the recruitment of downstream signaling effectors, such as GRB2, SRC, PI3K, STATs, and GRB2-associated-binding protein 1 (GAB1). Subsequently, the downstream c-MET cascade is activated through effector recruitment, thereby modulating cell fate (Fig. 2) [46, 47]. The regulation of this mechanism has been applied in anticancer therapy, with cabozantinib, a TK inhibitor, as an example [36, 43].

3.3 Downstream Signaling Pathways

c-MET-dependent signaling pathways play a crucial role in regulating cell proliferation, differentiation, and protection from apoptosis. Therefore, the HGF-induced c-MET signaling pathway is closely related to cancer cell regulation.

3.3.1 Ras Pathway

After HGF-induced Met dimerization, the GRB2 effector binds directly to the TK domain of c-MET, resulting in the recruitment of guanine nucleotide exchange factors [48]. Son of sevenless (SOS), a guanine nucleotide exchange factor, recruits RasGTP from the cytoplasm to the membrane. Subsequently, Ras-GTP activation enhances the activities of Raf, mitogen-activated protein kinases (MAPKs), extracellular signal-regulated kinase (ERK), Jun N-terminal kinase, and p38 [19]. Activated MAPKs translocate to the nucleus and phosphorylate transcription factors such as Elk1, Ets1, and c-Myc [49]. Additionally, the Ras cascade can be regulated by the scaffold adaptor GAB1 and the tyrosine phosphatase SHP2. The interaction between GAB1 and c-MET C-terminus recruits SHP2, leading to dephosphorylation of the p120-Ras-GAP binding site within GAB1. This disrupts p120, a negative regulator of Ras, thereby activating Ras signaling related to cell transformation, migration, and proliferation (Fig. 2) [19, 48].

3.3.2 PI3K Pathway

PI3K is involved in the HGF/c-MET signaling pathway in both Ras-dependent and Ras-independent manners [47]. The association with PI3K triggers the phosphorylation of various phosphatidylinositol intermediates. AKT kinase, which is phosphorylated through the Ras-dependent PI3K pathway, suppresses apoptosis by inactivating the proapoptotic protein and BCL-2 antagonist BAD (BCL2-associated agonist of cell death) and activating the E3 ubiquitin-protein ligase mouse double minute 2 (MDM2). AKT modulates cell growth and protein synthesis by inactivating glycogen synthase kinase 3β (GSK 3β) and activating mammalian target of rapamycin (mTOR) [19]. Independent of Ras, the PI3K-Akt pathway can activate IκB kinase (IKK) and cause IκB degradation through phosphorylation. The IKK-induced degradation of IκB leads to the translocation of NF-κB to the nucleus and stimulates the transcription of various genes, including mitogenic and antiapoptotic regulators (Fig. 2) [19, 39, 49].

4 HGF/c-MET as a Biomarker in Various Cancers

Although HGF/c-MET pathway has become a key actionable target across various solid tumors, the outcomes of therapies that target this pathway have been inconsistent in different tumor types. Although several clinical trials have shown promising results [48–52], others have failed to demonstrate sufficient clinical benefit from c-MET inhibitors [53–56]. Therefore, identifying diagnostic and prognostic biomarkers and predicting the responses to c-MET inhibitors in either tumor tissue or circulation are urgent needs.

4.1 Expression Patterns in Various Cancers

Numerous studies have established that HGF/c-MET is highly expressed across a wide range of cancers, including non-small cell lung cancer (NSCLC), breast cancer, colorectal cancer (CRC), gastric cancer, head and neck cancer, and kidney cancer [29, 57–61]. Overexpression rates (%) of c-MET and HGF are presented in Table 1 [33, 34, 62–195]. With the largest number of studies on lung cancer, the c-MET overexpression rate was found to be 45%, and the HGF overexpression rate was 50% [34, 62–89, 91–98, 102, 103, 195]. Many studies have investigated the expression of HGF/c-MET on breast and kidney cancers, with 42% and 64% expressions of c-MET and HGF in breast cancer [62, 70, 104–127] and 50% and 85% expressions of c-MET and HGF in kidney cancer [62, 70, 99, 129–139], respectively.

Elevated c-MET levels can lead to its constitutive activation and make cells responsive to even minimal amounts of HGF. Although many studies have not specified the level of c-MET receptor activity or phosphorylation, or compared its expression to that of normal tissue, an autocrine HGF/c-MET loop may drive increased cell proliferation and metastasis [24, 196–199]. In addition to

Table 1 High expression rates of HGF/c-MET in various cancers

Cancer site	c-MET (%)	HGF (%)	References
Lung	45	50	[34, 62–89, 91–98, 102, 103, 195]
Breast	42	64	[62, 70, 104–127]
Kidney	50	85	[62, 70, 99, 129–139]
Colon	65	95	[62, 70, 140–149]
Stomach	56	64	[70, 150–156, 158, 159]
Pancreas	51	53	[70, 90, 100, 160–164]
Head and neck	59	46	[33, 70, 101, 157, 165–177]
Liver	39	44	[70, 178–182]
Cervix	69	–	[70, 183–191]
Ovary	26	42	[62, 192, 193]
Thyroid	84	–	[194]

HGF hepatocyte growth factor

demonstrating constitutive activation of c-MET by an autocrine HGF/c-MET loop, HGF gene expression is upregulated by various cytokines and growth factors, including TNF-α, interleukin (IL)-1, epidermal growth factor (EGF), fibroblast growth factor (FGF), platelet-derived growth factor (PDGF), and prostaglandins, as well as through interactions with other RTKs such as epidermal growth factor receptor (EGFR) [200–202]. Besides tumor tissue levels of HGF being correlated with poor prognosis [203, 204], investigation on the prognostic value of circulating HGF levels has also revealed a negative correlation between HGF levels and patient survival across various cancer types [205]. These positive paracrine and autocrine loops of c-MET activation can, in turn, promote further Met expression [206].

Several studies have explored the potential of the soluble truncated ectodomain of the Met protein (sMet) as a biomarker across various cancer types in addition to HGF and c-MET, examining the correlation between sMet levels and Met protein expression in tissue. Most studies have reported a correlation between circulating Met levels and Met expression in tumor tissue [207–211].

4.2 Diagnostic Implications

Based on the increased distribution of HGF/c-MET in various cancers, expression of members of the HGF/c-MET axis could serve as diagnostic biomarkers.

Plasma HGF level was not sufficiently sensitive to detect early-stage NSCLC, although the sensitivity was increased in patients with lung squamous cell cancer (stages III and IV) [212]. In esophageal squamous cell carcinoma, serum HGF levels were elevated in patients compared with those in the control group [213]. In cervical cancer, serum HGF levels demonstrated the highest diagnostic value compared with other factors in distinguishing cervical squamous cell carcinoma in patients with cervical intraepithelial neoplasia and healthy controls [214]. The plasma HGF level was significantly different between patients with pancreatic cancer and normal controls and in patients with benign pancreatic tumor and those with chronic pancreatitis [215]. Similar findings have shown the diagnostic value of serum HGF levels in hepatocellular carcinoma (HCC) [216–218].

The diagnostic potential of sMet has been explored in metastatic uveal melanoma, where patients with metastatic cases had higher plasma sMet levels than those with nonmetastatic cases and healthy individuals [210]. In another study, the association between plasma sMet and prostate cancer was evaluated [209]. The sMet levels in plasma samples from patients with benign and malignant prostate disease were higher than in those from healthy controls [209]. Furthermore, sMet could differentiate malignant cases from healthy individuals [209]. Additionally, urinary Met levels have shown their potential for differentiating patients with invasive versus noninvasive bladder cancer [219].

4.3 Prognostic Significance

The study of the correlation of c-MET with disease prognosis will help distinguish the tumor types where the HGF/c-MET axis has the most significant impact on tumor behavior. Therefore, being not only a prognostic biomarker, c-MET can be used to identify patients with cancer who could potentially benefit from c-MET inhibitors. The primary goal of examining expression of the c-MET pathway across various cancer types is to identify predictive biomarkers that allow stratification of patients who will most likely benefit from c-MET inhibitors. This biomarker-driven approach, fundamental to precision medicine trials, may greatly enhance the effectiveness of targeted therapies [220].

Several studies have suggested that high c-MET and/or HGF expression levels are negative prognostic biomarkers in different cancers [163, 207, 221–224]. For NSCLC, several studies have reported that a high Met expression level was significantly associated with shorter overall survival (OS) [72]. Furthermore, the c-MET expression level was significantly higher in patients with non-squamous cell carcinoma with higher clinical stages [71]. Nearly half of lung adenocarcinomas exhibit high c-MET and HGF expression [65, 195, 225]. These high expression patterns have been associated with increased tumor growth rates, metastasis, poor prognosis, and radiotherapy resistance [196, 226, 227]. For breast cancer, multiple trials have shown conflicting results. However, recent meta-analyses have reported that patients with breast cancer with c-MET overexpression had poor prognosis, including shorter OS and relapse-free survival [104, 128, 141, 228]. Elevated c-MET/HGF levels in breast carcinoma have been associated with histological grade, poor prognosis, and a high proliferative cell index [123, 229, 230], as well as an increased incidence of metastases [120, 231]. The prognostic value of c-MET overexpression in CRC and head and neck cancer was similar, with shorter OS and disease-free survival (DFS) in patients with high c-MET expression levels [101, 141, 232, 233].

HGF overexpression in tumor tissue was observed in 67.1% of patients with tonsillar squamous cell carcinoma and was associated with shorter OS and progression-free survival (PFS) [174]. In endometrial cancer, patients with tumors positive for both HGF and FGF had a higher risk of recurrence than those negative for both markers [234]. Increased serum HGF levels were associated with a significantly worse prognosis in terms of DFS in patients with breast cancer [235]. Other studies have also demonstrated that elevated serum HGF levels are associated with increased lymph node involvement, a higher incidence of poorly differentiated tumors, more advanced cancer stages, and distant metastases [236–239]. For CRC, overexpression of serum HGF was associated with worse OS and DFS [222]. Furthermore, in patients undergoing CRC resection, elevated serum HGF levels were correlated with tumor size, lymph node metastasis, and distant metastasis [240]. Moreover, higher serum HGF levels were significantly correlated with disease stage and survival in esophageal cancer [213, 241].

The survival analysis of metastatic uveal melanoma indicated that patients with lower serum sMet levels had a longer median survival time than those with high expression [210]. A study of localized and metastatic prostate cancer has reported that soluble urinary sMet was significantly correlated with tumor metastasis [242].

5 Role of c-MET in Tumorigenesis

5.1 Aberrant c-MET Signaling

The c-MET gene, which encodes the c-MET receptor TK, plays an important role in various cell processes, including proliferation, survival, and migration. Under normal physiological conditions, c-MET is activated by its ligand, HGF, which promotes development and regeneration. However, in the case of cancer, c-MET can be abnormally activated through various mechanisms, such as sporadic or germline mutations, amplification, or protein overexpression, which can cause oncogenic activity through various signaling pathways [243]. This dysregulated c-MET signaling can lead to uncontrolled proliferation, migration, invasion, and angiogenesis, which may play an essential role in the initiation, progression, and survival of cancers. c-MET can activate several downstream signaling pathways, including PI3K/AKT, Ras/MAPK, JAK/STAT, SRC, and Wnt/β-catenin, which contribute to tumorigenesis and cancer progression. In addition, c-MET can interact with other receptor TKs to enhance oncogenic potential [19, 243]. Owing to its significant role in cancer biology, c-MET is considered an important target for cancer therapy, and various inhibitors have been developed to block its activities.

5.1.1 c-MET Mutations

Activating point mutations in c-MET occur in sporadic and hereditary papillary renal cancer, HCC, and other cancer types [62]. c-MET mutation sites are mainly located in the kinase domain and the JM domain. In recent years, novel c-MET mutations have been discovered in SEMA and IPT domains [244]. c-MET mutations can increase Met tyrosine phosphorylation and kinase activity, activate the Ras/Raf/MEK/ERK pathway, enhance cell survival, prevent apoptosis, and allow anchorage-independent growth. Recently, c-MET exon 14 splice site mutations have become a research hotspot [245]. These potentially targetable genetic defects, including exon 14 deletions and point mutations, induce exon 14 skipping, which delays c-MET ubiquitination and downregulation, leading to overexpression and subsequent persistent c-MET kinase activation [246]. The aberrant activation of c-MET by mutations leads to the development of various human malignancies and promotes cancer cell metastasis.

5.1.2 c-MET Amplification

Gene amplification refers to the presence of several gene copies that cause oncogene activation or resistance to therapy [247]. c-MET amplification promotes carcinogenesis in various cancers, such as gastric, colon, lung, and kidney cancers, as well as HCC, suggesting it as a potential new therapeutic target [248]. Amplified c-MET phosphorylates and activates Met protein, thereby activating the c-MET pathway and its downstream signaling pathways, which then promote tumor cell proliferation, migration, and invasion. A previous study examined primary CRC, CRC with liver metastases, normal colorectal mucosa, and normal liver tissue and reported that the copy number of the c-MET gene was increased in CRC samples with liver metastases and that the amplification rate of the c-MET gene was five times higher than in primary CRC. Overexpression of c-MET transcripts has also been detected, and its level of expression correlates with the depth of tumor invasion. c-MET gene amplification may promote cancer metastasis [249].

5.1.3 c-MET Overexpression

Studies over the past several decades have shown that c-MET overexpression is observed in lung, kidney, pancreatic, gastric, colon, breast, ovarian, cervical cancers and HCC, head and neck squamous cell carcinoma, and osteosarcoma. It is overexpressed in malignant tumor tissues compared with that in surrounding normal tissues. This overexpression and activation of c-MET occurs because of c-MET gene amplification, mutation, activation of K-RAS, insufficient degradation of c-MET, excessive HGF ligand secretion, hypoxia, inflammation, and other conditions [250]. c-MET overexpression promotes cancer initiation and development by activating the c-MET pathway and downstream signaling pathways, which may contribute to the acquisition of malignant and drug-resistant properties [251]. Previous studies have confirmed that c-MET overexpression is associated with poor prognosis or shorter survival in colorectal, gastric, breast, and pancreatic cancers and HCC and glioblastoma. In conclusion, c-MET overexpression plays an important role in the development and progression of most human cancers, and therapeutic strategies designed to inhibit c-MET overexpression have great clinical potential [250].

5.2 c-MET in Angiogenesis

Angiogenesis and lymphangiogenesis are important processes in cancer progression and metastasis. Vascular endothelial growth factor (VEGF) and the VEGF receptor (VEGFR) family play important roles in both processes. c-MET signaling is a potent inducer of endothelial cell growth and promotes angiogenesis and lymphangiogenesis. In addition, c-MET signaling can induce VEGFA expression and angiogenesis through signaling intermediates such as SHC and thrombospondin 1

[252]. Although c-MET and VEGFR do not physically associate or phosphorylate each other, they synergistically activate each other through common signaling intermediates: ERK/MAPK, AKT, and focal adhesion kinase (FAK) [253].

Another interesting aspect of c-MET in tumors is the regulation of c-MET expression and activation by hypoxia. Hypoxia induces expression of the transcription factor hypoxia-inducible factor 1α (HIF1α), and HIF-dependent Met expression occurs in several types of cancers [254, 255]. Although angiogenesis inhibitors can reduce tumor vascularization and induce tumor hypoxia, they can also promote c-MET-dependent cancer cell proliferation. Thus, combined therapies targeting both angiogenesis and c-MET are needed [243, 256].

5.3 c-MET Signaling in Metastasis

Previous studies have shown that HGF/c-MET signaling induces metastasis to different organs through the activation of multiple signal transduction pathways [257]. The c-MET bidentate docking site plays an essential role in metastasis. Single point mutations in this site can disrupt metastasis [258].

RAS/MAPK and RAC1 signaling are important in tumor growth and the early stages of metastasis. Signaling through the RAP1 and RAC1/CDC42 pathways regulates cadherin and integrin adhesion molecules to influence cell migration. c-MET controls these RAS and RAC1 activities [259, 260]. Lastly, although c-MET and Wnt/β-catenin signaling independently influence metastasis, they functionally cooperate in colon cancer [261].

5.4 Cross Talk Between c-MET and Other Signaling Pathways

Functional cross talk between c-MET and EGFR, ERBB2, or insulin-like growth factor 1 receptor (IGF1R) has been reported in several systems and is associated with cancer progression and therapeutic resistance [262, 263]. EGFR stimulation can lead to c-MET phosphorylation and activation [264]. c-MET amplification can activate ERBB3/PI3K/AKT signaling in cancer cells [265]. Thus, a combination of EGFR and Met inhibitors is required for cancer treatment. In addition, EGFR shares some downstream pathways with c-MET—MAPK (RAS/RAF/MEK/ERK) and PI3K/AKT/mTOR [266].

Cross talk with other developmental signaling pathways, including transforming growth factor beta (TGFβ) or Wnt/β-catenin, influences tumorigenesis and resistance to targeted therapies. Mutations or overexpression of the Wnt/β-catenin pathway can also lead to cancer. c-MET can contribute to the transcriptional activation of Wnt ligands or induce the activation of β-catenin/TCF/LEF target genes [267, 268]. In addition, HGF induces nuclear translocation of β-catenin/TCF and tyrosine phosphorylation of BCL9–2 in cancer cells [269, 270]. Mutational inactivation of the TGFβ receptor II gene results in upregulation of HGF and c-MET activation, leading to the development of epithelial tumors [271]. Altogether, these results show that c-MET cross talk with other signaling molecules can affect tumor development and progression.

6 c-MET as a Cancer Drug Target

The HGF/c-MET signaling pathway plays a central role in regulating various cellular processes involved in tumorigenesis, and dysregulation of this pathway has been implicated in multiple cancer types, making c-MET a highly promising target for therapeutic intervention. Two primary strategies have emerged for targeting c-MET: small-molecule inhibitors and monoclonal antibodies (mAbs) (Fig. 3) [272]. Each approach aims to inhibit key interactions or signaling steps within the HGF/c-MET axis, ultimately disrupting tumor-promoting signals. Small-molecule inhibitors offer direct inhibition of the TK domain, whereas monoclonal antibodies provide a means of interfering with ligand-receptor interactions or receptor dimerization. Each approach holds unique advantages and limitations, and ongoing clinical trials continue to refine these therapies to improve efficacy and safety across various cancer types. The completed or ongoing clinical trials targeting this axis are summarized in Table 2.

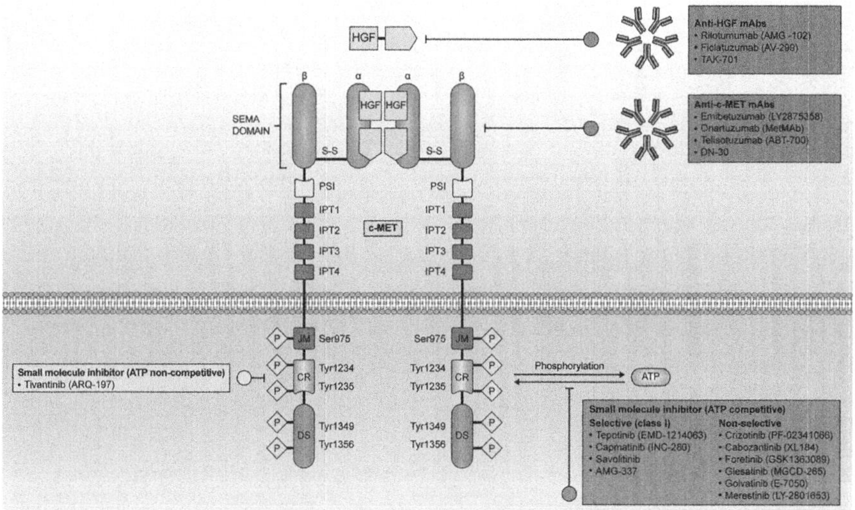

Fig. 3 Therapeutic strategies targeting the hepatocyte growth factor (HGF)/c-MET axis. Small-molecule inhibitors (ATP-competitive [selective/nonselective] inhibitor and adenosine triphosphate noncompetitive inhibitor) and monoclonal antibodies (mAbs)

Table 2 Summary of the drugs developed to target the HGF/c-MET pathway and their use in clinical trials (completed or ongoing)

Type of inhibitor		Name	Synonym(s)	Conditions	Phase	NCT number
Monoclonal Abs	Anti-HGF mAbs	Rilotumumab	AMG-102	NSCLC, OC, PPC, FTC, CRC, EC, GC	Phase 1,2	NCT01233687, NCT01039207, NCT00788957, NCT01443065, NCT01791374, NCT01233687
		Ficlatuzumab	AV-299	HNSCC, NSCLC	Phase 1,2,3	NCT06064877, NCT03422536, NCT01039948
	Anti-c-MET mAbs	Emibetuzumab	LY2875358	NSCLC, HCC, RCC, GC	Phase 1,2	NCT01900652, NCT02082210, NCT01287546, NCT01874938, NCT01602289
		Onartuzumab	MetMAb	NSCLC, GC, CRC, HCC, BrCa	Phase 1,2,3	NCT01456325, NCT01887886, NCT02031744, NCT01662869, NCT01632228, NCT01186991, NCT01418222, NCT00854308, NCT01897038,
		Telisotuzumab	ABT-700	NSCLC, SqCLC	Phase 2,3	NCT04928846, NCT03574753

Small-molecule inhibitor	ATP noncompetitive	Tivantinib		ARQ-197	HCC, NSCLC, CRC	Phase 1,2,3	NCT01755767, NCT02029157, NCT02049060, NCT01892527, NCT01075048
	ATP competitive	Nonselective	Crizotinib	PF-02341066	NSCLC, UM, NHL, MM	Phase 1,2,3	NCT04603807, NCT06140836, NCT05987332, NCT04322890, NCT03297606, NCT01712217, NCT01970865
			Cabozantinib	XL184	Thyroid cancer, HCC, RCC, PC, UC, MEL	Phase 1,2,3	NCT00704730, NCT01908426, NCT01865747, NCT06364631, NCT04820179, NCT05411081, NCT03866382, NCT05111574, NCT03299946, NCT05536141
			Foretinib	GSK1363089	HNC, BrCa, RCC, LC	Phase 1,2	NCT00725764, NCT01147484, NCT01138384, NCT00726323, NCT01068587
			Glesatinib	MGCD-265	NSCLC, Advanced Cancer	Phase 1,2	NCT02544633, NCT01930006, NCT00679133, NCT00697632
			Golvatinib	E-7050	HNC, GC, HCC	Phase 1,2	NCT01332266, NCT01428141, NCT01271504
			Merestinib	LY-2801653	AML, BTC, NHL, CRC, PC, BrCa	Phase 1	NCT03125239, NCT03027284, NCT02745769, NCT02791334
		Selective	Tepotinib	EMD-1214063	NSCLC, HM	Phase 1,2	NCT06083857, NCT06031688, NCT05159245, NCT05782361
			Capmatinib	INC-280	NSCLC, BrCa, MEL	Phase 1,2,3	NCT04427072, NCT04926831, NCT05243641, NCT03484923, NCT05642572
			Savolitinib		NSCLC, RCC, GC, CNS tumors	Phase 1,2,3	NCT05009836, NCT05043090, NCT05261399, NCT05015608, NCT05620628, NCT04322578, NCT04322890, NCT02117167, NCT03598244
				AMG-337	GC	Phase 1,2	NCT02096666

NSCLC non-small-cell lung cancer, *OC* ovarian carcinoma, *PPC* primary peritoneal carcinoma, *FTC* fallopian tube carcinoma, *CRC* colorectal cancer, *EC* esophageal cancer, *GC* gastric cancer, *HNSCC* head and neck squamous cell carcinoma, *HCC* hepatocellular cancer, *RCC* renal cell carcinoma, *BrCa* breast cancer, *SqCLC* squamous cell lung carcinoma, *UM* uveal melanoma, *NHL* non-Hodgkin's lymphoma, *MM* multiple myeloma, *PC* pancreatic cancer, *UC* urothelial carcinoma, *MEL* melanoma, *HNC* head and neck cancer, *LC* lung cancer, *AML* acute myeloid leukemia, *BTC* biliary tract carcinoma, *HM* hematological malignancy

6.1 Small-Molecule Inhibitors

Small-molecule inhibitors targeting c-MET are primarily designed to interfere with the receptor's TK domain, thereby blocking its phosphorylation and downstream signaling events critical for cancer cell survival and growth [273, 274]. These inhibitors can be categorized based on their binding mechanisms either in an ATP-competitive or noncompetitive manner. ATP-competitive inhibitors compete with ATP for the binding site on c-MET, effectively preventing phosphorylation [272, 273, 275]. Within this class, two main subtypes, class I and II inhibitors, can be further classified by their selectivity to c-MET. Class I inhibitors adopt a U-shaped conformation and interact specifically with the ATP-binding site of c-MET, often forming hydrogen bonds with key residues such as Met1160 [273, 275]. This binding specificity confers higher selectivity for c-MET, thus enhancing selectivity and reducing potential side effects. In contrast, class II inhibitors generally adopt an extended conformation exhibiting higher molecular weights and increased lipophilicity, which can impact their pharmacokinetic profiles and lead to multitargeted kinase inhibition [273, 275]. This leads to broader biological effects and, in some cases, undesirable off-target interactions. In addition to ATP-competitive inhibitors, atypical non-ATP-competitive inhibitors such as tivantinib have been developed. These bind in an unconventional manner, targeting alternative regions within the TK domain and modulating c-MET activity through distinct mechanisms [40].

6.1.1 Tivantinib (ARQ197)

ARQ197 is a non-ATP competitive inhibitor, developed by American ArQule Corporation and Japan's Daiichi Sankyo and Kyowa Hakko Kogyo. It binds to inactive c-MET receptors, blocking receptor activation and downstream signaling by inducing a "hydrophobic collapse" through direct interactions with A-loop and P-loop phenylalanines, as well as residues such as Arg1227 and Tyr1230 [276, 277]. Initially, tivantinib may exert its effects by directly inhibiting c-MET. However, recent studies have suggested that its action may instead involve microtubule depolymerization, as it inhibits tubulin polymerization, causing cell cycle arrest in the G2/M phase and promoting apoptosis [278, 279]. Combining tivantinib with erlotinib improves PFS and is well tolerated [280–282]. For patients with advanced solid tumors, tivantinib combined with sorafenib has demonstrated safety and enhanced efficacy, particularly in RCC, HCC, and melanoma with high c-MET levels. This combination enhances sorafenib's antitumor activity while potentially reducing resistance without off-target effects [283].

6.1.2 Crizotinib (PF-02341066)

PF-02341066 is a class II multi-target oral TKI of c-MET, derived from the first-generation c-MET inhibitor PHA-66752 (3-benzyloxy-2-amino). It also targets anaplastic lymphoma kinase (ALK) and ROS proto-oncogene 1 (ROS1) [284]. Crizotinib targets the TK domain of c-MET, inducing conformational changes in specific residues that disrupt ATP binding. In a phase I clinical trial for NSCLC with Met amplification, crizotinib demonstrated enhanced efficacy against c-MET-related cancers [285]. Crizotinib is a Food and Drug Administration (FDA)-approved drug for advanced NSCLC and is indicated for locally advanced or metastatic *ALK* fusion-positive NSCLC and *ROS1*-positive NSCLC [286, 287]. The application of crizotinib in combination with other treatments for other types of cancers is also common. In ovarian cancer, crizotinib and cisplatin induced G2/M cell cycle arrest and apoptosis [288]. Combining c-MET inhibitors and cytotoxic drugs in breast cancer cell lines inhibits growth, and the combination approach of crizotinib and EGFR-TKIs has greater efficacy than monotherapy in breast cancer [289].

6.1.3 Cabozantinib (XL184)

XL184 is a small-molecule inhibitor targeting c-MET, VEGFR1, VEGFR2, and VEGFR3, rearranged during transfection (RET), tyrosine–protein kinase receptor UFO (AXL), KIT, tyrosine receptor kinase B, fms-like TK3 (FLT-3), and TIE-2 [290]. Cabozantinib has demonstrated both preclinical and clinical activity in various tumors, including thyroid cancer [291, 292], RCC [293], prostate cancer [294], and NSCLC [295, 296]. The FDA has approved cabozantinib for two indications: advanced RCC [297] and progressive metastatic medullary thyroid cancer [291, 298]. For urological cancers other than RCC, cabozantinib effectively inhibits growth and invasion in bladder cancer cell lines driven by HGF (5637 and T24) by blocking HGF-MET signaling and reducing MMP1 expression, highlighting its potential in treating muscle-invasive bladder cancer [299]. Currently, cabozantinib is in clinical trials for prostate cancer and has been tested in phase II studies as a single agent or combined treatment in several other cancer types.

6.1.4 Capmatinib (INC280)

INC280, a selective c-MET inhibitor that blocks c-MET phosphorylation and activates downstream molecules in c-MET-dependent tumor cells. This leads to mitochondrial membrane depolarization and enhanced DNA repair [300, 301]. Capmatinib effectively inhibits cancer-associated fibroblast (CAF)-induced cell proliferation in ovarian cancer with HGF overexpression, overcoming CAF-induced drug resistance [300]. For NSCLC, capmatinib in combination with erlotinib could help overcome erlotinib resistance [301].

6.2 Monoclonal Antibodies (mAbs)

mAbs represent the second major approach to targeting the HGF/c-MET pathway. These antibodies inhibit c-MET either by preventing HGF binding to its receptor or by obstructing receptor dimerization, both of which are essential for receptor activation and downstream signaling [272]. Anti-HGF Abs block the ligand itself, preventing it from binding to c-MET and initiating receptor activation, whereas anti-c-MET Abs, such as onartuzumab, directly target the c-MET receptor [273]. By binding to the receptor, these antibodies inhibit ligand binding and can prevent receptor dimerization, effectively blocking c-MET activation [272, 273].

With an advancement in our understanding of the HGF/c-MET signaling pathway, therapeutic agents targeting this pathway have also been increasingly developed. This has led to numerous clinical trials exploring the safety and efficacy of c-MET as a cancer drug target [243]. Small-molecule inhibitors and mAbs have been tested across various phases of clinical development, each demonstrating potential benefits and limitations in different tumor types [302].

6.2.1 Onartuzumab (MetMab)

mAbs have shown promise in preclinical studies, leading to their evaluation in clinical trials [243]. Among anti-c-MET mAbs, onartuzumab (MetMab)—an mAb that targets the c-MET receptor—has been extensively studied [303, 304]. Onartuzumab functions by blocking the HGF a-chain and by forming a complex with the semi-PSI domain of c-MET without exerting subsequent signaling activation or triggering c-MET dimerization [305]. Onartuzumab has been applied for the treatment of patients with NSCLC and breast cancer [306]. Combining onartuzumab in combination with erlotinib (an EGFR inhibitor) significantly improved OS in patients with NSCLC in a phase II trial [307]. However, a subsequent phase III trial has reported decreased OS in patients who received combination therapy [56].

6.2.2 Rilotumumab (AMG102)

Rilotumumab (AMG102) is another type of mAbs—an anti-HGF Ab that interferes with c-MET's activation by HGF [308]. Rilotumumab has shown its efficacy by decreasing c-MET phosphorylation, which stabilizes the progression of certain solid tumors [309, 310].

6.3 Emerging Treatments

Although numerous mAbs and small-molecule inhibitors have been developed to treat HGF/c-MET axis-associated tumors, most of these therapies have struggled to advance into clinical use owing to concerns regarding their efficacy and associated complications. Recently, many clinical trials have sought combination approaches by combining c-MET inhibitors with other drugs, such as targeted therapies and immune checkpoint inhibitors, to avoid resistance induced by the utilization of a single drug.

6.3.1 Immune Checkpoint Inhibitors

Among combination drugs, immune checkpoint inhibitors such as PD-1 and CTLA-4 inhibitors show strong potential for combination therapies with c-MET inhibitors, as the HGF/c-MET axis plays a significant role in immunomodulation [34]. Research indicates that certain leukocyte subsets also express c-MET, and c-MET expression in cancer cells may be linked to the upregulation of immune suppressive molecules like programmed cell death ligand (PD-L1) and indoleamine-2,3-dioxygenase (IDO) [34, 311]. Thus, targeting c-MET could be a pivotal approach for reducing tumor cells' ability to evade antitumor immune responses. A study of 155 resected NSCLC tumor samples has reported that c-MET activation correlated with PD-L1 expression, suggesting that c-MET signaling can directly induce PD-L1 expression in NSCLC cell lines [312]. Furthermore, interferon (IFN)-γ-induced upregulation of PD-L1 can also be modulated through c-MET inhibition [313]. This evidence suggests that combining PD-L1 inhibitors with c-MET inhibitors could be effective for treating various cancers. Importantly, c-MET activation promotes PD-L1 expression via a distinct pathway that is independent of IFN-γ and the JAK–STAT pathway [314]. IDO, a key enzyme produced by tumor cells, also contributes to an immunosuppressive microenvironment, making IDO inhibitors a valuable therapeutic option for treating solid tumors with hot immune environments. Through the kynurenine pathway, IDO inhibitors directly impact T-cells and natural killer (NK) cells and deplete tryptophan, which is essential for T-cell function [315]. Inhibition of the c-MET receptor reduces IDO expression in a concentration-dependent manner, suggesting that a combined approach using c-MET inhibitors with IDO inhibitors may be promising for various tumors.

6.4 Antibody-Drug Conjugates

Antibody-drug conjugates, also known as smart chemotherapy, represent a promising new class of cancer therapeutics that leverage the target specificity of mAbs to deliver highly potent small molecules directly to tumors, minimizing damage to normal cells [35]. Currently, nine c-MET-targeting antibody-drug conjugates are undergoing various phases of clinical trials, and eight preclinical studies are focused on treating a range of solid tumors [35]. Telisotuzumab vedotin (Teliso-V; formerly known as ABBV-399) is a first-in-class c-MET-targeting antibody-drug conjugate that combines a humanized anti-c-MET mAb, telisotuzumab (formerly ABT-700), with the potent microtubule inhibitor monomethyl auristatin E (MMAE). This

MMAE is a synthetic derivative of dolastatin 10 that impedes cancer cell proliferation by inhibiting tubulin polymerization. The two are linked via a protease-cleavable valine-citrulline (vc) dipeptide linker, forming ABT-700-vcMMAE with an average drug-to-antibody ratio of 3:1 [316, 317]. Teliso-V selectively binds to c-MET-overexpressing cells, where it is internalized. Upon internalization, the linker undergoes proteolytic cleavage, releasing MMAE inside the cell. MMAE then binds to tubulin, blocking its polymerization, leading to G2/M cell cycle arrest and apoptosis [317, 318]. This targeted delivery mechanism allows Teliso-V to act independently of c-MET signaling [319]. Upon binding to c-MET, Teliso-V can inhibit both HGF-dependent and independent c-MET signaling. Although no c-MET-targeting antibody-drug conjugates have yet received global marketing approval, several antibody-drug conjugates are currently being researched and are at various clinical stages for treating c-MET-positive solid tumors. Several of these antibody-drug conjugates have shown promising antitumor activity in both clinical and preclinical settings, along with acceptable safety and tolerability profiles. Overall, anti-c-MET antibody-drug conjugates represent a promising strategy to combat c-MET-overexpressing solid tumors by selectively delivering highly potent small-molecule payloads directly into cancer cells.

7 Conclusions

RTKs, including c-MET, play a pivotal role in cellular communication and function. Understanding their mechanisms and roles in normal and pathological processes opens up opportunities for designing targeted therapies in cancer, in which signaling pathways are often dysregulated. Ongoing research and development of RTK inhibitors may improve outcomes in various diseases driven by abnormal RTK activity. The structure and function of c-MET are well-characterized, but certain aspects, such as its interactions with other tyrosine kinases and its functional integration, may still require further elucidation. The HGF/c-MET axis triggers signaling cascades critical to tumorigenesis, and extensive research on these pathways could elucidate carcinoma mechanisms while identifying promising targets for novel cancer therapies. Considering the complexity of the HGF/c-MET axis, further investigation is essential to elucidate how inhibiting c-MET activation impacts downstream pathways.

Over the past decade, novel selective c-MET inhibitors, encompassing small-molecule inhibitors and mAbs, have been developed. Although these targeted therapies have demonstrated preliminary clinical efficacy with manageable toxicity profiles, the majority of phase III trial results have shown unsatisfactory results. Currently, only crizotinib and cabozantinib are approved as c-MET inhibitors, and they have expanded treatment options for specific tumor types [272].

Recent clinical trials have highlighted the rapid acquisition of drug resistance with monotherapy, underscoring the necessity of exploring combination treatments involving c-MET inhibitors and other therapeutic agents, such as EGFR-TKIs. One study has demonstrated the effectiveness of simultaneously targeting two RTKs—c-MET and EGFR—as an approach to maximize the inhibition of cytoskeletal functions, cell viability, and cellular signaling, as well as to achieve complete regression of in vivo xenografts in cases of erlotinib resistance [320].

This approach may enhance treatment efficacy and improve outcomes for patients with c-MET-dependent tumors. Further research is imperative to elucidate the complex mechanisms underlying c-MET pathway activation and resistance, both primary and acquired. This understanding will be vital for refining targeted therapies and developing more effective treatment strategies. Emerging strategies, including the combination of c-MET inhibitors with immune checkpoint inhibitors or the development of antibody-drug conjugates, show promise in overcoming resistance and improving therapeutic outcomes.

However, several challenges persist in the field of c-MET-targeted therapies. These include significant variation in the reported prevalence of HGF/c-MET alterations across studies, the need for standardization of measurement protocols and scoring systems, and the identification of cancer types. Furthermore, in many c-MET-overexpressing cancer cells, this receptor is not the sole driver of carcinoma, as it often interacts with other TK receptors. Therefore, nonspecific interactions with other TK receptors result in various adverse side effects. Addressing these challenges will be crucial for improving patient selection and optimizing the balance between efficacy and toxicity.

In conclusion, although the HGF/c-MET axis remains a promising target in cancer therapy, overcoming the current limitations and challenges will be crucial for realizing its full therapeutic potential. Future research should focus on optimizing combination strategies, improving patient selection, and developing novel approaches to address the complexities of c-MET-dependent malignancies.

Acknowledgments This research was funded by the National Research Foundation of Korea (NRF) grants from the Korean government (MSIT), grant numbers 2019R1A2C1002863 and 2022R1A2C2003831. MID (Medical Illustration & Design), as a member of the Medical Research Support Services of Yonsei University College of Medicine, provided excellent support with medical illustration.

Conflicts of Interest none

References

1. Du Z, Lovly CM. Mechanisms of receptor tyrosine kinase activation in cancer. Mol Cancer. 2018;17:1–13.
2. Krause DS, Van Etten RA. Tyrosine kinases as targets for cancer therapy. N Engl J Med. 2005;353(2):172–87.
3. Vogelstein B, Papadopoulos N, Velculescu VE, Zhou S, Diaz LA Jr, Kinzler KW. Cancer genome landscapes. Science. 2013;339(6127):1546–58.
4. Drilon A, Cappuzzo F, Ou S-HI, Camidge DR. Targeting MET in lung cancer: will expectations finally be MET? J Thorac Oncol. 2017;12(1):15–26.

5. Goyal L, Muzumdar MD, Zhu AX. Targeting the HGF/c-MET pathway in hepatocellular carcinoma. Clin Cancer Res. 2013;19(9):2310–8.
6. Hack SP, Bruey J-M, Koeppen H. HGF/MET-directed therapeutics in gastroesophageal cancer: a review of clinical and biomarker development. Oncotarget. 2014;5(10):2866.
7. Peruzzi B, Bottaro DP. Targeting the c-met signaling pathway in cancer. Clin Cancer Res. 2006;12(12):3657–60.
8. Birchmeier C, Birchmeier W, Gherardi E, Vande Woude GF. Met, metastasis, motility and more. Nat Rev Mol Cell Biol. 2003;4(12):915–25.
9. Brand-Saberi B, Müller T, Wilting J, Christ B, Birchmeier C. Scatter factor/hepatocyte growth factor (SF/HGF) induces emigration of myogenic cells at interlimb levelin vivo. Dev Biol. 1996;179(1):303–8.
10. Heymann S, Koudrova M, Arnold H-H, Köster M, Braun T. Regulation and function of SF/HGF during migration of limb muscle precursor cells in chicken. Dev Biol. 1996;180(2):566–78.
11. Bladt F, Riethmacher D, Isenmann S, Aguzzi A, Birchmeier C. Essential role for the c-met receptor in the migration of myogenic precursor cells into the limb bud. Nature. 1995;376(6543):768–71.
12. Borowiak M, Garratt AN, Wüstefeld T, Strehle M, Trautwein C, Birchmeier C. Met provides essential signals for liver regeneration. Proc Natl Acad Sci. 2004;101(29):10608–13.
13. Huh C-G, Factor VM, Sánchez A, Uchida K, Conner EA, Thorgeirsson SS. Hepatocyte growth factor/c-met signaling pathway is required for efficient liver regeneration and repair. Proc Natl Acad Sci. 2004;101(13):4477–82.
14. Weidner KM, Arakaki N, Hartmann G, Vandekerckhove J, Weingart S, Rieder H, Fonatsch C, Tsubouchi H, Hishida T, Daikuhara Y. Evidence for the identity of human scatter factor and human hepatocyte growth factor. Proc Natl Acad Sci. 1991;88(16):7001–5.
15. Nakamura T, Nishizawa T, Hagiya M, Seki T, Shimonishi M, Sugimura A, Tashiro K, Shimizu S. Molecular cloning and expression of human hepatocyte growth factor. Nature. 1989;342(6248):440–3.
16. Stoker M, Gherardi E, Perryman M, Gray J. Scatter factor is a fibroblast-derived modulator of epithelial cell mobility. Nature. 1987;327(6119):239–42.
17. Rucki AA, Xiao Q, Muth S, Chen J, Che X, Kleponis J, Sharma R, Anders RA, Jaffee EM, Zheng L. Dual inhibition of Hedgehog and c-met pathways for pancreatic cancer treatment. Mol Cancer Ther. 2017;16(11):2399–409.
18. Organ SL, Tong J, Taylor P, St-Germain JR, Navab R, Moran MF, Tsao MS. Quantitative phospho-proteomic profiling of hepatocyte growth factor (HGF)-MET signaling in colorectal cancer. J Proteome Res. 2011;10(7):3200–11.
19. Trusolino L, Bertotti A, Comoglio PM. MET signalling: principles and functions in development, organ regeneration and cancer. Nat Rev Mol Cell Biol. 2010;11(12):834–48.
20. Seiden-Long I, Navab R, Shih W, Li M, Chow J, Zhu CQ, Radulovich N, Saucier C, Tsao MS. Gab1 but not Grb2 mediates tumor progression in Met overexpressing colorectal cancer cells. Carcinogenesis. 2008;29(3):647–55.
21. Peschard P, Park M. From Tpr-Met to Met, tumorigenesis and tubes. Oncogene. 2007;26(9):1276–85.
22. Sierra JR, Corso S, Caione L, Cepero V, Conrotto P, Cignetti A, Piacibello W, Kumanogoh A, Kikutani H, Comoglio PM. Tumor angiogenesis and progression are enhanced by Sema4D produced by tumor-associated macrophages. J Exp Med. 2008;205(7):1673–85.
23. Conrotto P, Valdembri D, Corso S, Serini G, Tamagnone L, Comoglio PM, Bussolino F, Giordano S. Sema4D induces angiogenesis through Met recruitment by Plexin B1. Blood. 2005;105(11):4321–9.
24. Yi S, Tsao MS. Activation of hepatocyte growth factor-met autocrine loop enhances tumorigenicity in a human lung adenocarcinoma cell line. Neoplasia. 2000;2(3):226–34.
25. Silvagno F, Follenzi A, Arese M, Prat M, Giraudo E, Gaudino G, Camussi G, Comoglio PM, Bussolino F. In vivo activation of met tyrosine kinase by heterodimeric hepatocyte growth factor molecule promotes angiogenesis. Arterioscler Thromb Vasc Biol. 1995;15(11):1857–65.

26. Zhu L, Xiong X, Kim Y, Okada N, Lu F, Zhang H, Sun H. Acid sphingomyelinase is required for cell surface presentation of Met receptor tyrosine kinase in cancer cells. J Cell Sci. 2016;129(22):4238–51.
27. Corso S, Comoglio PM, Giordano S. Cancer therapy: can the challenge be MET? Trends Mol Med. 2005;11(6):284–92.
28. Liu WT, Jing YY, Gf Y, Chen H, Han Zp YDD, Fan QM, Ye F, Li R, Gao L. Hepatic stellate cell promoted hepatoma cell invasion via the HGF/c-Met signaling pathway regulated by p53. Cell Cycle. 2016;15(7):886–94.
29. Zhang Y, Du Z, Zhang M. Biomarker development in MET-targeted therapy. Oncotarget. 2016;7:37370–89.
30. Benvenuti S, Comoglio PM. The MET receptor tyrosine kinase in invasion and metastasis. J Cell Physiol. 2007;213(2):316–25.
31. Danilkovitch-Miagkova A, Zbar B. Dysregulation of Met receptor tyrosine kinase activity in invasive tumors. J Clin Invest. 2002;109(7):863–7.
32. Tovar EA, Graveel CR. MET in human cancer: germline and somatic mutations. Ann Trans Med. 2017;5(10):205.
33. Anitha K, Dua K, Chellappan DK, Gupta G, Singh SK, Lakshmi SM, Bhatt S. HGF/c-MET: a potential target for the treatment of various cancers. Curr Enzym Inhib. 2023;19(2):71–80.
34. Titmarsh HF, O'connor R, Dhaliwal K, Akram AR. The emerging role of the c-MET-HGF axis in non-small cell lung cancer tumor immunology and immunotherapy. Front Oncol. 2020;10:54.
35. Mer AH, Mirzaei Y, Misamogooe F, Bagheri N, Bazyari A, Keshtkaran Z, Meyfour A, Shahedi A, Amirkhani Z, Jafari A. Progress of antibody–drug conjugates (ADCs) targeting c-met in cancer therapy; insights from clinical and preclinical studies. Drug Deliv Transl Res. 2024;14:1–26.
36. Raj S, Kesari KK, Kumar A, Rathi B, Sharma A, Gupta PK, Jha SK, Jha NK, Slama P, Roychoudhury S, et al. Molecular mechanism(s) of regulation(s) of c-MET/HGF signaling in head and neck cancer. Mol Cancer. 2022;21(1):31.
37. Uchikawa E, Chen Z, Xiao GY, Zhang X, Bai XC. Structural basis of the activation of c-MET receptor. Nat Commun. 2021;12(1):4074.
38. Zhang Y, Xia M, Jin K, Wang S, Wei H, Fan C, Wu Y, Li X, Li X, Li G, et al. Function of the c-met receptor tyrosine kinase in carcinogenesis and associated therapeutic opportunities. Mol Cancer. 2018;17(1):45.
39. Cecchi F, Rabe DC, Bottaro DP. Targeting the HGF/met signaling pathway in cancer therapy. Expert Opin Ther Targets. 2012;16(6):553–72.
40. Parikh PK, Ghate MD. Recent advances in the discovery of small molecule c-met kinase inhibitors. Eur J Med Chem. 2018;143:1103–38.
41. Silva Paiva R, Gomes I, Casimiro S, Fernandes I, Costa L. c-Met expression in renal cell carcinoma with bone metastases. J Bone Oncol. 2020;25:100315.
42. Tahira Y, Sakai K, Sato H, Imamura R, Matsumoto K. Dimer interface in natural variant NK1 Is dispensable for HGF-dependent met receptor activation. Int J Mol Sci. 2021;22(17):9240.
43. Fu J, Su X, Li Z, Deng L, Liu X, Feng X, Peng J. HGF/c-MET pathway in cancer: from molecular characterization to clinical evidence. Oncogene. 2021;40(28):4625–51.
44. de Nola G, Leclercq B, Mougel A, Taront S, Simonneau C, Forneris F, Adriaenssens E, Drobecq H, Iamele L, Dubuquoy L. Dimerization of kringle 1 domain from hepatocyte growth factor/scatter factor provides a potent MET receptor agonist. Life Sci Alliance. 2022;5(12)
45. Zhang Z, Li D, Yun H, Tong J, Liu W, Chai K, Zeng T, Gao Z, Xie Y. Opportunities and challenges of targeting c-met in the treatment of digestive tumors. Front Oncol. 2022;12:923260.
46. Shao Z, Pan H, Tu S, Zhang J, Yan S, Shao A. HGF/c-met axis: the advanced development in digestive system cancer. Front Cell Dev Biol. 2020;8:801.
47. Hass R, Jennek S, Yang Y, Friedrich K. C-met expression and activity in urogenital cancers – novel aspects of signal transduction and medical implications. Cell Commun Signal. 2017;15(1):10.

48. Bradley CA, Salto-Tellez M, Laurent-Puig P, Bardelli A, Rolfo C, Tabernero J, Khawaja HA, Lawler M, Johnston PG, Van Schaeybroeck S, et al. Targeting c-MET in gastrointestinal tumours: rationale, opportunities and challenges. Nat Rev Clin Oncol. 2017;14(9):562–76.
49. Organ SL, Tsao MS. An overview of the c-MET signaling pathway. Therap Adv Med Oncol. 2011;3(1_suppl):S7–S19.
50. Zhu M, Tang R, Doshi S, Oliner K, Dubey S, Jiang Y, Donehower R, Iveson T, Loh E, Zhang Y. Exposure-response analysis of rilotumumab in gastric cancer: the role of tumour MET expression. Br J Cancer. 2015;112(3):429–37.
51. Santoro A, Rimassa L, Borbath I, Daniele B, Salvagni S, Van Laethem JL, Van Vlierberghe H, Trojan J, Kolligs FT, Weiss A, et al. Tivantinib for second-line treatment of advanced hepatocellular carcinoma: a randomised, placebo-controlled phase 2 study. Lancet Oncol. 2013;14(1):55–63.
52. Iveson T, Donehower RC, Davidenko I, Tjulandin S, Deptala A, Harrison M, Nirni S, Lakshmaiah K, Thomas A, Jiang Y, et al. Rilotumumab in combination with epirubicin, cisplatin, and capecitabine as first-line treatment for gastric or oesophagogastric junction adenocarcinoma: an open-label, dose de-escalation phase 1b study and a double-blind, randomised phase 2 study. Lancet Oncol. 2014;15(9):1007–18.
53. Wakelee H, Zvirbule Z, De Braud F, Kingsley CD, Mekhail T, Lowe T, Schutte W, Lena H, Lawler W, Braiteh F, et al. Efficacy and safety of onartuzumab in combination with first-line bevacizumab- or pemetrexed-based chemotherapy regimens in advanced non-squamous non-small-cell lung cancer. Clin Lung Cancer. 2017;18(1):50–9.
54. Catenacci DVT, Tebbutt NC, Davidenko I, Murad AM, Al-Batran SE, Ilson DH, Tjulandin S, Gotovkin E, Karaszewska B, Bondarenko I, et al. Rilotumumab plus epirubicin, cisplatin, and capecitabine as first-line therapy in advanced MET-positive gastric or gastro-oesophageal junction cancer (RILOMET-1): a randomised, double-blind, placebo-controlled, phase 3 trial. Lancet Oncol. 2017;18(11):1467–82.
55. Rimassa L, Assenat E, Peck-Radosavljevic M, Zagonel V, Pracht M, Caremoli ER, Mathurin P, Harris WP, Bolondi L, Reig M, et al. Second-line tivantinib (ARQ 197) vs placebo in patients (Pts) with MET-high hepatocellular carcinoma (HCC): results of the METIV-HCC phase III trial. J Clin Oncol. 2017;35:4000.
56. Spigel DR, Edelman MJ, O'Byrne K, Paz-Ares L, Mocci S, Phan S, Shames DS, Smith D, Yu W, Paton VE, et al. Results from the phase III randomized trial of onartuzumab plus erlotinib versus erlotinib in previously treated stage IIIB or IV non-small-cell lung cancer: METLung. J Clin Oncol. 2017;35(4):412–20.
57. Finocchiaro G, Toschi L, Gianoncelli L, Baretti M, Santoro A. Prognostic and predictive value of MET deregulation in non-small cell lung cancer. Ann Transl Med. 2015;3(6):83.
58. Gambella M, Palumbo A, Rocci A. MET/HGF pathway in multiple myeloma: from diagnosis to targeted therapy? Expert Rev Mol Diagn. 2015;15(7):881–93.
59. Choi J, Lee HE, Lee HS, Han N, Kim MA, Kim WH. Evaluation of intratumoral and intertumoral heterogeneity of MET protein expression in gastric cancer. Appl Immunohistochem Mol Morphol. 2018;26(7):445–53.
60. Kim IH, Lee IH, Lee JE, Hong SH, Kim T-J, Lee KY, Kim YK, Kim SJ, Sung SW, Park JK. Prognostic impact of multiple clinicopathologic risk factors and c-MET overexpression in patients who have undergone resection of stage IB non–small-cell lung cancer. Clin Lung Cancer. 2016;17(5):e31–43.
61. Puvanenthiran S, Essapen S, Haagsma B, Bagwan I, Green M, Khelwatty SA, Seddon A, Modjtahedi H. Co-expression and prognostic significance of the HER family members, EGFRvIII, c-MET, CD44 in patients with ovarian cancer. Oncotarget. 2018;9(28):19662.
62. Ma PC, Tretiakova MS, MacKinnon AC, Ramnath N, Johnson C, Dietrich S, Seiwert T, Christensen JG, Jagadeeswaran R, Krausz T. Expression and mutational analysis of MET in human solid cancers. Genes Chromosom Cancer. 2008;47(12):1025–37.
63. Liu C, Tsao MS. In vitro and in vivo expressions of transforming growth factor-alpha and tyrosine kinase receptors in human non-small-cell lung carcinomas. Am J Pathol. 1993;142(4):1155.

64. Olivero M, Rizzo M, Madeddu R, Casadio C, Pennacchietti S, Nicotra M, Prat M, Maggi G, Arena N, Natali PG. Overexpression and activation of hepatocyte growth factor/scatter factor in human non-small-cell lung carcinomas. Br J Cancer. 1996;74(12):1862–8.
65. Takanami I, Tanana F, Hashizume T, Kikuchi K, Yamamoto Y, Yamamoto T, Kodaira S. Hepatocyte growth factor and c-met/hepatocyte growth factor receptor in pulmonary adenocarcinomas: an evaluation of their expression as prognostic markers. Oncology. 1996;53(5):392–7.
66. Tsao MS, Liu N, Chen JR, Pappas J, Ho J, To C, Viallet J, Park M, Zhu H. Differential expression of met/hepatocyte growth factor receptor in subtypes of non-small cell lung cancers. Lung Cancer. 1998;20(1):1–16.
67. Tokunou M, Niki T, Eguchi K, Iba S, Tsuda H, Yamada T, Matsuno Y, Kondo H, Saitoh Y, Imamura H. c-MET expression in myofibroblasts: role in autocrine activation and prognostic significance in lung adenocarcinoma. Am J Pathol. 2001;158(4):1451–63.
68. Ichimura E, Maeshima A, Nakajima T, Nakamura T. Expression of c-met/HGF receptor in human non-small cell lung carcinomas in vitro and in vivo and its prognostic significance. Jpn J Cancer Res. 1996;87(10):1063–9.
69. Hosoda H, Izumi H, Tukada Y, Takagiwa J, Chiaki T, Yano M, Arai H. Plasma hepatocyte growth factor elevation may be associated with early metastatic disease in primary lung cancer patients. Ann Thorac Cardiovasc Surg. 2012;18(1):1–7.
70. Moosavi F, Giovannetti E, Saso L, Firuzi O. HGF/MET pathway aberrations as diagnostic, prognostic, and predictive biomarkers in human cancers. Crit Rev Clin Lab Sci. 2019;56(8):533–66.
71. Pyo JS, Kang G, Cho WJ, Choi SB. Clinicopathological significance and concordance analysis of c-MET immunohistochemistry in non-small cell lung cancers: a meta-analysis. Pathol Res Pract. 2016;212(8):710–6.
72. Guo B, Cen H, Tan X, Liu W, Ke Q. Prognostic value of MET gene copy number and protein expression in patients with surgically resected non-small cell lung cancer: a meta-analysis of published literatures. PLoS One. 2014;9(6):e99399.
73. Feng Y, Minca EC, Lanigan C, Liu A, Zhang W, Yin L, Pennell NA, Farver C, Tubbs R, Ma PC. High MET receptor expression but not gene amplification in ALK 2p23 rearrangement positive non–small-cell lung cancer. J Thorac Oncol. 2014;9(5):646–53.
74. Huang L, An S-J, Chen Z-H, Su J, Yan H-H, Wu Y-L. MET expression plays differing roles in non–small-cell lung cancer patients with or without EGFR mutation. J Thorac Oncol. 2014;9(5):725–8.
75. Jin Y, Sun P-L, Kim H, Seo AN, Jheon S, Lee C-T, Chung J-H. MET gene copy number gain is an independent poor prognostic marker in Korean stage I lung adenocarcinomas. Ann Surg Oncol. 2014;21:621–8.
76. Kim MY, Koh J, Kim S, Go H, Jeon YK, Chung DH. Clinicopathological analysis of PD-L1 and PD-L2 expression in pulmonary squamous cell carcinoma: comparison with tumor-infiltrating T cells and the status of oncogenic drivers. Lung Cancer. 2015;88(1):24–33.
77. Murray S, Karavasilis V, Bobos M, Razis E, Papadopoulos S, Christodoulou C, Kosmidis P, Fountzilas G. Molecular predictors of response to tyrosine kinase inhibitors in patients with non-small-cell lung cancer. J Exp Clin Cancer Res. 2012;31:1–10.
78. Noro R, Seike M, Zou F, Soeno C, Matsuda K, Sugano T, Nishijima N, Matsumoto M, Kitamura K, Kosaihira S. MET FISH-positive status predicts short progression-free survival and overall survival after gefitinib treatment in lung adenocarcinoma with EGFR mutation. BMC Cancer. 2015;15:1–8.
79. Schildhaus HU, Schultheis AM, Rüschoff J, Binot E, Merkelbach-Bruse S, Fassunke J, Schulte W, Ko YD, Schlesinger A, Bos M. MET amplification status in therapy-naive adeno- and squamous cell carcinomas of the lung. Clin Cancer Res. 2015;21(4):907–15.
80. Tsuta K, Kozu Y, Mimae T, Yoshida A, Kohno T, Sekine I, Tamura T, Asamura H, Furuta K, Tsuda H. C-MET/phospho-MET protein expression and MET gene copy number in non-small cell lung carcinomas. J Thorac Oncol. 2012;7(2):331–9.

81. Wang X, Song N, Zhang Y, Cai Y, Liu Y, Qu X, Li Z, Li D, Hou K, Kang J. Coexpression of c-Met and Notch-1 correlates with poor prognosis in resected non-small-cell lung cancer. Tumour Biol. 2015;36:7053–9.
82. Watermann I, Schmitt B, Stellmacher F, Müller J, Gaber R, Kugler C, Reinmuth N, Huber R, Thomas M, Zabel P. Improved diagnostics targeting c-MET in non-small cell lung cancer: expression, amplification and activation? Diagn Pathol. 2015;10:1–12.
83. Xia N, An J, Qq J, Li M, Tan J, Cp H. Analysis of EGFR, EML4-ALK, KRAS, and c-MET mutations in Chinese lung adenocarcinoma patients. Exp Lung Res. 2013;39(8):328–35.
84. Yeung SF, Tong JH, Law PP, Chung LY, Lung RW, Tong CY, Chow C, Chan AW, Wan IY, Mok TS. Profiling of oncogenic driver events in lung adenocarcinoma revealed MET mutation as independent prognostic factor. J Thorac Oncol. 2015;10(9):1292–300.
85. Sun W, Song L, Ai T, Zhang Y, Gao Y, Cui J. Prognostic value of MET, cyclin D1 and MET gene copy number in non-small cell lung cancer. J Biomed Res. 2013;27(3):220.
86. Sun W, Ai T, Gao Y, Zhang Y, Cui J, Song L. Expression and prognostic relevance of MET and phospho-BAD in non-small cell lung cancer. Onco Targets Ther. 2013:1315–23.
87. Dziadziuszko R, Wynes MW, Singh S, Asuncion BR, Ranger-Moore J, Konopa K, Rzyman W, Szostakiewicz B, Jassem J, Hirsch FR. Correlation between MET gene copy number by silver in situ hybridization and protein expression by immunohistochemistry in non-small cell lung cancer. J Thorac Oncol. 2012;7(2):340–7.
88. Tachibana K, Minami Y, Shiba-Ishii A, Kano J, Nakazato Y, Sato Y, Goya T, Noguchi M. Abnormality of the hepatocyte growth factor/MET pathway in pulmonary adenocarcinogenesis. Lung Cancer. 2012;75(2):181–8.
89. Park S, Choi YL, Sung CO, An J, Seo J, Ahn MJ, Ahn JS, Park K, Shin YK, Erkin OC. High MET copy number and MET overexpression: poor outcome in non-small cell lung cancer patients. Histol Histopathol. 2012;27(2):197–207.
90. Park JK, Kim MA, Ryu JK, Yoon YB, Kim SW, Han HS, Kang GH, Kim H, Hwang JH, Kim YT. Postoperative prognostic predictors of pancreatic ductal adenocarcinoma: clinical analysis and immunoprofile on tissue microarrays. Ann Surg Oncol. 2012;19:2664–72.
91. Hu X, Fu X, Wen S, Zou X, Liu Y. Prognostic value of MACC1 and c-met expressions in non-small cell lung cancer. Chin J Lung Cancer. 2012;15(7)
92. Liu XH, Yang JQ, Wang HB. Expression of c-met and the prognosis in non-small cell lung cancer. Chin J Cell Mol Immunol. 2010:26.
93. Gallegos Ruiz M, Floor K, Ferreira J, Grunberg K, Thunnissen E, Belien J, Meijer G, Peters G, Smit E, Rodriguez J. Combined assessment of EGFR pathway-related molecular markers and prognosis of NSCLC patients. Cancer Res. 2008;68(9_Suppl):3542.
94. Nakamura Y, Niki T, Goto A, Morikawa T, Miyazawa K, Nakajima J, Fukayama M. c-Met activation in lung adenocarcinoma tissues: an immunohistochemical analysis. Cancer Sci. 2007;98(7):1006–13.
95. Ma G, Deng Y, Chen W, Liu Z, Ai C, Li X, Zhou Q. The prognostic role of MET protein expression among surgically resected non-small cell lung cancer patients: a meta-analysis. Front Oncol. 2019;9:1441.
96. Tsakonas G, Botling J, Micke P, Rivard C, LaFleur L, Mattsson J, Boyle T, Hirsch FR, Ekman S. c-MET as a biomarker in patients with surgically resected non-small cell lung cancer. Lung Cancer. 2019;133:69–74.
97. Zhang J, Sun J, Zhang Z, Liang X, Luo Y, Wu S, Liang Z. Protein overexpression and gene amplification of cellular mesenchymal-epithelial transition factor is associated with poor prognosis in micropapillary-predominant subtype pulmonary adenocarcinoma. Hum Pathol. 2018;72:59–65.
98. Kim IH, Lee IH, Lee JE, Hong SH, Kim T-J, Lee K-Y, Kim YK, Kim SJ, Sung SW, Park JK. Clinical significance of C-MET overexpression and epidermal growth factor receptor mutation in platinum-based adjuvant chemotherapy outcome in surgically resected lung adenocarcinoma. Ann Surg Oncol. 2017;24:770–7.
99. Kim JH, Kim BJ, Kim HS. Clinicopathological impacts of high c-Met expression in renal cell carcinoma: a meta-analysis and review. Oncotarget. 2017;8(43):75478.

100. Kim JH, Kim HS, Kim BJ, Lee J, Jang HJ. Prognostic value of c-met overexpression in pancreatic adenocarcinoma: a meta-analysis. Oncotarget. 2017;8(42):73098.
101. Kim JH, Kim BJ, Kim HS. Clinicopathological impacts of high c-met expression in head and neck squamous cell carcinoma: a meta-analysis and review. Oncotarget. 2017;8(68):113120.
102. Tong JH, Yeung SF, Chan AW, Chung LY, Chau SL, Lung RWM, Tong CY, Chow C, Tin EK, Yu YH. MET amplification and exon 14 splice site mutation define unique molecular subgroups of non–small cell lung carcinoma with poor prognosis. Clin Cancer Res. 2016;22(12):3048–56.
103. Masuya D, Huang C, Liu D, Nakashima T, Kameyama K, Haba R, Ueno M, Yokomise H. The tumour–stromal interaction between intratumoral c-met and stromal hepatocyte growth factor associated with tumour growth and prognosis in non-small-cell lung cancer patients. Br J Cancer. 2004;90(8):1555–62.
104. Zhao X, Qu J, Hui Y, Zhang H, Sun Y, Liu X, Zhao X, Zhao Z, Yang Q, Wang F. Clinicopathological and prognostic significance of c-met overexpression in breast cancer. Oncotarget. 2017;8(34):56758.
105. Ren X, Yuan L, Shen S, Wu H, Lu J, Liang Z. C-met and ERβ expression differences in basal-like and non-basal-like triple-negative breast cancer. Tumour Biol. 2016;37(8):11385–95.
106. Zagouri F, Brandstetter A, Moussiolis D, Chrysikos D, Dimitrakakis C, Tsigginou A, Marinopoulos S, Zografos GC, Sergentanis TN, Dimopoulos M-A. Low protein expression of MET in ER-positive and HER2-positive breast cancer. Anticancer Res. 2014;34(3):1227–31.
107. Koh YW, Lee HJ, Ahn JH, Lee JW, Gong G. MET expression is associated with disease-specific survival in breast cancer patients in the neoadjuvant setting. Pathol Res Pract. 2014;210(8):494–500.
108. Kim YJ, Choi JS, Seo J, Song JY, Eun Lee S, Kwon MJ, Kwon MJ, Kundu J, Jung K, Oh E. MET is a potential target for use in combination therapy with EGFR inhibition in triple-negative/basal-like breast cancer. Int J Cancer. 2014;134(10):2424–36.
109. Inanc M, Ozkan M, Karaca H, Berk V, Bozkurt O, Duran AO, Ozaslan E, Akgun H, Tekelioglu F, Elmali F. Cytokeratin 5/6, c-met expressions, and PTEN loss prognostic indicators in triple-negative breast cancer. Med Oncol. 2014;31:1–8.
110. Baccelli I, Stenzinger A, Vogel V, Pfitzner BM, Klein C, Wallwiener M, Scharpff M, Saini M, Holland-Letz T, Sinn H-P. Co-expression of MET and CD47 is a novel prognosticator for survival of luminal-type breast cancer patients. Oncotarget. 2014;5(18):8147.
111. Zagouri F, Bago-Horvath Z, Rössler F, Brandstetter A, Bartsch R, Papadimitriou C, Dimitrakakis C, Tsigginou A, Papaspyrou I, Giannos A. High MET expression is an adverse prognostic factor in patients with triple-negative breast cancer. Br J Cancer. 2013;108(5):1100–5.
112. Raghav KP, Wang W, Liu S, Chavez-MacGregor M, Meng X, Hortobagyi GN, Mills GB, Meric-Bernstam F, Blumenschein GR Jr, Gonzalez-Angulo AM. cMET and phospho-cMET protein levels in breast cancers and survival outcomes. Clin Cancer Res. 2012;18(8):2269–77.
113. Gisterek I, Lata E, Halon A, Matkowski R, Szelachowska J, Biecek P, Kornafel J. Prognostic role of c-met expression in breast cancer patients. Rep Pract Oncol Radiother. 2011;16(5):173–7.
114. Valente G, Nicotra G, Arrondini M, Castino R, Capparuccia L, Prat M, Kerim S, Tamagnone L, Isidoro C. Co-expression of plexin-B1 and met in human breast and ovary tumours enhances the risk of progression. Anal Cell Pathol. 2009;31(6):423–36.
115. Carracedo A, Egervari K, Salido M, Rojo F, Corominas JM, Arumi M, Corzo C, Tusquets I, Espinet B, Rovira A. FISH and immunohistochemical status of the hepatocyte growth factor receptor (c-met) in 184 invasive breast tumors. Breast Cancer Res. 2009;11:1–2.
116. Pozner-Moulis S, Cregger M, Camp RL, Rimm DL. Antibody validation by quantitative analysis of protein expression using expression of met in breast cancer as a model. Lab Investig. 2007;87(3):251–60.
117. Lindemann K, Resau J, Nährig J, Kort E, Leeser B, Annecke K, Welk A, Schäfer J, Vande Woude G, Lengyel E. Differential expression of c-met, its ligand HGF/SF and HER2/neu in DCIS and adjacent normal breast tissue. Histopathology. 2007;51(1):54–62.

118. Götte M, Kersting C, Radke I, Kiesel L, Wülfing P. An expression signature of syndecan-1 (CD138), E-cadherin and c-met is associated with factors of angiogenesis and lymphangiogenesis in ductal breast carcinoma in situ. Breast Cancer Res. 2007;9:1–12.
119. Chen HH, Su W-C, Lin P-W, Guo H-R, Lee W-Y. Hypoxia-inducible factor-1α correlates with MET and metastasis in node-negative breast cancer. Breast Cancer Res Treat. 2007;103:167–75.
120. Garcia S, Dalès J-P, Charafe-Jauffret E, Carpentier-Meunier S, Andrac-Meyer L, Jacquemier J, Andonian C, Lavaut M-N, Allasia C, Bonnier P. Poor prognosis in breast carcinomas correlates with increased expression of targetable CD146 and c-met and with proteomic basal-like phenotype. Hum Pathol. 2007;38(6):830–41.
121. Lengyel E, Prechtel D, Resau JH, Gauger K, Welk A, Lindemann K, Salanti G, Richter T, Knudsen B, Vande Woude GF. C-met overexpression in node-positive breast cancer identifies patients with poor clinical outcome independent of Her2/neu. Int J Cancer. 2005;113(4):678–82.
122. Tolgay Ocal I, Dolled-Filhart M, D'Aquila TG, Camp RL, Rimm DL. Tissue microarray-based studies of patients with lymph node negative breast carcinoma show that met expression is associated with worse outcome but is not correlated with epidermal growth factor family receptors. Cancer. 2003;97(8):1841–8.
123. Edakuni G, Sasatomi E, Satoh T, Tokunaga O, Miyazaki K. Expression of the hepatocyte growth factor/c-met pathway is increased at the cancer front in breast carcinoma. Pathol Int. 2001;51(3):172–8.
124. Nakopoulou, Gakiopoulou, Keramopoulos, Giannopoulou, Athanassiadou, Mavrommatis, Davaris. c-met tyrosine kinase receptor expression is associated with abnormal β-catenin expression and favourable prognostic factors in invasive breast carcinoma. Histopathology. 2000;36(4):313–25.
125. Camp RL, Rimm EB, Rimm DL. Met expression is associated with poor outcome in patients with axillary lymph node negative breast carcinoma. Cancer. 1999;86(11):2259–65.
126. Ghoussoub RA, Dillon DA, D'Aquila T, Rimm EB, Fearon ER, Rimm DL. Expression of c-met is a strong independent prognostic factor in breast carcinoma. Cancer. 1998;82(8):1513–20.
127. Narita T, Kimura N, Mitsuoka C, Toi M, Sato M, Matsuura N, Matsumoto K, Nakamura T, Kannagi R. Biological and clinical significance of hepatocyte growth factor in breast cancer. Int J Oncol. 1997;11(6):1305–10.
128. Wang F, Li S, Zhao Y, Yang K, Chen M, Niu H, Yang J, Luo Y, Tang W, Sheng M. Predictive role of the overexpression for CXCR4, C-Met, and VEGF-C among breast cancer patients: a meta-analysis. Breast. 2016;28:45–53.
129. Pisters LL, El-Naggar AK, Luo W, Malpica A, Lin SH. C-met proto-oncogene expression in benign and malignant human renal tissues. J Urol. 1997;158(3):724–8.
130. Inoue K, Karashima T, Chikazawa M, Iiyama T, Yoshikawa C, Furihata M, Ohtsuki Y, Shuin T. Overexpression of c-met proto-oncogene associated with chromophilic renal cell carcinoma with papillary growth. Virchows Arch. 1998;433(6):511–5.
131. Sweeney P, El-Naggar AK, Lin SH, Pisters LL. Biological significance of c-met over expression in papillary renal cell carcinoma. J Urol. 2002;168(1):51–5.
132. Miyata Y, Kanetake H, Kanda S. Presence of phosphorylated hepatocyte growth factor receptor/c-met is associated with tumor progression and survival in patients with conventional renal cell carcinoma. Clin Cancer Res. 2006;12(16):4876–81.
133. Gontero P, Ceratti G, Guglielmetti S, Andorno A, Terrone C, Bonvini D, Faggiano F, Tizzani A, Frea B, Valente G. Prognostic factors in a prospective series of papillary renal cell carcinoma. BJU Int. 2008;102(6):697–702.
134. Erlmeier F, Ivanyi P, Hartmann A, Autenrieth M, Wiedemann M, Weichert W, Steffens S. c-Met in chromophobe renal cell carcinoma. Med Oncol. 2017;34:1–6.
135. Chen S, Zhu Y, Cui J, Wang Y, Xia Y, Song J, Cheng S, Zhou C, Zhang D, Zhang B. The role of c-Met in prognosis and clinicopathology of renal cell carcinoma: results from a single-centre study and systematic review. In: Urologic oncology: seminars and original investigations. Elsevier; 2017.

136. Peltola KJ, Penttilä P, Rautiola J, Joensuu H, Hänninen E, Ristimäki A, Bono P. Correlation of c-met expression and outcome in patients with renal cell carcinoma treated with sunitinib. Clin Genitourin Cancer. 2017;15(4):487–94.
137. Macher-Goeppinger S, Keith M, Endris V, Penzel R, Tagscherer KE, Pahernik S, Hohenfellner M, Gardner H, Gruellich C, Schirmacher P. MET expression and copy number status in clear-cell renal cell carcinoma: prognostic value and potential predictive marker. Oncotarget. 2016;8(1):1046.
138. Kammerer-Jacquet S-F, Medane S, Bensalah K, Bernhard J-C, Yacoub M, Dupuis F, Ravaud A, Verhoest G, Mathieu R, Peyronnet B. Correlation of c-MET expression with PD-L1 expression in metastatic clear cell renal cell carcinoma treated by sunitinib first-line therapy. Target Oncol. 2017;12:487–94.
139. Oh RR, Park JY, Lee JH, Shin MS, Kim HS, Lee SK, Kim YS, Lee SH, Lee SN, Yang YM. Expression of HGF/SF and met protein is associated with genetic alterations of VHL gene in primary renal cell carcinomas. APMIS. 2002;110(3):229–38.
140. Di Renzo MF, Olivero M, Giacomini A, Porte H, Chastre E, Mirossay L, Nordlinger B, Bretti S, Bottardi S, Giordano S. Overexpression and amplification of the met/HGF receptor gene during the progression of colorectal cancer. Clin Cancer Res. 1995;1(2):147–54.
141. Liu Y, Yu XF, Zou J, Luo ZH. Prognostic value of c-met in colorectal cancer: a meta-analysis. World J Gastroenterol: WJG. 2015;21(12):3706.
142. Resnick MB, Routhier J, Konkin T, Sabo E, Pricolo VE. Epidermal growth factor receptor, c-MET, β-catenin, and p53 expression as prognostic indicators in stage II colon cancer: a tissue microarray study. Clin Cancer Res. 2004;10(9):3069–75.
143. Garouniatis A, Zizi-Sermpetzoglou A, Rizos S, Kostakis A, Nikiteas N, Papavassiliou AG. FAK, CD44v6, c-met and EGFR in colorectal cancer parameters: tumour progression, metastasis, patient survival and receptor crosstalk. Int J Color Dis. 2013;28:9–18.
144. Inno A, Di Salvatore M, Cenci T, Martini M, Orlandi A, Strippoli A, Ferrara AM, Bagalà C, Cassano A, Larocca LM. Is there a role for IGF1R and c-MET pathways in resistance to cetuximab in metastatic colorectal cancer? Clin Colorectal Cancer. 2011;10(4):325–32.
145. De Oliveira ATT, Matos D, Logullo AF, Da Silva SRM, Neto RA, Longatto Filho A, Saad SS. MET Is highly expressed in advanced stages of colorectal cancer and indicates worse prognosis and mortality. Anticancer Res. 2009;29(11):4807–11.
146. Kishiki T, Ohnishi H, Masaki T, Ohtsuka K, Ohkura Y, Furuse J, Watanabe T, Sugiyama M. Overexpression of MET is a new predictive marker for anti-EGFR therapy in metastatic colorectal cancer with wild-type KRAS. Cancer Chemother Pharmacol. 2014;73:749–57.
147. Ginty F, Adak S, Can A, Gerdes M, Larsen M, Cline H, Filkins R, Pang Z, Li Q, Montalto MC. The relative distribution of membranous and cytoplasmic met is a prognostic indicator in stage I and II colon cancer. Clin Cancer Res. 2008;14(12):3814–22.
148. Lee CT, Chow NH, Su PF, Lin SC, Lin PC, Lee JC. The prognostic significance of RON and MET receptor coexpression in patients with colorectal cancer. Dis Colon Rectum. 2008;51(8):1268–74.
149. Umeki K, Shiota G, Kawasaki H. Clinical significance of c-met oncogene alterations in human colorectal cancer. Oncology. 1999;56(4):314–21.
150. Peng Z, Zhu Y, Wang Q, Gao J, Li Y, Li Y, Ge S, Shen L. Prognostic significance of MET amplification and expression in gastric cancer: a systematic review with meta-analysis. PLoS One. 2014;9(1):e84502.
151. Li Y, Chen CQ, He YL, Cai SR, Yang DJ, He WL, Xu JB, Zan WH. Abnormal expression of E-cadherin in tumor cells is associated with poor prognosis of gastric carcinoma. J Surg Oncol. 2012;106(3):304–10.
152. Huang T-J, Wang J-Y, Lin S-R, Lian S-T, Hsieh J-S. Overexpression of the c-met protooncogene in human gastric carcinoma-correlation to clinical features. Acta Oncol. 2001;40(5):638–43.
153. Drebber U, Baldus SE, Nolden B, Grass G, Bollschweiler E, Dienes HP, Hölscher AH, Mönig SP. The overexpression of c-met as a prognostic indicator for gastric carcinoma compared to p53 and p21 nuclear accumulation. Oncol Rep. 2008;19(6):1477–83.

154. Nakajima M, Sawada H, Yamada Y, Watanabe A, Tatsumi M, Yamashita J, Matsuda M, Sakaguchi T, Hirao T, Nakano H. The prognostic significance of amplification and overexpression of c-met and c-erb B-2 in human gastric carcinomas. Cancer. 1999;85(9):1894–902.
155. Taniguchi K, Yonemura Y, Nojima N, Hirono Y, Fushida S, Fujimura T, Miwa K, Endo Y, Yamamoto H, Watanabe H. The relation between the growth patterns of gastric carcinoma and the expression of hepatocyte growth factor receptor (c-met), autocrine motility factor receptor, and urokinase-type plasminogen activator receptor. Cancer. 1998;82(11):2112–22.
156. Zhao J, Zhang X, Xin Y. Up-regulated expression of Ezrin and c-met proteins are related to the metastasis and prognosis of gastric carcinomas. Histol Histopathol. 2011;26(9):2011.
157. Zhao D, Wang SH, Feng Y, Hua CG, Zhao J, Tang XF. Intratumoral c-Met expression is associated with vascular endothelial growth factor C expression, lymphangiogenesis, and lymph node metastasis in oral squamous cell carcinoma: implications for use as a prognostic marker. Hum Pathol. 2011;42(10):1514–23.
158. Kubicka S, Claas C, Staab S, Kühnel F, Zender L, Trautwein C, Wagner S, Rudolph KL, Manns M. p53 mutation pattern and expression of c-erbB2 and c-met in gastric cancer: relation to histological subtypes, Helicobacter pylori infection, and prognosis. Dig Dis Sci. 2002;47:114–21.
159. Catenacci DV, Cervantes G, Yala S, Nelson EA, El-Hashani E, Kanteti R, El Dinali M, Hasina R, Brägelmann J, Seiwert T. RON (MST1R) is a novel prognostic marker and therapeutic target for gastroesophageal adenocarcinoma. Cancer Biol Ther. 2011;12(1):9–46.
160. Furukawa T, Duguid WP, Kobari M, Matsuno S, Tsao M-S. Hepatocyte growth factor and Met receptor expression in human pancreatic carcinogenesis. Am J Pathol. 1995;147(4):889.
161. Ide T, Kitajima Y, Miyoshi A, Ohtsuka T, Mitsuno M, Ohtaka K, Miyazaki K. The hypoxic environment in tumor-stromal cells accelerates pancreatic cancer progression via the activation of paracrine hepatocyte growth factor/c-met signaling. Ann Surg Oncol. 2007;14:2600–7.
162. Zhu GH, Huang C, Qiu ZJ, Liu J, Zhang ZH, Zhao N, Feng ZZ, Lv XH. Expression and prognostic significance of CD151, c-met, and integrin alpha3/alpha6 in pancreatic ductal adenocarcinoma. Dig Dis Sci. 2011;56:1090–8.
163. Neuzillet C, Couvelard A, Tijeras-Raballand A, de Mestier L, de Gramont A, Bédossa P, Paradis V, Sauvanet A, Bachet JB, Ruszniewski P. High c-Met expression in stage I–II pancreatic adenocarcinoma: proposal for an immunostaining scoring method and correlation with poor prognosis. Histopathology. 2015;67(5):664–76.
164. Tomihara H, Yamada D, Eguchi H, Iwagami Y, Noda T, Asaoka T, Wada H, Kawamoto K, Gotoh K, Takeda Y. MicroRNA-181b-5p, ETS1, and the c-met pathway exacerbate the prognosis of pancreatic ductal adenocarcinoma after radiation therapy. Cancer Sci. 2017;108(3):398–407.
165. Chau NG, Perez-Ordonez B, Zhang K, Pham NA, Ho J, Zhang T, Ludkovski O, Wang L, Chen EX, Tsao M-S. The association between EGFR variant III, HPV, p16, c-MET, EGFR gene copy number and response to EGFR inhibitors in patients with recurrent or metastatic squamous cell carcinoma of the head and neck. Head Neck Oncol. 2011;3:1–11.
166. Sawatsubashi M, Sasatomi E, Mizokami H, Tokunaga O, Shin T. Expression of c-met in laryngeal carcinoma. Virchows Arch. 1998;432:331–5.
167. Muzio LL, Leonardi R, Mignogna MD, Pannone G, Rubini C, Pieramici T, Trevisiol L, Ferrari F, Serpico R, Testa N. Scatter factor receptor (c-met) as possible prognostic factor in patients with oral squamous cell carcinoma. Anticancer Res. 2004;24(2C):1063–70.
168. Endo K, Shirai A, Furukawa M, Yoshizaki T. Prognostic value of cell motility activation factors in patients with tongue squamous cell carcinoma. Hum Pathol. 2006;37(8):1111–6.
169. Kim CH, Moon SK, Bae JH, Ho Lee J, Ho Han J, Kim K, Chang CE. Expression of hepatocyte growth factor and c-met in hypopharyngeal squamous cell carcinoma. Acta Otolaryngol. 2006;126(1):88–94.
170. Kim CH, Koh YW, Han JH, Kim JW, Lee JS, Baek SJ, Hwang HS, Choi EC. C-met expression as an indicator of survival outcome in patients with oral tongue carcinoma. Head Neck. 2010;32(12):1655–64.

171. Freudlsperger C, Alexander D, Reinert S, Hoffmann J. Prognostic value of c-met expression in oral squamous cell carcinoma. Exp Ther Med. 2010;1(1):69–72.
172. Lim YC, Han JH, Kang HJ, Kim YS, Lee BH, Choi EC, Kim CH. Overexpression of c-met promotes invasion and metastasis of small oral tongue carcinoma. Oral Oncol. 2012;48(11):1114–9.
173. Choe JY, Yun JY, Nam SJ, Kim JE. Expression of c-met is different along the location and associated with lymph node metastasis of head and neck carcinoma. Korean J Pathol. 2012;46(6):515.
174. Kwon MJ, Kim DH, Park HR, Shin HS, Kwon JH, Lee DJ, Kim JH, Cho SJ, Nam ES. Frequent hepatocyte growth factor overexpression and low frequency of c-met gene amplification in human papillomavirus–negative tonsillar squamous cell carcinoma and their prognostic significances. Hum Pathol. 2014;45(7):1327–38.
175. Baschnagel AM, Williams L, Hanna A, Chen PY, Krauss DJ, Pruetz BL, Akervall J, Wilson GD. c-Met expression is a marker of poor prognosis in patients with locally advanced head and neck squamous cell carcinoma treated with chemoradiation. Int J Radiat Oncol Biol Phys. 2014;88(3):701–7.
176. Li Y, Li W, He Q, Xu Y, Ren X, Tang X, Wen X, Yang X, Sun Y, Zeng J. Prognostic value of MET protein overexpression and gene amplification in locoregionally advanced nasopharyngeal carcinoma. Oncotarget. 2015;6(15):13309.
177. Lo Muzio L, Farina A, Rubini C, Coccia E, Capogreco M, Colella G, Leonardi R, Campisi G, Carinci F. Effect of c-met expression on survival in head and neck squamous cell carcinoma. Tumour Biol. 2006;27(3):115–21.
178. Ueki T, Fujimoto J, Suzuki T, Yamamoto H, Okamoto E. Expression of hepatocyte growth factor and its receptor c-met proto-oncogene in hepatocellular carcinoma. Hepatology. 1997;25(4):862–6.
179. Ke AW, Shi GM, Zhou J, Wu FZ, Ding ZB, Hu MY, Xu Y, Song ZJ, Wang ZJ, Wu JC. Role of overexpression of CD151 and/or c-met in predicting prognosis of hepatocellular carcinoma. Hepatology. 2009;49(2):491–503.
180. Lee SJ, Lee J, Sohn I, Mao M, Kai W, Park CK, Lim HY. A survey of c-MET expression and amplification in 287 patients with hepatocellular carcinoma. Anticancer Res. 2013;33(11):5179–86.
181. Kondo S, Ojima H, Tsuda H, Hashimoto J, Morizane C, Ikeda M, Ueno H, Tamura K, Shimada K, Kanai Y. Clinical impact of c-met expression and its gene amplification in hepatocellular carcinoma. Int J Clin Oncol. 2013;18:207–13.
182. Koh YW, Park YS, Kang HJ, Shim JH, Yu E. MET is a predictive factor for late recurrence but not for overall survival of early stage hepatocellular carcinoma. Tumour Biol. 2015;36:4993–5000.
183. Peng J, Qi S, Wang P, Li W, Liu C, Li F. Diagnosis and prognostic significance of c-met in cervical cancer: a meta-analysis. Dis Markers. 2016;2016(1):6594016.
184. Baykal C, yşe Ayhan A, Al A, Yüce K, Ayhan A. Overexpression of the c-met/HGF receptor and its prognostic significance in uterine cervix carcinomas. Gynecol Oncol. 2003;88(2):123–9.
185. Shu C, Fei JW, Xu P, Tan Y. Expression of c-met protein in cervical cancer and its relationship with angiogenesis. Matern Child Health Care China. 2008:24.
186. Wu H, Zhou C, Ji X, Ma D. The expression and relation of c-met and tissue inhibitor of metalloproteinase 2 in cervical carcinoma. Chin Prim Health Care. 2009:4.
187. Wang Z, Zhang X, Bao S, Ha CF. Relationship between expression of HGF/C-met and clinicopathological properties in uterine cervix carcinomas. J Ningxia Med Univ. 2010:5.
188. Zhang XY, Wang ZY, Huang LY, Ma YJ, Chen GQ. The clinical research of serum levels HGF and expression of its receptor c-met in cervical cancer. Ningxia Med J. 2010:8.
189. Kim BW, Cho H, Chung J-Y, Conway C, Ylaya K, Kim J-H, Hewitt SM. Prognostic assessment of hypoxia and metabolic markers in cervical cancer using automated digital image analysis of immunohistochemistry. J Transl Med. 2013;11:1–11.

190. Yin HW. Clinical significance of HGF, c-met and VEGF expression in uterine cancer tissue. China Pract Med. 2013:14.
191. Zhou N, Wu Y. Expression of MACCl and c-met proteins in cervical carcinoma and its correlation with pelvic lymph node metastasis. J Pract Obstet Gynecol. 2014;3:199–203.
192. Yamamoto S, Tsuda H, Miyai K, Takano M, Tamai S, Matsubara O. Gene amplification and protein overexpression of MET are common events in ovarian clear-cell adenocarcinoma: their roles in tumor progression and prognostication of the patient. Mod Pathol. 2011;24(8):1146–55.
193. Aune G, Lian A-M, Tingulstad S, Torp SH, Forsmo S, Reseland JE, Stunes AK, Syversen U. Increased circulating hepatocyte growth factor (HGF): a marker of epithelial ovarian cancer and an indicator of poor prognosis. Gynecol Oncol. 2011;121(2):402–6.
194. Ramirez R, Hsu D, Patel A, Fenton C, Dinauer C, Tuttle RM, Francis GL. Over-expression of hepatocyte growth factor/scatter factor (HGF/SF) and the HGF/SF receptor (cMET) are associated with a high risk of metastasis and recurrence for children and young adults with papillary thyroid carcinoma. Clin Endocrinol. 2000;53(5):635–44.
195. Tsao MS, Yang Y, Marcus A, Liu N, Mou L. Hepatocyte growth factor is predominantly expressed by the carcinoma cells in non–small-cell lung cancer. Hum Pathol. 2001;32(1):57–65.
196. Navab R, Liu J, Seiden-Long I, Shih W, Li M, Bandarchi B, Chen Y, Lau D, Zu Y-F, Cescon D. Co-overexpression of met and hepatocyte growth factor promotes systemic metastasis in NCI-H460 non-small cell lung carcinoma cells. Neoplasia. 2009;11(12):1292-IN6.
197. Yi S, Chen J, Viallet J, Schwall R, Nakamura T, Tsao M. Paracrine effects of hepatocyte growth factor/scatter factor on non-small-cell lung carcinoma cell lines. Br J Cancer. 1998;77(12):2162–70.
198. Rong S, Segal S, Anver M, Resau JH, Vande Woude GF. Invasiveness and metastasis of NIH 3T3 cells induced by met-hepatocyte growth factor/scatter factor autocrine stimulation. Proc Natl Acad Sci. 1994;91(11):4731–5.
199. Tsao MS, Zhu H, Giaid A, Viallet J, Nakamura T, Park M. Hepatocyte growth factor/scatter factor is an autocrine factor for human normal bronchial epithelial and lung carcinoma cells. Cell Growth Differ. 1993;4:571.
200. Weng J, Mohan RR, Li Q, Wilson SE. IL-1 upregulates keratinocyte growth factor and hepatocyte growth factor mRNA and protein production by cultured stromal fibroblast cells: interleukin-1β expression in the cornea. Cornea. 1997;16(4):465–71.
201. Wheeler DL, Yarden Y. Receptor tyrosine kinases: family and subfamilies. Springer; 2015.
202. Garnett J, Chumbalkar V, Vaillant B, Gururaj AE, Hill KS, Latha K, Yao J, Priebe W, Colman H, Elferink LA. Regulation of HGF expression by DEGFR-mediated c-met activation in glioblastoma cells. Neoplasia. 2013;15(1):73-IN21.
203. Bharti A, Ma PC, Maulik G, Singh R, Khan E, Skarin AT, Salgia R. Haptoglobin α-subunit and hepatocyte growth factor can potentially serve as serum tumor biomarkers in small cell lung cancer. Anticancer Res. 2004;24(2C):1031–8.
204. Siegfried JM, Weissfeld LA, Luketich JD, Weyant RJ, Gubish CT, Landreneau RJ. The clinical significance of hepatocyte growth factor for non–small cell lung cancer. Ann Thorac Surg. 1998;66(6):1915–8.
205. Dreikhausen L, Blank S, Sisic L, Heger U, Weichert W, Jäger D, Bruckner T, Giese N, Grenacher L, Falk C. Association of angiogenic factors with prognosis in esophageal cancer. BMC Cancer. 2015;15:1–11.
206. Boccaccio C, Gaudino G, Gambarotta G, Galimi F, Comoglio PM. Hepatocyte growth factor (HGF) receptor expression is inducible and is part of the delayed-early response to HGF. J Biol Chem. 1994;269(17):12846–51.
207. Gao HF, Li AN, Yang JJ, Chen ZH, Xie Z, Zhang XC, Su J, Lou NN, Yan HH, Han JF. Soluble c-met levels correlated with tissue c-met protein expression in patients with advanced non–small-cell Lung cancer. Clin Lung Cancer. 2017;18(1):85–91.

208. Lv H, Shan B, Tian Z, Li Y, Zhang Y, Wen S. Soluble c-met is a reliable and sensitive marker to detect c-met expression level in lung cancer. Biomed Res Int. 2015;2015(1):626578.
209. Kaye DR, Pinto PA, Cecchi F, Reilly J, Semerjian A, Rabe DC, Gupta G, Choyke PL, Bottaro DP. Tumor and plasma met levels in non-metastatic prostate cancer. PLoS One. 2016;11(6):e0157130.
210. Barisione G, Fabbi M, Gino A, Queirolo P, Orgiano L, Spano L, Picasso V, Pfeffer U, Mosci C, Jager MJ. Potential role of soluble c-met as a new candidate biomarker of metastatic uveal melanoma. JAMA Ophthalmol. 2015;133(9):1013–21.
211. Wader KF, Fagerli UM, Holt RU, Børset M, Sundan A, Waage A. Soluble c-met in serum of patients with multiple myeloma: correlation with clinical parameters. Eur J Haematol. 2011;87(5):394–9.
212. Fang MY, Wang SY, Zheng YB, Gong LY, Bao WL, Gu DL, Mao WM. Prognostic and predictive significance of plasma hepatocyte growth factor and carcinoembryonic antigen in non-small lung cancer after surgery. Eur Rev Med Pharm Sci. 2014;18(3):398.
213. Ren Y, Cao B, Law S, Xie Y, Lee PY, Cheung L, Chen Y, Huang X, Chan HM, Zhao P. Hepatocyte growth factor promotes cancer cell migration and angiogenic factors expression: a prognostic marker of human esophageal squamous cell carcinomas. Clin Cancer Res. 2005;11(17):6190–7.
214. Zhang Y, Wu JZ, Yang YQ, Ma R, Zhang JY, Feng JF. Expression of growth-regulated oncogene-1, hepatocyte growth factor, platelet-derived growth factor-AA and soluble E-selectin and their association with high-risk human papillomavirus infection in squamous cell carcinoma of the uterine cervix. Mol Med Rep. 2014;10(2):1013–24.
215. Barakat O, Rodriguez GC, Raijman I, Allison PM, Nieto J, Ozaki CF, Wood RP, Engler DA. Clinical value of plasma hepatocyte growth factor measurement for the diagnosis of periampullary cancer and prognosis after pancreaticoduodenectomy. J Surg Oncol. 2010;102(7):816–20.
216. Karabulut S, Tas F, Akyüz F, Ormeci A, Serilmez M, Soydinç H, Vatansever S, Yasasever V. Clinical significance of serum hepatocyte growth factor (HGF) levels in hepatocellular carcinoma. Tumour Biol. 2014;35:2327–33.
217. Alkozai EM, Porte RJ, Adelmeijer J, Zanetto A, Simioni P, Senzolo M, Lisman T. Levels of angiogenic proteins in plasma and platelets are not different between patients with hepatitis B/C-related cirrhosis and patients with cirrhosis and hepatocellular carcinoma. Platelets. 2015;26(6):577–82.
218. Yamagamim H, Moriyama M, Matsumura H, Aoki H, Shimizu T, Saito T, Kaneko M, Shioda A, Tanaka N, Arakawa Y. Serum concentrations of human hepatocyte growth factor is a useful indicator for predicting the occurrence of hepatocellular carcinomas in C-viral chronic liver diseases. Cancer. 2002;95(4):824–34.
219. McNeil BK, Sorbellini M, Grubb RL, Apolo AB, Cecchi F, Athauda G, Cohen B, Giubellino A, Simpson H, Agarwal PK. Preliminary evaluation of urinary soluble met as a biomarker for urothelial carcinoma of the bladder. J Transl Med. 2014;12:1–6.
220. Rashed WM. C-MET as a potential target therapy toward personalized therapy in some pediatric tumors: an overview. Crit Rev Oncol Hematol. 2018;131:7–15.
221. Kim HS, Chon HJ, Kim H, Shin SJ, Wacheck V, Gruver AM, Kim JS, Rha SY, Chung HC. MET in gastric cancer with liver metastasis: the relationship between MET amplification and met overexpression in primary stomach tumors and liver metastasis. J Surg Oncol. 2018;117(8):1679–86.
222. Huang C, Zhou Q, Hu Y, Wen Y, Qiu Z, Liang M, Mo J, Xu J, Sun C, Liu F. Hepatocyte growth factor is a prognostic marker in patients with colorectal cancer: a meta-analysis. Oncotarget. 2017;2017(8):23459–69.
223. Wang M, Liang L, Lei X, Multani A, Meric-Bernstam F, Tripathy D, Wu Y, Chen H, Zhang H. Evaluation of cMET aberration by immunohistochemistry and fluorescence in situ hybridization (FISH) in triple negative breast cancers. Ann Diagn Pathol. 2018;35:69–76.

224. Rimassa L, Abbadessa G, Personeni N, Porta C, Borbath I, Daniele B, Salvagni S, Van Laethem JL, Van Vlierberghe H, Trojan J. Tumor and circulating biomarkers in patients with second-line hepatocellular carcinoma from the randomized phase II study with tivantinib. Oncotarget. 2016;7(45):72622.
225. Siegfried JM, Weissfeld LA, Singh-Kaw P, Weyant RJ, Testa JR, Landreneau RJ. Association of immunoreactive hepatocyte growth factor with poor survival in resectable non-small cell lung cancer. Cancer Res. 1997;57(3):433–9.
226. De Bacco F, Luraghi P, Medico E, Reato G, Girolami F, Perera T, Gabriele P, Comoglio PM, Boccaccio C. Induction of MET by ionizing radiation and its role in radioresistance and invasive growth of cancer. JNCI J Natl Cancer Inst. 2011;103(8):645–61.
227. Matsui S, Osada S, Tomita H, Komori S, Mori R, Sanada Y, Takahashi T, Yamaguchi K, Yoshida K. Clinical significance of aggressive hepatectomy for colorectal liver metastasis, evaluated from the HGF/c-met pathway. Int J Oncol. 2010;37(2):289–97.
228. Yan S, Jiao X, Zou H, Li K. Prognostic significance of c-met in breast cancer: a meta-analysis of 6010 cases. Diagn Pathol. 2015;10:1–10.
229. Garcia S, Dales J-P, Charafe-Jauffret E, Carpentier-Meunier S, Andrac-Meyer L, Jacquemier J, Andonian C, Lavaut M-N, Allasia C, Bonnier P. Overexpression of c-met and of the transducers PI3K, FAK and JAK in breast carcinomas correlates with shorter survival and neoangiogenesis. Int J Oncol. 2007;31(1):49–58.
230. Ji Y, Ogawa M, Si Y, Nomura K, Kuramoto M, Saishoji T, Shin S. Immunoreactive hepatocyte growth factor is a strong and independent predictor of recurrence and survival in human breast cancer. Cancer Res. 1994;54(7):1630–3.
231. Chen HJ, Mok TS, Chen ZH, Guo AL, Zhang XC, Su J, Wu YL. Clinicopathologic and molecular features of epidermal growth factor receptor T790M mutation and c-MET amplification in tyrosine kinase inhibitor-resistant Chinese non-small cell lung cancer. Pathol Oncol Res. 2009;15:651–8.
232. Gao H, Guan M, Sun Z, Bai C. High c-met expression is a negative prognostic marker for colorectal cancer: a meta-analysis. Tumour Biol. 2015;36:515–20.
233. Szturz P, Budíková M, Vermorken JB, Horova I, Gál B, Raymond E, de Gramont A, Faivre S. Prognostic value of c-MET in head and neck cancer: a systematic review and meta-analysis of aggregate data. Oral Oncol. 2017;74:68–76.
234. Felix A, Edwards R, Stone R, Chivukula M, Parwani A, Bowser R, Linkov F, Weissfeld J. Associations between hepatocyte growth factor, c-met, and basic fibroblast growth factor and survival in endometrial cancer patients. Br J Cancer. 2012;106(12):2004–9.
235. Toi M, Taniguchi T, Ueno T, Asano M, Funata N, Sekiguchi K, Iwanari H, Tominaga T. Significance of circulating hepatocyte growth factor level as a prognostic indicator in primary breast cancer. Clin Cancer Res. 1998;4(3):659–64.
236. Ahmed H, Metwally F, Mahdy E, Shosha W, Ramadan S. Clinical value of serum hepatocyte growth factor, B-cell lymphoma-2 and nitric oxide in primary breast cancer patients. Eur Rev Med Pharm Sci. 2012;16(7):958.
237. Sheen-Chen SM, Liu YW, Eng HL, Chou FF. Serum levels of hepatocyte growth factor in patients with breast cancer. Cancer Epidemiol Biomarkers Prev. 2005;14(3):715–7.
238. Taniguchi T, Toi M, Inada K, Imazawa T, Yamamoto Y, Tominaga T. Serum concentrations of hepatocyte growth factor in breast cancer patients. Clin Cancer Res. 1995;1(9):1031–4.
239. El-Attar HA, Sheta MI. Hepatocyte growth factor profile with breast cancer. Indian J Pathol Microbiol. 2011;54(3):509–13.
240. Toiyama Y, Miki C, Inoue Y, Okugawa Y, Tanaka K, Kusunoki M. Serum hepatocyte growth factor as a prognostic marker for stage II or III colorectal cancer patients. Int J Cancer. 2009;125(7):1657–62.
241. Kilic-Baygutalp N, Ozturk N, Orsal-Ibisoglu E, Gundogdu B, Ozgeris F, Bakan N, Bakan E, Kilic A. Evaluation of serum HGF and CK18 levels in patients with esophageal cancer. Genet Mol Res. 2016;15(3)

242. Russo AL, Jedlicka K, Wernick M, McNally D, Kirk M, Sproull M, Smith S, Shankavaram U, Kaushal A, Figg WD. Urine analysis and protein networking identify met as a marker of metastatic prostate cancer. Clin Cancer Res. 2009;15(13):4292–8.
243. Gherardi E, Birchmeier W, Birchmeier C, Woude GV. Targeting MET in cancer: rationale and progress. Nat Rev Cancer. 2012;12(2):89–103.
244. Navis AC, van Lith SA, van Duijnhoven SM, de Pooter M, Yetkin-Arik B, Wesseling P, Hendriks WJ, Venselaar H, Timmer M, van Cleef P. Identification of a novel MET mutation in high-grade glioma resulting in an auto-active intracellular protein. Acta Neuropathol. 2015;130:131–44.
245. Awad MM, Oxnard GR, Jackman DM, Savukoski DO, Hall D, Shivdasani P, Heng JC, Dahlberg SE, Jänne PA, Verma S. MET exon 14 mutations in non–small-cell lung cancer are associated with advanced age and stage-dependent MET genomic amplification and c-met overexpression. J Clin Oncol. 2016;34(7):721–30.
246. Salgia R, Sattler M, Scheele J, Stroh C, Felip E. The promise of selective MET inhibitors in non-small cell lung cancer with MET exon 14 skipping. Cancer Treat Rev. 2020;87:102022.
247. Sesboüé R, Le Pessot F, Di Fiore F, Frebourg T. EGFR alterations and response to anti-EGFR therapy: is it a matter of gene amplification or gene copy number gain? Br J Cancer. 2012;106(2):426–7.
248. Mazzone M, Comoglio PM. The met pathway: master switch and drug target in cancer progression. FASEB J. 2006;20(10):1611–21.
249. Zeng ZS, Weiser MR, Kuntz E, Chen CT, Khan SA, Forslund A, Nash GM, Gimbel M, Yamaguchi Y, Culliford AT IV. C-met gene amplification is associated with advanced stage colorectal cancer and liver metastases. Cancer Lett. 2008;265(2):258–69.
250. Yang X, Liao HY, Zhang HH. Roles of MET in human cancer. Clin Chim Acta. 2022;525:69–83.
251. Sun S, Wang Z. Head neck squamous cell carcinoma c-met+ cells display cancer stem cell properties and are responsible for cisplatin-resistance and metastasis. Int J Cancer. 2011;129(10):2337–48.
252. Zhang YW, Su Y, Volpert OV, Woude GFV. Hepatocyte growth factor/scatter factor mediates angiogenesis through positive VEGF and negative thrombospondin 1 regulation. Proc Natl Acad Sci. 2003;100(22):12718–23.
253. Sulpice E, Ding S, Muscatelli-Groux B, Bergé M, Han ZC, Plouet J, Tobelem G, Merkulova-Rainon T. Cross-talk between the VEGF-A and HGF signalling pathways in endothelial cells. Biol Cell. 2009;101(9):525–39.
254. Hara S, Nakashiro K-i, Klosek SK, Ishikawa T, Shintani S, Hamakawa H. Hypoxia enhances c-met/HGF receptor expression and signaling by activating HIF-1α in human salivary gland cancer cells. Oral Oncol. 2006;42(6):593–8.
255. Scarpino S, Cancellario d'Alena F, Di Napoli A, Pasquini A, Marzullo A, Ruco LP. Increased expression of met protein is associated with up-regulation of hypoxia inducible factor-1 (HIF-1) in tumour cells in papillary carcinoma of the thyroid. J Pathol. 2004;202(3):352–8.
256. You WK, McDonald DM. The hepatocyte growth factor/c-met signaling pathway as a therapeutic target to inhibit angiogenesis. BMB Rep. 2008;41(12):833.
257. Gallego MI, Bierie B, Hennighausen L. Targeted expression of HGF/SF in mouse mammary epithelium leads to metastatic adenosquamous carcinomas through the activation of multiple signal transduction pathways. Oncogene. 2003;22(52):8498–508.
258. Giordano S, Bardelli A, Zhen Z, Menard S, Ponzetto C, Comoglio PM. A point mutation in the MET oncogene abrogates metastasis without affecting transformation. Proc Natl Acad Sci. 1997;94(25):13868–72.
259. Furge KA, Kiewlich D, Le P, Vo MN, Faure M, Howlett AR, Lipson KE, Woude GFV, Webb CP. Suppression of Ras-mediated tumorigenicity and metastasis through inhibition of the met receptor tyrosine kinase. Proc Natl Acad Sci. 2001;98(19):10722–7.

260. Royal I, Lamarche-Vane N, Lamorte L, Kaibuchi K, Park M. Activation of cdc42, rac, PAK, and rho-kinase in response to hepatocyte growth factor differentially regulates epithelial cell colony spreading and dissociation. Mol Biol Cell. 2000;11(5):1709–25.
261. Vermeulen L, De Sousa E, Melo F, Van Der Heijden M, Cameron K, De Jong JH, Borovski T, Tuynman JB, Todaro M, Merz C, Rodermond H. Wnt activity defines colon cancer stem cells and is regulated by the microenvironment. Nat Cell Biol. 2010;12(5):468–76.
262. Bauer TW, Somcio RJ, Fan F, Liu W, Johnson M, Lesslie DP, Evans DB, Gallick GE, Ellis LM. Regulatory role of c-met in insulin-like growth factor-I receptor–mediated migration and invasion of human pancreatic carcinoma cells. Mol Cancer Ther. 2006;5(7):1676–82.
263. Khoury H, Naujokas MA, Zuo D, Sangwan V, Frigault MM, Petkiewicz S, Dankort DL, Muller WJ, Park M. HGF converts ErbB2/Neu epithelial morphogenesis to cell invasion. Mol Biol Cell. 2005;16(2):550–61.
264. Yamamoto N, Mammadova G, Song RXD, Fukami Y, Ki S. Tyrosine phosphorylation of p145met mediated by EGFR and Src is required for serum-independent survival of human bladder carcinoma cells. J Cell Sci. 2006;119(22):4623–33.
265. Engelman JA, Zejnullahu K, Mitsudomi T, Song Y, Hyland C, Park JO, Lindeman N, Gale C-M, Zhao X, Christensen J. MET amplification leads to gefitinib resistance in lung cancer by activating ERBB3 signaling. Science. 2007;316(5827):1039–43.
266. Hsieh AC, Moasser MM. Targeting HER proteins in cancer therapy and the role of the non-target HER3. Br J Cancer. 2007;97(4):453–7.
267. Klaus A, Birchmeier W. Wnt signalling and its impact on development and cancer. Nat Rev Cancer. 2008;8(5):387–98.
268. Liu Y, Chattopadhyay N, Qin S, Szekeres C, Vasylyeva T, Mahoney ZX, Taglienti M, Bates CM, Chapman HA, Miner JH. Coordinate integrin and c-Met signaling regulate Wnt gene expression during epithelial morphogenesis. Development. 2009;136:843.
269. Monga SP, Mars WM, Pediaditakis P, Bell A, Mulé K, Bowen WC, Wang X, Zarnegar R, Michalopoulos GK. Hepatocyte growth factor induces Wnt-independent nuclear translocation of β-catenin after met-β-catenin dissociation in hepatocytes. Cancer Res. 2002;62(7):2064–71.
270. Brembeck FH, Schwarz-Romond T, Bakkers J, Wilhelm S, Hammerschmidt M, Birchmeier W. Essential role of BCL9-2 in the switch between β-catenin's adhesive and transcriptional functions. Genes Dev. 2004;18(18):2225–30.
271. Bhowmick NA, Chytil A, Plieth D, Gorska AE, Dumont N, Shappell S, Washington MK, Neilson EG, Moses HL. TGF-ß signaling in fibroblasts modulates the oncogenic potential of adjacent epithelia. Science. 2004;303(5659):848–51.
272. Puccini A, Marín-Ramos NI, Bergamo F, Schirripa M, Lonardi S, Lenz H-J, Loupakis F, Battaglin F. Safety and tolerability of c-MET inhibitors in cancer. Drug Saf. 2019;42:211–33.
273. Mo HN, Liu P. Targeting MET in cancer therapy. Chron Dis Transl Med. 2017;3(03):148–53.
274. Goździk-Spychalska J, Szyszka-Barth K, Spychalski Ł, Ramlau K, Wójtowicz J, Batura-Gabryel H, Ramlau R. C-MET inhibitors in the treatment of lung cancer. Curr Treat Options in Oncol. 2014;15:670–82.
275. Bouattour M, Raymond E, Qin S, Cheng AL, Stammberger U, Locatelli G, Faivre S. Recent developments of c-met as a therapeutic target in hepatocellular carcinoma. Hepatology. 2018;67(3):1132–49.
276. Zhao S, Ammanamanchi S, Brattain M, Cao L, Thangasamy A, Wang J, Freeman JW. Smad4-dependent TGF-β signaling suppresses RON receptor tyrosine kinase-dependent motility and invasion of pancreatic cancer cells. J Biol Chem. 2008;283(17):11293–301.
277. Okusaka T, Aramaki T, Inaba Y, Nakamura S, Morimoto M, Moriguchi M, Sato T, Ikawa Y, Ikeda M, Furuse J. Phase I study of tivantinib in Japanese patients with advanced hepatocellular carcinoma: distinctive pharmacokinetic profiles from other solid tumors. Cancer Sci. 2015;106(5):611–7.
278. Xiang Q, Zhen Z, Deng DY, Wang J, Chen Y, Li J, Zhang Y, Wang F, Chen N, Chen H. Tivantinib induces G2/M arrest and apoptosis by disrupting tubulin polymerization in hepatocellular carcinoma. J Exp Clin Cancer Res. 2015;34:1–13.

279. Best J, Schotten C, Lohmann G, Gerken G, Dechêne A. Tivantinib for the treatment of hepatocellular carcinoma. Expert Opin Pharmacother. 2017;18(7):727–33.
280. Sequist LV, Von Pawel J, Garmey EG, Akerley WL, Brugger W, Ferrari D, Chen Y, Costa DB, Gerber DE, Orlov S. Randomized phase II study of erlotinib plus tivantinib versus erlotinib plus placebo in previously treated non–small-cell lung cancer. J Clin Oncol. 2011;29(24):3307–15.
281. Yoshioka H, Azuma K, Yamamoto N, Takahashi T, Nishio M, Katakami N, Ahn M, Hirashima T, Maemondo M, Kim S. A randomized, double-blind, placebo-controlled, phase III trial of erlotinib with or without a c-met inhibitor tivantinib (ARQ 197) in Asian patients with previously treated stage IIIB/IV nonsquamous nonsmall-cell lung cancer harboring wild-type epidermal growth factor receptor (ATTENTION study). Ann Oncol. 2015;26(10):2066–72.
282. Scagliotti G, Von Pawel J, Novello S, Ramlau R, Favaretto A, Barlesi F, Akerley W, Orlov S, Santoro A, Spigel D. Phase III multinational, randomized, double-blind, placebo-controlled study of tivantinib (ARQ 197) plus erlotinib versus erlotinib alone in previously treated patients with locally advanced or metastatic nonsquamous non–small-cell lung cancer. J Clin Oncol. 2015;33(24):2667–74.
283. Puzanov I, Sosman J, Santoro A, Saif MW, Goff L, Dy GK, Zucali P, Means-Powell JA, Ma WW, Simonelli M. Phase 1 trial of tivantinib in combination with sorafenib in adult patients with advanced solid tumors. Investig New Drugs. 2015;33:159–68.
284. Ou SHI, Kwak EL, Siwak-Tapp C, Dy J, Bergethon K, Clark JW, Camidge DR, Solomon BJ, Maki RG, Bang YJ. Activity of crizotinib (PF02341066), a dual mesenchymal-epithelial transition (MET) and anaplastic lymphoma kinase (ALK) inhibitor, in a non-small cell lung cancer patient with de novo MET amplification. J Thorac Oncol. 2011;6(5):942–6.
285. Camidge DR, Ou S-HI, Shapiro G, Otterson GA, Villaruz LC, Villalona-Calero MA, Iafrate AJ, Varella-Garcia M, Dacic S, Cardarella S. Efficacy and safety of crizotinib in patients with advanced c-MET-amplified non-small cell lung cancer (NSCLC). Am Soc Clin Oncol. 2014;32:8001.
286. Shea M, Costa DB, Rangachari D. Management of advanced non-small cell lung cancers with known mutations or rearrangements: latest evidence and treatment approaches. Ther Adv Respir Dis. 2016;10(2):113–29.
287. Facchinetti F, Rossi G, Bria E, Soria J-C, Besse B, Minari R, Friboulet L, Tiseo M. Oncogene addiction in non-small cell lung cancer: focus on ROS1 inhibition. Cancer Treat Rev. 2017;55:83–95.
288. Huang XX, Xie FF, Hou LJ, Chen XX, Ou RY, Yu JT, Qiu JG, Zhang WJ, Jiang QW, Yang Y. Crizotinib synergizes with cisplatin in preclinical models of ovarian cancer. Am J Transl Res. 2017;9(4):1667.
289. Stanley A, Ashrafi GH, Seddon AM, Modjtahedi H. Synergistic effects of various her inhibitors in combination with IGF-1R, C-MET and Src targeting agents in breast cancer cell lines. Sci Rep. 2017;7(1):3964.
290. Yakes FM, Chen J, Tan J, Yamaguchi K, Shi Y, Yu P, Qian F, Chu F, Bentzien F, Cancilla B. Cabozantinib (XL184), a novel MET and VEGFR2 inhibitor, simultaneously suppresses metastasis, angiogenesis, and tumor growth. Mol Cancer Ther. 2011;10(12):2298–308.
291. Bentzien F, Zuzow M, Heald N, Gibson A, Shi Y, Goon L, Yu P, Engst S, Zhang W, Huang D. In vitro and in vivo activity of cabozantinib (XL184), an inhibitor of RET, MET, and VEGFR2, in a model of medullary thyroid cancer. Thyroid. 2013;23(12):1569–77.
292. Kurzrock R, Sherman SI, Ball DW, Forastiere AA, Cohen RB, Mehra R, Pfister DG, Cohen EE, Janisch L, Nauling F. Activity of XL184 (Cabozantinib), an oral tyrosine kinase inhibitor, in patients with medullary thyroid cancer. J Clin Oncol. 2011;29(19):2660–6.
293. Choueiri T, Pal SK, McDermott D, Morrissey S, Ferguson K, Holland J, Kaelin W, Dutcher J. A phase I study of cabozantinib (XL184) in patients with renal cell cancer. Ann Oncol. 2014;25(8):1603–8.
294. Nguyen HM, Ruppender N, Zhang X, Brown LG, Gross TS, Morrissey C, Gulati R, Vessella RL, Schimmoller F, Aftab DT. Cabozantinib inhibits growth of androgen-sen-

sitive and castration-resistant prostate cancer and affects bone remodeling. PLoS One. 2013;8(10):e78881.
295. Sadiq AA, Salgia R. MET as a possible target for non–small-cell lung cancer. J Clin Oncol. 2013;31(8):1089–96.
296. Drilon A, Rekhtman N, Arcila M, Wang L, Ni A, Albano M, Van Voorthuysen M, Somwar R, Smith RS, Montecalvo J. Cabozantinib in patients with advanced RET-rearranged non-small-cell lung cancer: an open-label, single-centre, phase 2, single-arm trial. Lancet Oncol. 2016;17(12):1653–60.
297. Choueiri TK, Escudier B, Powles T, Tannir NM, Mainwaring PN, Rini BI, Hammers HJ, Donskov F, Roth BJ, Peltola K. Cabozantinib versus everolimus in advanced renal cell carcinoma (METEOR): final results from a randomised, open-label, phase 3 trial. Lancet Oncol. 2016;17(7):917–27.
298. Velpula KK, Dasari VR, Asuthkar S, Gorantla B, Tsung AJ. EGFR and c-met cross talk in glioblastoma and its regulation by human cord blood stem cells. Transl Oncol. 2012;5(5):379-IN18.
299. Shintani T, Kusuhara Y, Daizumoto K, Dondoo TO, Yamamoto H, Mori H, Fukawa T, Nakatsuji H, Fukumori T, Takahashi M. The involvement of hepatocyte growth factor-MET-matrix metalloproteinase 1 signaling in bladder cancer invasiveness and proliferation. Effect of the MET inhibitor, cabozantinib (XL184), on bladder cancer cells. Urology. 2017;101:169. e7–. e13
300. Wang J, Cheng JX. C-met inhibition enhances chemosensitivity of human ovarian cancer cells. Clin Exp Pharmacol Physiol. 2017;44(1):79–87.
301. Liu X, Wang Q, Yang G, Marando C, Koblish HK, Hall LM, Fridman JS, Behshad E, Wynn R, Li Y. A novel kinase inhibitor, INCB28060, blocks c-MET–dependent signaling, neoplastic activities, and cross-talk with EGFR and HER-3. Clin Cancer Res. 2011;17(22):7127–38.
302. Falchook GS, Kurzrock R, Amin HM, Xiong W, Fu S, Piha-Paul SA, Janku F, Eskandari G, Catenacci DV, Klevesath M. First-in-man phase I trial of the selective MET inhibitor tepotinib in patients with advanced solid tumors. Clin Cancer Res. 2020;26(6):1237–46.
303. Jin H, Yang R, Zheng Z, Romero M, Ross J, Bou-Reslan H, Carano RA, Kasman I, Mai E, Young J. MetMAb, the one-armed 5D5 anti-c-met antibody, inhibits orthotopic pancreatic tumor growth and improves survival. Cancer Res. 2008;68(11):4360–8.
304. Martens T, Schmidt NO, Eckerich C, Fillbrandt R, Merchant M, Schwall R, Westphal M, Lamszus K. A novel one-armed anti-c-met antibody inhibits glioblastoma growth in vivo. Clin Cancer Res. 2006;12(20):6144–52.
305. Merchant M, Ma X, Maun HR, Zheng Z, Peng J, Romero M, Huang A, Ny Y, Nishimura M, Greve J. Monovalent antibody design and mechanism of action of onartuzumab, a MET antagonist with anti-tumor activity as a therapeutic agent. Proc Natl Acad Sci. 2013;110(32):E2987–E96.
306. Gelsomino F, Facchinetti F, Haspinger E, Garassino M, Trusolino L, De Braud F, Tiseo M. Targeting the MET gene for the treatment of non-small-cell lung cancer. Crit Rev Oncol Hematol. 2014;89(2):284–99.
307. Spigel DR, Ervin TJ, Ramlau RA, Daniel DB, Goldschmidt JH Jr, Blumenschein GR Jr, Krzakowski MJ, Robinet G, Godbert B, Barlesi F. Randomized phase II trial of onartuzumab in combination with erlotinib in patients with advanced non–small-cell lung cancer. J Clin Oncol. 2013;31(32):4105–14.
308. Giordano S. Rilotumumab, a mAb against human hepatocyte growth factor for the treatment of cancer. Curr Opin Mol Ther. 2009;11(4):448–55.
309. Wen PY, Schiff D, Cloughesy TF, Raizer JJ, Laterra J, Smitt M, Wolf M, Oliner KS, Anderson A, Zhu M. A phase II study evaluating the efficacy and safety of AMG 102 (rilotumumab) in patients with recurrent glioblastoma. Neuro-Oncology. 2011;13(4):437–46.
310. Gordon MS, Sweeney CJ, Mendelson DS, Eckhardt SG, Anderson A, Beaupre DM, Branstetter D, Burgess TL, Coxon A, Deng H. Safety, pharmacokinetics, and pharmacodynamics of AMG 102, a fully human hepatocyte growth factor–neutralizing monoclonal

antibody, in a first-in-human study of patients with advanced solid tumors. Clin Cancer Res. 2010;16(2):699–710.
311. Sharpe AH, Pauken KE. The diverse functions of the PD1 inhibitory pathway. Nat Rev Immunol. 2018;18(3):153–67.
312. Saigi M, Alburquerque-Bejar JJ, Mc Leer-Florin A, Pereira C, Pros E, Romero OA, Baixeras N, Esteve-Codina A, Nadal E, Brambilla E. MET-oncogenic and JAK2-inactivating alterations are independent factors that affect regulation of PD-L1 expression in lung cancer. Clin Cancer Res. 2018;24(18):4579–87.
313. Demuth C, Andersen MN, Jakobsen KR, Madsen AT, Sørensen BS. Increased PD-L1 expression in erlotinib-resistant NSCLC cells with MET gene amplification is reversed upon MET-TKI treatment. Oncotarget. 2017;8(40):68221.
314. Garcia-Diaz A, Shin DS, Moreno BH, Saco J, Escuin-Ordinas H, Rodriguez GA, Zaretsky JM, Sun L, Hugo W, Wang X. Interferon receptor signaling pathways regulating PD-L1 and PD-L2 expression. Cell Rep. 2017;19(6):1189–201.
315. Frumento G, Rotondo R, Tonetti M, Damonte G, Benatti U, Ferrara GB. Tryptophan-derived catabolites are responsible for inhibition of T and natural killer cell proliferation induced by indoleamine 2, 3-dioxygenase. J Exp Med. 2002;196(4):459–68.
316. Strickler JH, Weekes CD, Nemunaitis J, Ramanathan RK, Heist RS, Morgensztern D, Angevin E, Bauer TM, Yue H, Motwani M. First-in-human phase I, dose-escalation and-expansion study of telisotuzumab vedotin, an antibody–drug conjugate targeting c-met, in patients with advanced solid tumors. J Clin Oncol. 2018;36(33):3298–306.
317. Wang J, Anderson MG, Oleksijew A, Vaidya KS, Boghaert ER, Tucker L, Zhang Q, Han EK, Palma JP, Naumovski L. ABBV-399, a c-met antibody–drug conjugate that targets both MET–amplified and c-met–overexpressing tumors, irrespective of MET pathway dependence. Clin Cancer Res. 2017;23(4):992–1000.
318. Camidge DR, Morgensztern D, Heist RS, Barve M, Vokes E, Goldman JW, Hong DS, Bauer TM, Strickler JH, Angevin E. Phase I study of 2-or 3-week dosing of telisotuzumab vedotin, an antibody–drug conjugate targeting c-met, monotherapy in patients with advanced non–small cell lung carcinoma. Clin Cancer Res. 2021;27(21):5781–92.
319. Horinouchi H, Shibata Y, Looman J, Sui Y, Noon E, Lu S. Phase 2 study of telisotuzumab vedotin (Teliso-V) monotherapy in patients with previously untreated MET-amplified locally advanced/metastatic non-squamous non-small cell lung cancer (NSQ NSCLC). Am Soc Clin Oncol. 2023;41:TPS9149.
320. Tang Z, Du R, Jiang S, Wu C, Barkauskas D, Richey J, Molter J, Lam M, Flask C, Gerson S. Dual MET–EGFR combinatorial inhibition against T790M-EGFR-mediated erlotinib-resistant lung cancer. Br J Cancer. 2008;99(6):911–22.

Receptor Tyrosine Kinase Signaling and Mitochondrial Dynamics in Cancer

Sanam Kumari, Kunj Bihari Gupta, Sonu Kumar Gupta, Muthuswamy Thangaraju, and Kunal Kumar

1 Introduction

Receptor tyrosine kinases (RTKs) are a subclass of tyrosine kinases that regulate various intricate biological processes, such as cell proliferation, motility, differentiation, metabolism, and cell-to-cell communications. A diverse array of cell surface receptors that respond to various signaling molecules, such as growth factors, cytokines, and hormones, are members of the RTK family. The human genome encodes 58 RTKs out of ninety unique tyrosine kinases, further distributed among 20 subfamilies [1–3]. All RTKs share a similar protein structure comprising an extracellular ligand-binding domain, a single transmembrane helix, an intracellular region with the juxtamembrane regulatory region, a tyrosine kinase domain (TKD), and a carboxyl(C-) terminal tail [4]. The extracellular domain of RTKs is a dynamic region that regulates ligand binding, receptor activation, and subsequent signaling cascades, which make it a crucial factor in RTK activity and cellular responses [4].

S. Kumari
Department of Mathematics, Science and Technology, Paine College, Augusta, GA, USA
e-mail: skumari@paine.edu

K. B. Gupta
Georgia Cancer Center, Augusta University, Augusta, GA, USA
e-mail: kugupta@augusta.edu

S. K. Gupta
Department of Physiology and Endocrinology, Medical College of Georgia, Augusta University, Augusta, GA, USA
e-mail: sgupta2@augusta.edu

M. Thangaraju · K. Kumar (✉)
Department of Biochemistry and Molecular Biology, Medical College of Georgia, Augusta University, Augusta, GA, USA
e-mail: mthangaraju@augusta.edu; kukumar@augusta.edu

A. Sabarwal et al. (eds.), *Receptor Tyrosine Kinases in Cancer*, Cancer Drug Discovery and Development, https://doi.org/10.1007/978-3-031-93894-8_7

It has various structural elements such as EGF-like domains, immunoglobulin-like domains, fibronectin type III-like repeats, and cysteine-rich regions, which are the basis of the RTK classification into different families [5]. The specificity of each receptor's ligand binding and regulatory properties depends on these domains' number, arrangement, and combination [6]. The intracellular domain of RTKs forms an alpha-helical structure from a sequence of amino acids contributing to the kinase's overall function and regulation [7]. In the kinase domain, this helix helps maintain the enzymatic conformation and is also involved in adenosine triphosphate (ATP) binding. It interacts with other parts of the kinase domain, such as the activation loop, and regulates the kinases' activity [8]. Once the kinase is activated, the intracellular helix often undergoes a shift in position to acquire the alignment of required residues and becomes catalytically active. This helix influences the substrate access to the active site and might play a role in the interaction of regulatory molecules. The intracellular domain of RTK is considered a cornerstone of cellular signal transduction. The tyrosine kinase domain, which lies in the heart of the intracellular domain, is an enzymatic center that catalyzes the phosphorylation of specific tyrosine residues on the target protein via ATP [5]. The intracellular domain also contains regulatory regions, such as the juxtamembrane domain, which inhibits the kinase activity without ligand, and the C-terminal tail, often containing multiple tyrosine residues [9]. These residues are crucial for signal propagation, as they act as a docking site for the adaptor and effector proteins once phosphorylated.

The signaling process starts with the ligand binding to the RTK extracellular domain, which induces receptor dimerization and subsequent autophosphorylation. Autophosphorylation of a specific tyrosine residue within the intracellular domain creates a binding site for proteins with Src homology (SH2) or phosphoserine binding (PTB) domains [10, 11].

RTKs are shown to regulate normal cellular processes, such as cell growth, survival, differentiation, migration, metabolism, and the cell cycle. However, genomic studies have identified several alterations in genes encoding RTKs, such as *EGFR*, *HER2/ErbB2*, and *MET* [12].

The double-membrane organelles, such as mitochondria, execute several vital cellular functions, including energy production, redox status, generation of reactive oxygen species, cytosolic calcium level, and initiation of apoptosis [13]. Mitochondrial dysfunction is associated with many human disorders, including cardiac disorders [14], metabolic diseases, aging, nervous system diseases [15], and cancer progression [13]. Through mitochondrial dynamics, damaged or dysfunctional mitochondria can be eliminated via mitophagy to avert further cellular damage. Maintaining a proper balance of mitochondrial dynamics is crucial for the optimal functioning of mitochondria and cellular fate [16]. Fission aids in the quality control of mitochondria by removing impaired or malfunctioning mitochondria and promoting apoptosis in the face of significant cellular stress. At the same time, fusion facilitates the mixing and exchanging of intramitochondrial contents among mitochondria, which is vital for preserving mitochondrial function. Additionally, mitophagy plays a critical role in mitochondrial quality control by selectively

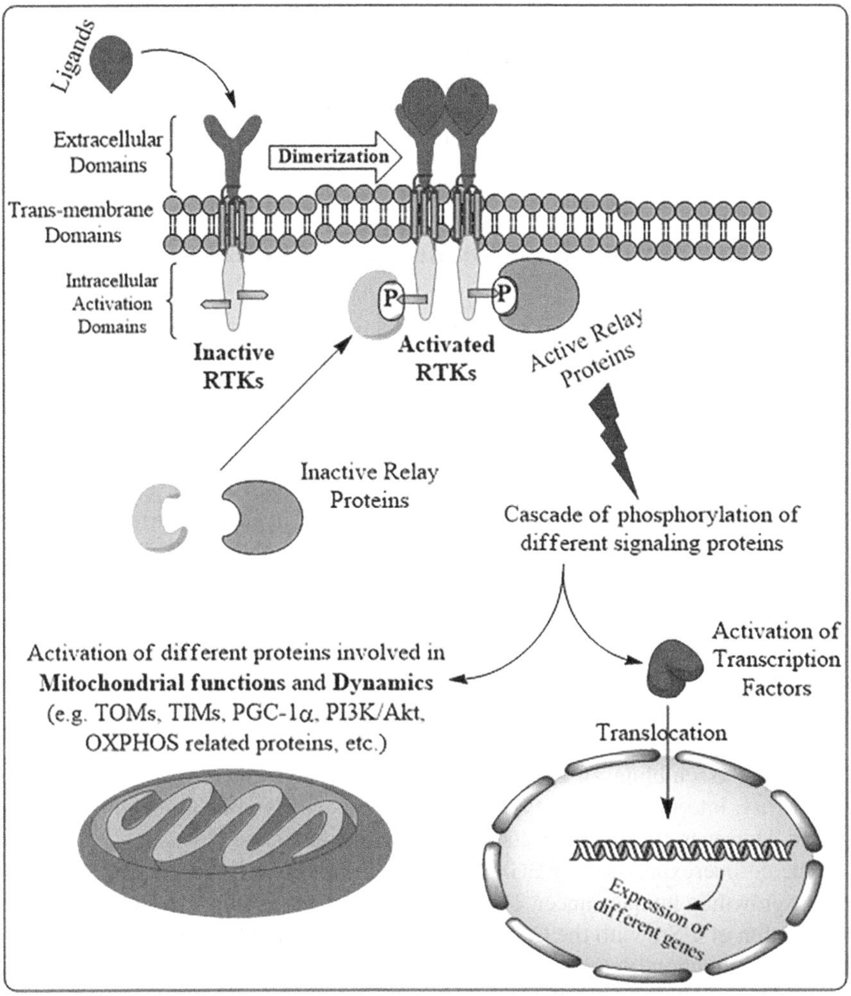

Fig. 1 Flow chart illustrating the activation of receptor tyrosine kinases (RTKs). The process begins with ligands binding to extracellular domains, leading to dimerization and activation of RTKs. This triggers phosphorylation, activating relay proteins. A cascade of phosphorylation events follows, involving different signaling proteins. The chart shows activation of transcription factors, translocation to the nucleus, and expression of genes. It also highlights the activation of proteins involved in mitochondrial functions and dynamics, such as TOMs (translocase of outer membranes), TIMs (translocase of inner membranes), PGC-1α (peroxisome proliferator-activated receptor gamma coactivator 1 alpha), PI3K/Akt (phosphatidylinositol 3-kinase/Akt), and OXPHOS (mitochondrial oxidative phosphorylation)-related proteins. Key elements include inactive and active RTKs, relay proteins, and a mitochondrion.

clearing damaged mitochondria and inhibiting the release of pro-apoptotic proteins [16]. Figure 1 summarizes the above discussion [17].

It has been demonstrated that mitochondrial dynamics are regulated at each step through carefully controlled upstream signaling pathways. This chapter explores how receptor tyrosine kinases (RTKs) are activated in normal physiological conditions. We also review the mechanisms by which different RTKs can influence mitochondrial dynamics in overly active human cancers.

2 Epidermal Growth Factor Receptor (EGFR)

Epidermal growth factor receptor (EGFR) is a tyrosine kinase expressed on the cell surface to trigger downstream kinase signaling pathway via ligand binding-mediated phosphorylation, and its autoactivation is correlated with tumor progression [18]. Upon ligand (EGF) binding, some EGF-EGFR complexes are internalized through endocytosis and degraded [19] after fusing with the endosome, and some complexes are recycled back to the cell surface [20, 21]. Although EGFR lacks a specific organelle-targeting sequence, the translocated EGFR in subcellular organelles has been reported [19, 22–24]. Previously, it was reported that EGF-induced EGFR translocates to the mitochondria, leading to mitochondrial fission and redistribution in the lamellipodia, upregulating cellular ATP production in highly invasive non-small-lung cancer cells. They showed that EGFR interacts with Mitofusin 1 (Mfn1) (one of the key components of mitochondrial dynamics) to induce mitochondrial fragmentation. Mfn1 has a GTPase domain, two transmembrane segments (TM), and two hydrophobic heptad repeats (HR). Mfn1 polymerizes through the HR2 domain by assembling an antiparallel-coiled coil. Mitochondrial EGFR interacts with Mfn1, interferes with Mfn1 polymerization, and inhibits mitochondrial elongation. However, the precise mechanism was not elucidated [25–27].

The co-overexpression of EGFR and c-Src is frequently linked with enhanced tumor growth in human cancer. This synergistic growth requires an EGF-dependent association of c-Src with the EGFR and phosphorylation of Tyr-845 of the receptor by c-Src. Mitochondrial cytochrome *c* oxidase subunit II (CoxII) binds to EGFR at Tyr(P)-845 following EGF binding. This association helps EGFR to translocate to the mitochondria. Michelle et al. demonstrated that c-Src translocates to the mitochondria with similar kinetics as EGFR, depending on the catalytic activity of EGFR and c-Src and the mitochondrial localization signal required. Both EGFR and c-Src can phosphorylate CoxII; furthermore, stimulation with EGF leads to a decrease in CoxII activity and ATP levels within the cell, a process significantly reliant on the presence of EGFR in the mitochondria. These results indicate that EGFR uniquely influences mitochondrial activity through interaction with and alteration of CoxII [28].

The EGFR is overexpressed in various human epithelial tumors, often because of gene amplification. In tumor, EGFR gene amplification frequently contains EGFR gene rearrangements that leads to mutation in the EGFR extracellular domains, and the most common mutated EGFR is EGFRvIII.

Additionally, it was reported the presence of EGFRvIII in the tumor cell mitochondria for the first time. They demonstrated that both EGFR and EGFRvIII are consistently present within the mitochondria. Notably, the mitochondrial transport of both EGFR and EGFRvIII significantly increased after treatments with apoptosis inducers, staurosporine, and anisomycin and with the EGFR kinase inhibitor, Iressa. By utilizing engineered mutant EGFR/EGFRvIII receptors designed for enhanced intracellular trafficking specifically to the mitochondria (and not to the cell nucleus), they found that glioblastoma cells with a higher mitochondrial presence of EGFRvIII demonstrated increased resistance to growth suppression and apoptosis caused by staurosporine and anisomycin. Tumor cells with accumulated mitochondrial EGFRvIII showed a strong resistance to growth inhibition induced by Iressa. These results suggest that apoptosis inducers and EGFR-targeted inhibitors promote the mitochondrial translocation of both EGFR and EGFRvIII and that the mitochondrial presence of these receptors plays a role in drug resistance in tumors. Additionally, this data supports a possible connection between the mitochondrial EGFR/EGFRvIII pathway and apoptosis [29].

ErbB2, a member of the EGFR family, also referred to as Her2/neu, is an oncogene that exhibits overexpression in various cancer types and is associated with an unfavorable prognosis [30]. Multiple reports have indicated that ErbB2 enhances the transformation and/or metastatic capabilities of human breast cancer [31, 32]. Furthermore, ErbB2 has been shown to activate signaling molecules that influence bioenergetic metabolism. It is reported that ErbB2 drives cancer cell proliferation and glycolysis by increasing lactate dehydrogenase isoform A (LDH-A) expression [33]. However, it remains uncertain if ErbB2 modulates cell metabolism through other mechanisms beyond enhancing glycolysis. It is well recognized that ErbB2 is situated at the plasma membrane, where it phosphorylates downstream targets on tyrosine residues in response to external signals [33]. Recent research has also indicated that ErbB2 can move into the nucleus and may act as a transcription factor [34].

One study demonstrated that Heregulin β1, a ligand for ErbB family receptor tyrosine kinases, influences cytochrome *c* oxidase subunit II (CoxII) in mammary epithelial cells [35]. Another piece of research noted that inhibiting the ErbB2 receptor causes cardiomyocyte death via mitochondrial and reactive oxygen species-dependent pathways [36]. Moreover, several significant kinases and phosphatases, such as PKC and Abl, Src, MAPK, and Shp-2, have been reported to mediate tyrosine phosphorylation and dephosphorylation of proteins within mitochondria [37]. These findings imply that the subcellular distribution of ErbB2 could determine signaling specificity, and ErbB2 might play a role in regulating mitochondrial dynamics. In conclusion, many ErbB2 translocate to the mitochondria of cancer cells and patient samples via a well-established mitochondrial translocation route aided by mtHSP70.

Furthermore, mtErbB2 influences mitochondrial respiratory functions and transitions the energy reliance of cancer cells from oxidative phosphorylation toward aerobic glycolysis. Notably, the migration of ErbB2 into mitochondria serves as a new mechanism for trastuzumab resistance. However, several questions need to be

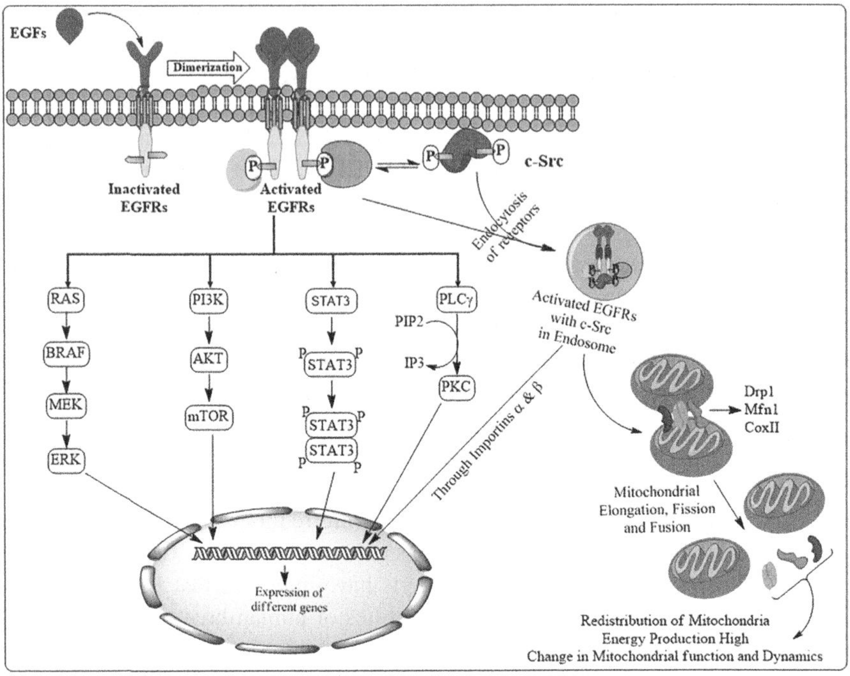

Fig. 2 Flow chart illustrating the signaling pathway of Epidermal Growth Factor Receptors (EGFRs). The process begins with the dimerization of inactivated EGFRs upon binding with EGFs, leading to activation. Activated EGFRs interact with c-Src, triggering endocytosis. The pathway branches into several signaling cascades: RAS to BRAF to MEK to ERK, PI3K to AKT to mTOR, and STAT3 phosphorylation. Another branch involves PLCγ leading to PIP2 conversion to IP3 and PKC activation. These pathways culminate in gene expression and mitochondrial changes, including elongation, fission, fusion, and redistribution, affecting energy production and dynamics. Key components include Drp1, Mfn1, and CoxII.

addressed, such as the following: What is the specific mechanism by which mtErbB2 modulates mitochondrial respiratory function? What are the potential implications of mtErbB2 in cancer progression and treatment? Addressing these queries will not only resolve a fascinating issue within the field of cancer biology, but more crucially, it may pave the way for identifying a new biomarker and therapeutic target for breast and other types of cancers [38]. The above discussion has been summarized in Fig. 2.

3 Insulin-like Growth Factor-1 Receptor (IGF-1R)

Insulin-like growth factor 1(IGF-1) receptor (IGF-1R) signaling plays a crucial role in growth and promotes tumor development, cancer cell survival, and proliferation. It also has a well-conserved and significant impact on mitochondrial function and homeostasis [39–41]. The absence of IGF-1 signals, which leads to decreased mitochondrial ability to generate ATP and reduced antioxidant capacity, can be reversed through IGF-1 therapy, indicating that IGF-1 signaling has protective effects on mitochondria [41]. However, anticancer treatments that target the IGF-1 receptor (IGF-1R) have been largely unsuccessful in clinical trials [42, 43].

Mitochondrial dysfunction frequently characterizes cancer cells, which can result from altered dynamics of fusion and fission, oxidative stress, metabolic reprogramming, and various other mitochondrial modifications, as discussed by Vyas and Haigis [44]. Growing evidence indicates that mitophagy dysregulation is prevalent in cancer, with the accumulation of dysfunctional mitochondria contributing to cancer advancement [45]. A precise balance between mitochondrial biogenesis and mitophagy is essential to maintain a sustainable population of mitochondria in healthy cells [46]. However, it has been suggested that mitophagy and mitochondrial biogenesis are disconnected in cancer due to increased mitochondrial biogenesis and decreased mitophagy [45].

Various reports have demonstrated that IGF-1 signaling impacts mitochondrial function and health, including maintaining mitochondrial DNA and RNA, biogenesis, oxidative phosphorylation, and reducing reactive oxygen species (ROS) production [41, 47, 48]. IGF-1 stimulates the expression of the highly conserved mitochondrial carrier for pyrimidine nucleotides, solute carrier family 25 (pyrimidine nucleotide carrier 1), member 33 (PNC1/SLC25A33), which is a homolog of the essential yeast carrier RIM2 and is critical for maintaining mitochondrial DNA and RNA. Notably, suppressing PNC1/SLC25A33 expression in cancer cell lines leads to severe mitochondrial dysfunction, heightened ROS production, and epithelial-mesenchymal transition (EMT) triggering [47]. This finding implies a significant role for IGF-1 signaling in sustaining healthy mitochondria and curbing ROS-induced EMT in cancer cells. Subsequently, it was reported that IGF-1 enhances mitochondrial biogenesis by upregulating the transcriptional co-activators peroxisome proliferator-activated receptor gamma coactivator 1-β (PGC-1β) and PGC-1-related coactivator (PRC), as well as promoting the expression of the mitophagy receptor BNIP3 (Bcl-2/adenovirus E1B 19 kDa protein-interacting protein 3) [49].

BNIP3 was initially characterized as a BH3-only member of the Bcl-2 family that can interact and inhibit pro-survival Bcl-2 proteins to trigger apoptosis [50]. It is closely related to BNIP3L/Nix, with both proteins featuring an LIR domain that allows interaction with LC3II on autophagosomes, thereby promoting mitophagy [51]. Although BNIP3 and BNIP3L exhibit distinct patterns of tissue-specific expression, both are upregulated by hypoxia to facilitate mitophagy independent of PINK1/Parkin [52]. Notably, the expression of BNIP3 has been shown to rise during the early phases of breast cancer but declines as tumors develop hypoxia [53].

Riis et al. (2020) [54] demonstrated that IGF-1 facilitates a conserved signal via Nrf2 to enhance BNIP3 expression in cancer cells and mouse embryonic fibroblasts (MEFs). Nrf2 is recognized for its cytoprotective role in responding to oxidative stress and detoxification. In the context of cancer, this could help cells thrive in challenging environments and promote chemoresistance [55, 56]. They found that the

IGF-1-mediated increase of BNIP3 [49] can occur through HIF-1α or NRF1, both of which are induced by Nrf2 [57–59], indicating that BNIP3 may not be a direct target gene of Nrf2 but rather a secondary one. The observation that NRF1 may play a role in inducing BNIP3 is noteworthy, as NRF1 is generally recognized for promoting mitochondrial biogenesis rather than mitophagy [60]. It has been reported that IGF-1 stimulates NRF1 during mitochondrial biogenesis [49]. NRF1 functioning downstream of IGF-1 and Nrf2 supports the notion that the synthesis and degradation of mitochondria are interconnected through an IGF-1-Nrf2-BNIP3 pathway.

Mitophagy is vital for maintaining overall cellular health, and its dysregulation is frequently observed in numerous diseases [61]. Moreover, mitophagy is critical in cell differentiation, which necessitates precise regulation of mitochondrial mass and distribution to match the cell's specific energy production and reactive oxygen species (ROS) clearance needs [62, 63]. BNIP3 has been shown to enhance mitochondrial function in stem cells under hypoxic conditions and to aid in the differentiation of epidermal keratinocytes [64, 65]. However, mitophagy's essential role in differentiating various cell types allows these processes to induce the dedifferentiation of fully developed cells during cancer progression, enabling cancer cells to attain stem-like characteristics [66]. Notably, it has been suggested that the IGF-1 signaling pathway maintains stem-like features and boosts cell survival during differentiation [67, 68].

The downregulation of either Nrf2 or BNIP3 in MCF-7 cells led to a significant change in mitochondrial morphology, resulting in more fused and elongated mitochondria. This difference was markedly more pronounced in R− (IGF1R KO) cells compared to R+ (IGF1R WT) cells. Furthermore, reducing BNIP3 expression was connected to the lower levels of Drp1, a mediator of mitochondrial fission known to be recruited by BNIP3 to facilitate mitophagy [69]. Therefore, IGF-induced BNIP3 can influence mitochondrial dynamics. Interestingly, the punctate structure of mitochondria and increased Drp1 levels observed in control MCF-7 cells compared to those with suppressed Nrf2 or BNIP3 have been similarly noted as characteristics associated with stem cells [62].

Furthermore, BNIP3 protein may have at least dual functions, either they can promote cell death under certain conditions or induce mitophagy, especially in hypoxic environments [70]. It has also been implicated in the regulation of cell proliferation [71]. As IGF-1 promotes BNIP3 in serum-deprived cells, while IGF-1 is recognized for its protective effect against apoptosis, it is plausible that BNIP3 initiates mitophagy in such conditions. The findings regarding autophagic flux in R− and R+ cells support this conclusion. R− cells displayed higher baseline levels of autophagic flux than R+ cells, which aligns with the expected activation of mTORC1 by IGF-1, inhibiting general autophagy [72]. However, R+ cells demonstrated a significantly stronger ability to induce autophagosomal turnover in response to metabolic or mitochondrial stress than R− cells, indicating that IGF-1 signaling enhances the capability of cells to trigger both autophagy and mitophagy. Additionally, there is a notable difference in the sensitivity of R− and R+ cells to cellular stress caused by serum starvation or DFP. R+ cells exhibited markedly lower levels of apoptosis under these stress conditions than R− cells, implying that the IGF-1 pathway boosts cellular resistance to such stresses. The increased levels of BNIP3 in R+ cells are anticipated to facilitate efficient mitophagy clearance in response to DFP. Moreover, BNIP3 has been shown to promote cell survival during nutrient deprivation [73], which may play a role in the enhanced survival seen with R+ cells in serum-deprived environments. It is also noteworthy that neither R− nor R+ cells experienced cell death in response to CCCP (carbonyl cyanide 3-chlorophenylhydrazone), which is known to initiate mitophagy through the PINK1-parkin pathway [74, 75] even though CCCP induced PINK1 expression. This indicates that R− cells lacking IGF-1 signaling may be particularly deficient in BNIP3-mediated mitophagy. Consequently, while alternative roles of BNIP3 cannot be entirely dismissed, the expression levels of BNIP3 are pivotal to the significant differences observed in autophagic flux and cell survival between R− and R+ cells when exposed to either serum starvation or DFP. In summary, these findings show that IGF-1 signaling plays a crucial role in mitochondrial dynamics and turnover through the actions of Nrf2 and BNIP3. This protective function for mitochondria may represent a critical vulnerability of this vital growth pathway and present opportunities for innovative therapeutic strategies in cancer [54].

4 Vascular Endothelial Growth Factor Receptors (VEGFR)

Mitochondrial Biogenesis Angiogenesis demands higher levels of ATP, which is produced in mitochondria by the oxidative phosphorylation process that needs higher biogenesis of mitochondria. The biogenesis of mitochondria requires close coordination between the expression of mitochondrial and nuclear genes, as well as synthesis, import, and assembling of mitochondrial proteins in a regulated way. One of many such regulators is PGC-1α (peroxisome proliferator-activated receptor γ coactivator-1 α), which acts upstream and regulates mtDNA transcription. Most mitochondrial proteins are synthesized by nuclear genes, which are translocated through the outer membranes (TOMs), which recognize N-terminal signal sequences of the preprotein and enter into the matrix through the inner membrane (TIMs). Researchers found that about 30 different proteins are involved in this process [76, 77].

Angiogenesis (i.e., formation of new blood vessels) and vascularization (denotes the formation of blood vessels within the tumor) are the essential processes of tumor growth and metastasis. This process helps in the constant supply of nutrients and oxygen to the tumor cells and helps to grow beyond a specific size (more than 2 mm^3) [78]. The vascular endothelial growth factor (VEGF) family of proteins, which includes VEGF-A, VEGF-B, VEGF-C, VEGF-D, and placenta growth factor (PIGF), are the major proteins associated with angiogenesis. VEGF are 40–45 kDa homodimer proteins secreted by a wide variety of cells and exert their biological functions after binding to the class of vascular endothelial growth factor receptors (VEGFRs) [79]. VEGFRs are a type of receptor tyrosine kinase (RTK). When

VEGF binds to it, receptors get homo-dimerized, and the intracellular tyrosine kinase domain gets phosphorylated, initiates a cascade of downstream signaling, and promotes angiogenesis. VEGFR mainly includes three types of receptors, viz., VEGFR-1 (also known as Flt-1, localized on blood vascular endothelial cells and critical for hematopoietic cell development), VEGFR-2 (KDR/Flk-1, expressed on both blood vascular and growing lymphatic vessels and critical for vascular endothelial cell development) and VEGFR-3 (Flt-4, expressed on blood vascular and critical for lymphatic endothelial cells development) [80] along with these neuropilins-1 and -2 (NRP-1 and NRP-2) act as a co-receptor for action of VEGF [81]. VEGFRs are crucial for normal physiological conditions as well as pathological conditions, including cancer, where they promote new blood vessel formation. As with other RTKs, VEGFRs also consist of three domains: the extracellular domain, which consists of 7 immunoglobulin-like domain, a 20-amino-acid-long α-helix transmembrane domain, and the intracellular tyrosine kinase domain containing key phosphorylation sites [82]. Downstream of VEGFR, three core signaling pathways occur: PI3K (phosphatidylinositol 3-kinase)-Akt (protein kinase B) pathway, the Raf-MEK (mitogen-activated or extracellular signal-regulated protein kinase kinase)-MAPK [83]. More specifically, this VEGF/VEGFR signaling regulate the expression of Akt3 protein through the sphin-gosine-1-phosphate S1P3/Edg3 [84] and Akt3 regulate the PGC-1α a master regulator of mitochondrial biogenesis. Specifically inhibiting the Akt3 results in the reduction of mtDNA and gene expression and affects mitochondrial biogenesis [85]. The PI3K/Akt signaling pathways also regulate the activation of the mammalian target of rapamycin (mTOR). This kinase regulates cell growth, survival, and size and is a vital component in the energy and nutrient pathways. mTOR is necessary for maintaining mitochondrial oxidative function and positively regulates the expression of PGC-1α [86, 87]. These findings point out the role of VEGF/VEGFR signaling in mitochondrial biogenesis and cancer progression.

5 Fibroblast Growth Factor Receptor (FGFR)

This is another subfamily of RTK that plays a significant role in various cellular processes, including regulating cell proliferation, differentiation, migration, survival, etc. Dysregulation of fibroblast growth factor receptors (FGFRs) is associated with different types of cancers, including urothelial carcinoma, hepatocellular carcinoma, ovarian cancer, lung adenocarcinoma, and glioblastoma [88]. Along with cancer, other developmental diseases like achondroplasia and thanatophoric dysplasia are also associated with its mutation. The human FGFR family consists of four members, viz., FGFR-1 to 4 [88], which act as a receptor for about 22 different members of fibroblast growth factor proteins (FGF; largest family of growth factor ligands divided into seven subfamilies) [89]. One FGFR can bind to multiple FGFs, and vice versa, allowing a wide range of signaling possibilities and highlighting the diversity and complexity of FDF-FDFR signaling. As with other RTKs, FGFRs also share a canonical structure which consists of a large extracellular domain, which consists of three immunoglobulin-like domains (viz., D-1 to 3). In between the D-1 and D-2 regions, there is a conserved, aspartic acid-rich motif, known as the acid box, a single-span transmembrane α-helix domain, and an intracellular tyrosine kinase domain having phosphorylation and dephosphorylation sites [88, 90]. A genetic sequencing study of more than four thousand cancer patients showed that about 7.1% of patients have some irregularities in the FGFR genes. These aberrations include an increase in gene number (53.7–66%), point mutations (26–38.8%), gene rearrangements (5.6–8%), etc. [91].

When FGFs (typical molecular weight is about 17–34 kDa) interact with the FGFRs, this binding causes dimerization of receptors followed by autophosphorylation of intracellular kinase, which triggers the cascade of downstream intracellular signaling pathways. Depending on the FGFs, this pathway may be mitogen-activated protein kinase (MAPK), extracellular signal-regulated kinase 1/2 (ERK1/2) pathway, signal transducer and activator of transcription (STAT) pathways, PI3K/Akt pathways, etc., and finally regulate various normal and pathophysiological processes, like angiogenesis, wound healing, embryonic development, metabolic regulation, etc. [92, 93]. Especially, FGFR-1 and FGFR-4, when interacting with FGF-21 and FGF-19, respectively, play a critical role in enhancing mitochondrial function and biogenesis. They activate the AMPK/MAPK and AMPK/SIRT-1/PGC-1α pathway, which expresses the nuclear respiratory factors (NRFs), and these are all essential for mitochondrial DNA replication and biogenesis. FGFR signaling has also been linked to metabolic reprogramming by increasing oxidative phosphorylation (OXPHOS), promoting high ATP production. These findings indicate that FGFRs help maintain mitochondrial health and regulate cellular functions by ensuring a constant energy supply.

6 Hepatocyte Growth Factor Receptor (HGFR)

It is a receptor protein with a 145 kDa size that belongs to the RTK subfamily. In humans, it is encoded by the MET gene, popularly known as MET or c-MET. It is a type of proto-oncogene and helps maintain cell growth, migration, and survival [94]. The aberrant expression of HGFR through overexpression, gene amplification, mutation, etc., leads to many disease conditions, including different cancers like breast, liver, gastric, lung, etc. [95]. As with other RTK, it also consists of an extracellular domain having two (α and β) Sema domains, four [1–4] immunoglobulin domains, and, in between these two, one cysteine-rich domain followed by one transmembrane domain and intracellular tyrosine kinase domain [96]. Hepatocyte growth factor (HGF) or scatter factor (SF) and its spliced variant (NK-1 and NK-2) are the ligands that interact with HGFR and initiate a cascade of pathways and regulate critical physiological processes. When SF binds with HGFR, the intracellular tyrosine, especially at sites 1234 and 1235, gets phosphorylated, which induces second messenger proteins (like PI3K, GAP, NCK, etc.) and promotes oncogenic signaling leading to tumor growth and metastasis [97]. At the downstream signaling, HGF/HGFR activates the PI3K/Akt, RAS/MAPK, STAT3, and PLCγ pathways and promotes cancers through angiogenesis, motility, invasion, metastasis, etc., and promotes resistance to traditional anticancer drugs, making it an important target for cancer therapy [98, 99]. HGF/HGFR signaling induces canonical oncogenic pathways and mitochondrial function and biogenesis to support energy demands during cell proliferation and rapid growth. It increases the expression of PGC-1α, NRFs, mitochondrial transcription factor A (TFAM), etc., required for mtDNA replications, transcriptions, and translocation of materials through TOMs and TIMs mechanism. The HFG/HFGR signaling and mitochondrial biogenesis are relevant to cancers, where rapid demands of energy are needed to support neoplastic growth.

7 Colon Carcinoma Kinase (CCK)

It is a novel, evolutionarily conserved, non-active member of the RTK family, isolated initially from colon carcinoma tissues [100]. It is also known as protein tyrosine kinase-like 7 (PTK7) and colon carcinoma kinase-4 (CCK4). In humans, it was encoded by the PTK7 gene. With other RTKs, it also has an extracellular domain consisting of seven immunoglobulin-like domains, one transmembrane α-helix domain, and an intracellular domain that lacks detectable catalytic kinase activity; instead, it retains its signal transduction ability, which makes it unique. Many reports show that the level of PTK7 expression is high in many cancers, including colon, lung, gastric, and acute myeloid leukemia [101].

Wound healing and regeneration require precise and coordinated movement of cells, accomplished by cell-cell communication. Wnt family proteins tightly regulate this process. PTK7 is a co-receptor of Wnt, which regulates canonical (β-catenin dependent) and noncanonical (β-catenin independent) Wnt signaling and regulates planar polarity of cells (through PCP pathways), cell migration, and adhesions [102]. PTK7 also acts as a co-receptor for other receptors like VEGFR-1 and Plexin A1, controlling angiogenesis, cell migration, etc. [103, 104]. PTK7 may also influence mitochondrial health and biogenesis by regulating the key transcription factors and related enzymes that interfere with mtDNA replication, thus affecting energy pool and total mitochondrial functions. This is important in the context of cancer progression and therapy, highlighting the probability of PTK7 as a therapeutic target, especially in the case of cancer treatment.

8 Tyrosine Kinase Receptor in Endothelial Cells (TIE)

Its isolation and characterization started in the 1990s and constituent a distinct part of the RTK subfamily, very unique extracellular receptor structures having two immunoglobulin-like domains, one epidermal growth factor, and three fibronectin type III domain [105]. The TIE receptor family consists of two members, viz., TIE-1 and TIE-2; TIE-2 is also known as TEK, interacting with closely related angiopoietin growth factors (ANGPT or ANG). ANG legends mainly interact with TIE-2 and promote angiogenic capillary growth and blood vessel maturity; this process also requires VEGF-dependent signaling, whereas TIE-1 is an orphan receptor (i.e., showing both agonist and antagonist action depending on context). The TIE-ANG signaling is important for cell-to-cell communications, cardiovascular, including blood and lymphatic vessel development and stability [106, 107]. Dysregulation of the TIE-ANG system contributes to many disease conditions, like vascular malformations and diverse types of cancers. The role of TIE in mitochondrial biogenesis is not fully elucidated; however, reports suggest it can regulate PI3K/Akt pathways that may lead to regulation of mitochondrial functioning and biogenesis in cancer cells, which requires further investigation.

9 Rearranged During Transfection (RET)

RET is an important member of the RTK subfamily and was first discovered in the year 1985 by Takahashi et al. [108]. It is important for the normal development of cells, and mutation leads to many disease conditions, including various cancers. It plays a significant role in kidney and nervous system development. With other RTK family proteins, it also has an extracellular domain having four cadherin-like domains (CLD1–4) and a conserved cysteine-rich domain, followed by a transmembrane α-helix domain, an intracellular catalytic kinase domain, and isoform-specific C-terminus; there is a juxtamembrane segment also that lies between the transmembrane domain and intracellular domain [109]. Based on this C-terminus structure, which is the result of alternate splicing, RET is of three types based on the number of amino acids: RET9 (having nine amino acids), RET41 (having 41 amino acids) and RET51 (having 51 amino acids) [110]. Although these isoforms share common sequences and are co-expressed in many tissues but may activate different pathways in normal and disease conditions, RET9 has a proactive role in promoting cancer, whereas RET51 has a proactive role in cell proliferation, migration, and anchorage-dependent growth. In some cases, like multiple endocrine neoplasia type-2 (MEN2), the abundance of RET51 is more than RET9 [111]; in the case of breast cancer, estrogen increases the expression of RET51 at a much higher level [112], and the level of RET51 is high in the stage IIB pancreatic tumor [113]. Based on these findings and other available literature, it is suggested that RET51 is more aggressive than RET9 in promoting cell proliferation, migration, and neoplastic growth of different cancers.

RET gets activated after binding with the coreceptor ligand complex, which belongs to glial cell line-derived neurotrophic factor (GDNF) family ligands (GFLs). These GFLs didn't directly interact with RET receptors but bind to GDNF family receptor-α (GFRα), which is a coreceptor. The GFL family consists of four types of members: GDNF, neurturin (NRTN), artemin (ARTN), and persephin (PSPN) [114,

115]. When these ligands and coreceptors interact with the RET receptor, it dimerizes and sends a signal to the intracellular catalytic site for autophosphorylation and starts downstream signaling. The RET intracellular site has 12 phosphorylation sites, out of which Tyr1062 is the main docking site for several adaptor proteins, such as RAS/ERK, RAS/MAPK, PI3K/AKT, JNK, etc. [116], which modulate the cellular proliferation, differentiation survival, and migration. RET is a proto-oncogene, and it gains in function through any mutation like gene fusion, chromosomal rearrangement, or missense mutation. Elevated RET expression leads to a different type of cancers in humans, like chronic myelomonocytic leukemia, colorectal, breast, ovarian, head and neck tumors, lung, etc. [117]. Although the direct role of RET receptor-related pathways is not linked to mitochondrial biogenesis, fusion, and fission of mitochondria, it can affect the metabolic pathways, especially energy metabolic pathways like glycolysis and oxidative phosphorylation, which are crucial for mitochondrial health and cancer propagation.

10 Lemur (LMR)

In the last decade, LMR receptor kinase family of proteins has gained much attention as it is involved in many human disease conditions, including neurodegenerative diseases, cystic fibrosis, and different types of cancer. The lemur kinase receptor, which is also known as apoptosis-associated tyrosine kinase (AATK or AATYK), is a type of RTK subfamily type XVIII. It consists of four members, named LMTK-1α, LMTK-1β, LMTK-2, and LMTK-3 which are encoded by three genes [118]. LMTK-1α and LMTK-1β are encoded by the same gene through alternate splicing [119]. Like other RTK family members, it also has an extracellular short domain, a transmembrane domain, followed by a long, extended intracellular kinase domain, taking its name from the long tail of Madagascar lemurs. Although it shows characteristics similar to the RTKs, a worldwide group of senior researchers recognizes it as a serine/threonine-protein kinase instead of a tyrosine kinase [118]. These receptor proteins are involved in many cellular processes, including signal transduction, apoptosis, and recent studies insinuating their role in cancer biology [120]. LMTK-2 is one of the most susceptible genes in the case of prostate cancer [121], whereas studies suggest that expression levels of LMTK-2 and LMTK-3 can be used as a predictive marker in many human cancers, like colorectal, gastric, breast, lung, etc. [122]. The direct role of LMTK on mitochondrial biogenesis is not clear yet, but it can affect the mitochondrial pathways in many ways to enhance cancer progression.

11 Platelet-Derived Growth Factor Receptor (PDGFR)

PDGFs are a growth factor family, each member containing one of four different polypeptide chains: PDGF-A, PDGF-B, PDGF-C, and PDGF-D [123–125]. Each chain is encoded by an individual gene located on chromosomes 7, 22, 4, and 11, respectively [126, 127]. These individual chains can form five subtypes of PDGF dimers through a disulfide bond, including four homodimers (PDGF-AA, PDGF-BB, PDGF-CC, and PDGF-DD) and one heterodimer (PDGF-AB) [128]. There are two types of receptors for PDGFs, PDGFRα and PDGFRβ, which belong to the class III receptor tyrosine kinases (RTKs), and have different expression patterns and physiological roles. PDGFRα signaling controls gastrulation and the development of several organs, such as the lung, intestine, skin, testis, kidney, bones, and neuroprotective tissues. PDGFRβ signaling is better recognized as an essential regulator of early hematopoiesis and blood vessel formation [129].

The PDGFR signaling can significantly influence mitochondrial dynamics in various cancers. Its role in cancer progression and regulation is an important aspect of mitochondrial dynamics, and it is a rapidly growing area of research. The size, shape, quality, quantity, and distribution of mitochondria within cells are determined by several processes that they go through, including fission, fusion, mitophagy, and energy transport during cancer. Mitochondrial fragmentation during cancer is a general phenomenon, and it is shown by the upregulation of the Drp1 gene and downregulation of the mfn2 gene in lung cancer [130], metastatic breast cancer [131], glioblastoma [132], neuroblastoma [133], colorectal cancer cells [134], pancreatic cancers [135], and melanoma [136]. Moreover, inhibition of Drp1 using mdivi-1 or Mfn2 overexpression using an adenoviral vector resulted in mitochondrial hyper-fusion in NSCLC, reduced cell proliferation, and enhanced apoptosis [130]. Furthermore, treatment with mdivi-1 or overexpression of the Mfn2 gene decreased tumor growth in a mouse xenograft NSCLC model. This suggests that aberrant fission-favoring mitochondrial dynamics stimulate the formation of NSCLC tumors by increasing mitotic fission and speeding up cell cycle progression through a Drp1/Mfn2 imbalance [130]. Several reports confirmed that mitochondrial fission stimulates autophagy, cell migration, tumor-associated macrophage infiltration, and tumor growth in hepatocellular carcinoma (HCC) [137, 138]. Mitochondrial fission was upregulated in HCC, similar to lung cancer, with increased expression of Drp1 and decreased expression of Mfn1 [139]. The overall survival rate was considerably lower for HCC patients with higher Drp1 expression and lower Mfn1 expression, or a higher Drp1:Mfn1 ratio than for those with a lower Drp1:Mfn1 ratio [139]. Additionally, Drp1 plays an important role in the development of primary and chronic epithelial ovarian cancer (EOC) as well as in the control of mitosis [140]. Similarly, in the case of breast cancer, Drp1 exhibited an increase in expression while mfn1 showed a decrease in triple-negative breast cancer (TNBC). This increase in the mitochondrial fission process is associated with a negative prognosis for patients, as it contributed to improved survival rates of cancer cells [141].

12 Nerve Growth Factor Receptor (NGFR)

NGF belongs to the neurotrophin family [142, 143] having a molecular weight of 130 KDa and is made up of 118 amino acids [17]. It is linked with neurotrophins in the peripheral nervous system and the central nervous system [144]. NGF is made up of three subunits: α, β, and γ [144]. NGF activates two types of receptors: the high-affinity tropomyosin-related kinase A receptor (TrkA) and the low-affinity p75 neurotrophin receptor (p75NTR) [145]. Vertebrates are known to have three distinct Trk receptors that exhibit a high degree of specificity and a high, picomolar affinity for their ligands. TrkA is the receptor for NGF, TrkB is the receptor for BDNF and NT-4/5, and the ideal receptor for NT-3 is TrkC. However, NT-3 can also bind to and signal through TrkA and TrkB receptors [146]. All three neurotrophin receptors TrkA, TrkB, and TrkC have a common architecture with an extracellular part comprising a cysteine-rich cluster (domain 1), three leucine-rich repeats (domain 2), a second cysteine-rich cluster (domain 3), and two immunoglobulin (Ig)-like domains (domains 4 and 5) [147]. The TrkA receptor is a transmembrane protein having a molecular weight of 140-kDa encoded by a proto-oncogene on chromosome 1. NGF promotes cell survival, differentiation, and proliferation when TrkA is activated. Furthermore, TrkA prevents apoptosis, increases the excitability of neurons, and induces mediator release. p75NTR receptor is a 75-kDa glycoprotein encoded by the *NGFR/p75*^{NTR} gene on chromosome 17. There are two types of NGFR/p75^{NTR} isoforms known, both of which are expressed in nerve cells: a full-length isoform and a short isoform (s-NGFR/p75^{NTR}) [148]. The full-length isoform consists of six exons and five introns [149]. The alternative splicing of exon III of the NGFR/p75NTR locus, which encodes for the CDR 2–4, results in s-NGFR/p75NTR isoform [150]. However, Sabry and his coworkers also reported a human NGFR/p75NTR isoform having a molecular weight of 36 kDa, but not yet been fully characterized.

NGFR has been primarily discovered in gliomas, neuroblastomas, and medulloblastomas [151]. It is found in many solid cancers, including breast cancer [152, 153]. lung cancer [154], colon cancer [155], pancreas [156], and prostate cancer [157]. Myelomas and lymphoid tumors are also sensitive to NGF and express high levels of NGFRs [158]. Previous research has demonstrated that TrkA and NGF are highly expressed in non-small cell lung cancer (NSCLC) and studied in detail. However, it is still unknown how TrkA and NGF behave in other subtypes of lung cancer [159]. Furthermore, the role of low-affinity nerve growth factor receptor p75NTR in enhanced activity of Trk and involvement in apoptosis are quite complex, and p75NTR-mediated signaling is yet to be established [160]. However, the overexpression of NGF and hypoxia-inducible factor-1α (HIF-1α) enhances angiogenesis in tumor tissues [161]. Additionally, a recent study discovered that NGF and HIF-1α are highly expressed in NSCLC tissues compared with para-cancerous lung tissues [162]. The expression of NGF and HIF-1α is closely associated with the microvascular mass in NSCLC. These findings suggest that NGF may have a significant impact on lung cancer and suggest that NGF offers a novel approach to treating lung cancer [162].

13 Muscle-Specific Kinase (MuSK)

Muscle-specific kinase (MuSK) belongs to the receptor tyrosine kinase having a molecular weight of 100 kDa, which was first isolated from the synapse-rich electric organ of the ray, Torpedo californica [162]. The *Musk* gene is located on mouse chromosome 4, and the human orthologue is on chromosome 9. The extracellular domain of MuSK is composed of three immunoglobin (Ig)-like domains and one cysteine-rich/frizzled domain, whereas the cytoplasmic domain is made up of tyrosine kinase domain [10]. MuSK is essentially present at the neuromuscular junction (NMJ) and plays an important role in its differentiation and development [163]. In addition, it is present in the liver, brain, and lungs [163]. MuSK is the central component of a multi-protein signaling complex activated by N-agrin, which is a heparan-sulfate proteoglycan secreted from the motor nerve terminal [164]. N-agrin interacts with MuSK through low-density lipoprotein receptor-related protein 4 (LPR4) binding [165]. In addition, the MuSK activation is mediated by another key player, downstream of tyrosine kinase 7 (Dok7), a tyrosine-phosphorylated motif in the JM domain of MuSK [166]. The cytosolic region of MuSK is constitutively bound by the tumorous imaginal discs protein (Tid1) [167]. Dishevelled (Dvl) connects the JM and kinase domains of MuSK and thus pairs MuSK to p21-activated kinase (PAK1) [168]. All these factors of the MuSK signaling complex are necessary to mediate AChR clustering in response to N-agrin. Other than this, the role of MuSK in cancer pathogenesis has not been well established in the literature, with one exception: hepatocellular carcinoma (HCC) [169]. The MuSK-LPR4 complex promotes endocytosis in HCC cells, and such events are often linked with signal amplification in several cancers [170]. However, MuSK is also involved in a muscle-related autoimmune disease of the neuromuscular junction, known as myasthenia gravis (MG) [171]. MG resulted from the production of autoantibodies directed against nicotinic acetylcholine receptors (nAchRs) that are present at these locations and it can be regarded as a type II hypersensitivity reaction [172]. nAchRs-directed autoantibodies are found in most cases; however, some are recognized as MuSK-specific [171].

14 Receptor-like Tyrosine Kinase (RYK)

The receptor-like protein tyrosine kinase was first defined in 1992 [173]. It is encoded by the RYK gene and has a vital role in numerous biological processes, such as tissue homeostasis and development. It is a co-receptor of the Wnt signaling pathway [174]. The structural components of RYK include an extracellular Wnt inhibitory factor (WIF) domain that mediates RYK high-affinity binding to Wnt ligands, an intracellular PDZ binding domain, and an unusual intracellular kinase domain that harbors mutations in evolutionarily conserved tyrosine kinase residues [174]. Moreover, RYK may bind to different WNT ligands, such as *WNT1*, *WNT3*, *WNT3A*, *WNT5B*, *WNT5A*, and *WNT11* through its extracellular WIF domain in humans. This causes the initiation of β-catenin-dependent or independent signaling cascade [175]. The *RYK* is involved in different biological processes by interacting with other WNT receptors or coreceptors, such as *ROR* [176], *FZD7* [177], *PTK7* [178], *LRP5/6*, and *VANGL* [179]. The components of the WNT pathway have been confirmed to directly regulate the DNA damage response (DDR) [180]. It has also been observed that *RYK* has a pathological connotation with the nervous system in diseases. In the *C. elegans* model, the intracellular domain of the *RYK* gene was found to inhibit the neuroprotective property of the transcription factor FOXO during the early phases of Huntington's disease [181]. Furthermore, it was noticed that *RYK* was enhanced in the motor neurons and ventral white matter of a mouse model of amyotrophic lateral sclerosis [182]. Moreover, *RYK* is highly expressed in different types of cancer comprising lung cancer [183], small cell lung cancer [184], breast cancer [185], gastric cancer, liver cancer [186], pancreatic cancer [187], chronic leukemia [188], acute leukemias [189], glioma [190], glioblastoma [191], mesothelioma [192], and head and neck squamous cell carcinoma [193]. However, little is known about the role of RYK in the molecular mechanism of tumor issues. Therefore, it can be a valuable target for further research and clinical applications.

15 Conclusions

Recent advancements in genomics, proteomics, and related fields have led to identifying various molecular changes associated with mitochondrial dynamics in cancer. Many of these changes have been observed in receptor tyrosine kinases (RTKs), which disrupt cellular signaling and play a crucial role in cancer progression. Understanding the structure of RTKs and their localization within different cellular compartments that influence their biological functions is essential. This knowledge aims to enhance treatment responses, minimize adverse effects, and prevent recurrences. The long-term efficacy of available treatments is often hindered by the development of drug resistance. Future research initiatives should prioritize unraveling the molecular mechanisms underlying these resistance phenomena. Additionally, exploring how RTK compartmentalization regulates the tumor microenvironment is vital. This could include pioneering new therapeutic agents and innovative drug formulations, such as nanocarriers, that specifically target different RTKs or their associated signaling pathways, thereby reducing the risks of side effects and resistance.

Acknowledgments The authors are thankful to Augusta University, Augusta, GA, USA, for providing infrastructure and research facilities. All figures were created with ChemBioDraw Ultra 11.0 software.

Conflict of Interest The authors declare that there is no conflict of interest.

References

1. Manning G, Whyte DB, Martinez R, Hunter T, Sudarsanam S. The protein kinase complement of the human genome. Science. 2002;298(5600):1912–34.
2. Blume-Jensen P, Hunter T. Oncogenic kinase signalling. Nature. 2001;411(6835):355–65.
3. Robinson DR, Wu YM, Lin SF. The protein tyrosine kinase family of the human genome. Oncogene. 2000;19(49):5548–57.
4. Hubbard SR. Structural analysis of receptor tyrosine kinases. Prog Biophys Mol Biol. 1999;71(3–4):343–58.
5. Gonzalez-Magaldi M, McCabe JM, Cartwright HN, Sun N, Leahy DJ. Conserved roles for receptor tyrosine kinase extracellular regions in regulating receptor and pathway activity. Biochem J. 2020;477(21):4207–20.
6. Yao Z, Stagljar I. Multiple functions of protein phosphatases in receptor tyrosine kinase signaling revealed by interactome analysis. Mol Cell Oncol. 2017;4(3):e1297101.
7. Diwanji D, Thaker T, Jura N. More than the sum of the parts: toward full-length receptor tyrosine kinase structures. IUBMB Life. 2019;71(6):706–20.
8. Matsushita C, Tamagaki H, Miyazawa Y, Aimoto S, Smith SO, Sato T. Transmembrane helix orientation influences membrane binding of the intracellular juxtamembrane domain in Neu receptor peptides. Proc Natl Acad Sci USA. 2013;110(5):1646–51.
9. Hedger G, Sansom MSP, Koldsø H. The juxtamembrane regions of human receptor tyrosine kinases exhibit conserved interaction sites with anionic lipids. Sci Rep. 2015;5(1):9198.
10. Lemmon MA, Schlessinger J. Cell signaling by receptor tyrosine kinases. Cell. 2010;141(7):1117–34.
11. Tomuleasa C, Tigu A-B, Munteanu R, Moldovan C-S, Kegyes D, Onaciu A, et al. Therapeutic advances of targeting receptor tyrosine kinases in cancer. Signal Transduct Target Ther. 2024;9(1):201.
12. Vogelstein B, Papadopoulos N, Velculescu VE, Zhou S, Diaz LA Jr, Kinzler KW. Cancer genome landscapes. Science. 2013;339(6127):1546–58.
13. Wallace DC. Mitochondria and cancer. Nat Rev Cancer. 2012;12(10):685–98.
14. Frohman MA. Mitochondria as integrators of signal transduction and energy production in cardiac physiology and disease. J Mol Med (Berl). 2010;88(10):967–70.
15. Andreux PA, Houtkooper RH, Auwerx J. Pharmacological approaches to restore mitochondrial function. Nat Rev Drug Discov. 2013;12(6):465–83.
16. Passarella S, Atlante A, Valenti D, de Bari L. The role of mitochondrial transport in energy metabolism. Mitochondrion. 2003;2(5):319–43.
17. Aarão TLS, de Sousa JR, Falcão ASC, Falcão LFM, Quaresma JAS. Nerve growth factor and pathogenesis of leprosy: review and update. Front Immunol. 2018;9:939.
18. Sharma SV, Bell DW, Settleman J, Haber DA. Epidermal growth factor receptor mutations in lung cancer. Nat Rev Cancer. 2007;7(3):169–81.
19. Bitler BG, Goverdhan A, Schroeder JA. MUC1 regulates nuclear localization and function of the epidermal growth factor receptor. J Cell Sci. 2010;123(Pt 10):1716–23.
20. Chi S, Cao H, Wang Y, McNiven MA. Recycling of the epidermal growth factor receptor is mediated by a novel form of the clathrin adaptor protein Eps15. J Biol Chem. 2011;286(40):35196–208.

21. Goh LK, Sorkin A. Endocytosis of receptor tyrosine kinases. Cold Spring Harb Perspect Biol. 2013;5(5):a017459.
22. Lo HW, Ali-Seyed M, Wu Y, Bartholomeusz G, Hsu SC, Hung MC. Nuclear-cytoplasmic transport of EGFR involves receptor endocytosis, importin beta1 and CRM1. J Cell Biochem. 2006;98(6):1570–83.
23. Du Y, Shen J, Hsu JL, Han Z, Hsu MC, Yang CC, et al. Syntaxin 6-mediated Golgi translocation plays an important role in nuclear functions of EGFR through microtubule-dependent trafficking. Oncogene. 2014;33(6):756–70.
24. Boerner JL, Demory ML, Silva C, Parsons SJ. Phosphorylation of Y845 on the epidermal growth factor receptor mediates binding to the mitochondrial protein cytochrome c oxidase subunit II. Mol Cell Biol. 2004;24(16):7059–71.
25. Che TF, Lin CW, Wu YY, Chen YJ, Han CL, Chang YL, et al. Mitochondrial translocation of EGFR regulates mitochondria dynamics and promotes metastasis in NSCLC. Oncotarget. 2015;6(35):37349–66.
26. Ramírez S, Gómez-Valadés AG, Schneeberger M, Varela L, Haddad-Tóvolli R, Altirriba J, et al. Mitochondrial dynamics mediated by mitofusin 1 is required for POMC neuron glucose-sensing and insulin release control. Cell Metab. 2017;25(6):1390–9.e6.
27. Lee JY, Kapur M, Li M, Choi MC, Choi S, Kim HJ, et al. MFN1 deacetylation activates adaptive mitochondrial fusion and protects metabolically challenged mitochondria. J Cell Sci. 2014;127(Pt 22):4954–63.
28. Demory ML, Boerner JL, Davidson R, Faust W, Miyake T, Lee I, et al. Epidermal growth factor receptor translocation to the mitochondria. J Biol Chem. 2009;284(52):36592–604.
29. Cao X, Zhu H, Ali-Osman F, Lo HW. EGFR and EGFRvIII undergo stress- and EGFR kinase inhibitor-induced mitochondrial translocalization: a potential mechanism of EGFR-driven antagonism of apoptosis. Mol Cancer. 2011;10:26.
30. Slamon DJ, Godolphin W, Jones LA, Holt JA, Wong SG, Keith DE, et al. Studies of the HER-2/neu proto-oncogene in human breast and ovarian cancer. Science. 1989;244(4905):707–12.
31. Guy CT, Webster MA, Schaller M, Parsons TJ, Cardiff RD, Muller WJ. Expression of the neu protooncogene in the mammary epithelium of transgenic mice induces metastatic disease. Proc Natl Acad Sci USA. 1992;89(22):10578–82.
32. Tan M, Li P, Sun M, Yin G, Yu D. Upregulation and activation of PKC alpha by ErbB2 through Src promotes breast cancer cell invasion that can be blocked by combined treatment with PKC alpha and Src inhibitors. Oncogene. 2006;25(23):3286–95.
33. Zhao YH, Zhou M, Liu H, Ding Y, Khong HT, Yu D, et al. Upregulation of lactate dehydrogenase A by ErbB2 through heat shock factor 1 promotes breast cancer cell glycolysis and growth. Oncogene. 2009;28(42):3689–701.
34. Elizalde PV, Cordo Russo RI, Chervo MF, Schillaci R. ErbB-2 nuclear function in breast cancer growth, metastasis and resistance to therapy. Endocr Relat Cancer. 2016;23(12):T243–T57.
35. Sun Y, Lin H, Zhu Y, Ma C, Ye J, Luo J. Induction or suppression of expression of cytochrome C oxidase subunit II by heregulin beta 1 in human mammary epithelial cells is dependent on the levels of ErbB2 expression. J Cell Physiol. 2002;192(2):225–33.
36. Gordon LI, Burke MA, Singh AT, Prachand S, Lieberman ED, Sun L, et al. Blockade of the erbB2 receptor induces cardiomyocyte death through mitochondrial and reactive oxygen species-dependent pathways. J Biol Chem. 2009;284(4):2080–7.
37. Lim S, Smith KR, Lim S-TS, Tian R, Lu J, Tan M. Regulation of mitochondrial functions by protein phosphorylation and dephosphorylation. Cell Biosci. 2016;6(1):25.
38. Ding Y, Liu Z, Desai S, Zhao Y, Liu H, Pannell LK, et al. Receptor tyrosine kinase ErbB2 translocates into mitochondria and regulates cellular metabolism. Nat Commun. 2012;3(1):1271.
39. Hakuno F, Takahashi SI. IGF1 receptor signaling pathways. J Mol Endocrinol. 2018;61(1):T69–t86.

40. Denduluri SK, Idowu O, Wang Z, Liao Z, Yan Z, Mohammed MK, et al. Insulin-like growth factor (IGF) signaling in tumorigenesis and the development of cancer drug resistance. Genes Dis. 2015;2(1):13–25.
41. Sádaba MC, Martín-Estal I, Puche JE, Castilla-Cortázar I. Insulin-like growth factor 1 (IGF-1) therapy: mitochondrial dysfunction and diseases. Biochim Biophys Acta (BBA) - Mol Basis Dis. 2016;1862(7):1267–78.
42. Qu X, Wu Z, Dong W, Zhang T, Wang L, Pang Z, et al. Update of IGF-1 receptor inhibitor (ganitumab, dalotuzumab, cixutumumab, teprotumumab and figitumumab) effects on cancer therapy. Oncotarget. 2017;8(17):29501–18.
43. Osher E, Macaulay VM. Therapeutic targeting of the IGF Axis. Cells. 2019;8(8)
44. Vyas S, Zaganjor E, Haigis MC. Mitochondria and cancer. Cell. 2016;166(3):555–66.
45. Drake LE, Springer MZ, Poole LP, Kim CJ, Macleod KF. Expanding perspectives on the significance of mitophagy in cancer. Semin Cancer Biol. 2017;47:110–24.
46. Pickles S, Vigié P, Youle RJ. Mitophagy and quality control mechanisms in mitochondrial maintenance. Curr Biol. 2018;28(4):R170–R85.
47. Favre C, Zhdanov A, Leahy M, Papkovsky D, O'Connor R. Mitochondrial pyrimidine nucleotide carrier (PNC1) regulates mitochondrial biogenesis and the invasive phenotype of cancer cells. Oncogene. 2010;29(27):3964–76.
48. Ribeiro M, Rosenstock TR, Oliveira AM, Oliveira CR, Rego AC. Insulin and IGF-1 improve mitochondrial function in a PI-3K/Akt-dependent manner and reduce mitochondrial generation of reactive oxygen species in Huntington's disease knock-in striatal cells. Free Radic Biol Med. 2014;74:129–44.
49. Lyons A, Coleman M, Riis S, Favre C, O'Flanagan CH, Zhdanov AV, et al. Insulin-like growth factor 1 signaling is essential for mitochondrial biogenesis and mitophagy in cancer cells. J Biol Chem. 2017;292(41):16983–98.
50. Ray R, Chen G, Vande Velde C, Cizeau J, Park JH, Reed JC, et al. BNIP3 heterodimerizes with Bcl-2/Bcl-XL and induces cell death independent of a Bcl-2 homology 3 (BH3) domain at both mitochondrial and nonmitochondrial sites. J Biol Chem. 2000;275(2):1439–48.
51. Ney PA. Mitochondrial autophagy: origins, significance, and role of BNIP3 and NIX. Biochimica et Biophysica Acta (BBA). Mol Cell Res. 2015;1853(10, Part B):2775–83.
52. Clague MJ, Urbé S. Diverse routes to mitophagy governed by ubiquitylation and mitochondrial import. Trends Cell Biol. 2025;
53. Chourasia AH, Tracy K, Frankenberger C, Boland ML, Sharifi MN, Drake LE, et al. Mitophagy defects arising from BNip3 loss promote mammary tumor progression to metastasis. EMBO Rep. 2015;16(9):1145–63.
54. Riis S, Murray JB, O'Connor R. IGF-1 signalling regulates mitochondria dynamics and turnover through a conserved GSK-3β–Nrf2–BNIP3 pathway. Cells. 2020;9(1):147.
55. Rojo de la Vega M, Chapman E, Zhang DD. NRF2 and the hallmarks of cancer. Cancer Cell. 2018;34(1):21–43.
56. Wang X-J, Sun Z, Villeneuve NF, Zhang S, Zhao F, Li Y, et al. Nrf2 enhances resistance of cancer cells to chemotherapeutic drugs, the dark side of Nrf2. Carcinogenesis. 2008;29(6):1235–43.
57. Lee S, Hallis SP, Jung K-A, Ryu D, Kwak M-K. Impairment of HIF-1α-mediated metabolic adaption by NRF2-silencing in breast cancer cells. Redox Biol. 2019;24:101210.
58. Zhang H-S, Du G-Y, Zhang Z-G, Zhou Z, Sun H-L, Yu X-Y, et al. NRF2 facilitates breast cancer cell growth via HIF1α-mediated metabolic reprogramming. Int J Biochem Cell Biol. 2018;95:85–92.
59. Ma Q. Role of Nrf2 in oxidative stress and toxicity. Ann Rev Pharmacol Toxicol. 2013;53:401–26.
60. Shao D, Liu Y, Liu X, Zhu L, Cui Y, Cui A, et al. PGC-1 beta-regulated mitochondrial biogenesis and function in myotubes is mediated by NRF-1 and ERR alpha. Mitochondrion. 2010;10(5):516–27.

61. Um JH, Yun J. Emerging role of mitophagy in human diseases and physiology. BMB Rep. 2017;50(6):299–307.
62. Fu W, Liu Y, Yin H. Mitochondrial dynamics: biogenesis, fission, fusion, and mitophagy in the regulation of stem cell behaviors. Stem Cells Int. 2019;2019(1):9757201.
63. Esteban-Martínez L, Boya P. BNIP3L/NIX-dependent mitophagy regulates cell differentiation via metabolic reprogramming. Autophagy. 2018;14(5):915–7.
64. Lee HJ, Jung YH, Choi GE, Ko SH, Lee SJ, Lee SH, et al. BNIP3 induction by hypoxia stimulates FASN-dependent free fatty acid production enhancing therapeutic potential of umbilical cord blood-derived human mesenchymal stem cells. Redox Biol. 2017;13:426–43.
65. Moriyama M, Moriyama H, Uda J, Kubo H, Nakajima Y, Goto A, et al. BNIP3 upregulation via stimulation of ERK and JNK activity is required for the protection of keratinocytes from UVB-induced apoptosis. Cell Death Dis. 2017;8(2):e2576.
66. Reya T, Morrison SJ, Clarke MF, Weissman IL. Stem cells, cancer, and cancer stem cells. Nature. 2001;414(6859):105–11.
67. Teng CF, Jeng LB, Shyu WC. Role of insulin-like growth factor 1 receptor signaling in stem cell stemness and therapeutic efficacy. Cell Transplant. 2018;27(9):1313–9.
68. Huat TJ, Khan AA, Pati S, Mustafa Z, Abdullah JM, Jaafar H. IGF-1 enhances cell proliferation and survival during early differentiation of mesenchymal stem cells to neural progenitor-like cells. BMC Neurosci. 2014;15:91.
69. Lee Y, Lee HY, Hanna RA, Gustafsson ÅB. Mitochondrial autophagy by Bnip3 involves Drp1-mediated mitochondrial fission and recruitment of Parkin in cardiac myocytes. Am J Physiol Heart Circ Physiol. 2011;301(5):H1924–31.
70. Zhang J, Ney PA. Role of BNIP3 and NIX in cell death, autophagy, and mitophagy. Cell Death Differ. 2009;16(7):939–46.
71. Singh A, Azad M, Shymko MD, Henson ES, Katyal S, Eisenstat DD, et al. The BH3 only Bcl-2 family member BNIP3 regulates cellular proliferation. PLoS One. 2018;13(10):e0204792.
72. Jung CH, Ro SH, Cao J, Otto NM, Kim DH. mTOR regulation of autophagy. FEBS Lett. 2010;584(7):1287–95.
73. Macher-Goeppinger S, Keith M, Hatiboglu G, Hohenfellner M, Schirmacher P, Roth W, et al. Expression and functional characterization of the BNIP3 protein in renal cell carcinomas. Transl Oncol. 2017;10(6):869–75.
74. Matsuda N, Sato S, Shiba K, Okatsu K, Saisho K, Gautier CA, et al. PINK1 stabilized by mitochondrial depolarization recruits Parkin to damaged mitochondria and activates latent Parkin for mitophagy. J Cell Biol. 2010;189(2):211–21.
75. Kondapalli C, Kazlauskaite A, Zhang N, Woodroof HI, Campbell DG, Gourlay R, et al. PINK1 is activated by mitochondrial membrane potential depolarization and stimulates Parkin E3 ligase activity by phosphorylating Serine 65. Open Biol. 2012;2(5):120080.
76. Neupert W, Herrmann JM. Translocation of proteins into mitochondria. Annu Rev Biochem. 2007;76(1):723–49.
77. Baker MJ, Frazier AE, Gulbis JM, Ryan MT. Mitochondrial protein-import machinery: correlating structure with function. Trends Cell Biol. 2007;17(9):456–64.
78. Folkman J. Angiogenesis. Annu Rev Med. 2006;57(1):1–18.
79. Patel SA, Nilsson MB, Le X, Cascone T, Jain RK, Heymach JV. Molecular mechanisms and future implications of VEGF/VEGFR in cancer therapy. Clin Cancer Res. 2023;29(1):30–9.
80. Holmes K, Roberts OL, Thomas AM, Cross MJ. Vascular endothelial growth factor receptor-2: structure, function, intracellular signalling and therapeutic inhibition. Cell Signal. 2007;19(10):2003–12.
81. Soker S, Takashima S, Miao HQ, Neufeld G, Klagsbrun M. Neuropilin-1 is expressed by endothelial and tumor cells as an isoform-specific receptor for vascular endothelial growth factor. Cell. 1998;92(6):735–45.

82. Liu Y, Li Y, Wang Y, Lin C, Zhang D, Chen J, et al. Recent progress on vascular endothelial growth factor receptor inhibitors with dual targeting capabilities for tumor therapy. J Hematol Oncol. 2022;15(1):89.
83. Claesson-Welsh L. VEGF receptor signal transduction—a brief update. Vasc Pharmacol. 2016;86:14–7.
84. Fieber CB, Eldridge J, Taha TA, Obeid LM, Muise-Helmericks RC. Modulation of total Akt kinase by increased expression of a single isoform: requirement of the sphingosine-1-phosphate receptor, Edg3/S1P3, for the VEGF-dependent expression of Akt3 in primary endothelial cells. Exp Cell Res. 2006;312(7):1164–73.
85. Wright GL, Maroulakou IG, Eldridge J, Liby TL, Sridharan V, Tsichlis PN, et al. VEGF stimulation of mitochondrial biogenesis: requirement of AKT3 kinase. FASEB J. 2008;22(9):3264.
86. Cunningham JT, Rodgers JT, Arlow DH, Vazquez F, Mootha VK, Puigserver P. mTOR controls mitochondrial oxidative function through a YY1–PGC-1α transcriptional complex. Nature. 2007;450(7170):736–40.
87. Guo D, Wang Q, Li C, Wang Y, Chen X. VEGF stimulated the angiogenesis by promoting the mitochondrial functions. Oncotarget. 2017;8(44):77020.
88. Dai S, Zhou Z, Chen Z, Xu G, Chen Y. Fibroblast growth factor receptors (FGFRs): structures and small molecule inhibitors. Cells. 2019;8(6):614.
89. Ornitz DM, Itoh N. Fibroblast growth factors. Genome Biol. 2001;2:1–12.
90. Farrell B, Breeze AL. Structure, activation and dysregulation of fibroblast growth factor receptor kinases: perspectives for clinical targeting. Biochem Soc Trans. 2018;46(6):1753–70.
91. Zhang P, Yue L, Leng Q, Chang C, Gan C, Ye T, et al. Targeting FGFR for cancer therapy. J Hematol Oncol. 2024;17(1):39.
92. Beenken A, Mohammadi M. The FGF family: biology, pathophysiology and therapy. Nat Rev Drug Discov. 2009;8(3):235–53.
93. Edirisinghe O, Ternier G, Alraawi Z, Suresh Kumar TK. Decoding FGF/FGFR signaling: insights into biological functions and disease relevance. Biomol Ther. 2024;14(12):1622.
94. Bottaro DP, Rubin JS, Faletto DL, Chan AM-L, Kmiecik TE, Vande Woude GF, et al. Identification of the hepatocyte growth factor receptor as the c-met proto-oncogene product. Science. 1991;251(4995):802–4.
95. Mungunsukh O, McCart EA, Day RM. Hepatocyte growth factor isoforms in tissue repair, cancer, and fibrotic remodeling. Biomedicines. 2014;2(4):301–26.
96. Birchmeier C, Birchmeier W, Gherardi E, Vande Woude GF. Met, metastasis, motility and more. Nat Rev Mol Cell Biol. 2003;4(12):915–25.
97. Johnson M, Kochhar K, Nakamura T, Iyer A. Hepatocyte growth factor-induced signal transduction in two normal mouse epithelial cell lines. Biochem Mol Biol Int. 1995;36(3):465–74.
98. Meng W, Chen T. Association between the HGF/c-MET signaling pathway and tumorigenesis, progression and prognosis of hepatocellular carcinoma. Oncol Rep. 2021;46(3):1–10.
99. Liu X, Newton RC, Scherle PA. Developing c-MET pathway inhibitors for cancer therapy: progress and challenges. Trends Mol Med. 2010;16(1):37–45.
100. Mossie K, Jallal B, Alves F, Sures I, Plowman GD, Ullrich A. Colon carcinoma kinase-4 defines a new subclass of the receptor tyrosine kinase family. Oncogene. 1995;11(10):2179–84.
101. Meng L, Sefah K, O'Donoghue MB, Zhu G, Shangguan D, Noorali A, et al. Silencing of PTK7 in colon cancer cells: caspase-10-dependent apoptosis via mitochondrial pathway. PLoS One. 2010;5(11):e14018.
102. Peradziryi H, Tolwinski NS, Borchers A. The many roles of PTK7: a versatile regulator of cell–cell communication. Arch Biochem Biophys. 2012;524(1):71–6.
103. Lee HK, Chauhan SK, Kay E, Dana R. Flt-1 regulates vascular endothelial cell migration via a protein tyrosine kinase-7–dependent pathway. Blood. 2011;117(21):5762–71.
104. Wagner G, Peradziryi H, Wehner P, Borchers A. PlexinA1 interacts with PTK7 and is required for neural crest migration. Biochem Biophys Res Commun. 2010;402(2):402–7.

105. Partanen J, Armstrong E, Mäkelä TP, Korhonen J, Sandberg M, Renkonen R, et al. A novel endothelial cell surface receptor tyrosine kinase with extracellular epidermal growth factor homology domains. Mol Cell Biol. 1992;12(4):1698–707.
106. D'Amico G, Korhonen EA, Anisimov A, Zarkada G, Holopainen T, Hägerling R, et al. Tie1 deletion inhibits tumor growth and improves angiopoietin antagonist therapy. J Clin Invest. 2014;124(2):824–34.
107. Qu X, Zhou B, Baldwin HS. Tie1 is required for lymphatic valve and collecting vessel development. Dev Biol. 2015;399(1):117–28.
108. Takahashi M, Ritz J, Cooper GM. Activation of a novel human transforming gene, ret, by DNA rearrangement. Cell. 1985;42(2):581–8.
109. Saha D, Ryan KR, Lakkaniga NR, Acharya B, Garcia NG, Smith EL, et al. Targeting rearranged during transfection in cancer: a perspective on small-molecule inhibitors and their clinical development. J Med Chem. 2021;64(16):11747–73.
110. Subbiah V, Yang D, Velcheti V, Drilon A, Meric-Bernstam F. State-of-the-art strategies for targeting RET-dependent cancers. J Clin Oncol. 2020;38(11):1209–21.
111. Le Hir H, Charlet-Berguerand N, Gimenez-Roqueplo A-P, Mannelli M, Plouin P-F, de Franciscis V, et al. Relative expression of the RET9 and RET51 isoforms in human pheochromocytomas. Oncology. 2000;58(4):311–8.
112. Griseri P, Garrone O, Sardo AL, Monteverde M, Rusmini M, Tonissi F, et al. Genetic and epigenetic factors affect RET gene expression in breast cancer cell lines and influence survival in patients. Oncotarget. 2016;7(18):26465.
113. Veit C, Genze F, Menke A, Hoeffert S, Gress TM, Gierschik P, et al. Activation of phosphatidylinositol 3-kinase and extracellular signal-regulated kinase is required for glial cell line-derived neurotrophic factor-induced migration and invasion of pancreatic carcinoma cells. Cancer Res. 2004;64(15):5291–300.
114. Ibáñez CF, Andressoo J-O. Biology of GDNF and its receptors—relevance for disorders of the central nervous system. Neurobiol Dis. 2017;97:80–9.
115. Mulligan LM. RET revisited: expanding the oncogenic portfolio. Nat Rev Cancer. 2014;14(3):173–86.
116. Mahato AK, Sidorova YA. RET receptor tyrosine kinase: role in neurodegeneration, obesity, and cancer. Int J Mol Sci. 2020;21(19):7108.
117. Zhou L, Li J, Zhang X, Xu Z, Yan Y, Hu K. An integrative pan cancer analysis of RET aberrations and their potential clinical implications. Sci Rep. 2022;12(1):13913.
118. Mórotz GM, Bradbury NA, Caluseriu O, Hisanaga S-i, Miller CC, Swiatecka-Urban A, et al. A revised nomenclature for the lemur family of protein kinases. Commun Biol. 2024;7(1):57.
119. Wendler F, Purice T-M, Simon T, Stebbing J, Giamas G. The LMTK-family of kinases: emerging important players in cell physiology and pathogenesis. Biochim Biophys Acta Mol Basis Dis. 2021;1867(9):165372.
120. Ditsiou A, Gagliano T, Samuels M, Vella V, Tolias C, Giamas G. The multifaceted role of lemur tyrosine kinase 3 in health and disease. Open Biol. 2021;11(9):210218.
121. Shah K, Bradbury NA. Lemur tyrosine kinase 2, a novel target in prostate cancer therapy. Oncotarget. 2015;6(16):14233.
122. Cruz DF, Farinha CM, Swiatecka-Urban A. Unraveling the function of lemur tyrosine kinase 2 network. Front Pharmacol. 2019;10:24.
123. Li X, Pontén A, Aase K, Karlsson L, Abramsson A, Uutela M, et al. PDGF-C is a new protease-activated ligand for the PDGF α-receptor. Nat Cell Biol. 2000;2(5):302–9.
124. Bergsten E, Uutela M, Li X, Pietras K, Östman A, Heldin C-H, et al. PDGF-D is a specific, protease-activated ligand for the PDGF β-receptor. Nat Cell Biol. 2001;3(5):512–6.
125. Fredriksson L, Li H, Eriksson U. The PDGF family: four gene products form five dimeric isoforms. Cytokine Growth Factor Rev. 2004;15(4):197–204.
126. Betsholtz C, Johnsson A, Heldin C-H, Westermark B, Lind P, Urdea MS, et al. cDNA sequence and chromosomal localization of human platelet-derived growth factor A-chain and its expression in tumour cell lines. Nature. 1986;320(6064):695–9.

127. Uutela M, Laurén J, Bergsten E, Li X, Horelli-Kuitunen N, Eriksson U, et al. Chromosomal location, exon structure, and vascular expression patterns of the human PDGFC and PDGFD genes. Circulation. 2001;103(18):2242–7.
128. Manzat Saplacan RM, Balacescu L, Gherman C, Chira RI, Craiu A, Mircea PA, et al. The role of PDGFs and PDGFRs in colorectal cancer. Mediat Inflamm. 2017;2017(1):4708076.
129. Andrae J, Gallini R, Betsholtz C. Role of platelet-derived growth factors in physiology and medicine. Genes Dev. 2008;22(10):1276–312.
130. Rehman J, Zhang HJ, Toth PT, Zhang Y, Marsboom G, Hong Z, et al. Inhibition of mitochondrial fission prevents cell cycle progression in lung cancer. FASEB J. 2012;26(5):2175.
131. Zhao J, Zhang J, Yu M, Xie Y, Huang Y, Wolff DW, et al. Mitochondrial dynamics regulates migration and invasion of breast cancer cells. Oncogene. 2013;32(40):4814–24.
132. Wan Y-Y, Zhang J-F, Yang Z-J, Jiang L-P, Wei YF, Lai Q-N, et al. Involvement of Drp1 in hypoxia-induced migration of human glioblastoma U251 cells. Oncol Rep. 2014;32(2):619–26.
133. Hagenbuchner J, Kuznetsov A, Obexer P, Ausserlechner M. BIRC5/Survivin enhances aerobic glycolysis and drug resistance by altered regulation of the mitochondrial fusion/fission machinery. Oncogene. 2013;32(40):4748–57.
134. Inoue-Yamauchi A, Oda H. Depletion of mitochondrial fission factor DRP1 causes increased apoptosis in human colon cancer cells. Biochem Biophys Res Commun. 2012;421(1):81–5.
135. Kashatus JA, Nascimento A, Myers LJ, Sher A, Byrne FL, Hoehn KL, et al. Erk2 phosphorylation of Drp1 promotes mitochondrial fission and MAPK-driven tumor growth. Mol Cell. 2015;57(3):537–51.
136. Serasinghe MN, Wieder SY, Renault TT, Elkholi R, Asciolla JJ, Yao JL, et al. Mitochondrial division is requisite to RAS-induced transformation and targeted by oncogenic MAPK pathway inhibitors. Mol Cell. 2015;57(3):521–36.
137. Sun X, Cao H, Zhan L, Yin C, Wang G, Liang P, et al. Mitochondrial fission promotes cell migration by Ca2+/CaMKII/ERK/FAK pathway in hepatocellular carcinoma. Liver Int. 2018;38(7):1263–72.
138. Bao D, Zhao J, Zhou X, Yang Q, Chen Y, Zhu J, et al. Mitochondrial fission-induced mtDNA stress promotes tumor-associated macrophage infiltration and HCC progression. Oncogene. 2019;38(25):5007–20.
139. Huang Q, Zhan L, Cao H, Li J, Lyu Y, Guo X, et al. Increased mitochondrial fission promotes autophagy and hepatocellular carcinoma cell survival through the ROS-modulated coordinated regulation of the NFKB and TP53 pathways. Autophagy. 2016;12(6):999–1014.
140. Tanwar DK, Parker DJ, Gupta P, Spurlock B, Alvarez RD, Basu MK, et al. Crosstalk between the mitochondrial fission protein, Drp1, and the cell cycle is identified across various cancer types and can impact survival of epithelial ovarian cancer patients. Oncotarget. 2016;7(37):60021.
141. Chen L, Zhang J, Lyu Z, Chen Y, Ji X, Cao H, et al. Positive feedback loop between mitochondrial fission and notch signaling promotes survivin-mediated survival of TNBC cells. Cell Death Dis. 2018;9(11):1050.
142. Hoyle GW. Neurotrophins and lung disease. Cytokine Growth Factor Rev. 2003;14(6):551–8.
143. Frossard N, Freund V, Advenier C. Nerve growth factor and its receptors in asthma and inflammation. Eur J Pharmacol. 2004;500(1–3):453–65.
144. Wise BL, Seidel MF, Lane NE. The evolution of nerve growth factor inhibition in clinical medicine. Nat Rev Rheumatol. 2021;17(1):34–46.
145. Stabile A, Pistilli A, Crispoltoni L, Montagnoli C, Tiribuzi R, Casali L, et al. A role for NGF and its receptors TrKA and p75NTR in the progression of COPD. Biol Chem. 2016;397(2):157–63.
146. Berkemeier LR, Winslow JW, Kaplan DR, Nikolics K, Goeddel DV, Rosenthal A. Neurotrophin-5: a novel neurotrophic factor that activates trk and trkB. Neuron. 1991;7(5):857–66.

147. Schneider R, Schweiger M. A novel modular mosaic of cell adhesion motifs in the extracellular domains of the neurogenic trk and trkB tyrosine kinase receptors. Oncogene. 1991;6(10):1807–11.
148. von Schack D, Casademunt E, Schweigreiter R, Meyer M, Bibel M, Dechant G. Complete ablation of the neurotrophin receptor p75NTR causes defects both in the nervous and the vascular system. Nat Neurosci. 2001;4(10):977–8.
149. Wang Y-T, Lu X-M, Shu Y-H, Xiao L, Chen K-T. Selection of human p75NTR tag SNPs and its biological significance for clinical association studies. Biomed Mater Eng. 2014;24(6):3833–9.
150. Casaccia-Bonnefil P, Gu C, Khursigara G, Chao MV. p75 neurotrophin receptor as a modulator of survival and death decisions. Microsc Res Tech. 1999;45(4–5):217–24.
151. Aloe L, Rocco ML, Balzamino BO, Micera A. Nerve growth factor: role in growth, differentiation and controlling cancer cell development. J Exp Clin Cancer Res. 2016;35:1–7.
152. Descamps S, Toillon R-A, Adriaenssens E, Pawlowski V, Cool SM, Nurcombe V, et al. Nerve growth factor stimulates proliferation and survival of human breast cancer cells through two distinct signaling pathways. J Biol Chem. 2001;276(21):17864–70.
153. Pundavela J, Roselli S, Faulkner S, Attia J, Scott RJ, Thorne RF, et al. Nerve fibers infiltrate the tumor microenvironment and are associated with nerve growth factor production and lymph node invasion in breast cancer. Mol Oncol. 2015;9(8):1626–35.
154. Vaishnavi A, Capelletti M, Le AT, Kako S, Butaney M, Ercan D, et al. Oncogenic and drug-sensitive NTRK1 rearrangements in lung cancer. Nat Med. 2013;19(11):1469–72.
155. Anagnostopoulou V, Pediaditakis I, Alkahtani S, Alarifi SA, Schmidt E-M, Lang F, et al. Differential effects of dehydroepiandrosterone and testosterone in prostate and colon cancer cell apoptosis: the role of nerve growth factor (NGF) receptors. Endocrinology. 2013;154(7):2446–56.
156. Miknyoczki SJ, Lang D, Huang L, Klein-Szanto AJ, Dionne CA, Ruggeri BA. Neurotrophins and Trk receptors in human pancreatic ductal adenocarcinoma: expression patterns and effects on in vitro invasive behavior. Int J Cancer. 1999;81(3):417–27.
157. Pundavela J, Demont Y, Jobling P, Lincz LF, Roselli S, Thorne RF, et al. ProNGF correlates with Gleason score and is a potential driver of nerve infiltration in prostate cancer. Am J Pathol. 2014;184(12):3156–62.
158. Thiele CJ, Li Z, McKee AE. On Trk—the TrkB signal transduction pathway is an increasingly important target in cancer biology. Clin Cancer Res. 2009;15(19):5962–7.
159. Ricci A, Greco S, Mariotta S, Felici L, Bronzetti E, Cavazzana A, et al. Neurotrophins and neurotrophin receptors in human lung cancer. Am J Respir Cell Mol Biol. 2001;25(4):439–46.
160. Roux PP, Colicos MA, Barker PA, Kennedy TE. p75 neurotrophin receptor expression is induced in apoptotic neurons after seizure. J Neurosci. 1999;19(16):6887–96.
161. Eggert A, Grotzer MA, Ikegaki N, Liu XG, Evans AE, Brodeur GM. Expression of neurotrophin receptor TrkA inhibits angiogenesis in neuroblastoma. Med Pediatr Oncol. 2000;35(6):569–72.
162. Lu Q-l, Liu J, Zhu X-l, Xu W-j. Expression of nerve growth factor and hypoxia inducible factor-1α and its correlation with angiogenesis in non-small cell lung cancer. J Huazhong Univ Sci Technolog Med Sci. 2014;34:359–62.
163. Li L, Xiong W-C, Mei L. Neuromuscular junction formation, aging, and disorders. Annu Rev Physiol. 2018;80(1):159–88.
164. Mittaud P, Camilleri AA, Willmann R, Erb-Vögtli S, Burden SJ, Fuhrer C. A single pulse of agrin triggers a pathway that acts to cluster acetylcholine receptors. Mol Cell Biol. 2004;24:7841.
165. Kim N, Stiegler AL, Cameron TO, Hallock PT, Gomez AM, Huang JH, et al. Lrp4 is a receptor for Agrin and forms a complex with MuSK. Cell. 2008;135(2):334–42.
166. Bergamin E, Hallock PT, Burden SJ, Hubbard SR. The cytoplasmic adaptor protein Dok7 activates the receptor tyrosine kinase MuSK via dimerization. Mol Cell. 2010;39(1):100–9.

167. Linnoila J, Wang Y, Yao Y, Wang Z-Z. A mammalian homolog of drosophila tumorous imaginal discs, Tid1, mediates agrin signaling at the neuromuscular junction. Neuron. 2008;60(4):625–41.
168. Luo ZG, Wang Q, Zhou JZ, Wang J, Luo Z, Liu M, et al. Regulation of AChR clustering by Dishevelled interacting with MuSK and PAK1. Neuron. 2002;35(3):489–505.
169. Chakraborty S, Lakshmanan M, Swa HL, Chen J, Zhang X, Ong YS, et al. An oncogenic role of Agrin in regulating focal adhesion integrity in hepatocellular carcinoma. Nat Commun. 2015;6(1):6184.
170. Casaletto JB, McClatchey AI. Spatial regulation of receptor tyrosine kinases in development and cancer. Nat Rev Cancer. 2012;12(6):387–400.
171. Borges LS, Richman DP. Muscle-specific kinase myasthenia gravis. Front Immunol. 2020;11:707.
172. Dresser L, Wlodarski R, Rezania K, Soliven B. Myasthenia gravis: epidemiology, pathophysiology and clinical manifestations. J Clin Med. 2021;10(11):2235.
173. Hovens CM, Stacker SA, Andres A-C, Harpur AG, Ziemiecki A, Wilks AF. RYK, a receptor tyrosine kinase-related molecule with unusual kinase domain motifs. Proc Natl Acad Sci. 1992;89(24):11818–22.
174. Roy JP, Halford MM, Stacker SA. The biochemistry, signalling and disease relevance of RYK and other WNT-binding receptor tyrosine kinases. Growth Factors. 2018;36(1–2):15–40.
175. Bovolenta P, Rodriguez J, Esteve P. Frizzled/RYK mediated signalling in axon guidance, vol. 133; 2006. p. 4399.
176. Green J, Nusse R, van Amerongen R. The role of Ryk and Ror receptor tyrosine kinases in Wnt signal transduction. Cold Spring Harb Perspect Biol. 2014;6(2):a009175.
177. Zhang Z, Gao S, Xu Y, Zhao C. Regulation of ABCG2 expression by Wnt5a through FZD7 in human pancreatic cancer cells. Mol Med Rep. 2021;23(1):1.
178. Liu K, Song G, Zhang X, Li Q, Zhao Y, Zhou Y, et al. PTK7 is a novel oncogenic target for esophageal squamous cell carcinoma. World J Surg Oncol. 2017;15:1–8.
179. Macheda ML, Sun WW, Kugathasan K, Hogan BM, Bower NI, Halford MM, et al. The Wnt receptor Ryk plays a role in mammalian planar cell polarity signaling. J Biol Chem. 2012;287(35):29312–23.
180. Zhang X, Yu X. Crosstalk between Wnt/β-catenin signaling pathway and DNA damage response in cancer: a new direction for overcoming therapy resistance. Front Pharmacol. 2023;14:1230822.
181. Liu Y, Wang X, Lu C-C, Sherman-Kermen R, Steward O, Xu X-M, et al. Repulsive Wnt signaling inhibits axon regeneration after CNS injury. J Neurosci. 2008;28(33):8376–82.
182. Tury A, Tolentino K, Zou Y. Altered expression of atypical PKC and Ryk in the spinal cord of a mouse model of amyotrophic lateral sclerosis. Dev Neurobiol. 2014;74(8):839–50.
183. Ohara S, Suda K, Fujino T, Hamada A, Koga T, Nishino M, et al. Dose-dependence in acquisition of drug tolerant phenotype and high RYK expression as a mechanism of osimertinib tolerance in lung cancer. Lung Cancer. 2021;154:84–91.
184. Hamilton G, Rath B, Klameth L, Hochmair M. Receptor tyrosine kinase expression of circulating tumor cells in small cell lung cancer. Oncoscience. 2015;2(7):629.
185. Liang Z, Liu L, Gao R, Che C, Yang G. Downregulation of exosomal miR-7-5p promotes breast cancer migration and invasion by targeting RYK and participating in the atypical WNT signalling pathway. Cell Mol Biol Lett. 2022;27(1):88.
186. Fu Y, Chen Y, Huang J, Cai Z, Wang Y. RYK, a receptor of noncanonical Wnt ligand Wnt5a, is positively correlated with gastric cancer tumorigenesis and potential of liver metastasis. Am J Physiol-Gastrointest Liver Physiol. 2020;318(2):G352–G60.
187. Okura R, Fujihara S, Iwama H, Morishita A, Chiyo T, Watanabe M, et al. MicroRNA profiles during galectin-9-induced apoptosis of pancreatic cancer cells. Oncol Lett. 2018;15(1):407–14.

188. Micci F, Panagopoulos I, Haugom L, Andersen HK, Tjønnfjord GE, Beiske K, et al. t (3; 21)(q22; q22) leading to truncation of the RYK gene in atypical chronic myeloid leukemia. Cancer Lett. 2009;277(2):205–11.
189. Alvarez-Zavala M, Riveros-Magaña AR, García-Castro B, Barrera-Chairez E, Rubio-Jurado B, Garcés-Ruíz OM, et al. WNT receptors profile expression in mature blood cells and immature leukemic cells: RYK emerges as a hallmark receptor of acute leukemia. Eur J Haematol. 2016;97(2):155–65.
190. Habu M, Koyama H, Kishida M, Kamino M, Iijima M, Fuchigami T, et al. Ryk is essential for Wnt-5a-dependent invasiveness in human glioma. J Biochem. 2014;156(1):29–38.
191. Adamo A, Fiore D, De Martino F, Roscigno G, Affinito A, Donnarumma E, et al. RYK promotes the stemness of glioblastoma cells via the WNT/β-catenin pathway. Oncotarget. 2017;8(8):13476.
192. Ventura E, Belfiore A, Iozzo RV, Giordano A, Morrione A. Progranulin and EGFR modulate receptor-like tyrosine kinase sorting and stability in mesothelioma cells. Am J Phys Cell Phys. 2023;325(2):C391–405.
193. Reddy RB, Khora SS, Suresh A. Molecular prognosticators in clinically and pathologically distinct cohorts of head and neck squamous cell carcinoma—a meta-analysis approach. PLoS One. 2019;14(7):e0218989.

Molecular Mechanisms of Acquired Therapeutic Resistance to Tyrosine Kinase Targeted Therapy

Liwu Fu and Min Luo

1 Introduction

Protein tyrosine kinases (PTKs) are a group of enzymes that catalyze phosphorylation of specific tyrosine residues within target proteins. Aberrant activation of PTKs due to gain-of-function mutation, genomic amplification, chromosomal rearrangements, or autocrine activation can lead to dysregulated kinase activity, ultimately leading to uncontrolled cell proliferation and tumorigenesis [1]. Given the critical role of PTK and its widespread aberrant activation in various cancers, PTK has emerged as a key target for cancer therapy. Tyrosine kinase inhibitors (TKIs) are target-specific inhibitors of PTKs that have been employed in the treatment of malignancies and play an important role in targeted therapies for cancer [2]. As of 2024, hundreds of TKIs have been studied, 76 of which have been approved by the United States Food and Drug Administration (FDA) for clinical cancer treatment [2–4]. Despite the considerable success of TKIs, drug resistance has developed over time and has become a major obstacle, limiting the efficacy of TKIs. TKI resistance can be basically classified into two main categories: intrinsic and acquired. The intrinsic resistance is usually due to initial genetic alteration in a patient's tumor that affects drug metabolism, bio-distribution, and pharmacodynamics. Acquired resistance to TKIs manifests as resistance to the development of resistance after an initial positive response to TKI treatment. Compared with intrinsic resistance, acquired resistance is frequently intricate and multifaceted, typically classified into tyrosine kinase-dependent and independent mechanisms, depending on whether they are associated with the tyrosine kinase domain [5]. Tyrosine kinase-dependent

L. Fu (✉) · M. Luo
State Key Laboratory of Oncology in South China, Collaborative Innovation Center for Cancer Medicine, Guangdong Esophageal Cancer Institute, Sun Yat-sen University Cancer Center, Guangzhou, People's Republic of China
e-mail: Fulw@mail.sysu.edu.cn; luomin2@sysucc.org.cn

A. Sabarwal et al. (eds.), *Receptor Tyrosine Kinases in Cancer*, Cancer Drug Discovery and Development, https://doi.org/10.1007/978-3-031-93894-8_8

mechanisms include acquired mutations in the targeted tyrosine kinase, amplification or fusion of the target proteins/genes, activation of alternative receptor tyrosine kinase signaling, and downstream pathways. The tyrosine kinase domain exhibits common structural elements, such as the ATP-binding pocket located between the N-terminal lobe containing the αC helix and the C-terminal lobe containing the activation loop. Mutations in these structural components can lead to resistance to inhibitors. And tyrosine kinase-independent mechanisms include epigenetic modification, histologic transformation, cancer stem cell, cell metabolism reprogramming, cell death inhibition, tumor microenvironment, drug delivery abnormality, and so on. These two modalities may work synergistically to lead to the accumulation of key events at the DNA, RNA, and protein levels, resulting in abnormal cell proliferation and apoptosis failure, ultimately leading to drug resistance. In this chapter, we discussed the current understanding of the mechanisms of resistance to TKIs. We highlight drug resistance caused by different intracellular and extracellular changes in tumor cells. An in-depth understanding of the mechanisms of TKI resistance provides a broader perspective for the development of more effective small-molecule inhibitors or antibodies and also provides a theoretical basis for improving the efficacy of TKIs.

2 Tyrosine Kinase-Dependent Mechanism

2.1 Second Mutation of Target

Genetic mutations in the tyrosine kinase are a very common driving oncogene and thus constitutes a core druggable target in the context of cancer treatment. Unfortunately, PTKs develop secondary mutations during TKI therapy, leading to acquired drug resistance and posing a huge barrier to cancer management. The presence of other mutations in the gene alters tyrosine kinase affinity for ATP, thereby restoring ATP affinity to the wild-type level, which represents a common key mechanism of EGFR-TKIs resistance [6]. Depending on the location and function of the mutant gene in the target kinase domain, secondary targeted mutations can be classified into five types: gatekeeper mutations, solvent-front mutations, covalent binding site mutations, compound mutations, and other site mutations [7]. The most commonly reported secondary mutations associated with TKI resistance are summarized in Table 1.

2.1.1 Gatekeeper Mutation

Of all types of targeted mutations, gatekeeper mutation is the most common and widely studied mechanism of TKI resistance. The gatekeeper residue is located in the hinge region of the tyrosine kinase ATP-binding pocket [29]. They play a pivotal

Table 1 Representative mutations associated with TKI resistance

RTKs	Gatekeeper mutations	Other mutations	Solvent-front mutations	Covalent binding site mutations	Ref
EGFR	T790M	L692V, E709A/K, F716L, L718Q/V, G719X, G724S, V726M, V738G, L747S/P, Δexon19, E758D, D761Y, M766Q, S768I, V786M, T790M loss, G796R/S, L798I, Exon20 insertions, L792X/F/H/R/Y/V/P, V802I/F, G824D, V834L, V843I, T844V, T854A, L858R, L861X	L792X	C797S/X	[5, 8, 9]
FGFR1	V561F/M	N546K/H	NA	NA	[10]
FGFR2	V564F/I	M536I, M538I, I548V, N549K/H, N550H, E565A, E584G, L617V, L618M, V652L, K659M, K660E/M, K678M, H682L, K714R, E719G, K641R,	NA	NA	[11]
FGFR3	V355M	V555M	NA	NA	[12]
FGFR4	V550L/E/M	C552A	NA	NA	[13]
ALK	L1196M	L1198F, F1245V, G1269A/S, Q1146K, L1152V/Q, C1156Y, L1174V, V1180L, S1206Y, E1210K	G1202R, D1203N, I1171T	NA	[14, 15]
ROS1	L2026M	L2086F, L2155S, L1174F, L1951R, S1986F/Y, L1986F/L, F2004C/I/V S1896F, L2000V	G2032R/K, D2033N	NA	[16, 17]
RET	V804L/M	E732K, V738A, G807V, G810S, M918T, S904F, F998V, I788N, V804L/M/E, Y806N/C/E, V871I, L881V, A883F, S891A	L730V/I, G810A/C/S/R	NA	[18–20]
PDGFR*α*	T674I	NA	NA	NA	[21]
NTRK1	F589L	G667C/S, A608D	G595R/A	NA	[22]
NTRK2	F633L	G709C	G639R	NA	[23]
NTRK3	F617L	G696A	G623R	NA	[24]
BTK	T474I	NA	C481S	NA	[25]
ABL	T315I	Q252H, Y253F, E255K, V299L, F317I/L, M351T, H396P	NA	NA	[26]
FLT	F691I/L	D835Y/V, Y842C/H, D200N, K429F, Y572C, L601F	NA	NA	[27]
KIT	T670I	D816H, D820A, N822K, Y823D, A829P, V653A, V654A, T669I	NA	NA	[28]

RTK receptor tyrosine kinase, *EGFR* epidermal growth factor receptor, *FGFR* fibroblast growth factor receptors, *ALK* anaplastic lymphoma kinase, *ROS1* c-ROS proto-oncogene 1 receptor tyrosine kinase, *RET* rearranged during transfection proto-oncogene, *PDGFR* platelet-derived growth factor receptor, *NTRK* non-receptor tyrosine kinase, *BTK* Bruton's tyrosine kinase, *ABL* ABL proto-oncogene, *FLT* Fms-related receptor tyrosine kinase, *KIT* kinase insert domain receptor role in controlling the accessibility of TKIs to the ATP-binding pocket [29]. Mutations in gatekeeper residues can affect the interaction between inhibitors and their targeted kinases, reducing the efficacy of TKIs and leading to drug resistance. For example, the EGFR T790M mutation is the most common resistance mechanism for first- and second-generation TKIs, especially in exons 18, 20, and 21 [7, 30–38] (Fig. 1), accounting for approximately one-third of third-generation EGFR-TKI resistance mechanisms [39].

2.1.2 Solvent-Front Mutation

Another major targeted mutation mechanism that leads to TKI resistance is solvent-front mutation, which confers acquired resistance to anaplastic lymphoma kinase (ALK), ROS1, rearranged during transfection proto-oncogene (RET) fusion-targeted therapy [14, 40, 41]. Solvent-front mutations occur in amino acid residues located near the surface of the solvent-accessible in the ATP-binding pocket [7].

2.1.3 Covalent Binding Site Mutation

A less common yet formidable mechanism of resistance in targeted therapy is a covalent binding site mutation. Osimertinib is a third-generation EGFR-TKI with increased potency and selectivity specifically designed to overcome the EGFR-T790M resistance mutation through efficient covalent binding to the ATP-binding

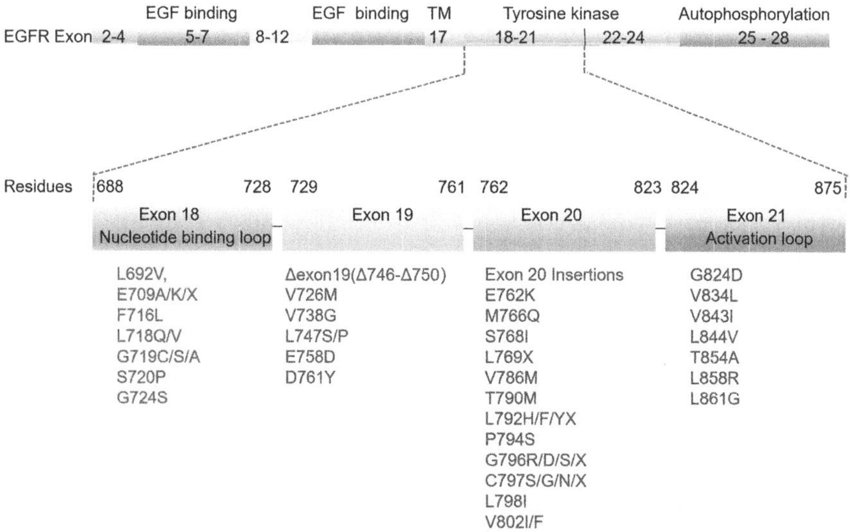

Fig. 1 Summary of EGFR mutations associated with resistance to EGFR-TKIs. The location of different mutations associated with resistance on various domains of the EGFR protein and exons was depicted. *TM* transmembrane domain

pocket [42]. While osimertinib effectively counteracts the EGFR T790M resistance mutation, its covalent binding to cysteine 797 (C797) residues in the ATP-binding pocket renders the C797 residue susceptible to mutations [8]. EGFR-C797 mutation occurs at the EGFR C797 codon in exon 20, which causes cysteine to be replaced by serine, resulting in the loss of the covalent bond between osimertinib and EGFR [36]. C797 is the most common mutation hotspot, occurring in 10–26% of patients undergoing osimertinib treatment [9].

2.1.4 Compound Mutations

Compound mutations are when there are two or more mutations within the same gene or across different genes in the same molecule. Compound mutations have the potential to confer cross-resistance to multiple TKIs through steric hindrance mechanisms [7]. Patients with EGFR compound mutations have been observed to experience worse clinical outcomes than patients with EGFR-TKI sensitive mutations (ex19del, L858R) but have a better prognosis than patients with resistant mutations (T790M, ex20ins). Non-C797 EGFR mutations often coexist with other EGFR mutations [39].

2.1.5 Other Site Mutations

In addition to the classic secondary mutations discussed above, mutations at other sites of the target kinase domain have also been reported to confer TKI resistance. Typically, mutations at other sites occur in kinase residues that have important functions. They are usually less common. For example, EGFR-T854A, EGFR-D761Y, EGFR-L747S, EGFR-L792F/H have been found to confer resistance to EGFR-TKIs [43–46]. D276G mutation, located on the β3-αC loop of BCR-ABL kinase domain, was reported to cause resistance to imatinib [47]. Similarly, ALK-G1269A and ALK-C1156Y, ALK-L1152R, ALK-F1174C, and ALK-1151 T insertion were identified in ALK-rearranged cancers, conferring resistance [48–51].

2.2 *Amplification of Target Genes*

Genomic amplification and overexpression are the second major type of mutation. Cancer cells may upregulate the expression of target proteins, allowing them to maintain signaling despite TKI inhibition. For example, amplification or overexpression of wild-type or truncated and constitutively activated EGFR is postulated to be a common mechanism of resistance in EGFR-TKI treatment. In addition, the upregulation of its ligand EGF and the inferior degradation of EGFR in lysosomes lead to sustained activation of the EGFR pathway and also induce resistance to TKIs [52]. Increased or expanded ALK copy number has been reported in non-small cell lung carcinoma (NSCLC) patients with crizotinib resistance [53]. B-cell receptor (BCR)-ABL gene amplification has also been observed in ABL-inhibitor-resistant cells [54].

2.3 Oncogenic Fusions

Tumors harboring sensitizing mutations are sensitive to TKIs, whereas tumors with acquired RTK fusions are resistant due to activated bypass signaling pathways [55]. RET fusion is the most prevalent acquired RTK fusion, particularly in the context of third-generation EGFR-TKIs, which encompass a range of fusions, including the following: RET-KIF5B, RET-CCDC6, RET-NCOA4, RET-MYH9, RET-TRIM33, RET-PRKAR1A, RET-GOLGA5, RET-TRIM24, RET-KTN1, RET-RFG9, RET-ELKS, RET-CPM1, and RET-ERC1 [18, 31, 56, 57]. Additionally, anaplastic lymphoma kinase (ALK) fusions have also been demonstrated to potentially mediate osimertinib resistance. They include SPTBN1-ALK, STRN-ALK, EML4-ALK, DCTN1-ALK, and PLEKHA7-ALK [58, 59]. The fusions of the BRAF gene were AGK-BRAF, ZDHHC20-BRAF, PCBP2-BRAF, and ESYT2-BRAF [60]. Other fusions include FGFR3-TACC3, FGFR2-CCDC6, NTRK1-TPM3, and GOPC-ROS1 [56, 61, 62].

2.4 Other Nontarget Gene Alterations

Receptor tyrosine kinase (RTK) is a membrane protein that affects cellular functions by activating downstream signaling pathways; however, molecular alterations of other membrane proteins or multiple oncogenes that can bypass repressed signaling pathways may also cause resistance to single-target TKIs [63] (see Table 2). Met oncogene amplification accounts for 5–20% of acquired resistance to EGFR-targeted agents, and this resistance form occurs in both first- and second-line osimertinib treatments [118]. Herein, EGFR mutations with MET amplification attenuate the efficacy of a single TKI in inhibiting proliferation caused by aberrant activation of other PTKs. In addition to MET, another frequently amplified gene is HER2, which encodes the ErbB2 receptor tyrosine kinase that activates the MAPK and PI3K pathways. Its aberrant amplification also leads to secondary resistance to EGFR-TKIs [119]. EGFR and KIT amplification have been reported in patients resistant to ALK inhibitors [41]. FGFR (FGFR1, FGFR2, and FGFR3) amplification, as well as IGF/IGFR amplification and overactivation, are mechanisms of resistance to EGFR-TKIs [120]. Moreover, nerve growth factor (NGF)-mediated activation of the RAS/MAPK pathway can lead to resistance to multiple TKIs in neuroblastoma [121]. The alterations in different genes associated with resistance to TKIs are shown in Table 2.

Table 2 The alterations in different genes associated with acquired resistance to TKIs

Resistance to	Gene	Alteration types	Ref
EGFR-I	MET	Amplification	[64]
		METex14 mutation, D1228Y/N, Y1230H/C, K957N	[5, 65]
	HER2	Amplification	[36]
		Exon 16 skipping mutation, C805S, exon 20 insertion	[66]
	HER3	Heterodimers	[67]
	KRAS	Amplification	[68]
		Mutation: G12V/D/S/A, A146T, G13D, Q61K/H, R68S	[55, 65]
	NRAS	Amplification Mutation: E63K, Q61K	[68]
	BRAF	Amplification	[68]
		Mutation: V600E/F, G466R, G469A, G467A/V, D549G	[69, 70]
		Fusion	[37]
	PIK3CA	Amplification	[71]
		Mutation: R88Q, N345K, E545K, E418K, E542K	[58, 68]
	MAPK1	Amplification	[72]
	FGF/FGFR	Amplification	[73]
		Mutation	[68]
		FGFR3 fusion	
	RB1	Mutation	[74]
	KIT	Mutation	[36]
	ALK	Mutation	[58]
		Fusion	[37]
	RET	Mutation/fusion	[57]
	IGFBP3	Loss	[68]
	IGFBP7	Overexpression	[68]
	IGF1R	Amplification	[65]
	AXL	Overexpression	[68]
	TP53	Mutation: R273C	[5, 68]
	MEK	Mutation	[70]
	ROS1	Rearrangement	[75]
	PTEN	Loss	[76]
	CCND1/2	Amplification	[58]
	NTRK	Fusion	[37]
	mTOR	Mutation: L1433S	[77]
	CCNE1	Amplification	[58]
	CDK4/6	Overexpression	[58]
	CDKN2A	Mutation: E27fs	[58]
	NF2	Loss	[74]
	EPHA2	Overexpression	[78]

continued

Table 2 continued

Resistance to	Gene	Alteration types	Ref
ALK-I	MET	Amplification Mutation: E1012	[79]
	IGF-1R	Amplification	[80]
	KIT	Mutation: D820E	[81]
	EGFR	Mutation: P26_C291del, P753S	[81]
	MAP2K1	Mutation	[17]
	KRAS	Amplification Mutation: G12C	[17]
	AXL	Overexpression	[82]
	PIM	Overexpression	[83]
	BRAF	Fusion	[84]
	NRAS	Mutation	[17]
	PTPN1/ PTPN2	Loss	[85]
	SRC	Activation	[86]
HER2-I	HER1/HER3	Amplification	[87]
	IGF1R	Amplification	[88]
	AXL	Heterodimerization	[89]
	VEGFR	Amplification	[90]
	EphB	Overexpression	[91]
	EPHA2	Overexpression	[92]
	FGFR	Amplification	[93]
	PTEN	Loss	[94]
	PIK3CA	Mutation: exon 9 or exon 20	[95]
	CDKN2A	Mutation: E27fs	[96]
	CCND1/2	Amplification	[96]
	CDK4/6	Amplification	[96]
	CCNE1	Amplification	[97]
FGFR-I	PIK3CA	Mutation	[98].
	KRAS	Mutation: G12V	[99]
	MYC	Amplification	[100]
RET-I	MDM2	Amplification	[101]
	NRAS	Mutation: Q61K	[101]
	MET	Amplification	[18]
		Mutation	[40]
	AXL	Overexpression	[102]
	BRAF	Mutation: V600E	[101]
	YAP	Overexpression	[18]
ABL-I	KIT	Mutation: V654A, T670I, D816V, D820A	[103]
	AXL	Overexpression	[104]
	PDGFRA	Mutation: D842V	[101]
	FGFR	Mutation: V564F	[105]
	RET	Mutation: V804M	[106]
	RUNX1/ PMRD16	Deletion	[107]
	GATA2	Mutation	[107]

continued

Table 2 continued

Resistance to	Gene	Alteration types	Ref
VEGFR-I	JAK	Mutation	[108]
	PKC	Overexpression	[109]
	MET	Amplification	[110]
	IGF1R	Amplification	[80]
	AXL	Overexpression	[111]
	PTEN	Loss	[112]
	EphB	Overexpression	[113]
	EphA2	Overexpression	[114]
KIT-I	CTNNB1	Mutation: S33C	[115]
MET-I	PI3KCA	Mutation	[116]
ROS-I	MEK1/MEKK1	In-frame deletion	[117]

2.5 *Activation of Aberrant Bypassing Pathways*

Inhibition of a specific target frequently activates horizontal or longitudinal pathways in an attempt to evade cancer cell death. The PI3K/AKT, RAS/MAPK/ERK, and JAK/STAT signaling pathways are the most important pathways for RTK activation. Upregulated receptors or ligands can enhance signaling, while point mutations and gene rearrangements result in sustained activation of RTKs in the absence of ligands. Sometimes multiple RTKs activate the same downstream pathway, and the inability of a TKI to inhibit all of them leads to persistent pathway activation and drug resistance. Furthermore, activation of a single RTK due to a dominant ligand may lead to cascading activation of other RTKs, resulting in a signal transduction scaffold with a network potential greater than that of the initiating dominant kinase. This phenomenon is known as pathway co-option [107]. This indicates that cells can survive through signaling pathways such as JAK/STAT, RAS/MAPK, and PI3K/AKT pathways which are caused by HER2, MET, FGFR mutations, oncogene fusions (e.g., ALK, NTRK, BRAF/ROS1), upregulation (e.g., AXL, HER2, MET), gene amplification (e.g., HER2, MET, FGFR, IGF-1R [122–125]). In addition, crosstalk between different RTKs, such as between MET, EGFR, and HER3, can lead to more complex aberrant signal transduction [126]. Here, we summarized the aberrant activation of several types of PTK-related signaling pathways that may be resistant to TKIs (Fig. 2).

2.5.1 EGFR Pathway

EGFR signaling pathway activation is the most common mechanism of resistance to non-EGFR TKIs. It can be triggered by ligands or dimers with HER2 or AXL, activated PI3K/AKT, RAS/MAPK, and JAK/STAT3 signaling pathways [127, 128]. In most cases of aberrant forms of EGFR in malignancies, manifest as mutations and/or gene amplification, which can lead to sustained activation of the EGFR pathway to promote cell proliferation, differentiation, survival, and migration [129].

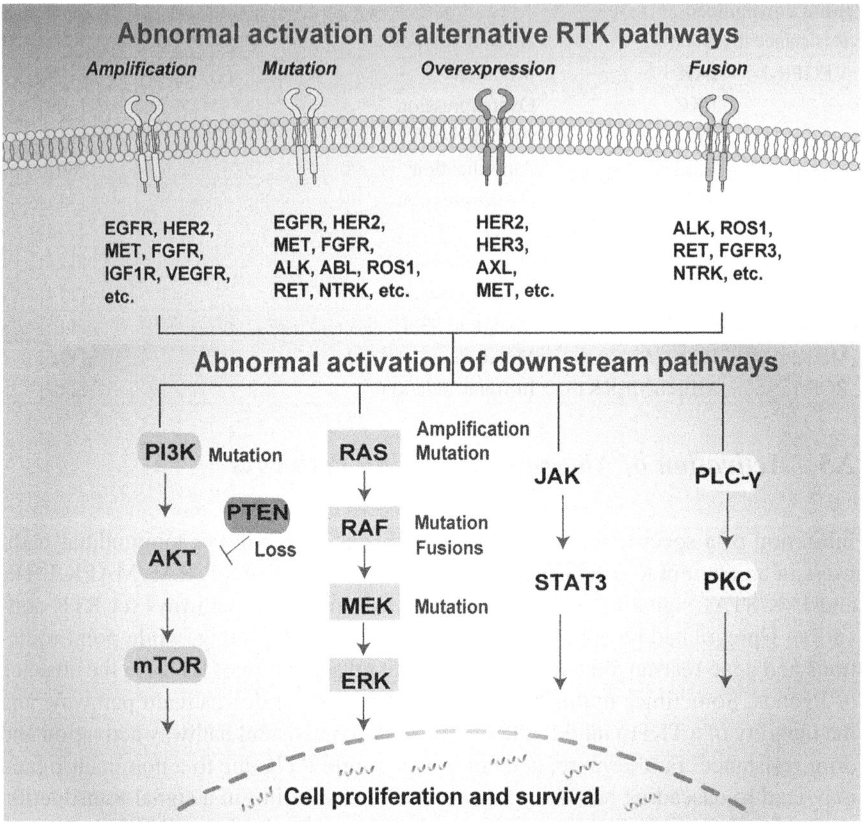

Fig. 2 Overview of tyrosine kinase-dependent mechanisms to TKI resistance. Acquired mutations or amplification of TKs, activation of bypass signals, and mutational activation of downstream molecules are thought to be mechanisms for inducing resistance to TKIs

2.5.2 HER2/3 Pathway

The HER family consists of EGFR, HER2, HER3, and HER4 and has a cytoplasmic TK domain that can be activated by ligand binding, followed by dimerization. It has been documented that the upregulation of other members of the HER family serves as a mechanism of resistance to TKIs. Although HER2 does not appear to have an intrinsic ligand binding capacity, it is capable of reversibly interacting with ligand-activated EGFR or HER3 to form active heterodimers that activate downstream signals to control cell proliferation and migration. HER2 overexpression and exon 16-skipping mutations were involved in resistance to EGFR-TKIs [66]. HER3 can form heterodimers with other members of the HER family, such as EGFR and HER2, or it can be amplified via the proto-oncogene c-Met *[130]*. HER3 is a key activator of the downstream PI3K/AKT and RAS/MAPK pathways, contributing to the survival of most tumor cells, thereby further promoting resistance to third-generation EGFR-TKIs [67].

2.5.3 HGF/c-MET Pathway

Hepatocyte growth factor (HGF) or MET gene mutation, amplification, or protein overexpression leads to sustained activation of downstream signaling cascades, such as PI3K/Akt and RAS/RAF/ERK/MAPK, which promote resistance to TKIs [36, 58]. The MET exon 14 skipping mutation mediated resistance to osimertinib in NSCLC [131]. MET amplification has been proposed as a driver of resistance to EGFR inhibitor (gefitinib), RET inhibitors (selpercatinib), ALK inhibitors (alectinib), and KRAS inhibitors (sotorasib) through the activation of the HER3-PI3K/Akt signaling pathway in NSCLC [132–134].

2.5.4 IGF/IGF1R Pathway

Aberrant expression of insulin-like growth factor (IGF)/IGF1 receptor (IGF1R) triggers continuous activation of RAS/MAPK and PI3K/AKT signaling pathways, leading to the development of secondary resistance to TKIs [125]. Heterodimerization of IGF1R and HER family members activates IGF1R and its downstream signaling, resulting in the development of TKI resistance [135]. IGF1R and EMT cross talk may be sufficient to cause erlotinib resistance in EGFR-mutated NSCLC [38]. Nuclear accumulation of IGF1R by amphiregulin induces cell cycle arrest via p21WAF1/cyclin-dependent kinase-interacting protein 1 (CIP1) upregulation and prevents gefitinib from inducing cell apoptosis [136].

2.5.5 GAS6/AXL Amplification

Anexelekto (AXL), activated by its ligand growth arrested-specific gene 6 (GAS6), triggers activation of MEK/ERK and PI3K/AKT pathways, thereby enhancing resistance to third-generation EGFR-TKIs [137]. AXL degradation is slower in osimertinib-resistant cells than in drug-sensitive cells [138]. Additionally, activation of AXL has been observed to correlate with EMT signatures and to confer acquired resistance to erlotinib in NSCLC cell lines [139]. AXL co-clusters with and activates EGFR, HER2, and c-MET, resulting in downstream ERK phosphorylation and temporal response, motility, and invasion [128, 140].

2.5.6 ALK/ROS1 Translocations

The kinase domains of ROS1 and anaplastic lymphoma kinase (ALK) share a large number of homologies. Chromosomal rearrangements of ALK or ROS1 genes have been shown to define two distinct molecular subsets of non-small-cell lung cancer (NSCLC) [141]. In both cases, gene rearrangements result in the expression of constitutively activated fusion kinases and are involved in tumorigenesis in multiple tumor types. ROS1 G2032R is the most common resistance substitution, similar to

ALK G1202R [16]. ALK gene rearrangement (EML4-ALK) has been observed in 3–6% of lung adenocarcinomas, and EML4-ALK overexpression activates ERK and STAT3 [142]. Genetic rearrangements of the ROS1 gene account for 1–2% of NSCLC [143]. To date, various ROS1 fusion variants have been identified in NSCLC, which result in the constitutive activity of the ROS1 kinase [144]. CD74-ROS1 is the most common variant, occurring in up to 50% of cases [145]. Other frequently occurring ROS1 fusion chaperones are SDC4, SLC34A2, LRIG3, EZR, and TPM3, which exhibit distinctively activation of MAPK signaling [144]. Due to the structural homology between the ALK and ROS1 kinase domains, ALK-TKIs can be used to cross-inhibit ROS1 kinase activity [146].

2.5.7 Other RTK Pathways

In addition to the RTK signaling pathways described above, there are a number of key players that can similarly be involved in TKI resistance. FGF2 is a ligand for the FGF receptor (FGFR) belonging to the TKR family, and stimulation of the FGF pathway promotes the reactivation of the MAPK pathway, thus consequently attenuating the efficacy of imatinib [147]. Inappropriate expression of FGF ligands and FGFRs (FGFR1–4) or FGFR mutations results in receptor dimerization and activation of PI3K/AKT/mTOR and RAS/MAPK pathways [130]. FGFR activation is an escape mechanism for TKI resistance that compensates for the loss of EGFR-driven signal transduction pathways [72]. The fusion of FGFR3-TACC3 can be used as a bypass resistance mechanism to pretreated EGFR-TKIs, such as naquotinib, afatinib, erlotinib, and osimertinib in these lung cancer patients [62, 148]. The vascular endothelial growth factor (VEGF)/VEGF receptor 2 (VEGFR2) signaling cascade via VEGFR2/PI3K/mTOR induces regulation of VEGF secretion and has been implicated in TKI resistance [130]. Mutations within the activation loop of members of the class III RTK subfamily, including KIT, platelet-derived growth factor receptor (PDGFRA), and FLT3, have been observed in a variety of human tumor types [149]. These mutations confer constitutive activation as well as resistance to type II TKIs, such as imatinib and sunitinib [150]. KIT or PDGFRA kinase activation leads to activation of PI3K, PLCγ, and MAPK signaling pathways, which in turn mediate EGFR transactivation and evade EGFR-TKIs [151]. In vitro studies, alterations in common signaling pathway mediators, such as Src family kinases, AXL receptor tyrosine kinases, and transmembrane protein CUB domain-containing protein 1 (CDCP1), have also been identified as promoting TKI resistance [152, 153].

2.6 *Abnormal Activation of Downstream Pathways*

There are four well-elucidated signaling networks downstream of the RTK, namely, RAS/MAPK, PI3K/AKT, JAK/STAT3, and PLC-γ/PKC, that facilitate a myriad of biological outcomes (reviewed in [154]). Cancer cells can bypass inhibitor-mediated oncogene blockade by altering downstream components of inhibitory kinases in the absence of upstream stimulation, leading to drug resistance. All of these genes encode key signaling molecules in the RTK downstream signaling pathway, and their mutations cause continuous RTK pathway activation and transcription of downstream genes, leading to sustained cell proliferation.

2.6.1 Activation of MAPK/ERK Signaling Pathway

Several alterations, such as NRAS mutations (G12V, G12R, Q61K, and E63K), KRAS mutations (G12S, G12A, Q61H, G12D, G13D, and A146T), BRAF mutations (G469A, V599E, and V600E), and MEK1 mutations, make NSCLC resistant to EGFR, MET, or ALK-TKIs [64, 120, 155]. In these cases, constitutive activation of the RAS/MAPK pathway can be directly induced by altered KRAS and BRAF, independent of upstream stimulation [64]. Deletion of neurofibromin 1 (NF1), a RAS GTPase-activating protein that negatively regulates the RAS/MAPK pathway, is associated with erlotinib resistance [156]. In addition, Src homology 2 domain-containing phosphatase (SHP2) is a key cascade of activation of the RAS/MAPK signaling pathway [157].

2.6.2 Activation of PI3K/AKT Signaling Pathway

Similarly, mutations encoding PI3K have been observed to lead to constitutive activation of the PI3K/Akt signaling pathway, mediating resistance to MET inhibitors and EGFR-TKIs [158]. Deletion or mutation of PTEN increases PI3K levels and induces overactivation of AKT, resulting in resistance to the third-generation EGFR-TKIs [73, 159]. PIK3CA is one of the PI3K isoforms, and PIK3CA mutations, specifically E545K, E542K, R88Q, N345K, and E418K, have been reported to aberrantly activate AKT and confer resistance to osimertinib or rociletinib [31, 58].

2.6.3 Activation of JAK-STAT3 Signaling Pathway

Inhibition of various RTKs, including EGFR, MEK, FGFR, and HER2, triggers feedback activation of the IL-6/JAK/STAT3 signaling pathway [160]. Nerve growth factor (NGF)/tropomyosin receptor kinase A (TrkA) selectively activates the JAK/STAT3 pathway, thereby inducing EMT and erlotinib resistance in head and neck squamous cell carcinoma (HNSCC) [161]. Aberrant FLT3 induction activates FLT3-JAK/STAT3 signaling pathway, thereby conferring resistance to multiple BCR/ABL-TKIs [162].

2.6.4 Activation of PLCγ/PKCδ Signaling Pathway

TKI-induced dimerization of EGFR with other RTKs, such as AXL and HER2, has been observed to promote nuclear translocation of PKCδ and the activation of AKT and NF-κB signaling pathways, ultimately leading to resistance to third-generation EGFR-TKIs [83, 163].

3 Tyrosine Kinase-Independent Mechanism

3.1 Epigenetic Alterations

3.1.1 Noncoding RNAs (ncRNAs)

ncRNAs do not code for proteins but are involved in posttranscriptional modifications. ncRNAs mainly include microRNAs (miRNAs), long non-coding RNAs (lncRNAs), and circular RNAs (circRNAs). By sharing a common signaling molecule, different ncRNA regulatory axes form a broad ncRNA regulatory network that regulates TKI resistance. A preliminary circRNA/lncRNA/miRNA/mRNA network has been constructed to elucidate the role of ncRNAs in regulating signaling pathways and drug resistance [164]. The ncRNAs associated with resistance to TKIs are presented in Table 3.

3.1.2 Other Epigenetic Modifications

Other types of epigenetic modifications with TKI resistance, such as DNA hypermethylation, histone acetylation, RNA methylation, and chromatin remodeling, are presented in Table 4. mRNA N6-methyladenosine (m6A) modification can affect gene transcription by affecting mRNA stability, and m6A methyltransferase (METTL) plays an important role in TKI resistance. METTL3 can alter MET levels to modulate the PI3K/AKT pathway, leading to resistance to gefitinib [209]. In addition, METTL3 upregulates AXL expression in ovarian cancer and induces EMT [210]. There is now ample evidence that mutations in epigenetic regulatory genes, such as DNA methyltransferase 3A (DNMT3A), ten-eleven translocation 2 (TET2), histone-lysine N-methyltransferase (EZH2), and additional sex combs-like 1 (ASXL1), isocitrate dehydrogenase 1 (IDH1), and SET binding protein 1 (SETBP1), increase during disease progression and accelerates leukemia stem cell production, maintenance, and chronic myelogenous leukemia (CML) progression [122, 211]. Genomic instability results in the accumulation of mutations and other molecular or chromosomal aberrations, which ultimately lead to acquired resistance to EGFR-TKIs in cancer [212].

Table 3 The ncRNAs associated with resistance to TKIs

ncRNAs	Target	Resistant to	Ref
miR-200	ZEB1/2	EGFR-TKI	[165]
miR-34a	Snail	EGFR-TKI	[82]
miR-451	MDR1	EGFR-TKI	[166]
miR-21-3p	ADSL	EGFR-TKI	[167]
miR-153-3p	RPS6KB1	Cabozantinib	[168]
miR-186	TWIST	EGFR-TKI	[165]
miR-125-5p	Ataxin-1	Sorafenib	[169]
miR-210-5p	ATAD3A	Sorafenib	[170]
miR-424	CBX4	Sorafenib	[171]
miR-125d-5p	PUMA	Sorafenib	[172]
miR-494	PTEN	Sorafenib	[173]
miR-30a-5p	CLCF1	Sorafenib	[174]
miR-335	AXL	EGFR-TKI	[139]
miR-221	p27kip1	Lapatinib	[175]
miR-196a	GLTP	Gefitinib.	[176]
miR-92a-1-5p	MLKL	ABL-TKI	[177]
miR-212	ABCG2	Sunitinib	[178]
miR-452-5p	SMAD4	Sunitinib	[179]
miR-25-3p	PTEN	EGFR-TKI	[180]
lncRNA-RP11-390F4.3	Snail, Twist1, ZEB1, ZEB2	EGFR-TKI	[181]
lncRNA-ECVSR	ERβ/HIF2-α	Sunitinib	[182]
lncRNA-CASC9	EZH2/DUSP1	Gefitinib	[183]
LncRNA-PCAT-1	PI3K/AKT pathway	Gefitinib	[184]
LncRNA-HIF1A-AS2	miR-146b-5p/IL-6/STAT3	EGFR-TKI	[181]
circFN1	miR-1205/E2F1	Sorafenib	[185]
circFOXM1	miR-1324/MECP2	Sorafenib	[186]
circRNA-SORE	YBX1	Sorafenib	[187]
circRNA-ASK1	ASK1-272a.a/PI3K/AKT	Gefitinib	[188]
circRNA hsa-circ-0005576	miR-512-5p/IGF-1R	Osimertinib	[189]
circ-E-Cad	C-E-Cad/EGFR	EGFR-TKI	[190]
ncRNA-RP11616M22.7	RASSF	Imatinib	[191]

3.2 Histologic and Phenotypic Transformations

3.2.1 Epithelial-Mesenchymal Transition

Epithelial to mesenchymal transition (EMT) is a reversible process in which the absence of epithelial cell markers, like E-cadherin, and upregulation of mesenchymal cell markers, such as N-cadherin and vimentin, facilitates cancer cell migration and invasion of surrounding tissues [165]. Cancer cells receiving EMT are less dependent on targeted pathways and more resistant to TKI therapy [165]. Furthermore, EMT-related molecules, such as SNAIL/2, ZEB1/2, CLDN1, and

Table 4 The epigenetic modifications of genes associated with TKI resistance

Genes	Resistance to	Ref
GABBR2	Erlotinib	[192]
ADAMTSL5	Sorafenib/lenvatinib/regorafenib	[193]
SETD2	Imatinib	[194]
EZH2	Sorafenib	[195]
KMT5C	EGFR-TKI	[196]
SMYD2	Sunitinib	[197]
BORIS	ALK-inhibitor	[198]
HDAC5	Sorafenib	[199]
KDM6A	Imatinib	[200]
METTL3	Gefitinib	[201]
METTL7B	EGFR-TKI	[202]
METTL-KIAA1429	Gefitinib	[203]
YTHDF2	Gefitinib	[188]
YTHDF	Erlotinib	[204]
IGFBP2	Selumetinib	[205]
RBP MUSASHI-2 (MSI2)	Osimertinib	[206]
DNMT-1	Cabozantinib	[207]
RBP2	ABL-inhibitor	[208]

CD44, were significantly upregulated in TKI-resistant tumor biopsies [213]. TGF-β secretion triggered by EGFR inhibition is a canonical inducer of EMT [214]. It binds to TGFR and induces the expression of EMT transcription factors, which downregulates epithelial-related proteins and initiates EMT via SMAD pathways or non-SMAD pathways, such as PI3K/AKT and RAS/MAPK pathways [214]. Hypoxia can interact with EMT-related pathways (e.g., TGFβ, WNT, NOTCH, and hedgehog) through HIF, directly leading to EMT or accelerated EMT [215]. Some PTKs and their own ligands, such as FGF/FGFR1, HGF/MET, IGF/IGFR, GAS6/AXL, and NGF/TrkA, have been implicated in EMT and TKI resistance [161, 216, 217].

3.2.2 Small Cell Lung Cancer Transformation

In addition to EMT, the transformation of NSCLC into small cell lung cancer (SCLC) is another mechanism by which cancer cells escape from TKI-targeted therapy [218]. Approximately 3–14% of EGFR-TKI-resistant patients experience conversion from NSCLC to SCLC [120]. Decreased levels of EGFR protein and mRNA have been observed in SCLC. After effective TKI treatment, SCLC became dominant and retained mutant EGFR, indicating that the SCLC samples are most likely transformed from EGFR-mutant NSCLC. SCLC transformation may be associated with the inactivation of Rb and p53 [219, 220].

3.3 *Quiescent Stem Cells*

Stem cells are a group of cells that exist in tumor tissue with the capacity for unlimited proliferation and multidirectional differentiation potential, which are closely associated with tumor recurrence, metastasis, and treatment resistance to [221]. Resistant cells exhibit stem cell-like attributes, such as increased ALDH1A1 expression, increased percentage of side population (SP) cells, and enhanced self-renewal capacity [222]. Cancer stem cell-like cell (CSLC)-mediated therapeutic resistance in ALK-rearranged lung cancer represents one of the numerous different mechanisms of ALK-TKIs [222]. CML stem cells are resistant to TKIs, and it is hypothesized that resistance/relapse is predominantly due to these cells [223]. Leukemia stem cells in some patients can survive for a long time even after achieving complete molecular biological remission [223].

3.4 *Tumor Microenvironment Heterogeneity*

Tumor growth and development is a dynamic process, and interactions between tumor cells and their surrounding components elucidate the mechanisms of resistance [224]. Uncontrolled expression of bioactive substances by cellular components of the tumor microenvironment (TME) can compensate for inhibited signal transduction in tumor cells, induce immunosuppression, and promote EMT and stemness of tumor cells, all of which contribute to TKI resistance [225]. TKIs, especially antiangiogenesis-related TKIs (VEGFR-TKI, PDGFR-TKI, etc.), have the potential to remodel TME by modulating various components, including tumor-infiltrating immune cells (e.g., T cells, NK cells, and macrophages), immunomodulatory stromal cells, cytokines, chemokines, and other factors, which play a dual role in tumor progression and drug resistance [226].

3.4.1 TAMs

Tumor-associated macrophages (TAMs) promote angiogenesis, tumor cell invasion, metastasis, and immunosuppression in TME [227]. TAMs are abundant in TME and are associated with adverse clinical outcomes [227]. TAMs can be divided into two subgroups, M1 and M2 [228]. M1 macrophages typically exhibit an antitumor phenotype, while M2 macrophages are often considered as pro-tumor cells [228]. Tumor cell-derived HGF, VEGF, IL-6, and TNF-α promote TAM M2 polarization, thereby activating a multitude of tumor cell signaling pathways involved in TKI resistance, including MAPK, PI3K, YAP, NF-κB, Wnt, and RAS pathways [224, 229–232]. High TAM-derived epiregulin (EREG) induces the formation of EGFR/HER2 heterodimers on cancer cells, leading to AKT pathway activation and erlotinib resistance in NSCLC patients [233, 234].

3.4.2 TANs

Similar to TAMs, tumor-associated neutrophils (TANs) can also be polarized in response to TME signaling. VEGF-TKI induces IL-17A secretion by γδ T cells to promote N2 polarization of TANs and depletion of CD8+ T cells, creating an immunosuppressive microenvironment and leading to VEGFR-TKI resistance [235]. Sorafenib induces hypoxia and then activates HIF-1α/NF-κB/Chemokine C-X-C Motif ligand 5 (CXCL5) signaling to recruit TAN and inhibit apoptosis, further triggering the infiltration of TAM and Treg cells [236], which subsequently renders tumor cells insensitive to VEGFR-TKIs.

3.4.3 Tregs

Regulatory T cells secrete immunosuppressive cytokines, like TGF-β and interleukin-6 (IL-6), which inhibit effector T cell function, thereby contributing to TKI resistance [237]. Treg cells deliver sorafenib resistance through the immunosuppressive landscape by upregulating T cell factor 1 (TCF1), programmed cell death protein 1 (PD-1), and cytotoxic T lymphocyte-associated protein 4 (CTLA-4) and releasing immunosuppressive cytokines, like IL-10 [14, 238]. Elevated levels of TNF-α have been demonstrated to lead to EMT and can also induce NF-κB pathway activation to avoid sorafenib-mediated cell death [239].

3.4.4 MDSCs

Myeloid-derived suppressor cells (MDSCs) release IL-6, granulocyte-macrophage colony-stimulating factor (GM-CSF), and granulocyte colony-stimulating factor (G-CSF) through STAT5 activation, leading to VEGFR-TKIs resistance [240]. MDSCs-derived IL-6 induces FGF1 secretion by HCC cells to activate CAFs and induce sorafenib resistance [241]. Fibroblast growth factor 2 (FGF-2) released from the bone marrow matrix also drives imatinib resistance by activating downstream MAPK signaling [147].

3.4.5 Stromal Cells

Stromal cells within the TME, such as cancer-associated fibroblasts (CAFs) and tumor-associated endothelial cells, play a key role in regulating immune responses and promoting tumor survival [242]. CAFs are capable of secreting a variety of cytokines, including IGF-1, IL-6, IL-8, TGF-β, and extracellular matrix proteins that support tumor growth and metastasis [225, 242, 243]. For instance, fibroblasts can mediate resistance to RAF inhibitors in BRAF-mutant melanoma by secreting

HGF and activating the MAPK and PI3K signaling pathways in tumor cells [244]. CAF-derived secreted phosphoprotein 1 (SSP1) was found to be a candidate molecule responsible for HCC resistance to sorafenib/lenvatinib in HCC [245]. Furthermore, CAFs can induce EMT in tumor cells, leading to increased aggressiveness and resistance to TKI therapy [243]. CAFs express more CAF markers like α-smooth muscle agonists-α (α-SMA), fibroblast activation protein (FAP), and PDGFR, which induce activation of ERK, JAK1/STAT3, PI3K-AKT-mROR pathways, and subsequently TKI resistance [243, 246]. Additionally, CAFs-derived cardiotrophin-like cytokine factor 1 (CLCF1) and chemokines such as chemokine (C-C motif) ligand 2 CCL2 can form TME suitable for cellular stemness and immunosuppression, further exacerbating immune evasion and TKI resistance [247, 248]. Moreover, CAFs can release NRG1β, which binds to HER3 and stabilizes HER2-HER3 dimers to maintain permanent activation of downstream pathways, resulting in lapatinib resistance [249]. In addition, upregulation of VEGFA, FGF2, FGFR2, and PDGFRα in stromal cells leads to acquired resistance to TKIs [250–252].

3.5 Cellular Metabolism

The role of tumor metabolic reprogramming in the induction of TKI resistance provides new insight into the improvement of drug resistance. Resistant cell lines significantly induce hypoxic signaling and have higher levels of glycolytic pathway and fatty acid metabolic pathways compared to TKI-sensitive cells [253]. Hexokinase (HK) is upregulated due to hypoxia and is associated with HCC resistance to regorafenib [254]. PKM2 overexpression is associated with erlotinib resistance in EGFR-mutant lung cancer [255]. Sterol regulatory element binding protein 1 (SREBP1) is responsible for de novo synthesis, which is highly expressed in gefitinib-resistant cells [256]. Gefitinib downregulates the expression of peroxisome proliferator-activated receptor alpha (PPARα), liver X receptor alpha (LXRα), and ATP-binding cassette subfamily B member 1 (ABCA1), leading to cholesterol efflux dysregulation and resistance induction in NSCLC cell lines [257]. Cholesterol accumulation in lipid rafts promotes EGFR-Src interactions, activates ERK/SP1 signaling via Src, and enhances estrogen-related receptor alpha (ERRα) expression, thereby regulating ROS levels and maintaining cell proliferation, leading to drug resistance [258]. Furthermore, phospholipid and methionine metabolism have also been implicated in drug resistance [259, 260]. Phosphoglycerate dehydrogenase (PHGDH) is a major driver of sorafenib resistance in HCC [261]. Fumarate reacts with PTEN and blocks the binding of PTEN to cell membranes, thereby decreasing its inhibitory effect on the PI3K/AKT pathway, resulting in sustained pathway activation and further resistance to sunitinib [262].

3.6 *Cell Death Inhibition*

3.6.1 Apoptosis Inhibition

In cancer, dysregulation of apoptosis signaling, particularly activation of the anti-apoptotic pathway, allows cancer cells to evade apoptosis, leading to uncontrolled proliferation and TKI resistance. Bcl-2 interacting mediator of cell death (BIM) deletion polymorphisms is associated with apoptotic resistance to EGFR-TKIs (e.g., gefitinib and erlotinib) in NSCLC harboring EGFR mutations [263]. Hypoxia upregulates FGFR to activate the RAS/MAPK pathway, thereby downregulating BIM expression and inducing osimertinib resistance [72]. In addition to BIM, some cell lines overexpress antiapoptotic proteins such as BCL-2 [264], due to overactivation of the PI3K/AKT and RAS/MAPK pathways.

3.6.2 Non-Apoptotic Cell Death Inhibition

In addition to apoptosis, there are several forms of programmed cell death, such as ferroptosis, autophagy, necroptosis, and pyroptosis, which are closely associated with TKI resistance. Activation of the nuclear factor erythroid 2-related factor 2 (NRF2) pathway prevents ferroptosis and increases TKI resistance [265]. ABCC5 stabilizes SCL7A11, thereby increasing the uptake of GSH precursors and enhancing sorafenib efflux, which ultimately leads to TKI resistance [266]. In addition to the NRF2 pathways, the transcription complex yes-associated protein (YAP)/tafazzin (TAZ)-TEA domain transcription factor (TEAD) downstream of the Hippo pathway is a key factor in sorafenib resistance in HCC [267]. Additionally, tumor cell-derived stearoyl-CoA desaturase-1 (SCD1), tumor endothelial cells (TECs), and adipocyte-derived fatty acid binding protein 4 (FABP4) induce ferroptosis resistance in sunitinib or sorafenib-treated breast cancer cells [268]. GPX4 overexpression has been observed in EMT-induced hypermesenchymal state and is associated with erlotinib resistance [269]. In addition, lipid metabolism reprogramming is intimately related to ferroptosis, and they can reinforce each other to improve the survival rate of tumor cells. In FLT-TKI-resistant AML cells with FLT3-ITD, FLT3-ITD-mediated aberrant activation of FLT3 induced autophagy and resistance to FLT-TKIs [270]. Additionally, aberrant AXL activation also increases autophagic flux, leading to resistance to the first- and third-generation EGFR-TKIs erlotinib and rociletinib [271]. Serine- and arginine-rich splicing factor 1 (SRSF1) is an autophagy suppressor, and its knockout may activate autophagy and inhibit the proliferation of gefitinib-resistant cancer cells [272]. HIF induces HSP90α expression, which binds the necrosome complex, induces autophagolysosomal degradation of MLKL, blocks necroptosis, and induces HCC resistance to sorafenib [273].

3.7 Abnormal Metabolism of Drug Delivery

TKIs are predominantly administered orally, and their absorption is inevitably influenced by first-pass metabolism, drug transporters, and intestinal enzymes [274]. For example, approximately 95% of imatinib binds to plasma proteins (primarily albumin) and α-1 acid glycoprotein (AGP, a hepatic acute phase protein), which reduces accessibility [275, 276]. Decreased or increased influx of drug transporter-mediated TKIs in cancer cell membranes has been associated with drug resistance [277]. Decreased expression and activity of solute carrier (SLC) transporter proteins like SLC22A1, SLC16A1, and octamer binding transcription factor 1 (OCT1) can reduce the intracellular concentration of TKIs [277–279]. The efflux of TKIs is mainly dependent on the ATP-binding cassette (ABC) family proteins, and overexpression of ABC transporters can impair TKI uptake, leading to drug resistance [280]. When TKI is administered concomitantly with hepatic enzyme inducers (e.g., rifampin, phenytoin sodium, and carbamazepine), the plasma concentration of TKIs decreases [281]. The efficacy of trastuzumab may be diminished by distribution disorders at the tissue and vessel levels, including alterations in vascular density and patency [90].

3.8 Other Mechanisms

Cell-cycle-related genes were altered, including increases in G1/S-specific cyclin D (CCND), G1/S-specific cyclin E1 (CCNE1), and cyclin-dependent kinases 4/6 (CDK4/6), as well as cyclin-dependent kinase inhibitor 2A (CDKN2A) E27fs, in 12% and 10% of NSCLC patients treated with first- and second-line osimertinib [58]. Aurora kinase A drives the evolution of resistance to third-generation EGFR inhibitors in lung cancer [282]. Amplification of FGF2/FGFR, RB1 mutations, and KIT mutations also creates resistance to third-generation EGFR-TKIs [70, 73, 74]. The overview of the mechanisms of acquired resistance to TKIs is summarized in Fig. 3.

4 Conclusion

Tumors are constantly changing during targeted therapy, which may further induce heterogeneity and affect therapeutic outcome. Therefore, it is essential to analyze the changes comprehensively in tumor cells and microenvironment to gain an in-depth understanding of the molecular mechanisms underlying the development of drug resistance. The future treatment will be genotype-dependent, with a strictly selected combination of drugs, ensuring precise targeting and robust killing of

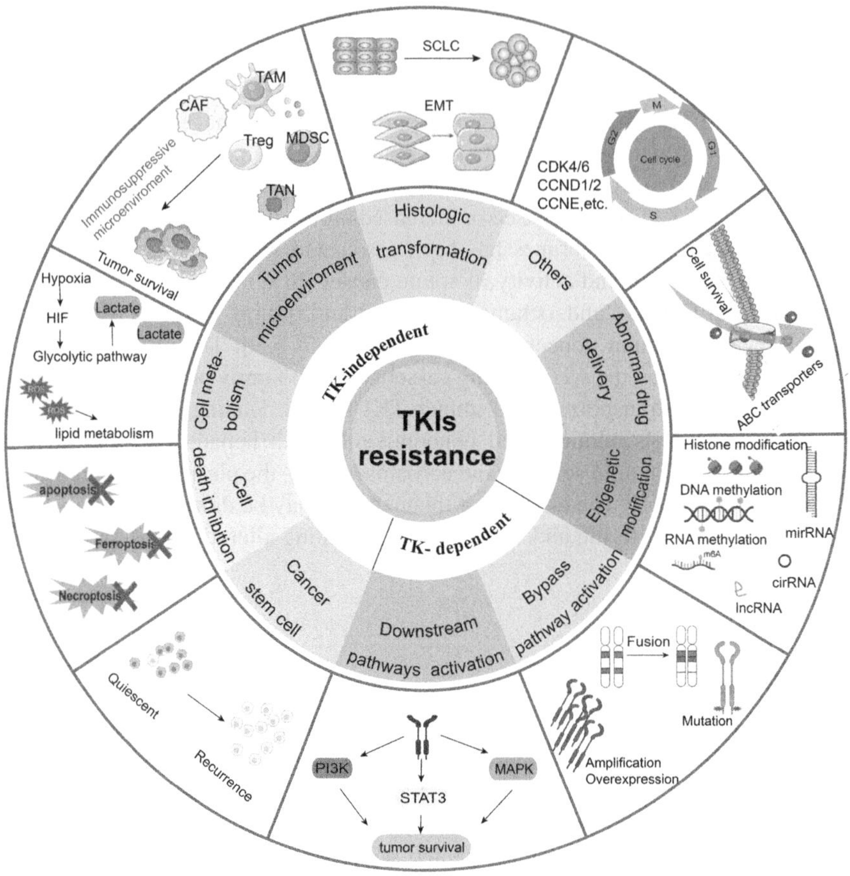

Fig. 3 Overview of the mechanisms of TKI resistance, dependent and independent of tyrosine kinase

tumor cells, enhancing the antitumor immune response, inhibiting angiogenesis, and blocking the interaction between tumor cells and other noncellular components simultaneously. Besides, understanding the molecular mechanisms of the disease and relevant pharmacokinetic information is pivotal for choosing the best therapeutic approach. And the dynamic changes in tumor growth during treatment should also be monitored for the development of effective, individualized treatment regimens. To appropriately combat TKI resistance, we still have a long way to go.

Acknowledgments The figures were created using Adobe Illustrator 2022 software.

Conflicts of Interest The authors have no conflicts of interest to declare.

References

1. Du Z, Lovly CM. Mechanisms of receptor tyrosine kinase activation in cancer. Mol Cancer. 2018;17(1):58.
2. Cohen P, Cross D, Jänne PA. Kinase drug discovery 20 years after imatinib: progress and future directions. Nat Rev Drug Discov. 2021;20(7):551–69.
3. Huang L, Jiang S, Shi Y. Tyrosine kinase inhibitors for solid tumors in the past 20 years (2001–2020). J Hematol Oncol. 2020;13(1):143.
4. Roskoski R. Properties of FDA-approved small molecule protein kinase inhibitors: a 2024 update. Pharmacol Res. 2024;200:107059.
5. Rotow J, Bivona TG. Understanding and targeting resistance mechanisms in NSCLC. Nat Rev Cancer. 2017;17(11):637–58.
6. Du X, Yang B, An Q, Assaraf YG, Cao X, Xia J. Acquired resistance to third-generation EGFR-TKIs and emerging next-generation EGFR inhibitors. Innovation (Camb). 2021;2(2):100103.
7. Ou X, Gao G, Habaz IA, Wang Y. Mechanisms of resistance to tyrosine kinase inhibitor-targeted therapy and overcoming strategies. MedComm. 2020;5(9):e694.
8. Cooper AJ, Sequist LV, Lin JJ. Third-generation EGFR and ALK inhibitors: mechanisms of resistance and management. Nat Rev Clin Oncol. 2022;19(8):499–514.
9. Dong RF, Zhu ML, Liu MM, Xu YT, Yuan LL, Bian J, et al. EGFR mutation mediates resistance to EGFR tyrosine kinase inhibitors in NSCLC: from molecular mechanisms to clinical research. Pharmacol Res. 2021;167:105583.
10. Cowell JK, Qin H, Hu T, Wu Q, Bhole A, Ren M. Mutation in the FGFR1 tyrosine kinase domain or inactivation of PTEN is associated with acquired resistance to FGFR inhibitors in FGFR1-driven leukemia/lymphomas. Int J Cancer. 2017;141(9):1822–9.
11. Goyal L, Saha SK, Liu LY, Siravegna G, Leshchiner I, Ahronian LG, et al. Polyclonal secondary FGFR2 mutations drive acquired resistance to FGFR inhibition in patients with FGFR2 fusion-positive Cholangiocarcinoma. Cancer Discov. 2017;7(3):252–63.
12. Yue S, Li Y, Chen X, Wang J, Li M, Chen Y, et al. FGFR-TKI resistance in cancer: current status and perspectives. J Hematol Oncol. 2021;14(1):23.
13. Hatlen MA, Schmidt-Kittler O, Sherwin CA, Rozsahegyi E, Rubin N, Sheets MP, et al. Acquired on-target clinical resistance validates FGFR4 as a driver of hepatocellular carcinoma. Cancer Discov. 2019;9(12):1686–95.
14. Gainor JF, Dardaei L, Yoda S, Friboulet L, Leshchiner I, Katayama R, et al. Molecular mechanisms of resistance to first- and second-generation ALK inhibitors in ALK-rearranged lung cancer. Cancer Discov. 2016;6(10):1118–33.
15. Yanagitani N, Uchibori K, Koike S, Tsukahara M, Kitazono S, Yoshizawa T, et al. Drug resistance mechanisms in Japanese anaplastic lymphoma kinase-positive non-small cell lung cancer and the clinical responses based on the resistant mechanisms. Cancer Sci. 2020;111(3):932–9.
16. Gainor JF, Tseng D, Yoda S, Dagogo-Jack I, Friboulet L, Lin JJ, et al. Patterns of metastatic spread and mechanisms of resistance to Crizotinib in ROS1-positive non-small-cell lung cancer. JCO Precis Oncologia. 2017;2017:1.
17. Lin JJ, Choudhury NJ, Yoda S, Zhu VW, Johnson TW, Sakhtemani R, et al. Spectrum of mechanisms of resistance to Crizotinib and Lorlatinib in ROS1 fusion-positive lung cancer. Clin Cancer Res. 2021;27(10):2899–909.
18. Román-Gil MS, Pozas J, Rosero-Rodríguez D, Chamorro-Pérez J, Ruiz-Granados Á, Caracuel IR, et al. Resistance to RET targeted therapy in thyroid cancer: molecular basis and overcoming strategies. Cancer Treat Rev. 2022;105:102372.
19. Dagogo-Jack I, Stevens SE, Lin JJ, Nagy R, Ferris L, Shaw AT, et al. Emergence of a RET V804M gatekeeper mutation during treatment with Vandetanib in RET-rearranged NSCLC. J Thorac Oncol. 2018;13(11):e226–e7.

20. Solomon BJ, Tan L, Lin JJ, Wong SQ, Hollizeck S, Ebata K, et al. RET solvent front mutations mediate acquired resistance to selective RET inhibition in RET-driven malignancies. J Thorac Oncol. 2020;15(4):541–9.
21. Lierman E, Smits S, Appleby N, Conneally E, Michaux L, Vandenberghe P. FIP1L1-PDGFRα p.T674I-D842L: a novel and Ponatinib resistant compound mutation in FIP1L1-PDGFRα positive leukemia. Hemasphere. 2019;3(2):e182.
22. Drilon A, Laetsch TW, Kummar S, DuBois SG, Lassen UN, Demetri GD, et al. Efficacy of Larotrectinib in TRK fusion-positive cancers in adults and children. N Engl J Med. 2018;378(8):731–9.
23. Cocco E, Scaltriti M, Drilon A. NTRK fusion-positive cancers and TRK inhibitor therapy. Nat Rev Clin Oncol. 2018;15(12):731–47.
24. Drilon A, Li G, Dogan S, Gounder M, Shen R, Arcila M, et al. What hides behind the MASC: clinical response and acquired resistance to entrectinib after ETV6-NTRK3 identification in a mammary analogue secretory carcinoma (MASC). Ann Oncol. 2016;27(5):920–6.
25. Woyach JA, Ruppert AS, Guinn D, Lehman A, Blachly JS, Lozanski A, et al. BTK(C481S)-mediated resistance to Ibrutinib in chronic lymphocytic leukemia. J Clin Oncol. 2017;35(13):1437–43.
26. Amarante-Mendes GP, Rana A, Datoguia TS, Hamerschlak N, Brumatti G. BCR-ABL1 tyrosine kinase complex signaling transduction: challenges to overcome resistance in chronic myeloid leukemia. Pharmaceutics. 2022;14(1)
27. McMahon CM, Ferng T, Canaani J, Wang ES, Morrissette JJD, Eastburn DJ, et al. Clonal selection with RAS pathway activation mediates secondary clinical resistance to selective FLT3 inhibition in acute myeloid leukemia. Cancer Discov. 2019;9(8):1050–63.
28. Tamborini E, Bonadiman L, Greco A, Albertini V, Negri T, Gronchi A, et al. A new mutation in the KIT ATP pocket causes acquired resistance to imatinib in a gastrointestinal stromal tumor patient. Gastroenterology. 2004;127(1):294–9.
29. Zhou Y, Xiang S, Yang F, Lu X. Targeting gatekeeper mutations for kinase drug discovery. J Med Chem. 2022;65(23):15540–58.
30. Sequist LV, Waltman BA, Dias-Santagata D, Digumarthy S, Turke AB, Fidias P, et al. Genotypic and histological evolution of lung cancers acquiring resistance to EGFR inhibitors. Sci Transl Med. 2011;3(75):75ra26.
31. Ricordel C, Friboulet L, Facchinetti F, Soria JC. Molecular mechanisms of acquired resistance to third-generation EGFR-TKIs in EGFR T790M-mutant lung cancer. Ann Oncol. 2018;29(suppl_1):i28–37.
32. Gazdar AF. Activating and resistance mutations of EGFR in non-small-cell lung cancer: role in clinical response to EGFR tyrosine kinase inhibitors. Oncogene. 2009;28(Suppl 1):S24–31.
33. Lynch TJ, Bell DW, Sordella R, Gurubhagavatula S, Okimoto RA, Brannigan BW, et al. Activating mutations in the epidermal growth factor receptor underlying responsiveness of non-small-cell lung cancer to gefitinib. N Engl J Med. 2004;350(21):2129–39.
34. Kuan FC, Kuo LT, Chen MC, Yang CT, Shi CS, Teng D, et al. Overall survival benefits of first-line EGFR tyrosine kinase inhibitors in EGFR-mutated non-small-cell lung cancers: a systematic review and meta-analysis. Br J Cancer. 2015;113(10):1519–28.
35. Vyse S, Huang PH. Targeting EGFR exon 20 insertion mutations in non-small cell lung cancer. Signal Transduct Target Ther. 2019;4:5.
36. Leonetti A, Sharma S, Minari R, Perego P, Giovannetti E, Tiseo M. Resistance mechanisms to osimertinib in EGFR-mutated non-small cell lung cancer. Br J Cancer. 2019;121(9):725–37.
37. Passaro A, Jänne PA, Mok T, Peters S. Overcoming therapy resistance in EGFR-mutant lung cancer. Nat Cancer. 2021;2(4):377–91.
38. Tripathi SK, Pandey K, Rengasamy KRR, Biswal BK. Recent updates on the resistance mechanisms to epidermal growth factor receptor tyrosine kinase inhibitors and resistance reversion strategies in lung cancer. Med Res Rev. 2020;40(6):2132–76.
39. Tan CS, Kumarakulasinghe NB, Huang YQ, Ang YLE, Choo JR, Goh BC, et al. Third generation EGFR TKIs: current data and future directions. Mol Cancer. 2018;17(1):29.

40. Lin JJ, Liu SV, McCoach CE, Zhu VW, Tan AC, Yoda S, et al. Mechanisms of resistance to selective RET tyrosine kinase inhibitors in RET fusion-positive non-small-cell lung cancer. Ann Oncol. 2020;31(12):1725–33.
41. Russo M, Misale S, Wei G, Siravegna G, Crisafulli G, Lazzari L, et al. Acquired resistance to the TRK inhibitor Entrectinib in colorectal cancer. Cancer Discov. 2016;6(1):36–44.
42. Jänne PA, Yang JC, Kim DW, Planchard D, Ohe Y, Ramalingam SS, et al. AZD9291 in EGFR inhibitor-resistant non-small-cell lung cancer. N Engl J Med. 2015;372(18):1689–99.
43. Bean J, Riely GJ, Balak M, Marks JL, Ladanyi M, Miller VA, et al. Acquired resistance to epidermal growth factor receptor kinase inhibitors associated with a novel T854A mutation in a patient with EGFR-mutant lung adenocarcinoma. Clin Cancer Res. 2008;14(22):7519–25.
44. Balak MN, Gong Y, Riely GJ, Somwar R, Li AR, Zakowski MF, et al. Novel D761Y and common secondary T790M mutations in epidermal growth factor receptor-mutant lung adenocarcinomas with acquired resistance to kinase inhibitors. Clin Cancer Res. 2006;12(21):6494–501.
45. Costa DB, Halmos B, Kumar A, Schumer ST, Huberman MS, Boggon TJ, et al. BIM mediates EGFR tyrosine kinase inhibitor-induced apoptosis in lung cancers with oncogenic EGFR mutations. PLoS Med. 2007;4(10):1669–79. discussion 80.
46. Fairclough SR, Kiedrowski LA, Lin JJ, Zelichov O, Tarcic G, Stinchcombe TE, et al. Identification of osimertinib-resistant EGFR L792 mutations by cfDNA sequencing: oncogenic activity assessment and prevalence in large cfDNA cohort. Exp Hematol Oncol. 2019;8:24.
47. Piazza RG, Magistroni V, Gasser M, Andreoni F, Galietta A, Scapozza L, et al. Evidence for D276G and L364I Bcr-Abl mutations in Ph+ leukaemic cells obtained from patients resistant to Imatinib. Leukemia. 2005;19(1):132–4.
48. Friboulet L, Li N, Katayama R, Lee CC, Gainor JF, Crystal AS, et al. The ALK inhibitor ceritinib overcomes crizotinib resistance in non-small cell lung cancer. Cancer Discov. 2014;4(6):662–73.
49. Choi YL, Soda M, Yamashita Y, Ueno T, Takashima J, Nakajima T, et al. EML4-ALK mutations in lung cancer that confer resistance to ALK inhibitors. N Engl J Med. 2010;363(18):1734–9.
50. Sasaki T, Okuda K, Zheng W, Butrynski J, Capelletti M, Wang L, et al. The neuroblastoma-associated F1174L ALK mutation causes resistance to an ALK kinase inhibitor in ALK-translocated cancers. Cancer Res. 2010;70(24):10038–43.
51. Katayama R, Shaw AT, Khan TM, Mino-Kenudson M, Solomon BJ, Halmos B, et al. Mechanisms of acquired crizotinib resistance in ALK-rearranged lung cancers. Sci Transl Med. 2012;4(120):120ra17.
52. Ménard L, Floc'h N, Martin MJ, Cross DAE. Reactivation of mutant-EGFR degradation through Clathrin inhibition overcomes resistance to EGFR tyrosine kinase inhibitors. Cancer Res. 2018;78(12):3267–79.
53. Pailler E, Oulhen M, Borget I, Remon J, Ross K, Auger N, et al. Circulating tumor cells with aberrant ALK copy number predict progression-free survival during Crizotinib treatment in ALK-rearranged non-small cell lung cancer patients. Cancer Res. 2017;77(9):2222–30.
54. Poudel G, Tolland MG, Hughes TP, Pagani IS. Mechanisms of resistance and implications for treatment strategies in chronic myeloid Leukaemia. Cancers (Basel). 2022;14(14)
55. Chmielecki J, Mok T, Wu YL, Han JY, Ahn MJ, Ramalingam SS, et al. Analysis of acquired resistance mechanisms to osimertinib in patients with EGFR-mutated advanced non-small cell lung cancer from the AURA3 trial. Nat Commun. 2023;14(1):1071.
56. Zhu VW, Klempner SJ, Ou SI. Receptor tyrosine kinase fusions as an actionable resistance mechanism to EGFR TKIs in EGFR-mutant non-small-cell lung cancer. Trends Cancer. 2019;5(11):677–92.
57. Piotrowska Z, Isozaki H, Lennerz JK, Gainor JF, Lennes IT, Zhu VW, et al. Landscape of acquired resistance to Osimertinib in EGFR-mutant NSCLC and clinical validation of combined EGFR and RET inhibition with Osimertinib and BLU-667 for acquired RET fusion. Cancer Discov. 2018;8(12):1529–39.

58. Papadimitrakopoulou VA, Wu YL, Han JY, Ahn MJ, Ramalingam SS, John T, et al. LBA51 – analysis of resistance mechanisms to osimertinib in patients with EGFR T790M advanced NSCLC from the AURA3 study. Ann Oncol. 2018;29:viii741.
59. Lin CC, Shih JY, Yu CJ, Ho CC, Liao WY, Lee JH, et al. Outcomes in patients with non-small-cell lung cancer and acquired Thr790Met mutation treated with osimertinib: a genomic study. Lancet Respir Med. 2018;6(2):107–16.
60. Schrock AB, Zhu VW, Hsieh WS, Madison R, Creelan B, Silberberg J, et al. Receptor tyrosine kinase fusions and BRAF kinase fusions are rare but actionable resistance mechanisms to EGFR tyrosine kinase inhibitors. J Thorac Oncol. 2018;13(9):1312–23.
61. Zeng L, Yang N, Zhang Y. GOPC-ROS1 rearrangement as an acquired resistance mechanism to Osimertinib and responding to Crizotinib combined treatments in lung adenocarcinoma. J Thorac Oncol. 2018;13(7):e114–e6.
62. Ou SI, Horn L, Cruz M, Vafai D, Lovly CM, Spradlin A, et al. Emergence of FGFR3-TACC3 fusions as a potential by-pass resistance mechanism to EGFR tyrosine kinase inhibitors in EGFR mutated NSCLC patients. Lung Cancer. 2017;111:61–4.
63. Westover D, Zugazagoitia J, Cho BC, Lovly CM, Paz-Ares L. Mechanisms of acquired resistance to first- and second-generation EGFR tyrosine kinase inhibitors. Ann Oncol. 2018;29(suppl_1):i10–i9.
64. Suzawa K, Offin M, Lu D, Kurzatkowski C, Vojnic M, Smith RS, et al. Activation of KRAS mediates resistance to targeted therapy in MET exon 14-mutant non-small cell lung cancer. Clin Cancer Res. 2019;25(4):1248–60.
65. Bahcall M, Awad MM, Sholl LM, Wilson FH, Xu M, Wang S, et al. Amplification of wild-type KRAS imparts resistance to Crizotinib in MET exon 14 mutant non-small cell lung cancer. Clin Cancer Res. 2018;24(23):5963–76.
66. Hsu CC, Liao BC, Liao WY, Markovets A, Stetson D, Thress K, et al. Exon 16-skipping HER2 as a novel mechanism of Osimertinib resistance in EGFR L858R/T790M-positive non-small cell lung cancer. J Thorac Oncol. 2020;15(1):50–61.
67. Romaniello D, Marrocco I, Belugali Nataraj N, Ferrer I, Drago-Garcia D, Vaknin I, et al. Targeting HER3, a catalytically defective receptor tyrosine kinase, prevents resistance of lung cancer to a third-generation EGFR kinase inhibitor. Cancers (Basel). 2020;12(9)
68. Zalaquett Z, Catherine Rita Hachem M, Kassis Y, Hachem S, Eid R, Raphael Kourie H, et al. Acquired resistance mechanisms to osimertinib: the constant battle. Cancer Treat Rev. 2023;116:102557.
69. Ho CC, Liao WY, Lin CA, Shih JY, Yu CJ, Yang JC. Acquired BRAF V600E mutation as resistant mechanism after treatment with Osimertinib. J Thorac Oncol. 2017;12(3):567–72.
70. Leonetti A, Facchinetti F, Rossi G, Minari R, Conti A, Friboulet L, et al. BRAF in non-small cell lung cancer (NSCLC): pickaxing another brick in the wall. Cancer Treat Rev. 2018;66:82–94.
71. Ham JS, Kim S, Kim HK, Byeon S, Sun JM, Lee SH, et al. Two cases of small cell lung cancer transformation from EGFR mutant adenocarcinoma during AZD9291 treatment. J Thorac Oncol. 2016;11(1):e1–4.
72. Lu Y, Liu Y, Oeck S, Zhang GJ, Schramm A, Glazer PM. Hypoxia induces resistance to EGFR inhibitors in lung cancer cells via upregulation of FGFR1 and the MAPK pathway. Cancer Res. 2020;80(21):4655–67.
73. Nagano T, Tachihara M, Nishimura Y. Mechanism of resistance to epidermal growth factor receptor-tyrosine kinase inhibitors and a potential treatment strategy. Cells. 2018;7(11)
74. Nilsson MB, Sun H, Robichaux J, Pfeifer M, McDermott U, Travers J, et al. A YAP/FOXM1 axis mediates EMT-associated EGFR inhibitor resistance and increased expression of spindle assembly checkpoint components. Sci Transl Med. 2020;12(559)
75. Roskoski R Jr. ROS1 protein-tyrosine kinase inhibitors in the treatment of ROS1 fusion protein-driven non-small cell lung cancers. Pharmacol Res. 2017;121:202–12.

76. Wang YT, Pan SH, Tsai CF, Kuo TC, Hsu YL, Yen HY, et al. Phosphoproteomics reveals HMGA1, a CK2 substrate, as a drug-resistant target in non-small cell lung cancer. Sci Rep. 2017;7:44021.
77. Park HR, Kim TM, Lee Y, Kim S, Park S, Ju YS, et al. Acquired resistance to third-generation EGFR tyrosine kinase inhibitors in patients with De novo EGFR(T790M)-mutant NSCLC. J Thorac Oncol. 2021;16(11):1859–71.
78. Lin L, Sabnis AJ, Chan E, Olivas V, Cade L, Pazarentzos E, et al. The hippo effector YAP promotes resistance to RAF- and MEK-targeted cancer therapies. Nat Genet. 2015;47(3):250–6.
79. Dagogo-Jack I, Yoda S, Lennerz JK, Langenbucher A, Lin JJ, Rooney MM, et al. MET alterations are a recurring and actionable resistance mechanism in ALK-positive lung cancer. Clin Cancer Res. 2020;26(11):2535–45.
80. Lovly CM, McDonald NT, Chen H, Ortiz-Cuaran S, Heukamp LC, Yan Y, et al. Rationale for co-targeting IGF-1R and ALK in ALK fusion-positive lung cancer. Nat Med. 2014;20(9):1027–34.
81. Lin YT, Chiang CL, Hung JY, Lee MH, Su WC, Wu SY, et al. Resistance profiles of anaplastic lymphoma kinase tyrosine kinase inhibitors in advanced non-small-cell lung cancer: a multicenter study using targeted next-generation sequencing. Eur J Cancer. 2021;156:1–11.
82. Yun MR, Lim SM, Kim SK, Choi HM, Pyo KH, Kim SK, et al. Enhancer remodeling and MicroRNA alterations are associated with acquired resistance to ALK inhibitors. Cancer Res. 2018;78(12):3350–62.
83. Trigg RM, Lee LC, Prokoph N, Jahangiri L, Reynolds CP, Amos Burke GA, et al. The targetable kinase PIM1 drives ALK inhibitor resistance in high-risk neuroblastoma independent of MYCN status. Nat Commun. 2019;10(1):5428.
84. Kang J, Deng QM, Peng KC, Li P, Zhu BT, Wang P, et al. Clinicopathological features and resistance mechanisms in HIP1-ALK-rearranged lung cancer: a multicenter study. Genes Chromosomes Cancer. 2022;61(4):177–86.
85. Karaca Atabay E, Mecca C, Wang Q, Ambrogio C, Mota I, Prokoph N, et al. Tyrosine phosphatases regulate resistance to ALK inhibitors in ALK+ anaplastic large cell lymphoma. Blood. 2022;139(5):717–31.
86. Yoshida R, Sasaki T, Minami Y, Hibino Y, Okumura S, Sado M, et al. Activation of Src signaling mediates acquired resistance to ALK inhibition in lung cancer. Int J Oncol. 2017;51(5):1533–40.
87. Lee HJ, Seo AN, Kim EJ, Jang MH, Kim YJ, Kim JH, et al. Prognostic and predictive values of EGFR overexpression and EGFR copy number alteration in HER2-positive breast cancer. Br J Cancer. 2015;112(1):103–11.
88. Gianni L, Pienkowski T, Im YH, Tseng LM, Liu MC, Lluch A, et al. 5-year analysis of neoadjuvant pertuzumab and trastuzumab in patients with locally advanced, inflammatory, or early-stage HER2-positive breast cancer (NeoSphere): a multicentre, open-label, phase 2 randomised trial. Lancet Oncol. 2016;17(6):791–800.
89. Adam-Artigues A, Arenas EJ, Martínez-Sabadell A, Brasó-Maristany F, Cervera R, Tormo E, et al. Targeting HER2-AXL heterodimerization to overcome resistance to HER2 blockade in breast cancer. Sci Adv. 2022;8(20):eabk2746.
90. Hori A, Shimoda M, Naoi Y, Kagara N, Tanei T, Miyake T, et al. Vasculogenic mimicry is associated with trastuzumab resistance of HER2-positive breast cancer. Breast Cancer Res. 2019;21(1):88.
91. Ding J, Yao Y, Huang G, Wang X, Yi J, Zhang N, et al. Targeting the EphB4 receptor tyrosine kinase sensitizes HER2-positive breast cancer cells to Lapatinib. Cancer Lett. 2020;475:53–64.
92. Chen Z, Liu Z, Zhang M, Huang W, Li Z, Wang S, et al. EPHA2 blockade reverses acquired resistance to afatinib induced by EPHA2-mediated MAPK pathway activation in gastric cancer cells and avatar mice. Int J Cancer. 2019;145(9):2440–9.

93. Fernández-Nogueira P, Mancino M, Fuster G, López-Plana A, Jauregui P, Almendro V, et al. Tumor-associated fibroblasts promote HER2-targeted therapy resistance through FGFR2 activation. Clin Cancer Res. 2020;26(6):1432–48.
94. Ebbesen SH, Scaltriti M, Bialucha CU, Morse N, Kastenhuber ER, Wen HY, et al. Pten loss promotes MAPK pathway dependency in HER2/neu breast carcinomas. Proc Natl Acad Sci USA. 2016;113(11):3030–5.
95. Belli V, Matrone N, Napolitano S, Migliardi G, Cottino F, Bertotti A, et al. Combined blockade of MEK and PI3KCA as an effective antitumor strategy in HER2 gene amplified human colorectal cancer models. J Exp Clin Cancer Res. 2019;38(1):236.
96. Goel S, Wang Q, Watt AC, Tolaney SM, Dillon DA, Li W, et al. Overcoming therapeutic resistance in HER2-positive breast cancers with CDK4/6 inhibitors. Cancer Cell. 2016;29(3):255–69.
97. DiPeri TP, Evans KW, Raso MG, Zhao M, Rizvi YQ, Zheng X, et al. Adavosertib enhances antitumor activity of Trastuzumab Deruxtecan in HER2-expressing cancers. Clin Cancer Res. 2023;29(21):4385–98.
98. Wang L, Šuštić T, Leite de Oliveira R, Lieftink C, Halonen P, van de Ven M, et al. A functional genetic screen identifies the phosphoinositide 3-kinase pathway as a determinant of resistance to fibroblast growth factor receptor inhibitors in FGFR mutant urothelial cell carcinoma. Eur Urol. 2017;71(6):858–62.
99. Liu H, Ai J, Shen A, Chen Y, Wang X, Peng X, et al. C-Myc alteration determines the therapeutic response to FGFR inhibitors. Clin Cancer Res. 2017;23(4):974–84.
100. Mahe M, Dufour F, Neyret-Kahn H, Moreno-Vega A, Beraud C, Shi M, et al. An FGFR3/MYC positive feedback loop provides new opportunities for targeted therapies in bladder cancers. EMBO Mol Med. 2018;10(4)
101. Subbiah V, Yang D, Velcheti V, Drilon A, Meric-Bernstam F. State-of-the-art strategies for targeting RET-dependent cancers. J Clin Oncol. 2020;38(11):1209–21.
102. Nelson-Taylor SK, Le AT, Yoo M, Schubert L, Mishall KM, Doak A, et al. Resistance to RET-inhibition in RET-rearranged NSCLC is mediated by reactivation of RAS/MAPK signaling. Mol Cancer Ther. 2017;16(8):1623–33.
103. Zeng C, Zhu L, Jia X, Pang Y, Li Z, Lu X, et al. Spectrum of activity of dasatinib against mutant KIT kinases associated with drug-sensitive and drug-resistant gastrointestinal stromal tumors. Gastric Cancer. 2020;23(5):837–47.
104. Dumas PY, Naudin C, Martin-Lannerée S, Izac B, Casetti L, Mansier O, et al. Hematopoietic niche drives FLT3-ITD acute myeloid leukemia resistance to quizartinib via STAT5-and hypoxia-dependent upregulation of AXL. Haematologica. 2019;104(10):2017–27.
105. Serrano C, Wang Y, Mariño-Enríquez A, Lee JC, Ravegnini G, Morgan JA, et al. KRAS and KIT gatekeeper mutations confer polyclonal primary Imatinib resistance in GI stromal tumors: relevance of concomitant phosphatidylinositol 3-kinase/AKT dysregulation. J Clin Oncol. 2015;33(22):e93–6.
106. Subbiah V, Velcheti V, Tuch BB, Ebata K, Busaidy NL, Cabanillas ME, et al. Selective RET kinase inhibition for patients with RET-altered cancers. Ann Oncol. 2018;29(8):1869–76.
107. Bavaro L, Martelli M, Cavo M, Soverini S. Mechanisms of disease progression and resistance to tyrosine kinase inhibitor therapy in chronic myeloid leukemia: an update. Int J Mol Sci. 2019;20(24)
108. Rummelt C, Gorantla SP, Meggendorfer M, Charlet A, Endres C, Döhner K, et al. Activating JAK-mutations confer resistance to FLT3 kinase inhibitors in FLT3-ITD positive AML in vitro and in vivo. Leukemia. 2021;35(7):2017–29.
109. Minciacchi VR, Kumar R, Krause DS. Chronic myeloid leukemia: a model disease of the past, present and future. Cells. 2021;10(1)
110. Cascone T, Xu L, Lin HY, Liu W, Tran HT, Liu Y, et al. The HGF/c-MET pathway is a driver and biomarker of VEGFR-inhibitor resistance and vascular remodeling in non-small cell lung cancer. Clin Cancer Res. 2017;23(18):5489–501.

111. Pinato DJ, Brown MW, Trousil S, Aboagye EO, Beaumont J, Zhang H, et al. Integrated analysis of multiple receptor tyrosine kinases identifies Axl as a therapeutic target and mediator of resistance to sorafenib in hepatocellular carcinoma. Br J Cancer. 2019;120(5):512–21.
112. Sanchez-Martin A, Sanchon-Sanchez P, Romero MR, Marin JJG, Briz O. Impact of tumor suppressor genes inactivation on the multidrug resistance phenotype of hepatocellular carcinoma cells. Biomed Pharmacother. 2023;165:115209.
113. Leung HW, Leung CON, Lau EY, Chung KPS, Mok EH, Lei MML, et al. EPHB2 activates β-catenin to enhance cancer stem cell properties and drive Sorafenib resistance in hepatocellular carcinoma. Cancer Res. 2021;81(12):3229–40.
114. Ruan H, Li S, Bao L, Zhang X. Enhanced YB1/EphA2 axis signaling promotes acquired resistance to sunitinib and metastatic potential in renal cell carcinoma. Oncogene. 2020;39(38):6113–28.
115. Cho J, Kim SY, Kim YJ, Sim MH, Kim ST, Kim NKD, et al. Emergence of CTNNB1 mutation at acquired resistance to KIT inhibitor in metastatic melanoma. Clin Transl Oncol. 2017;19(10):1247–52.
116. Kim S, Kim TM, Kim DW, Kim S, Kim M, Ahn YO, et al. Acquired resistance of MET-amplified non-small cell lung cancer cells to the MET inhibitor Capmatinib. Cancer Res Treat. 2019;51(3):951–62.
117. Sato H, Schoenfeld AJ, Siau E, Lu YC, Tai H, Suzawa K, et al. MAPK pathway alterations correlate with poor survival and drive resistance to therapy in patients with lung cancers driven by ROS1 fusions. Clin Cancer Res. 2020;26(12):2932–45.
118. Wang Q, Yang S, Wang K, Sun SY. MET inhibitors for targeted therapy of EGFR TKI-resistant lung cancer. J Hematol Oncol. 2019;12(1):63.
119. Ren S, Wang J, Ying J, Mitsudomi T, Lee DH, Wang Z, et al. Consensus for HER2 alterations testing in non-small-cell lung cancer. ESMO Open. 2022;7(1):100395.
120. Oxnard GR, Hu Y, Mileham KF, Husain H, Costa DB, Tracy P, et al. Assessment of resistance mechanisms and clinical implications in patients with EGFR T790M-positive lung cancer and acquired resistance to Osimertinib. JAMA Oncol. 2018;4(11):1527–34.
121. Lebedev T, Vagapova E, Spirin P, Rubtsov P, Astashkova O, Mikheeva A, et al. Growth factor signaling predicts therapy resistance mechanisms and defines neuroblastoma subtypes. Oncogene. 2021;40(44):6258–72.
122. Braun TP, Eide CA, Druker BJ. Response and resistance to BCR-ABL1-targeted therapies. Cancer Cell. 2020;37(4):530–42.
123. Chaib I, Karachaliou N, Pilotto S, Codony Servat J, Cai X, Li X, et al. Co-activation of STAT3 and YES-associated protein 1 (YAP1) pathway in EGFR-mutant NSCLC. J Natl Cancer Inst. 2017;109(9)
124. Wu SG, Shih JY. Management of acquired resistance to EGFR TKI-targeted therapy in advanced non-small cell lung cancer. Mol Cancer. 2018;17(1):38.
125. Manabe T, Yasuda H, Terai H, Kagiwada H, Hamamoto J, Ebisudani T, et al. IGF2 autocrine-mediated IGF1R activation is a clinically relevant mechanism of Osimertinib resistance in lung cancer. Mol Cancer Res. 2020;18(4):549–59.
126. Stern YE, Al-Ghabkari A, Monast A, Fiset B, Aboualizadeh F, Yao Z, et al. Met-HER3 crosstalk supports proliferation via MPZL3 in MET-amplified cancer cells. Cell Mol Life Sci. 2022;79(3):178.
127. Gandullo-Sánchez L, Ocaña A, Pandiella A. HER3 in cancer: from the bench to the bedside. J Exp Clin Cancer Res. 2022;41(1):310.
128. Okamoto K, Ando T, Izumi H, Kobayashi SS, Shintani T, Gutkind JS, et al. AXL activates YAP through the EGFR-LATS1/2 axis and confers resistance to EGFR-targeted drugs in head and neck squamous cell carcinoma. Oncogene. 2023;42(39):2869–77.
129. Nastały P, Stoupiec S, Popęda M, Smentoch J, Schlomm T, Morrissey C, et al. EGFR as a stable marker of prostate cancer dissemination to bones. Br J Cancer. 2020;123(12):1767–74.
130. Huang L, Fu L. Mechanisms of resistance to EGFR tyrosine kinase inhibitors. Acta Pharm Sin B. 2015;5(5):390–401.

131. Suzawa K, Offin M, Schoenfeld AJ, Plodkowski AJ, Odintsov I, Lu D, et al. Acquired MET exon 14 alteration drives secondary resistance to epidermal growth factor receptor tyrosine kinase inhibitor in EGFR-mutated lung cancer. JCO precis. Oncologia. 2019:3.
132. Rosen EY, Johnson ML, Clifford SE, Somwar R, Kherani JF, Son J, et al. Overcoming MET-dependent resistance to selective RET inhibition in patients with RET fusion-positive lung cancer by combining Selpercatinib with Crizotinib. Clin Cancer Res. 2021;27(1):34–42.
133. Kogita A, Togashi Y, Hayashi H, Banno E, Terashima M, De Velasco MA, et al. Activated MET acts as a salvage signal after treatment with alectinib, a selective ALK inhibitor, in ALK-positive non-small cell lung cancer. Int J Oncol. 2015;46(3):1025–30.
134. Suzuki S, Yonesaka K, Teramura T, Takehara T, Kato R, Sakai H, et al. KRAS inhibitor resistance in MET-amplified KRAS (G12C) non-small cell lung cancer induced by RAS- and non-RAS-mediated cell signaling mechanisms. Clin Cancer Res. 2021;27(20):5697–707.
135. Lenz G, Hamilton A, Geng S, Hong T, Kalkum M, Momand J, et al. T-Darpp activates IGF-1R signaling to regulate glucose metabolism in Trastuzumab-resistant breast cancer cells. Clin Cancer Res. 2018;24(5):1216–26.
136. Guerard M, Robin T, Perron P, Hatat AS, David-Boudet L, Vanwonterghem L, et al. Nuclear translocation of IGF1R by intracellular amphiregulin contributes to the resistance of lung tumour cells to EGFR-TKI. Cancer Lett. 2018;420:146–55.
137. Taniguchi H, Yamada T, Wang R, Tanimura K, Adachi Y, Nishiyama A, et al. AXL confers intrinsic resistance to osimertinib and advances the emergence of tolerant cells. Nat Commun. 2019;10(1):259.
138. Kim D, Bach DH, Fan YH, Luu TT, Hong JY, Park HJ, et al. AXL degradation in combination with EGFR-TKI can delay and overcome acquired resistance in human non-small cell lung cancer cells. Cell Death Dis. 2019;10(5):361.
139. Safaric Tepes P, Pal D, Lindsted T, Ibarra I, Lujambio A, Jimenez Sabinina V, et al. An epigenetic switch regulates the ontogeny of AXL-positive/EGFR-TKi-resistant cells by modulating miR-335 expression. elife. 2021;10:10.
140. Antony J, Huang RY. AXL-driven EMT state as a targetable conduit in cancer. Cancer Res. 2017;77(14):3725–32.
141. Lin JJ, Ritterhouse LL, Ali SM, Bailey M, Schrock AB, Gainor JF, et al. ROS1 fusions rarely overlap with other oncogenic drivers in non-small cell lung cancer. J Thorac Oncol. 2017;12(5):872–7.
142. Qin Z, Sun H, Yue M, Pan X, Chen L, Feng X, et al. Phase separation of EML4-ALK in firing downstream signaling and promoting lung tumorigenesis. Cell Discov. 2021;7(1):33.
143. Singal G, Miller PG, Agarwala V, Li G, Kaushik G, Backenroth D, et al. Association of patient characteristics and tumor genomics with clinical outcomes among patients with non-small cell lung cancer using a clinicogenomic database. JAMA. 2019;321(14):1391–9.
144. Neel DS, Allegakoen DV, Olivas V, Mayekar MK, Hemmati G, Chatterjee N, et al. Differential subcellular localization regulates oncogenic signaling by ROS1 kinase fusion proteins. Cancer Res. 2019;79(3):546–56.
145. Liu Y, Liu T, Li N, Wang T, Pu Y, Lin R. Identification of a novel WNK1-ROS1 fusion in a lung adenocarcinoma sensitive to crizotinib. Lung Cancer. 2019;129:92–4.
146. Facchinetti F, Loriot Y, Kuo MS, Mahjoubi L, Lacroix L, Planchard D, et al. Crizotinib-resistant ROS1 mutations reveal a predictive kinase inhibitor sensitivity model for ROS1- and ALK-rearranged lung cancers. Clin Cancer Res. 2016;22(24):5983–91.
147. Li F, Huynh H, Li X, Ruddy DA, Wang Y, Ong R, et al. FGFR-mediated reactivation of MAPK signaling attenuates antitumor effects of Imatinib in gastrointestinal stromal tumors. Cancer Discov. 2015;5(4):438–51.
148. Desai A, Adjei AA. FGFR signaling as a target for lung cancer therapy. J Thorac Oncol. 2016;11(1):9–20.
149. Klug LR, Kent JD, Heinrich MC. Structural and clinical consequences of activation loop mutations in class III receptor tyrosine kinases. Pharmacol Ther. 2018;191:123–34.

150. Hemming ML, Heinrich MC, Bauer S, George S. Translational insights into gastrointestinal stromal tumor and current clinical advances. Ann Oncol. 2018;29(10):2037–45.
151. Bosbach B, Rossi F, Yozgat Y, Loo J, Zhang JQ, Berrozpe G, et al. Direct engagement of the PI3K pathway by mutant KIT dominates oncogenic signaling in gastrointestinal stromal tumor. Proc Natl Acad Sci USA. 2017;114(40):E8448–e57.
152. Ichihara E, Westover D, Meador CB, Yan Y, Bauer JA, Lu P, et al. SFK/FAK signaling attenuates Osimertinib efficacy in both drug-sensitive and drug-resistant models of EGFR-mutant lung cancer. Cancer Res. 2017;77(11):2990–3000.
153. Karachaliou N, Chaib I, Cardona AF, Berenguer J, Bracht JWP, Yang J, et al. Common co-activation of AXL and CDCP1 in EGFR-mutation-positive non-small cell lung cancer associated with poor prognosis. EBioMedicine. 2018;29:112–27.
154. Saraon P, Pathmanathan S, Snider J, Lyakisheva A, Wong V, Stagljar I. Receptor tyrosine kinases and cancer: oncogenic mechanisms and therapeutic approaches. Oncogene. 2021;40(24):4079–93.
155. Pailler E, Faugeroux V, Oulhen M, Mezquita L, Laporte M, Honoré A, et al. Acquired resistance mutations to ALK inhibitors identified by single circulating tumor cell sequencing in ALK-rearranged non-small-cell lung cancer. Clin Cancer Res. 2019;25(22):6671–82.
156. de Bruin EC, Cowell C, Warne PH, Jiang M, Saunders RE, Melnick MA, et al. Reduced NF1 expression confers resistance to EGFR inhibition in lung cancer. Cancer Discov. 2014;4(5):606–19.
157. Sun Y, Meyers BA, Czako B, Leonard P, Mseeh F, Harris AL, et al. Allosteric SHP2 inhibitor, IACS-13909, overcomes EGFR-dependent and EGFR-independent resistance mechanisms toward Osimertinib. Cancer Res. 2020;80(21):4840–53.
158. Jamme P, Fernandes M, Copin MC, Descarpentries C, Escande F, Morabito A, et al. Alterations in the PI3K pathway drive resistance to MET inhibitors in NSCLC harboring MET exon 14 skipping mutations. J Thorac Oncol. 2020;15(5):741–51.
159. To KKW, Wu WKK, Loong HHF. PPARgamma agonists sensitize PTEN-deficient resistant lung cancer cells to EGFR tyrosine kinase inhibitors by inducing autophagy. Eur J Pharmacol. 2018;823:19–26.
160. Wang Y, Zhou Q, Liu C, Zhang R, Xing B, Du J, et al. Targeting IL-6/STAT3 signaling abrogates EGFR-TKI resistance through inhibiting Beclin-1 dependent autophagy in HNSCC. Cancer Lett. 2024;586:216612.
161. Lin C, Ren Z, Yang X, Yang R, Chen Y, Liu Z, et al. Nerve growth factor (NGF)-TrkA axis in head and neck squamous cell carcinoma triggers EMT and confers resistance to the EGFR inhibitor erlotinib. Cancer Lett. 2020;472:81–96.
162. Shin JE, Kim SH, Kong M, Kim HR, Yoon S, Kee KM, et al. Targeting FLT3-TAZ signaling to suppress drug resistance in blast phase chronic myeloid leukemia. Mol Cancer. 2023;22(1):177.
163. Lee PC, Fang YF, Yamaguchi H, Wang WJ, Chen TC, Hong X, et al. Targeting PKCδ as a therapeutic strategy against heterogeneous mechanisms of EGFR inhibitor resistance in EGFR-mutant lung cancer. Cancer Cell. 2018;34(6):954–69.e4.
164. Dai C, Liu B, Li S, Hong Y, Si J, Xiong Y, et al. Construction of a circRNA-miRNA-mRNA regulated pathway involved in EGFR-TKI lung adenocarcinoma resistance. Technol Cancer Res Treat. 2021;20:15330338211056809.
165. Tulchinsky E, Demidov O, Kriajevska M, Barlev NA, Imyanitov E. EMT: a mechanism for escape from EGFR-targeted therapy in lung cancer. Biochim Biophys Acta Rev Cancer. 2019;1871(1):29–39.
166. Huang J, Pan B, Xia G, Zhu J, Li C, Feng J. LncRNA SNHG15 regulates EGFR-TKI acquired resistance in lung adenocarcinoma through sponging miR-451 to upregulate MDR-1. Cell Death Dis. 2020;11(7):525.
167. Zhang WC, Skiados N, Aftab F, Moreno C, Silva L, Corbilla PJA, et al. MicroRNA-21 guide and passenger strand regulation of adenylosuccinate lyase-mediated purine metabo-

lism promotes transition to an EGFR-TKI-tolerant persister state. Cancer Gene Ther. 2022;29(12):1878–94.
168. Joo LJS, Weiss J, Gill AJ, Clifton-Bligh R, Brahmbhatt H, MacDiarmid JA, et al. RET kinase-regulated MicroRNA-153-3p improves therapeutic efficacy in medullary thyroid carcinoma. Thyroid. 2019;29(6):830–44.
169. Hirao A, Sato Y, Tanaka H, Nishida K, Tomonari T, Hirata M, et al. MiR-125b-5p is involved in Sorafenib resistance through Ataxin-1-mediated epithelial-mesenchymal transition in hepatocellular carcinoma. Cancers (Basel). 2021;13(19)
170. Wu H, Wang T, Liu Y, Li X, Xu S, Wu C, et al. Mitophagy promotes sorafenib resistance through hypoxia-inducible ATAD3A dependent Axis. J Exp Clin Cancer Res. 2020;39(1):274.
171. Zhao W, Ma B, Tian Z, Han H, Tang J, Dong B, et al. Inhibiting CBX4 efficiently protects hepatocellular carcinoma cells against sorafenib resistance. Br J Cancer. 2021;124(7):1237–48.
172. Fernández-Tussy P, Rodríguez-Agudo R, Fernández-Ramos D, Barbier-Torres L, Zubiete-Franco I, Davalillo SL, et al. Anti-miR-518d-5p overcomes liver tumor cell death resistance through mitochondrial activity. Cell Death Dis. 2021;12(6):555.
173. Pollutri D, Patrizi C, Marinelli S, Giovannini C, Trombetta E, Giannone FA, et al. The epigenetically regulated miR-494 associates with stem-cell phenotype and induces sorafenib resistance in hepatocellular carcinoma. Cell Death Dis. 2018;9(1):4.
174. Zhang Z, Tan X, Luo J, Yao H, Si Z, Tong JS. The miR-30a-5p/CLCF1 axis regulates sorafenib resistance and aerobic glycolysis in hepatocellular carcinoma. Cell Death Dis. 2020;11(10):902.
175. Huynh TK, Huang CH, Chen JY, Yao JH, Yang YS, Wei YL, et al. miR-221 confers lapatinib resistance by negatively regulating p27(kip1) in HER2-positive breast cancer. Cancer Sci. 2021;112(10):4234–45.
176. Liu BJ, Li FF, Xie YX, Fan CY, Liu WJ, Qiu JG, et al. miR-196a upregulation contributes to Gefitinib resistance through inhibiting GLTP expression. Int J Mol Sci. 2022;23(3)
177. Huang X, Chen Z, Ni F, Ye X, Qian W. Shikonin overcomes drug resistance and induces necroptosis by regulating the miR-92a-1-5p/MLKL axis in chronic myeloid leukemia. Aging (Albany NY). 2020;12(17):17662–80.
178. Reustle A, Fisel P, Renner O, Büttner F, Winter S, Rausch S, et al. Characterization of the breast cancer resistance protein (BCRP/ABCG2) in clear cell renal cell carcinoma. Int J Cancer. 2018;143(12):3181–93.
179. Zhai W, Li S, Zhang J, Chen Y, Ma J, Kong W, et al. Sunitinib-suppressed miR-452-5p facilitates renal cancer cell invasion and metastasis through modulating SMAD4/SMAD7 signals. Mol Cancer. 2018;17(1):157.
180. Wang D, Wang X, Si M, Yang J, Sun S, Wu H, et al. Exosome-encapsulated miRNAs contribute to CXCL12/CXCR4-induced liver metastasis of colorectal cancer by enhancing M2 polarization of macrophages. Cancer Lett. 2020;474:36–52.
181. Peng PH, Chieh-Yu Lai J, Hsu KW, Wu KJ. Hypoxia-induced lncRNA RP11-390F4.3 promotes epithelial-mesenchymal transition (EMT) and metastasis through upregulating EMT regulators. Cancer Lett. 2020;483:35–45.
182. He M, Yang H, Shi H, Hu Y, Chang C, Liu S, et al. Sunitinib increases the cancer stem cells and vasculogenic mimicry formation via modulating the lncRNA-ECVSR/ERβ/Hif2-α signaling. Cancer Lett. 2022;524:15–28.
183. Chen Z, Chen Q, Cheng Z, Gu J, Feng W, Lei T, et al. Long non-coding RNA CASC9 promotes gefitinib resistance in NSCLC by epigenetic repression of DUSP1. Cell Death Dis. 2020;11(10):858.
184. Wang S, Liu C, Lei Q, Wu Z, Miao X, Zhu D, et al. Relationship between long non-coding RNA PCAT-1 expression and gefitinib resistance in non-small-cell lung cancer cells. Respir Res. 2021;22(1):146.
185. Yang C, Dong Z, Hong H, Dai B, Song F, Geng L, et al. circFN1 mediates Sorafenib resistance of hepatocellular carcinoma cells by sponging miR-1205 and regulating E2F1 expression. Mol Ther Nucleic Acids. 2020;22:421–33.

186. Weng H, Zeng L, Cao L, Chen T, Li Y, Xu Y, et al. circFOXM1 contributes to sorafenib resistance of hepatocellular carcinoma cells by regulating MECP2 via miR-1324. Mol Ther Nucleic Acids. 2021;23:811–20.
187. Xu J, Ji L, Liang Y, Wan Z, Zheng W, Song X, et al. CircRNA-SORE mediates sorafenib resistance in hepatocellular carcinoma by stabilizing YBX1. Signal Transduct Target Ther. 2020;5(1):298.
188. Wang T, Liu Z, She Y, Deng J, Zhong Y, Zhao M, et al. A novel protein encoded by circASK1 ameliorates gefitinib resistance in lung adenocarcinoma by competitively activating ASK1-dependent apoptosis. Cancer Lett. 2021;520:321–31.
189. Liu S, Jiang Z, Xiao P, Li X, Chen Y, Tang H, et al. Hsa_circ_0005576 promotes osimertinib resistance through the miR-512-5p/IGF1R axis in lung adenocarcinoma cells. Cancer Sci. 2022;113(1):79–90.
190. Harjes U. EGFR is going circular. Nat Rev Cancer. 2021;21(5):280.
191. Shao Y, Lian S, Zheng J, Tong H, Wang J, Xu J, et al. RP11-616M22.7 recapitulates imatinib resistance in gastrointestinal stromal tumor. Mol Ther Nucleic Acids. 2021;25:264–76.
192. Niu X, Liu F, Zhou Y, Zhou Z, Zhou D, Wang T, et al. Genome-wide DNA methylation analysis reveals GABBR2 as a novel epigenetic target for EGFR 19 deletion lung adenocarcinoma with induction Erlotinib treatment. Clin Cancer Res. 2017;23(17):5003–14.
193. Arechederra M, Bazai SK, Abdouni A, Sequera C, Mead TJ, Richelme S, et al. ADAMTSL5 is an epigenetically activated gene underlying tumorigenesis and drug resistance in hepatocellular carcinoma. J Hepatol. 2021;74(4):893–906.
194. Sheng Y, Ji Z, Zhao H, Wang J, Cheng C, Xu W, et al. Downregulation of the histone methyltransferase SETD2 promotes imatinib resistance in chronic myeloid leukaemia cells. Cell Prolif. 2019;52(4):e12611.
195. Wang Z, Dai J, Yan J, Zhang Y, Yin Z. Targeting EZH2 as a novel therapeutic strategy for sorafenib-resistant thyroid carcinoma. J Cell Mol Med. 2019;23(7):4770–8.
196. Pal AS, Agredo A, Lanman NA, Son J, Sohal IS, Bains M, et al. Loss of KMT5C promotes EGFR inhibitor resistance in NSCLC via LINC01510-mediated upregulation of MET. Cancer Res. 2022;82(8):1534–47.
197. Yan L, Ding B, Liu H, Zhang Y, Zeng J, Hu J, et al. Inhibition of SMYD2 suppresses tumor progression by down-regulating microRNA-125b and attenuates multi-drug resistance in renal cell carcinoma. Theranostics. 2019;9(26):8377–91.
198. Berko ER, Mossé YP. Thrown for a loop: awakening BORIS to evade ALK inhibition therapy. Cancer Cell. 2019;36(4):345–7.
199. Hu B, Xu Y, Li YC, Huang JF, Cheng JW, Guo W, et al. CD13 promotes hepatocellular carcinogenesis and sorafenib resistance by activating HDAC5-LSD1-NF-κB oncogenic signaling. Clin Transl Med. 2020;10(8):e233.
200. Zhang C, Shen L, Zhu Y, Xu R, Deng Z, Liu X, et al. KDM6A promotes imatinib resistance through YY1-mediated transcriptional upregulation of TRKA independently of its demethylase activity in chronic myelogenous leukemia. Theranostics. 2021;11(6):2691–705.
201. Liu S, Li Q, Li G, Zhang Q, Zhuo L, Han X, et al. The mechanism of m(6)A methyltransferase METTL3-mediated autophagy in reversing gefitinib resistance in NSCLC cells by β-elemene. Cell Death Dis. 2020;11(11):969.
202. Song H, Liu D, Wang L, Liu K, Chen C, Wang L, et al. Methyltransferase like 7B is a potential therapeutic target for reversing EGFR-TKIs resistance in lung adenocarcinoma. Mol Cancer. 2022;21(1):43.
203. Tang J, Han T, Tong W, Zhao J, Wang W. N(6)-methyladenosine (m(6)A) methyltransferase KIAA1429 accelerates the gefitinib resistance of non-small-cell lung cancer. Cell Death Discov. 2021;7(1):108.
204. Li K, Peng ZY, Gao S, Wang QS, Wang R, Li X, et al. M6A associated TSUC7 inhibition contributed to Erlotinib resistance in lung adenocarcinoma through a notch signaling activation dependent way. J Exp Clin Cancer Res. 2021;40(1):325.

205. Sa R, Liang R, Qiu X, He Z, Liu Z, Chen L. IGF2BP2-dependent activation of ERBB2 signaling contributes to acquired resistance to tyrosine kinase inhibitor in differentiation therapy of radioiodine-refractory papillary thyroid cancer. Cancer Lett. 2022;527:10–23.
206. Yiming R, Takeuchi Y, Nishimura T, Li M, Wang Y, Meguro-Horike M, et al. MUSASHI-2 confers resistance to third-generation EGFR-tyrosine kinase inhibitor osimertinib in lung adenocarcinoma. Cancer Sci. 2021;112(9):3810–21.
207. Wang JH, Zeng Z, Sun J, Chen Y, Gao X. A novel small-molecule antagonist enhances the sensitivity of osteosarcoma to cabozantinib in vitro and in vivo by targeting DNMT-1 correlated with disease severity in human patients. Pharmacol Res. 2021;173:105869.
208. Yin X, Zhou M, Fu Y, Yang L, Xu M, Sun T, et al. Histone demethylase RBP2 mediates the blast crisis of chronic myeloid leukemia through an RBP2/PTEN/BCR-ABL cascade. Cell Signal. 2019;63:109360.
209. Gao F, Wang Q, Zhang C, Zhang C, Qu T, Zhang J, et al. RNA methyltransferase METTL3 induces intrinsic resistance to gefitinib by combining with MET to regulate PI3K/AKT pathway in lung adenocarcinoma. J Cell Mol Med. 2021;25(5):2418–25.
210. Hua W, Zhao Y, Jin X, Yu D, He J, Xie D, et al. METTL3 promotes ovarian carcinoma growth and invasion through the regulation of AXL translation and epithelial to mesenchymal transition. Gynecol Oncol. 2018;151(2):356–65.
211. Meenakshi Sundaram DN, Jiang X, Brandwein JM, Valencia-Serna J, Remant KC, Uludağ H. Current outlook on drug resistance in chronic myeloid leukemia (CML) and potential therapeutic options. Drug Discov Today. 2019;24(7):1355–69.
212. Liu B, Duenas D, Zheng L, Reckamp K, Shen B. Genomic instability as a major mechanism for acquired resistance to EGFR tyrosine kinase inhibitors in cancer. Protein Cell. 2022;13(2):82–9.
213. Weng CH, Chen LY, Lin YC, Shih JY, Lin YC, Tseng RY, et al. Epithelial-mesenchymal transition (EMT) beyond EGFR mutations per se is a common mechanism for acquired resistance to EGFR TKI. Oncogene. 2019;38(4):455–68.
214. Soucheray M, Capelletti M, Pulido I, Kuang Y, Paweletz CP, Becker JH, et al. Intratumoral heterogeneity in EGFR-mutant NSCLC results in divergent resistance mechanisms in response to EGFR tyrosine kinase inhibition. Cancer Res. 2015;75(20):4372–83.
215. Tam SY, Wu VWC, Law HKW. Hypoxia-induced epithelial-mesenchymal transition in cancers: HIF-1α and beyond. Front Oncol. 2020;10:486.
216. Zhu C, Wei Y, Wei X. AXL receptor tyrosine kinase as a promising anti-cancer approach: functions, molecular mechanisms and clinical applications. Mol Cancer. 2019;18(1):153.
217. Lee YS, Kim SM, Kim BW, Chang HJ, Kim SY, Park CS, et al. Anti-cancer effects of HNHA and Lenvatinib by the suppression of EMT-mediated drug resistance in cancer stem cells. Neoplasia. 2018;20(2):197–206.
218. Oser MG, Niederst MJ, Sequist LV, Engelman JA. Transformation from non-small-cell lung cancer to small-cell lung cancer: molecular drivers and cells of origin. Lancet Oncol. 2015;16(4):e165–72.
219. Lee JK, Lee J, Kim S, Kim S, Youk J, Park S, et al. Clonal history and genetic predictors of transformation into small-cell carcinomas from lung adenocarcinomas. J Clin Oncol. 2017;35(26):3065–74.
220. Niederst MJ, Sequist LV, Poirier JT, Mermel CH, Lockerman EL, Garcia AR, et al. RB loss in resistant EGFR mutant lung adenocarcinomas that transform to small-cell lung cancer. Nat Commun. 2015;6:6377.
221. Huang T, Song X, Xu D, Tiek D, Goenka A, Wu B, et al. Stem cell programs in cancer initiation, progression, and therapy resistance. Theranostics. 2020;10(19):8721–43.
222. Honkanen T, Wilenius E, Koivunen P, Koivunen JP. HER2 regulates cancer stem-like cell phenotype in ALK translocated NSCLC. Int J Oncol. 2017;51(2):599–606.
223. Houshmand M, Simonetti G, Circosta P, Gaidano V, Cignetti A, Martinelli G, et al. Chronic myeloid leukemia stem cells. Leukemia. 2019;33(7):1543–56.

224. Binnewies M, Roberts EW, Kersten K, Chan V, Fearon DF, Merad M, et al. Understanding the tumor immune microenvironment (TIME) for effective therapy. Nat Med. 2018;24(5):541–50.
225. Patel SA, Nilsson MB, Yang Y, Le X, Tran HT, Elamin YY, et al. IL6 mediates suppression of T- and NK-cell function in EMT-associated TKI-resistant EGFR-mutant NSCLC. Clin Cancer Res. 2023;29(7):1292–304.
226. Isomoto K, Haratani K, Hayashi H, Shimizu S, Tomida S, Niwa T, et al. Impact of EGFR-TKI treatment on the tumor immune microenvironment in EGFR mutation-positive non-small cell lung cancer. Clin Cancer Res. 2020;26(8):2037–46.
227. Hwang I, Kim JW, Ylaya K, Chung EJ, Kitano H, Perry C, et al. Tumor-associated macrophage, angiogenesis and lymphangiogenesis markers predict prognosis of non-small cell lung cancer patients. J Transl Med. 2020;18(1):443.
228. Boutilier AJ, Elsawa SF. Macrophage polarization states in the tumor microenvironment. Int J Mol Sci. 2021;22(13)
229. Pistamaltzian NF, Georgoulias V, Kotsakis A. The role of immune checkpoint inhibitors in advanced non-small cell lung cancer. Expert Rev Respir Med. 2019;13(5):435–47.
230. Yin X, Han S, Song C, Zou H, Wei Z, Xu W, et al. Metformin enhances gefitinib efficacy by interfering with interactions between tumor-associated macrophages and head and neck squamous cell carcinoma cells. Cell Oncol (Dordr). 2019;42(4):459–75.
231. Dong N, Shi X, Wang S, Gao Y, Kuang Z, Xie Q, et al. M2 macrophages mediate sorafenib resistance by secreting HGF in a feed-forward manner in hepatocellular carcinoma. Br J Cancer. 2019;121(1):22–33.
232. Mantovani A, Allavena P, Marchesi F, Garlanda C. Macrophages as tools and targets in cancer therapy. Nat Rev Drug Discov. 2022;21(11):799–820.
233. Ma S, Zhang L, Ren Y, Dai W, Chen T, Luo L, et al. Epiregulin confers EGFR-TKI resistance via EGFR/ErbB2 heterodimer in non-small cell lung cancer. Oncogene. 2021;40(14):2596–609.
234. Liu S, Wang Y, Han Y, Xia W, Zhang L, Xu S, et al. EREG-driven oncogenesis of head and neck squamous cell carcinoma exhibits higher sensitivity to Erlotinib therapy. Theranostics. 2020;10(23):10589–605.
235. Zhang Z, Yang C, Li L, Zhu Y, Su K, Zhai L, et al. "γδT cell-IL17A-neutrophil" axis drives immunosuppression and confers breast cancer resistance to high-dose anti-VEGFR2 therapy. Front Immunol. 2021;12:699478.
236. Zhou SL, Zhou ZJ, Hu ZQ, Huang XW, Wang Z, Chen EB, et al. Tumor-associated neutrophils recruit macrophages and T-regulatory cells to promote progression of hepatocellular carcinoma and resistance to Sorafenib. Gastroenterology. 2016;150(7):1646–58.e17.
237. Jia Y, Li X, Jiang T, Zhao S, Zhao C, Zhang L, et al. EGFR-targeted therapy alters the tumor microenvironment in EGFR-driven lung tumors: implications for combination therapies. Int J Cancer. 2019;145(5):1432–44.
238. Dieci MV, Conte P, Bisagni G, Brandes AA, Frassoldati A, Cavanna L, et al. Association of tumor-infiltrating lymphocytes with distant disease-free survival in the ShortHER randomized adjuvant trial for patients with early HER2+ breast cancer. Ann Oncol. 2019;30(3):418–23.
239. Tan W, Luo X, Li W, Zhong J, Cao J, Zhu S, et al. TNF-α is a potential therapeutic target to overcome sorafenib resistance in hepatocellular carcinoma. EBioMedicine. 2019;40:446–56.
240. Bauer R, Udonta F, Wroblewski M, Ben-Batalla I, Santos IM, Taverna F, et al. Blockade of myeloid-derived suppressor cell expansion with all-trans retinoic acid increases the efficacy of antiangiogenic therapy. Cancer Res. 2018;78(12):3220–32.
241. Deng X, Li X, Guo X, Lu Y, Xie Y, Huang X, et al. Myeloid-derived suppressor cells promote tumor growth and sorafenib resistance by inducing FGF1 upregulation and fibrosis. Neoplasia. 2022;28:100788.
242. Chen X, Song E. Turning foes to friends: targeting cancer-associated fibroblasts. Nat Rev Drug Discov. 2019;18(2):99–115.
243. Yi Y, Zeng S, Wang Z, Wu M, Ma Y, Ye X, et al. Cancer-associated fibroblasts promote epithelial-mesenchymal transition and EGFR-TKI resistance of non-small cell

lung cancers via HGF/IGF-1/ANXA2 signaling. Biochim Biophys Acta Mol basis Dis. 2018;1864(3):793–803.

244. Straussman R, Morikawa T, Shee K, Barzily-Rokni M, Qian ZR, Du J, et al. Tumour microenvironment elicits innate resistance to RAF inhibitors through HGF secretion. Nature. 2012;487(7408):500–4.
245. Eun JW, Yoon JH, Ahn HR, Kim S, Kim YB, Lim SB, et al. Cancer-associated fibroblast-derived secreted phosphoprotein 1 contributes to resistance of hepatocellular carcinoma to sorafenib and lenvatinib. Cancer Commun (Lond). 2023;43(4):455–79.
246. Zervantonakis IK, Poskus MD, Scott AL, Selfors LM, Lin JR, Dillon DA, et al. Fibroblast-tumor cell signaling limits HER2 kinase therapy response via activation of MTOR and antiapoptotic pathways. Proc Natl Acad Sci USA. 2020;117(28):16500–8.
247. Modak RV, de Oliveira Rebola KG, McClatchy J, Mohammadhosseini M, Damnernsawad A, Kurtz SE, et al. Targeting CCL2/CCR2 signaling overcomes MEK inhibitor resistance in acute myeloid leukemia. Clin Cancer Res. 2024;30(10):2245–59.
248. Feng H, Liu K, Shen X, Liang J, Wang C, Qiu W, et al. Targeting tumor cell-derived CCL2 as a strategy to overcome Bevacizumab resistance in ETV5(+) colorectal cancer. Cell Death Dis. 2020;11(10):916.
249. Watson SS, Dane M, Chin K, Tatarova Z, Liu M, Liby T, et al. Microenvironment-mediated mechanisms of resistance to HER2 inhibitors differ between HER2+ breast cancer subtypes. Cell Syst. 2018;6(3):329–42.e6.
250. Hall RD, Le TM, Haggstrom DE, Gentzler RD. Angiogenesis inhibition as a therapeutic strategy in non-small cell lung cancer (NSCLC). Transl Lung Cancer Res. 2015;4(5):515–23.
251. Osude C, Lin L, Patel M, Eckburg A, Berei J, Kuckovic A, et al. Mediating EGFR-TKI resistance by VEGF/VEGFR autocrine pathway in non-small cell lung cancer. Cells. 2022;11(10)
252. Luo Z, Tian M, Yang G, Tan Q, Chen Y, Li G, et al. Hypoxia signaling in human health and diseases: implications and prospects for therapeutics. Signal Transduct Target Ther. 2022;7(1):218.
253. Apicella M, Giannoni E, Fiore S, Ferrari KJ, Fernández-Pérez D, Isella C, et al. Increased lactate secretion by cancer cells sustains non-cell-autonomous adaptive resistance to MET and EGFR targeted therapies. Cell Metab. 2018;28(6):848–65.e6.
254. Sofer S, Lamkiewicz K, Armoza Eilat S, Partouche S, Marz M, Moskovits N, et al. A genome-wide CRISPR activation screen reveals hexokinase 1 as a critical factor in promoting resistance to multi-kinase inhibitors in hepatocellular carcinoma cells. FASEB J. 2022;36(3):e22191.
255. Chen C, Liu WR, Zhang B, Zhang LM, Li CG, Liu C, et al. LncRNA H19 downregulation confers erlotinib resistance through upregulation of PKM2 and phosphorylation of AKT in EGFR-mutant lung cancers. Cancer Lett. 2020;486:58–70.
256. Xu C, Zhang L, Wang D, Jiang S, Cao D, Zhao Z, et al. Lipidomics reveals that sustained SREBP-1-dependent lipogenesis is a key mediator of gefitinib-acquired resistance in EGFR-mutant lung cancer. Cell Death Discov. 2021;7(1):353.
257. Sharma N, Bhushan A, He J, Kaushal G, Bhardwaj V. Metabolic plasticity imparts erlotinib-resistance in pancreatic cancer by upregulating glucose-6-phosphate dehydrogenase. Cancer Metab. 2020;8:19.
258. Pan Z, Wang K, Wang X, Jia Z, Yang Y, Duan Y, et al. Cholesterol promotes EGFR-TKIs resistance in NSCLC by inducing EGFR/Src/Erk/SP1 signaling-mediated ERRα re-expression. Mol Cancer. 2022;21(1):77.
259. Swinnen JV, Dehairs J, Talebi A. Membrane lipid remodeling takes center stage in growth factor receptor-driven cancer development. Cell Metab. 2019;30(3):407–8.
260. Xu Q, Li Y, Gao X, Kang K, Williams JG, Tong L, et al. HNF4α regulates sulfur amino acid metabolism and confers sensitivity to methionine restriction in liver cancer. Nat Commun. 2020;11(1):3978.

261. Wei L, Lee D, Law CT, Zhang MS, Shen J, Chin DW, et al. Genome-wide CRISPR/Cas9 library screening identified PHGDH as a critical driver for Sorafenib resistance in HCC. Nat Commun. 2019;10(1):4681.
262. Ge X, Li M, Yin J, Shi Z, Fu Y, Zhao N, et al. Fumarate inhibits PTEN to promote tumorigenesis and therapeutic resistance of type2 papillary renal cell carcinoma. Mol Cell. 2022;82(7):1249–60. e7.
263. Tanimoto A, Takeuchi S, Arai S, Fukuda K, Yamada T, Roca X, et al. Histone deacetylase 3 inhibition overcomes BIM deletion polymorphism-mediated Osimertinib resistance in EGFR-mutant lung cancer. Clin Cancer Res. 2017;23(12):3139–49.
264. Su Y, Li X, Ma J, Zhao J, Liu S, Wang G, et al. Targeting PI3K, mTOR, ERK, and Bcl-2 signaling network shows superior antileukemic activity against AML ex vivo. Biochem Pharmacol. 2018;148:13–26.
265. Sun X, Ou Z, Chen R, Niu X, Chen D, Kang R, et al. Activation of the p62-Keap1-NRF2 pathway protects against ferroptosis in hepatocellular carcinoma cells. Hepatology. 2016;63(1):173–84.
266. Huang W, Chen K, Lu Y, Zhang D, Cheng Y, Li L, et al. ABCC5 facilitates the acquired resistance of sorafenib through the inhibition of SLC7A11-induced ferroptosis in hepatocellular carcinoma. Neoplasia. 2021;23(12):1227–39.
267. Gao R, Kalathur RKR, Coto-Llerena M, Ercan C, Buechel D, Shuang S, et al. YAP/TAZ and ATF4 drive resistance to Sorafenib in hepatocellular carcinoma by preventing ferroptosis. EMBO Mol Med. 2021;13(12):e14351.
268. Luis G, Godfroid A, Nishiumi S, Cimino J, Blacher S, Maquoi E, et al. Tumor resistance to ferroptosis driven by Stearoyl-CoA Desaturase-1 (SCD1) in cancer cells and Fatty Acid Biding Protein-4 (FABP4) in tumor microenvironment promote tumor recurrence. Redox Biol. 2021;43:102006.
269. Viswanathan VS, Ryan MJ, Dhruv HD, Gill S, Eichhoff OM, Seashore-Ludlow B, et al. Dependency of a therapy-resistant state of cancer cells on a lipid peroxidase pathway. Nature. 2017;547(7664):453–7.
270. Heydt Q, Larrue C, Saland E, Bertoli S, Sarry JE, Besson A, et al. Oncogenic FLT3-ITD supports autophagy via ATF4 in acute myeloid leukemia. Oncogene. 2018;37(6):787–97.
271. Lotsberg ML, Wnuk-Lipinska K, Terry S, Tan TZ, Lu N, Trachsel-Moncho L, et al. AXL targeting abrogates Autophagic flux and induces immunogenic cell death in drug-resistant cancer cells. J Thorac Oncol. 2020;15(6):973–99.
272. Lv Y, Zhang W, Zhao J, Sun B, Qi Y, Ji H, et al. SRSF1 inhibits autophagy through regulating Bcl-x splicing and interacting with PIK3C3 in lung cancer. Signal Transduct Target Ther. 2021;6(1):108.
273. Liao Y, Yang Y, Pan D, Ding Y, Zhang H, Ye Y, et al. HSP90α mediates Sorafenib resistance in human hepatocellular carcinoma by necroptosis inhibition under hypoxia. Cancers (Basel). 2021;13(2)
274. Assaraf YG, Brozovic A, Gonçalves AC, Jurkovicova D, Linē A, Machuqueiro M, et al. The multi-factorial nature of clinical multidrug resistance in cancer. Drug Resist Updat. 2019;46:100645.
275. Rudich A, Garzon R, Dorrance A. Non-coding RNAs are implicit in chronic myeloid leukemia therapy resistance. Int J Mol Sci. 2022;23(20)
276. Talati C, Pinilla-Ibarz J. Resistance in chronic myeloid leukemia: definitions and novel therapeutic agents. Curr Opin Hematol. 2018;25(2):154–61.
277. Neul C, Schaeffeler E, Sparreboom A, Laufer S, Schwab M, Nies AT. Impact of membrane drug transporters on resistance to small-molecule tyrosine kinase inhibitors. Trends Pharmacol Sci. 2016;37(11):904–32.
278. Matheux A, Gassiot M, Fromont G, Leenhardt F, Boulahtouf A, Fabbrizio E, et al. PXR modulates the prostate cancer cell response to Afatinib by regulating the expression of the Monocarboxylate transporter SLC16A1. Cancers (Basel). 2021;13(14)

279. Branford S, Apperley JF. Measurable residual disease in chronic myeloid leukemia. Haematologica. 2022;107(12):2794–809.
280. Beretta GL, Cassinelli G, Pennati M, Zuco V, Gatti L. Overcoming ABC transporter-mediated multidrug resistance: the dual role of tyrosine kinase inhibitors as multitargeting agents. Eur J Med Chem. 2017;142:271–89.
281. Wind S, Schmid U, Freiwald M, Marzin K, Lotz R, Ebner T, et al. Clinical pharmacokinetics and pharmacodynamics of Nintedanib. Clin Pharmacokinet. 2019;58(9):1131–47.
282. Shah KN, Bhatt R, Rotow J, Rohrberg J, Olivas V, Wang VE, et al. Aurora kinase A drives the evolution of resistance to third-generation EGFR inhibitors in lung cancer. Nat Med. 2019;25(1):111–8.

Receptor Tyrosine Kinase Inhibitors in Combination Therapies: Integrating Chemotherapy and Radiation Therapy for Improved Cancer Treatment

Rohit Patel and Johnna Francis Varghese

1 Introduction

Chemotherapy is the standard first-line treatment for several types of cancer. Over the years, a combination of chemotherapy and radiation therapy, known as chemoradiotherapy, has been used to treat advanced solid tumors, despite the risk of damage to normal cells. However, a long-term treatment strategy is necessary to improve patient survival. This strategy should focus on targeted therapies with fewer side effects. Unfortunately, the development of resistance to chemotherapeutics makes their use in long-term treatment challenging. In recent decades, researchers have explored various combinatorial approaches, including chemotherapy or radiation alongside other selective inhibitors, such as receptor tyrosine kinase inhibitors (RTKIs). These combination therapies have shown increased therapeutic effectiveness compared to monotherapy. Moreover, these approaches have been tested in clinical trials and demonstrated significant benefits for patients. Receptor tyrosine kinases (RTKs) are a subset of tyrosine kinases (TKs) actively contributing to inter−/intracellular communication and varied biological functions such as cell growth, differentiation, and metabolism. Given their broad influence on these cellular functions, it is not surprising that RTK dysfunction is implicated in numerous diseases, particularly cancers [1]. A detailed understanding of specific RTKs, their signaling pathways, and their effects on cell function has enabled the development of targeted therapies, leading to notable improvements in clinical outcomes. A total of 90 tyrosine kinases have been identified in the human genome of which 58 are RTKs and 32 are cellular TKs [2].

R. Patel · J. F. Varghese (✉)
Department of Neurology, Brigham and Women's Hospital, Harvard Medical School, Boston, MA, USA
e-mail: rpatel84@bwh.harvard.edu; jvarghese4@bwh.harvard.edu

A. Sabarwal et al. (eds.), *Receptor Tyrosine Kinases in Cancer*, Cancer Drug Discovery and Development, https://doi.org/10.1007/978-3-031-93894-8_9

Despite sequence variability among human TKs, they generally share a conserved 3D structure, particularly around the ATP-binding pocket within their catalytic domain. The activation loop's starting sequence (Asp-Phe-Gly, or DFG) is often preserved, playing a crucial role in controlling access to the catalytic site [3]. The structure and signaling pathways of RTKs have been much better understood over the last three decades. Inhibiting these RTK signals can help prevent the pathological effects of mutated or dysfunctional RTKs [1, 4]. Inhibitors designed to target the TKs are defined as tyrosine kinase inhibitors (TKIs) which include RTKIs and cellular TKIs. Categorized as small molecules and macromolecules, the TKIs target a range of signaling pathways, including RTK signaling, that regulate cellular proliferation and tumor angiogenesis [5–7]. The interaction of TKIs with their targets plays a crucial role in determining their mechanism of action, potency, selectivity, and safety profiles [8]. While only a few TKIs are highly selective for specific protein kinases, the majority inhibit a broad range of kinases. This lack of selectivity increases the risk of toxicity [9]. Studies state that approximately 5% of multikinase inhibitors (MKIs) have been discontinued due to TKI-associated adverse effects [8]. Toxicity associated with TKIs can arise from excessive inhibition of RTK function or from off-target effects, which occur when multiple RTKs are inhibited due to limited selectivity of the TKI [8–11]. Several TKIs have now been approved by the US Food and Drug Administration (FDA). Moreover, the combination of various inhibitors with chemotherapeutic drugs and radiation therapy has been explored for a better treatment strategy. In this chapter, we explore a range of selective RTKIs targeting c-Met, Axl, vascular endothelial growth factor receptor (VEGFR), epidermal growth factor receptor (EGFR), and fibroblast growth factor receptor (FGFR) tyrosine kinase receptors. We focus on the use of these RTKIs alone and in combination with established radiation and chemotherapy treatments, highlighting their efficacy and mechanisms of action.

2 Receptor Tyrosine Kinase Inhibitors in Combination Therapy

RTKIs are low-molecular-weight organic compounds. High-throughput screening (HTS) methods are typically used to select effective RTKIs from large compound libraries. An ideal RTKI candidate should meet specific criteria to advance in development, particularly in terms of absorption, distribution, metabolism, excretion, and toxicity for in vivo applications [12]. These properties present several challenges in identifying the ideal inhibitors, including issues with membrane permeability, drug inactivation through metabolism, reduction in efficacy due to nonspecific intracellular interactions, rapid excretion, toxicity, and inadequate distribution to target cellular compartments [12].

RTKIs function by blocking critical pathways by targeting signaling molecules essential for cell survival. They penetrate the plasma membrane and bind to the cytoplasmic domains of RTKs, inhibiting the kinase domain's catalytic activity by obstructing adenosine triphosphate (ATP) or substrate binding [13] (Fig. 1). RTKIs are divided into irreversible or reversible, which can be further categorized mainly into type I, type II, and type III (Fig. 2). Irreversible inhibitors covalently bind to the ATP site and permanently block the kinase, while reversible inhibitors have various binding modes and interact with the kinase at different motifs [14, 15]. The majority of current RTKIs are ATP-competitive inhibitors, classified as type I. However, developing highly selective type I TKIs is challenging due to the conserved nature of ATP binding sites in kinase domains with the DFG motif positioned toward the catalytic site and significant competition with intracellular ATP. As a result, type I RTKIs may inadvertently target other kinases, suggesting that some anti-tumor effects could be attributed to interactions with additional signaling molecules. In contrast, type II and type III RTKIs are non-ATP-competitive inhibitors that induce structural changes in RTKs. These conformational shifts alter the kinase domain, leading to a loss of catalytic activity [16]. The type II inhibitors target inactive kinases, binding to the ATP site with the DFG motif rotated outward, which allows access to adjacent regions, and the type III inhibitors interact with allosteric pockets adjacent to the ATP site, avoiding direct interaction with ATP. Additionally, these

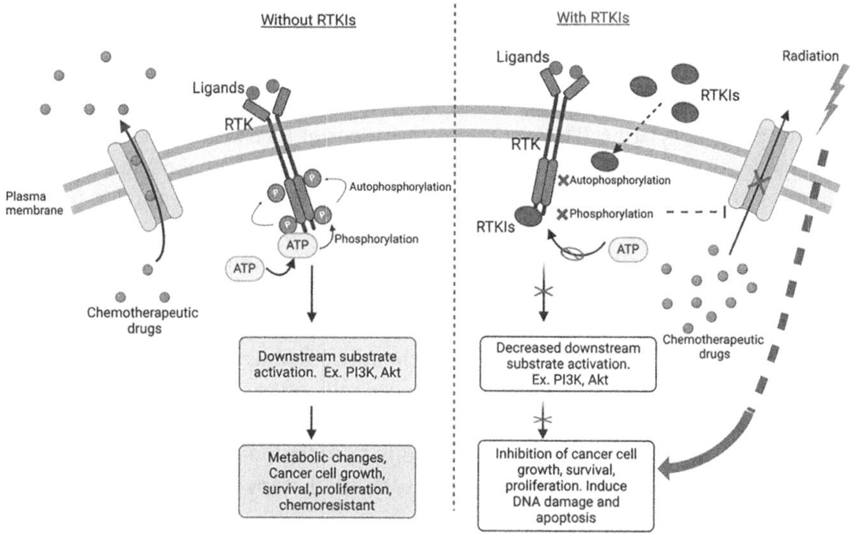

Fig. 1 Diagram illustrating RTK activation and its downstream effects, comparing scenarios with and without RTK inhibitors (RTKIs). In the absence of RTKIs, ligand binding to the receptor triggers ATP binding, leading to receptor phosphorylation, autophosphorylation, and activation. This cascade stimulates downstream signaling pathways, ultimately promoting cancer cell growth and development of chemoresistance. Conversely, in the presence of RTKIs, ATP binding to the receptor is blocked, preventing phosphorylation and autophosphorylation. This inhibition disrupts downstream signaling, effectively suppressing cancer cell growth. This also leads to inhibition of efflux of chemotherapeutic drugs when RTKIs are combined with radiation, making the therapy more effective

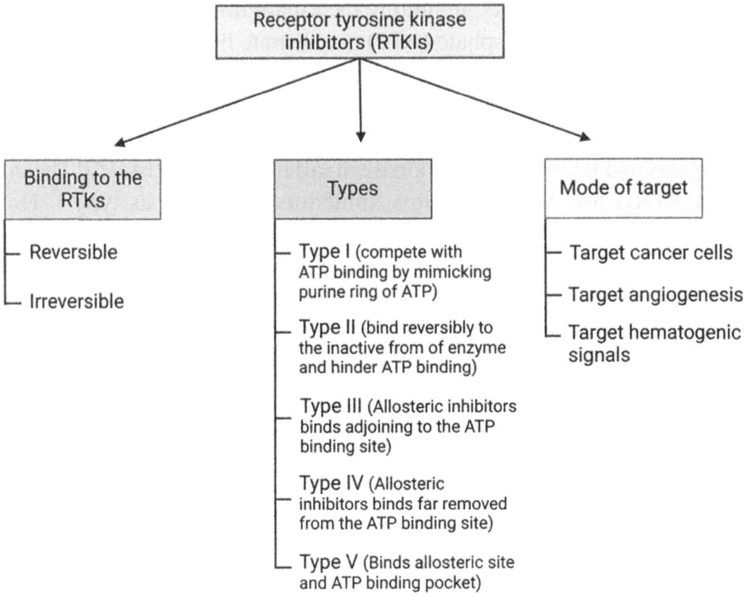

Fig. 2 Flow diagram categorizing RTKIs based on their binding affinities, target RTK binding sites, and modes of action

inhibitors can bind to residues within the kinase domain, effectively preventing tyrosine phosphorylation. There are two other types of RTKIs, the type IV inhibitors that bind to distant allosteric sites, far removed from the ATP pocket, and the type V inhibitors that encompass inhibitors exhibiting multiple binding modes (Fig. 2). RTKIs when combined with chemotherapy or radiation therapy boost their therapeutic effectiveness while overcoming challenges like drug resistance and long-term side effects. RTKIs function by blocking signaling pathways involved in cancer cell growth and survival, while chemotherapy and radiation therapy inhibit cancer cell growth by inducing DNA and mitochondrial damage, ultimately triggering apoptosis. In response to chemotherapy, cancer cells often adapt by activating alternative growth signaling pathways, which RTKIs help to block. These properties of RTKIs increase the sensitivity of cancer cells to radiation and chemotherapeutics, making them more vulnerable to radiation–/chemotherapy-induced damage.

3 RTKIs in Combination with Chemotherapy and Radiation Therapy

The recurrent dysregulation of RTKs in cancer, along with their association with disease progression and poor prognosis, has positioned these receptors as promising targets for therapeutic intervention. Recent innovations in the formulation of inhibitors that specifically target RTKs have significantly transformed the landscape of cancer therapy. Patients frequently exhibit a positive response to RTKIs at the outset; however, extended administration of these agents can lead to the development of resistance, culminating in treatment failure and disease progression. Tumor cells may adopt multiple survival mechanisms, including the acquisition of mutations or the activation of alternate signaling pathways, to counteract the inhibitory effects of RTKIs [17]. The efficacy of RTKIs in cancer therapy is well-established, as they focus on the key factors involved in tumorigenesis [18]. Over the past few decades, the clinical application of targeted therapies has become increasingly prominent, particularly with the use of RTKIs that inhibit a wide range of pro-tumorigenic signals. Mechanistically, RTKIs can be classified into three categories: (i) agents that directly attack cancer cells, (ii) those that normalize angiogenesis, and (iii) agents that influence hematologic lineage cells (Fig. 2). The emergence of interest in protein kinase inhibitors can be traced back to the FDA's endorsement of the TKI imatinib in 2001. This oral chemotherapy agent specifically targets the BCR-Abl hybrid protein, a TK signaling molecule associated with Philadelphia-chromosome-positive chronic myelogenous leukemia. Following the launch of imatinib, the utilization of TKIs has significantly broadened, especially in oncology, owing to the essential functions that TKs serve in cellular signaling pathways [14, 19].

FDA-approved small molecule RTKIs have widely been used to treat varied cancers including hematologic, endocrinologic, urogenital, gastrointestinal, and pulmonary cancers as first-line therapy. The FDA approved RTKIs are either highly selective to a specific RTK (e.g., afatinib and lapatinib) or nonselectively target multiple RTKs (e.g., imatinib and sorafenib) or have dual specificity. Few of these RTKIs specifically inhibit members of one family of RTK. We will discuss the critical RTKs and their associated RTKIs along with their inhibitory mode of action (Table 1). We will also elaborate on the combinatorial therapies available with selective RTKIs and their anticancer mode of action.

3.1 c-Met Inhibitors in Combination Therapy

Hepatocyte growth factor (HGF) activates the c-Met receptor by binding and inducing dimerization and autophosphorylation, which triggers downstream signaling through pathways like mitogen-activated protein kinase (MAPK), phosphoinositide 3-kinase (PI3K), Src, and signal transducers and activators of transcription (STAT) [20]. Constitutive activation of c-Met can occur via HGF-cMet autocrine loops, receptor overexpression, or activating mutations [21]. The c-Met is encoded by the MET gene and is composed of α- and β-subunits and is essential for normal development [22]. MET is expressed across various adult tissues; its levels are typically low and mainly limited to epithelial and mesenchymal cells [23]. Together with its ligand, HGF, c-Met supports several critical functions, such as embryogenesis, cell proliferation, survival, differentiation, invasion, and tissue repair [24]. The

Table 1 The table represents key cancer-associated RTKs, their corresponding RTKIs, binding sites, modes of inhibition, and known combinations with chemotherapy or radiation therapy; other denotes inhibitors targeting molecules other than RTKs

RTKs	RTKIs	Binding of RTKIs	Inhibition	Known combination therapy
c-Met	Cabozantinib	ATP binding site	Phosphorylation	Radiation therapy
	EMD1214063, EMD1204831	ATP binding site	Phosphorylation	Other
	LY2801653	ATP binding site or allosteric binding site	Phosphorylation and autophosphorylation	Other
	Tivantinib	Non-ATP binding site	Autophosphorylation	Chemotherapy
AXL	R428	ATP binding site	Phosphorylation	Chemotherapy
VEGFR	Sorafenib	Non-ATP binding site	Autophosphorylation	Chemotherapy
	Lenvatinib	ATP binding site	Phosphorylation	Chemotherapy, radiation therapy
	Apatinib	ATP binding site	Phosphorylation	Chemotherapy, radiation therapy
	Vatalanib	ATP binding site	Phosphorylation	Radiation therapy
	Tivozanib	ATP binding site	Phosphorylation	Other
	Cediranib	ATP binding site	Phosphorylation	Radiation therapy
EGFR	Gefitinib	ATP binding site	Autophosphorylation	Chemotherapy, radiation therapy
	Erlotinib	ATP binding site	Phosphorylation	Chemotherapy, radiation therapy
	Apatinib	ATP binding site	Phosphorylation	Radiation therapy
	Icotinib	ATP binding site	Phosphorylation	Unknown
	Afatinib	Non-ATP binding site	Autophosphorylation	Chemotherapy, radiation therapy
	Lapatinib	ATP binding site	Phosphorylation	Chemotherapy, other
	Dacomitinib	ATP binding site	Phosphorylation	Unknown
PDGFR	CP-673451	ATP binding site	Phosphorylation and autophosphorylation	Unknown
	Crenolanib	ATP binding site	Phosphorylation	Chemotherapy
FGFR	PD173074	ATP binding site	Phosphorylation and autophosphorylation	Chemotherapy

cytoplasmic domain of c-Met has unique docking sites that, when phosphorylated, interact with PI3K and SHC, promoting proliferation through the Src homology and collagen (SHC)-growth factor receptor-bound protein 2 (GRB2)-Son of Sevenless (SOS)-rat sarcoma (Ras) pathway and enhancing cell migration and survival [25]. Cabozantinib is an example of c-Met RTKIs that binds to the ATP binding site of c-Met and inhibits PI3K activation, thereby inhibiting the growth of cancer cells. In this section we will discuss well-known c-Met inhibitors and their efficacy in combination with chemotherapy or radiation therapy.

In cancer, c-Met dysregulation can occur through various mechanisms, which include mutations, gene amplifications, overexpression, and enhanced ligand stimulation [26]. High MET expression is found in breast cancer [27], non-small cell lung cancer (NSCLC) [28], gastric cancer [29], other solid tumors, and hematologic malignancies [30, 31]. Elevated MET expression and activity are associated with increased metastasis, greater tumor aggressiveness, and poor prognosis [32].

The ligand of c-Met binds to its extracellular domain resulting in dimerization of the receptor which induces transphosphorylation of Tyr 1234 and 1235 residues in the kinase area leading to structural transformations within the receptor. Further, Tyr 1349 phosphorylation and 1356 phosphorylation at the carboxy-terminal domain are also necessary for c-Met signaling [33]. The phosphorylation and activation of c-Met further utilize the PI3K/protein kinase B (Akt), Src/focal adhesion kinase (FAK), and Ras/rapidly accelerated fibrosarcoma (Raf)/mitogen-activated protein kinase (MEK)/extracellular signal-regulated kinase (ERK) pathways for signal transduction leading to augmented cell growth, survival, proliferation, motility, and angiogenesis [34]. The cabozantinib functions by inhibiting the activation of c-Met, thereby limiting cancer cell progression. In patients with renal cell carcinoma (RCC), the rate of grade $\geq$3 adverse events when stereotactic ablative body radiotherapy is administered concurrently and non-concurrently with cabozantinib is 3.6% and 9.1%, respectively [35]. In hepatocellular carcinoma (HCC) when cabozantinib is combined with pan-mammalian target of rapamycin (mTOR) inhibitor (MLN0218), it leads to tumor regression in c-Met/β-catenin mice. On the contrary, combining this RTKI with checkpoint inhibitor anti-programmed death-ligand 1 (PD-L1) antibody does not render any significant therapeutic benefit to treat HCC [36].

3.1.1 EMD1214063 and EMD1204831

These selective c-Met-RTKIs are ATP competitive [37] and are identified to limit HGF-induced c-Met phosphorylation in A549 lung cancer cells and reduce human-derived tumors in xenograft mice [37]. Both these RTKIs induce similar anti-tumor effects as shown in the preclinical studies, where they decrease the autophosphorylation of c-Met, leading to dephosphorylation and inactivation of Akt, ERK1/2, and phospholipase C gamma (PLCγ) in cancer cell lines [38, 39]. EMD1214063 induces autophagy in gastric cancer cells, and combining autophagy inhibitors with this RTKI leads to augmented cell death [40]. This RTKI has also been studied in neuroblastoma both in vitro and in vivo, where it induces apoptosis by inhibiting HGF-mediated c-Met phosphorylation followed by MEK dephosphorylation. A similar effect is observed in neuroblastoma immunocompromised xenograft mice [41]. No reports are available regarding utilizing these RTKIs in combination with other well-rooted cancer therapies.

3.1.2 LY2801653 (Merestinib)

This RTKI is a type II MET/RON RTKI that functions by inhibiting c-Met autophosphorylation and PI3K/STAT3 phosphorylation as shown in HGF-stimulated lung cancer and glioblastoma cells [42]. Functionally, this RTKI has also shown antiproliferative properties by inhibiting c-Met phosphorylation in vitro and xenograft mice models [42]. Further, LY2801653 inhibits NSCLC cell line growth and patient-derived tumor xenografts. It also prevents the constitutive activation of c-Met signal transduction and lung cancer cell proliferation, anchorage-independent migration, invasion, and growth. Treatment of the orthotopic model from lung cancer cells using LY2801653 significantly inhibits both tumor growth and metastasis [43]. This RTKI when administered in combination with trametinib, a MEK inhibitor, inhibits Akt phosphorylation and promotes apoptosis by inducing the cleavage of poly (ADP-ribose) polymerase (PARP) protein in metastatic uveal melanoma explants. Thus, it selectively blocks c-Met and disrupts the resistance developed against MEK inhibitors [44].

3.1.3 Tivantinib

Tivantinib, a non-ATP-competitive inhibitor is administered orally and is highly selective for c-Met. It functions by binding to dephosphorylated c-Met and prevents its autophosphorylation after ATP binding [45]. This RTKI is capable of inducing caspase-mediated apoptosis, and inhibiting cell proliferation and growth of tumor cells in xenograft mice models [46]. Tivantinib underwent the first clinical evaluation in patients with advanced and metastatic tumors and showed appreciable levels of tolerance with mild or moderate toxicities such as thrombocytopenia, leukopenia, dehydration, and vomiting [46]. Thus, this RTKI independently has shown promising results both in vitro and in vivo in a wide range of cancers [46]. Several chemotherapeutic agents have been explored in combination with tivantinib, where combining it with 5-FU/leucovorin/oxaliplatin (FOLFOX) in patients with gastroesophageal cancer displayed therapeutic response and a progression-free survival in comparison to patients administered with only FOLFOX [47]. Also addition of tivantinib to first-line therapy carboplatin and pemetrexed in patients with advanced non-squamous NSCLC or malignant pleural mesothelioma portrayed better safety and preliminary anti-tumor efficacy [48]. Similarly, antimetabolite gemcitabine has been approved as monotherapy and in combination to treat metastatic breast cancer, NSCLC, and advanced ovarian cancer. Combining tivantinib with gemcitabine leads to better safety, tolerability, and anti-tumor activity [49].

3.2 AXL Inhibitors in Combination Therapy

Axl derived from the Greek term *anexelekto,* synonymously known as uncontrolled, is an integral part of the Tyro3, Axl, and Mertk (TAM) family of RTKs and plays a vital role in promoting cell proliferation by activating key signaling pathways, including PI3K/Akt/mTOR, Ras/Raf/MEK/ERK, Janus kinases (JAK)/STAT, and nuclear factor kappa B (NF-κB) [50]. The TAM family receptors are defined by the presence of two Ig-like domains, two fibronectin type II repeats present in the extracellular region, and a cytoplasmic domain containing the kinase activity [51]. The vitamin-K-dependent protein, growth arrest-specific 6 (Gas6), is a critical ligand of Axl [52] inducing Axl autophosphorylation at the Tyr 702 and 703 and MER autophosphorylation at Tyr 749, 753, and 754 residues in the activation loop [50, 53]. The residues Tyr 779, 821, and 866 in the Axl and the MER Tyr 872 undergo phosphorylation and provide the docking sites for the interaction with intracellular downstream signaling molecules [54]. Axl is expressed in several cancers with a moderate expression in normal organs [55, 56]. The expression for Axl is correlated to poor cancer prognosis [53]. For example, it is associated with unfavorable outcomes in acute myeloid leukemia (AML). Studies indicate that Axl inhibition sensitizes AML stem/progenitor cells to the anti-apoptotic agent, venetoclax, yielding a strong synergistic effect in in vitro and xenograft models [57]. Axl has been shown as a promising drug target in advanced-stage cancers [58]. Further, a combination of the DNA methyltransferase inhibitor, decitabine, with the histone deacetylase inhibitor, vorinostat, inhibits AML cell viability and induces apoptosis via enhanced acetylation of histones and decreased expression of DNA methyltransferase 1(DNMT1). However, this combinatorial therapy also induces the expression of Axl. Supplementing this combinatorial therapy with an Axl-specific inhibitor, BGB324, has demonstrated greater efficacy in increasing AML cell sensitivity compared to decitabine and vorinostat alone [59].

3.2.1 R428

This Axl-RTKI is extremely selective in its function where it dephosphorylates Axl in a dose-dependent manner, thus preventing invasion of human and murine breast cancers (MDA-MB-231 and 4 T1, respectively) [60]. In mouse models, oral administration of R428 with pharmacologically effective doses demonstrates good plasma stability. It effectively inhibits breast cancer cell metastasis and angiogenesis by blocking Akt and ERK phosphorylation. Additionally, R428 showed strong efficacy in combination with cisplatin by reducing liver and lung metastases in a breast cancer model [60]. The effects of R428 have been evaluated on NSCLC cells in in vitro and xenograft models, both as a standalone treatment and in combination with other RTKIs or chemotherapeutic agents. In NCI-H1299 cells (mesenchymal, EGFR wild-type, and erlotinib-resistant), R428 combined with erlotinib showed a synergistic antiproliferative effect. In NCI-H1299 xenograft mice, R428 significantly enhanced the anti-tumor activity of docetaxel. Additionally, R428 treatment of A549 NSCLC cells, either alone or in combination with erlotinib or the anti-vascular endothelial growth factor (VEGF) antibody bevacizumab, demonstrates additive anti-tumor effects. In the HCC827 xenograft model, R428 treatment delays the development of resistance to erlotinib [61]. R428 in combination with auranofin, a thioredoxin reductase inhibitor, synergistically inhibits breast cancer cell proliferation. This combination promotes apoptosis by caspase-3 activation and PARP cleavage, and additionally inhibits cancer cell migration [62].

3.3 VEGFR Inhibitors in Combination Therapy

An array of VEGFs is expressed by solid tumors that bind to their cognate receptors on the vascular endothelium inducing angiogenesis. There are several vascular endothelial growth factor receptor (VEGFR)-TKIs identified till date and investigated to treat various solid tumors. First developed for treating renal cell cancer, sorafenib functions by targeting serine/threonine kinase Raf or by limiting the autophosphorylation of several other RTKs that include VEGFR, c-Kit proto-oncogene receptor tyrosine kinase (c-Kit), rearranged during transfection (RET), and platelet-derived growth factor receptor (PDGFR) [63]. This RTKI was the first FDA-approved treatment for advanced primary HCC in 2007 [64]. Recent studies on sorafenib-based combination therapies show promising outcomes. Combining sorafenib with transcatheter arterial chemoembolization (TACE) has significantly improved overall response rates (55.9% vs. 37.3%) and disease control rates (86.4% vs. 67.8%) over TACE alone in HCC [65]. Further research is essential to confirm these findings and clarify the combination's risk–benefit profile.

3.3.1 Lenvatinib

This is another RTKI that was approved by the FDA in 2018 and functions distinctly when compared to sorafenib. Lenvatinib targets angiogenesis and a wide range of receptors including PDGFRα and VEGFR1–3 [65]. Distinctly, it targets fibroblast growth factor receptors (FGFR) 1–4 that contribute to angiogenesis and promote tumor progression and metastasis [66]. In addition, it inhibits cell proliferation by limiting the proto-oncogenes KIT and RET [67]. Mechanistically, lenvatinib binds to VEGFR2 by targeting its ATP binding site through a unique interaction involving its quinoline moiety, which mimics ATP, and a cyclopropane ring engaging nearby regions. This binding approach distinguishes lenvatinib from other VEGFR2 kinase inhibitors, suggesting a novel mode of interaction that could contribute to its effectiveness and selectivity [68]. In combination therapies, lenvatinib paired with an anti-programmed cell death protein-1 (PD-1) antibody (pembrolizumab) has shown promising outcomes, notably improving objective response rate (ORR) and disease control rate (DCR) in unresectable HCC [69]. This led the FDA to designate this combination as a breakthrough therapy for first-line treatment for HCC [70]. Other combinations, such as lenvatinib with TACE or microwave ablation, also demonstrate enhanced clinical benefits, including increased ORR and improved survival outcomes. Lenvatinib when administered with a new regimen comprising of low-dose cisplatin combined with 5-fluorouracil and Lipiodol via the hepatic arterial infusion chemotherapy increased the efficacy of this RTKI in HCC [71]. Combining radiation therapy with lenvatinib proved to be effective against differentiated thyroid cancer where it synergistically inhibited cell replication and colony formation in vitro. This combination also limited tumor growth in nude mice, suppressing the expression of proliferation marker Ki67 and inducing a G2/M cell cycle arrest and apoptosis [72].

3.3.2 Apatinib

This is an RTKI that selectively blocks VEGFR2 by occupying its intracellular ATP binding site, thereby preventing angiogenesis in tumor tissues. It was approved as a third- and second-line treatment for gastric cancer and HCC, respectively, by the Chinese National Medical Products Administration (NMPA) [73]. In terms of combinatorial therapies, apatinib when combined with camrelizumab or TACE showed prominent results with reduced ORR and DCR and improved median progression-free survival (PFS), proving to be a potential therapeutic for cancer patients [74]. Apatinib when combined with S-1 plus oxaliplatin as a neoadjuvant therapy was effective against advanced gastric cancer with better safety profiles [75]. This RTKI when combined with intensity-modulated radiation therapy is better tolerated and effective in improving the PFS and DCR as well as has potential anti-tumor activity in patients with unresectable HCC [76].

3.3.3 Vatalanib

This VEGFR-TKI is selective and is capable of inhibiting class III kinases such as PDGFRβ and KIT. In subcutaneous xenograft mouse models, it efficiently inhibits growth and microvasculature [77]. It has also demonstrated anti-tumor and antiangiogenic activities in renal carcinoma cell lines. A combination of vatalanib with everolimus on gastric cancer cells in vitro shows that everolimus but not vatalanib was effective in reducing cell proliferation with no effects on apoptosis. Vatalanib eliminated angiogenic effect which was incomplete in presence of everolimus alone. Under in vivo conditions, the combination was effective in reducing the tumor size, proliferation, and vascular density when compared to their individual effects [78]. Administering vatalanib in combination with radiation and temozolomide in newly diagnosed glioblastoma patients was well tolerated with better safety records [79].

3.3.4 Tivozanib

This oral quinolone urea derivative selectively inhibits the VEGFR family signaling and suppresses angiogenesis. This RTKI has been shown effective against human lung, liver, breast, ovarian, pancreas, and prostate cancer as well as in brain xenograft models [80]. Tivozanib functions by inhibiting the VEGFR2 phosphorylation induced by VEGF along with limiting the migratory effects on human umbilical vein endothelial cells (HUVECs) in vitro. A similar effect is displayed against a wide range of human tumor xenografts [81]. Tivozanib effects were also evaluated using a rat colon cancer RCN-9 syngeneic model, where RCN-9 tumor cells were transplanted into the peritoneal cavity of F344 rats. Treatment with this RTKI reduced angiogenesis, tumor nodules, and accumulation of malignant ascites. Additionally, it also induced vascular regression in newly formed tumors and extended survival, even in rats with advanced-stage tumors. This RTKI currently is in phase I and II clinical trials, for patients with RCC and breast cancer [82, 83]. In combination with sorafenib, this RTKI is more effective in RCC compared to when administered individually [80, 84].

3.3.5 Cediranib

This RTKI is an indole-ether quinazoline and is an extremely potent inhibitor of VEGF signaling [85]. This RTKI inhibits VEGF-induced phosphorylation of VEGFR2 and VEGFR3 followed by limiting the activation of downstream signaling molecules such as ERK1/2, Akt, and CREB in human endothelial cells (ECs). Cediranib thus affects the homeostatic function of ECs and lymphangiogenesis. It thereby inhibits the proliferation, survival, migration, and angiogenic properties of the cells [85]. In vivo administration of cediranib in the xenograft model of breast cancer cells causes regression in tumor growth, along with depreciated blood flow, microvascular density, and proliferation [86]. Preclinical studies demonstrate this RTKI to reduce tumor growth in cancers such as lung, breast, ovarian, RCC, and NSCLC [87]. Most phase III trials have failed to show beneficial effects with cediranib as individual and combination therapy in NSCLC, kidney, colorectal cancer, and glioblastoma [88, 89]. A phase Ib study has shown that cediranib in combination with standard radiation therapy is well tolerated in patients with newly diagnosed glioblastoma [90]. However, a phase II, placebo-controlled double-blind clinical trial in similar patients shows no significant difference in survival when cediranib is administered in combination with radiation and temozolomide versus radiation with temozolomide [91].

3.4 EGFR Inhibitors in Combination Therapy

The EGFR family of receptors plays a pivotal role in cell proliferation, survival, and differentiation. Under normal physiological conditions, EGF, the soluble ligand, binds to the extracellular domain of the receptor following EGFR homo−/heterodimerization with other associated members subsequently leading to specific tyrosine residue phosphorylation present in the intracellular region [92].

EGFR-TKIs have become the standard treatment for patients with NSCLC who carry EGFR mutations and are recommended as the first-line therapy for advanced cases. Currently, three oral EGFR-TKIs—gefitinib (Iressa), erlotinib (Tarceva), and afatinib (Giotrif)—are approved for initial treatment in advanced NSCLC with EGFR mutations. Gefitinib and erlotinib are first-generation RTKIs that reversibly inhibit EGFR, whereas afatinib, a second-generation TKI, irreversibly inhibits EGFR. These RTKIs have demonstrated higher response rates and improved PFS compared to traditional platinum-based chemotherapy. In the LUX-Lung 7 phase III trial, afatinib showed a significant PFS advantage over gefitinib in NSCLC patients with EGFR mutations (HR 0.73; 95% CI 0.57–0.95). However, phase III trials have typically included patients with a good performance status, leaving the effectiveness of this RTKI in NSCLC patients with compromised performance status being uncertain in real-world settings. Moreover, chemotherapy following EGFR-TKI treatment shows better outcome in NSCLC patients [93]. Additionally, other medications like statins and metformin have shown potential anti-tumor effects in NSCLC, while antacids and corticosteroids may reduce the efficacy of EGFR-TKIs in advanced NSCLC. Mutation, receptor overexpression, ligand-dependent stimulation, and receptor dimerization are some reasons for abrogated EGFR activation which is associated with epithelial origin tumors such as NSCLC [94], colorectal cancer [95], breast cancer [96], and pancreatic cancer [97]. EGFR expression is also associated with poor prognosis and malignancies [98, 99]. Hence an extensively selective inhibition of EGFR is an ideal approach for treating cancer.

3.4.1 Gefitinib and Erlotinib

These are two specific EGFR-TKIs approved for treating NSCLC [94, 100]. Erlotinib in combination with gemcitabine has also been approved to treat patients with metastatic pancreatic cancer [101]. Structurally, mechanistically, and therapeutically similar to gefitinib and erlotinib, there is another selective EGFR-TKI, i.e., icotinib hydrochloride that has demonstrated anti-tumor effects. Icotinib is capable of inhibiting EGF-induced EGFR phosphorylation, thus blocking EGFR-mediated intracellular tyrosine phosphorylation; this was also observed in in vivo nude mice that received human tumor-derived xenografts [102]. Combining gefitinib with chemotherapy agents, pemetrexed and carboplatin, showed increased survival in patients with advanced lung adenocarcinoma compared to treatment involving pemetrexed and carboplatin individually [103]. The combination of gefitinib along with radiation therapy has also shown improved efficacy and is well tolerated in NSCLC patients who are not eligible to undergo surgery or chemoradiotherapy [104]. Further, the NSCLC patients when subjected to chemoradiation with erlotinib may have an improved overall survival, but not a PFS [105].

3.4.2 Afatinib and Neratinib

These two second-generation EGFR-TKIs have been studied both in monotherapy and in combinatorial therapy. They are dual inhibitors of EGFR and human epidermal growth factor receptor 2 (HER2). Binding to EGFR and HER2 receptors, these molecules decrease autophosphorylation in tumor cells by targeting the cysteine residue embedded in the ATP-binding pocket of the receptors. Afatinib was developed as an irreversible EGFR-TKI unlike its counterparts gefitinib and erlotinib, which led to acquired resistance in patients. It is also an irreversible inhibitor of HER4 where it functions by covalently binding to the cysteine at position 803 [106]. Afatinib has been examined as a first-line therapy for patients with NSCLC and for patients pre-treated with a different RTKI and as a second- and third-line treatment for NSCLC patients [107]. Neratinib has been effective against NSCLC and breast tumors. HER2-transformed cells have resistance to erlotinib; however, HER2-overexpressing cells such as Calu-3 and NSCLC are sensitive to neratinib. BT474 HER2-overexpressing cell xenografts are highly sensitive to neratinib [108]. This drug is currently being investigated both as an independent and combinatorial therapeutic molecule. Interestingly, afatinib, in combination with ionizing radiation, has shown reduced phosphorylation of EGFR and ERK, thereby reducing tumor progression in xenograft model of afatinib-resistant cell line derived from NSCLC. Further, afatinib has also shown clinical benefits in patients with cancer progression post erlotinib/gefitinib treatment. When combined with paclitaxel, afatinib shows cancer PFS in patients with stage IIIb or IV NSCLC [109].

3.4.3 Lapatinib

A reversible inhibitor of HER2/EGFR covalently binds to the Cys 773 and Cys 805 residues of EGFR and HER2, respectively. It can sensitize mutant cell lines that are resistant to erlotinib and gefitinib [110]. It blocks EGFR/HER2 by limiting the tyrosine phosphorylation at the TK domain, thereby preventing activation of ERK1/2 and Akt leading to apoptosis of tumor cells both under in vitro and in vivo conditions [111, 112]. Clinical trials showed that lapatinib has variable effects on the downregulation of ERK and Akt in patients with metastatic cancers with overexpressing EGFR/HER2 and has fewer side effects. In combination with capecitabine, this drug has shown effective anti-tumor effects in addition to prolonged time of progression of the disease in breast cancer patients that were previously subjected to treatment with trastuzumab, an anthracycline, and a taxane [113]. Lapatinib in 2007 was approved as the second-line treatment for breast cancer and as a first-line therapy in combination with an aromatase inhibitor for postmenopausal women with HER2-positive breast cancer.

3.4.4 Dacomitinib

A selective quinazoline irreversible pan-EGFR-TKI is an apt candidate for treating NSCLC. Preclinical studies state that it has a similar mode of action when compared to erlotinib, gefitinib, and canertinib [113]. This molecule inhibits EGFR phosphorylation and its downstream molecules Akt and ERK1/2 and induces apoptosis in the EGFR-expressing H3255 cell line. This RTKI is effective in limiting tumor growth in ovarian cancer cell (SKOV3) xenografts that overexpress EGFR and HER2-4. Dacomitinib inhibits HER2 phosphorylation at tyrosine residue 1248, similarly for A431 human squamous cell carcinoma xenografts [114]. Preliminary studies of phase II clinical trials show this RTKI to be a first-line therapy for head and neck cancer and a second- or third-line therapy for NSCLC compared to erlotinib with some side effects [113]. A phase II trial is ongoing to evaluate the effect of dacomitinib as a first-line treatment for lung cancer patients with EGFR mutations.

3.5 Platelet-Derived Growth Factor Receptor (PDGFR)-Specific TKI in Combination Therapy

PDGFR belongs to the type III tyrosine kinase family. These are exclusively expressed in fibroblasts, smooth muscle cells in the lungs, and vascular wall pericytes. PDGFR overexpression and activation are characteristic features of various malignancies including gliomas and NSCLC [115]. PDGFR activation leads to stimulation of MAPK, PI3K, and PLC-γ pathways involved in cellular and developmental responses [116].

3.5.1 CP-673451

CP-673451 is an extremely selective benzimidazole and a reversible ATP-competitive inhibitor of PDGFR. Under in vitro conditions, this RTKI inhibits cell growth of H1703, a mesenchymal NSCLC cell line highly expressing PDGFRα and H292, an epithelial-derived cell line. Its efficacy has also been documented in in vivo xenograft mice models with lung and colon cancer cells [117]. CP-673451 depreciates cell proliferation via dephosphorylating PDGFR, Akt, glycogen synthase kinase 3 alpha/beta (GSK-3α/β), and impaired rhabdosphere-forming capacity in both RD and RUCH2 rhabdomyosarcoma cells [118]. RD and RUCH2 cell line xenografts in the presence of this RTKI show decreased proliferation, tumor growth, and stromal infiltration [118]. It has been effective in preclinical settings; however further investigations are required to examine its effect on PDGFR-overexpressing tumors.

3.5.2 Crenolanib

It is a selective inhibitor of PDGFR with a 100-fold greater affinity for PDGFR than VEGFR2 and fibroblast growth factor receptor (FGFR)2 [119]. It has been evaluated in phase I trial with better tolerability. Crenolanib displayed strong antiproliferative effects in BaF3 D842V and EOL-1 cell lines that have a PDGFRα-dependent growth. This RTKI is more potent than imatinib in inhibiting the kinase activity in imatinib-resistant mutated PDGFRα gastric cancer cells [115]. Crenolanib shows better efficacy and safety when combined with chemotherapeutic drugs such as cytarabine and daunorubicin in patients with newly diagnosed FLT3 mutant acute myeloid leukemia [120].

3.6 FGFR Inhibitors in Combination Therapy

Fibroblast growth factor (FGF) binding to its cognate receptors activates the MAPK and the PI3K/Akt signaling pathways. These two critical pathways play an important role in tumor cell proliferation, angiogenesis, survival, and migration [121]. Most of the FGFR-TKIs are multi-targeted inhibitors.

3.6.1 PD173074

This RTKI has antiangiogenic properties. It binds to the TK domain of the FGFR family. It is capable of selectively inhibiting micro-capillary formation shown by HUVECs in Matrigel. In vivo studies in mammary tumor mice indicate that oral administration of PD173074 inhibits angiogenesis in a dose-dependent manner when administered in combination with photodynamic therapy [122]. This RTKI inhibits FGF2-mediated pro-oligodendrocyte differentiation into oligodendrocyte. Moreover, it prevents PDGF-mediated MAPK activation of the cells and downregulates the myelin genes 2′,3′-cyclic-nucleotide 3′-phosphodiesterase (CNP) and myelin basic protein (MBP) [123]. PD173074 also facilitates the accumulation of paclitaxel, a chemotherapeutic drug, by reducing its efflux. This RTKI also promotes therapeutic efficacy in chemoresistant cancer in patients [124]. A combination of PD173074 with paclitaxel and doxorubicin has also shown enhanced cytotoxic effect of these chemotherapeutic agents in endometrial cancer cell line [125].

4 Conclusion

As of November 2023, regulatory authorities, including the FDA and the European Medicines Agency (EMA), have approved more than 100 small molecules or antibodies targeting specific RTKs for clinical application. Prominent examples of these approved therapies include imatinib, gefitinib, and cetuximab, which are utilized in the treatment of different cancer types. The side effects of each RTKI are monitored to identify the desired outcomes of cancer therapy. Combination therapies incorporating RTKIs have emerged as a promising strategy to overcome the limitations of monotherapies in cancer treatment. By simultaneously targeting multiple signaling pathways involved in tumor growth, angiogenesis, and resistance mechanisms, these combinatorial approaches have demonstrated enhanced therapeutic efficacy across various malignancies. Preclinical and clinical studies suggest that pairing RTKIs with chemotherapy, immunotherapy, or other targeted agents can synergistically inhibit tumor progression, delay resistance, and improve patient outcomes. However, challenges such as toxicity, optimal dosing schedules, and biomarker-driven patient selection remain critical hurdles. Combination of RTKIs along with other chemotherapeutic agents and/or radiotherapy requires a risk–benefit assessment of patients and characterizing the diseases' state to optimize the therapy regime. Additionally, a significant number of other RTKIs are expected to gain approval in the near future, thereby broadening the range of therapeutic strategies available for cancer treatment. Most of the therapies available for cancer have significantly augmented patients' PFS; however, none stand out to cure the disease. Despite the benefits of RTKIs in combination therapies, cancer cells often develop resistance due to mutations and the upregulation of alternative signaling pathways. This remains a major challenge in combination therapy strategies. While RTKIs tend to have fewer side effects, the long-term use of combination therapies may lead to significant side effects that require attention. Moving forward, a deeper understanding of tumor biology, coupled with precision medicine approaches, will be essential to refine these strategies and maximize the clinical benefit of RTKI-based combination therapies. Ongoing clinical trials and research are essential to refine these combinations, optimize patient selection, and address the challenges posed by drug resistance. Ultimately, integrating RTKIs into rational, multimodal treatment regimens holds great potential to transform the therapeutic landscape of cancer.

Acknowledgments The figures were created with BioRender.com. (Created in BioRender. Patel, R. (2025) https://BioRender.com/8ba2bqp).

Conflict of Interest The authors declare that there are no conflicts of interest.

References

1. Kolibaba KS, Druker BJ. Protein tyrosine kinases and cancer. Biochim Biophys Acta. 1997;1333(3):F217–48.
2. Broekman F, Giovannetti E, Peters GJ. Tyrosine kinase inhibitors: multi-targeted or single-targeted? World J Clin Oncol. 2011;2(2):80–93.
3. Tong M, Seeliger MA. Targeting conformational plasticity of protein kinases. ACS Chem Biol. 2015;10(1):190–200.
4. Jiao Q, Bi L, Ren Y, Song S, Wang Q, Wang YS. Advances in studies of tyrosine kinase inhibitors and their acquired resistance. Mol Cancer. 2018;17(1):36.
5. Nair P, Prado M, Perea-Gil I, Karakikes I. Concise review: precision matchmaking: induced pluripotent stem cells meet cardio-oncology. Stem Cells Transl Med. 2019;8(8):758–67.
6. Wu F, Yang J, Liu J, Wang Y, Mu J, Zeng Q, et al. Signaling pathways in cancer-associated fibroblasts and targeted therapy for cancer. Signal Transduct Target Ther. 2021;6(1):218.
7. Zhong L, Li Y, Xiong L, Wang W, Wu M, Yuan T, et al. Small molecules in targeted cancer therapy: advances, challenges, and future perspectives. Signal Transduct Target Ther. 2021;6(1):201.
8. Moran G, Elaine A. Protein kinase inhibitors – selectivity or toxicity? In: Rajesh Kumar S, editor. Protein kinases. Rijeka: IntechOpen; 2021. p. Ch. 2.
9. Hantschel O. Unexpected off-targets and paradoxical pathway activation by kinase inhibitors. ACS Chem Biol. 2015;10(1):234–45.
10. Chen ZI, Ai DI. Cardiotoxicity associated with targeted cancer therapies. Mol Clin Oncol. 2016;4(5):675–81.
11. Miljković F, Bajorath J. Computational analysis of kinase inhibitors identifies promiscuity cliffs across the human Kinome. ACS Omega. 2018;3(12):17295–308.
12. Prueksaritanont T, Tang C. ADME of biologics-what have we learned from small molecules? AAPS J. 2012;14(3):410–9.
13. Johnson LN. Protein kinase inhibitors: contributions from structure to clinical compounds. Q Rev Biophys. 2009;42(1):1–40.
14. Wu P, Nielsen TE, Clausen MH. FDA-approved small-molecule kinase inhibitors. Trends Pharmacol Sci. 2015;36(7):422–39.
15. Norman RA, Toader D, Ferguson AD. Structural approaches to obtain kinase selectivity. Trends Pharmacol Sci. 2012;33(5):273–8.
16. Garuti L, Roberti M, Bottegoni G. Non-ATP competitive protein kinase inhibitors. Curr Med Chem. 2010;17(25):2804–21.
17. Yang Y, Li S, Wang Y, Zhao Y, Li Q. Protein tyrosine kinase inhibitor resistance in malignant tumors: molecular mechanisms and future perspective. Signal Transduct Target Ther. 2022;7(1):329.
18. Bhullar KS, Lagarón NO, McGowan EM, Parmar I, Jha A, Hubbard BP, et al. Kinase-targeted cancer therapies: progress, challenges and future directions. Mol Cancer. 2018;17(1):48.
19. Manning G, Whyte DB, Martinez R, Hunter T, Sudarsanam S. The protein kinase complement of the human genome. Science. 2002;298(5600):1912–34.
20. Bouattour M, Raymond E, Qin S, Cheng AL, Stammberger U, Locatelli G, et al. Recent developments of c-Met as a therapeutic target in hepatocellular carcinoma. Hepatology. 2018;67(3):1132–49.
21. Gentile A, Trusolino L, Comoglio PM. The Met tyrosine kinase receptor in development and cancer. Cancer Metastasis Rev. 2008;27(1):85–94.
22. Birchmeier C, Gherardi E. Developmental roles of HGF/SF and its receptor, the c-Met tyrosine kinase. Trends Cell Biol. 1998;8(10):404–10.
23. Sonnenberg E, Meyer D, Weidner KM, Birchmeier C. Scatter factor/hepatocyte growth factor and its receptor, the c-Met tyrosine kinase, can mediate a signal exchange between mesenchyme and epithelia during mouse development. J Cell Biol. 1993;123(1):223–35.

24. Birchmeier C, Birchmeier W, Gherardi E, Vande Woude GF. Met, metastasis, motility and more. Nat Rev Mol Cell Biol. 2003;4(12):915–25.
25. Trusolino L, Comoglio PM. Scatter-factor and semaphorin receptors: cell signalling for invasive growth. Nat Rev Cancer. 2002;2(4):289–300.
26. Ghiso E, Giordano S. Targeting MET: why, where and how? Curr Opin Pharmacol. 2013;13(4):511–8.
27. Shattuck DL, Miller JK, Carraway KL III, Sweeney C. Met receptor contributes to trastuzumab resistance of Her2-overexpressing breast cancer cells. Cancer Res. 2008;68(5):1471–7.
28. Park S, Choi Y-L, Sung CO, An J, Seo J, Ahn M-J, et al. High MET copy number and MET overexpression: poor outcome in non-small cell lung cancer patients. Histol Histopathol. 2012;27
29. Catenacci DV, Henderson L, Xiao S-Y, Patel P, Yauch RL, Hegde P, et al. Durable complete response of metastatic gastric cancer with anti-Met therapy followed by resistance at recurrence. Cancer Discov. 2011;1(7):573–9.
30. Eksioglu-Demiralp E, Akdeniz T, Bayik M. Aberrant expression of c-Met and HGF/c-Met pathway provides survival advantage in B-chronic lymphocytic leukemia. Cytometry B Clin Cytom. 2011;80(1):1–7.
31. Papadhimitriou SI, Abazis D, Repa C, Papaconstantinou C, Papanastasiou C, Pangalos C, et al. An unusual cytogenetic abnormality involving chromosomes 1 and 7 in a case of chronic myelomonocytic leukemia. Cancer Genet Cytogenet. 1995;85(1):75–7.
32. Boccaccio C, Comoglio PM. Invasive growth: a MET-driven genetic programme for cancer and stem cells. Nat Rev Cancer. 2006;6(8):637–45.
33. Graveel CR, Tolbert D, Woude GFV. MET: a critical player in tumorigenesis and therapeutic target. Cold Spring Harb Perspect Biol. 2013;5(7):a009209.
34. Rosen LS, Senzer N, Mekhail T, Ganapathi R, Chai F, Savage RE, et al. A phase I dose-escalation study of Tivantinib (ARQ 197) in adult patients with metastatic solid tumors. Clin Cancer Res. 2011;17(24):7754–64.
35. Loo Gan C, Huang J, Pan E, Xie W, Schmidt AL, Labaki C, et al. Real-world practice patterns and safety of concurrent radiotherapy and cabozantinib in metastatic renal cell carcinoma: results from the international metastatic renal cell carcinoma database consortium. Eur Urol Oncol. 2023;6(2):204–11.
36. Shang R, Song X, Wang P, Zhou Y, Lu X, Wang J, et al. Cabozantinib-based combination therapy for the treatment of hepatocellular carcinoma. Gut. 2021;70(9):1746–57.
37. Bladt F, Faden B, Friese-Hamim M, Knuehl C, Wilm C, Fittschen C, et al. EMD 1214063 and EMD 1204831 constitute a new class of potent and highly selective c-Met inhibitors. Clin Cancer Res. 2013;19(11):2941–51.
38. Medová M, Pochon B, Streit B, Blank-Liss W, Francica P, Stroka D, et al. The novel ATP-competitive inhibitor of the MET hepatocyte growth factor receptor EMD1214063 displays inhibitory activity against selected MET-mutated variants. Mol Cancer Ther. 2013;12(11):2415–24.
39. Leiser D, Pochon B, Blank-Liss W, Francica P, Glück AA, Aebersold DM, et al. Targeting of the MET receptor tyrosine kinase by small molecule inhibitors leads to MET accumulation by impairing the receptor downregulation. FEBS Lett. 2014;588(5):653–8.
40. Humbert M, Medová M, Aebersold DM, Blaukat A, Bladt F, Fey MF, et al. Protective autophagy is involved in resistance towards MET inhibitors in human gastric adenocarcinoma cells. Biochem Biophys Res Commun. 2013;431(2):264–9.
41. Scorsone K, Zhang L, Woodfield SE, Hicks J, Zage PE. The novel kinase inhibitor EMD1214063 is effective against neuroblastoma. Investig New Drugs. 2014;32:815–24.
42. Yan SB, Peek VL, Ajamie R, Buchanan SG, Graff JR, Heidler SA, et al. LY2801653 is an orally bioavailable multi-kinase inhibitor with potent activity against MET, MST1R, and other oncoproteins, and displays anti-tumor activities in mouse xenograft models. Investig New Drugs. 2013;31(4):833–44.

43. Wu W, Bi C, Credille KM, Manro JR, Peek VL, Donoho GP, et al. Inhibition of tumor growth and metastasis in non–small cell lung cancer by LY2801653, an inhibitor of several oncokinases, including MET. Clin Cancer Res. 2013;19(20):5699–710.
44. Cheng H, Chua V, Liao C, Purwin TJ, Terai M, Kageyama K, et al. Co-targeting HGF/cMET signaling with MEK inhibitors in metastatic uveal melanoma. Mol Cancer Ther. 2017;16(3):516–28.
45. Eathiraj S, Palma R, Hirschi M, Volckova E, Nakuci E, Castro J, et al. A novel mode of protein kinase inhibition exploiting hydrophobic motifs of autoinhibited kinases: discovery of ATP-independent inhibitors of fibroblast growth factor receptor. J Biol Chem. 2011;286(23):20677–87.
46. Munshi N, Jeay S, Li Y, Chen C-R, France DS, Ashwell MA, et al. ARQ 197, a novel and selective inhibitor of the human c-Met receptor tyrosine kinase with antitumor activity. Mol Cancer Ther. 2010;9(6):1544–53.
47. Pant S, Patel M, Kurkjian C, Hemphill B, Flores M, Thompson D, et al. A phase II study of the c-Met inhibitor Tivantinib in combination with FOLFOX for the treatment of patients with previously untreated metastatic adenocarcinoma of the distal esophagus, gastroesophageal junction, or stomach. Cancer Investig. 2017;35(7):463–72.
48. Zucali P, Simonelli M, De Vincenzo F, Fatuzzo G, Bertossi M, Perrino M, et al. Tivantinib in combination with carboplatin and Pemetrexed as first line treatment in patients with advanced non-squamous NSCLC or malignant pleural mesothelioma: results of phase I trial. Ann Oncol. 2015;26:vi78.
49. Pant S, Saleh M, Bendell J, Infante J, Jones S, Kurkjian C, et al. A phase I dose escalation study of oral c-MET inhibitor tivantinib (ARQ 197) in combination with gemcitabine in patients with solid tumors. Ann Oncol. 2014;25(7):1416–21.
50. Linger RM, Keating AK, Earp HS, Graham DK. TAM receptor tyrosine kinases: biologic functions, signaling, and potential therapeutic targeting in human cancer. Adv Cancer Res. 2008;100:35–83.
51. O'bryan JP, Frye RA, Cogswell PC, Neubauer A, Kitch B, Prokop C, et al. Axl, a transforming gene isolated from primary human myeloid leukemia cells, encodes a novel receptor tyrosine kinase. Mol Cell Biol. 1991;11(10):5016–31.
52. Stitt TN, Conn G, Goret M, Lai C, Bruno J, Radzlejewski C, et al. The anticoagulation factor protein S and its relative, Gas6, are ligands for the Tyro 3/Axl family of receptor tyrosine kinases. Cell. 1995;80(4):661–70.
53. Linger RM, Keating AK, Earp HS, Graham DK. Taking aim at Mer and Axl receptor tyrosine kinases as novel therapeutic targets in solid tumors. Expert Opin Ther Targets. 2010;14(10):1073–90.
54. Braunger J, Schleithoff L, Schulz AS, Kessler H, Lammers R, Ullrich A, et al. Intracellular signaling of the Ufo/Axl receptor tyrosine kinase is mediated mainly by a multi-substrate docking-site. Oncogene. 1997;14(22):2619–31.
55. Tsou A-P, Wu K-M, Tsen T-Y, Chi C-W, Chiu J-H, Lui W-Y, et al. Parallel hybridization analysis of multiple protein kinase genes: identification of gene expression patterns characteristic of human hepatocellular carcinoma. Genomics. 1998;50(3):331–40.
56. Ghosh AK, Secreto C, Boysen J, Sassoon T, Shanafelt TD, Mukhopadhyay D, et al. The novel receptor tyrosine kinase Axl is constitutively active in B-cell chronic lymphocytic leukemia and acts as a docking site of nonreceptor kinases: implications for therapy. Blood J Am Soc Hematol. 2011;117(6):1928–37.
57. Niu X, Rothe K, Chen M, Grasedieck S, Li R, Nam SE, et al. Targeting AXL kinase sensitizes leukemic stem and progenitor cells to venetoclax treatment in acute myeloid leukemia. Blood. 2021;137(26):3641–55.
58. Yadav M, Sharma A, Patne K, Tabasum S, Suryavanshi J, Rawat L, et al. AXL signaling in cancer: from molecular insights to targeted therapies. Signal Transduct Target Ther. 2025;10(1):37.

59. Young CS, Clarke KM, Kettyle LM, Thompson A, Mills KI. Decitabine-Vorinostat combination treatment in acute myeloid leukemia activates pathways with potential for novel triple therapy. Oncotarget. 2017;8(31):51429–46.
60. Holland SJ, Pan A, Franci C, Hu Y, Chang B, Li W, et al. R428, a selective small molecule inhibitor of Axl kinase, blocks tumor spread and prolongs survival in models of metastatic breast cancer. Cancer Res. 2010;70(4):1544–54.
61. Wnuk-Lipinska K, Gausdal G, Sandal T, Frink R, Hinz S, Hellesøy M, et al. Abstract B30: selective small molecule AXL inhibitor BGB324 overcomes acquired drug resistance in non-small cell lung carcinoma models. Clin Cancer Res. 2014;20(2_Supplement):B30-B.
62. Ryu YS, Shin S, An HG, Kwon TU, Baek HS, Kwon YJ, et al. Synergistic induction of apoptosis by the combination of an Axl inhibitor and Auranofin in human breast cancer cells. Biomol Ther (Seoul). 2020;28(5):473–81.
63. Wilhelm S, Carter C, Lynch M, Lowinger T, Dumas J, Smith RA, et al. Discovery and development of sorafenib: a multikinase inhibitor for treating cancer. Nat Rev Drug Discov. 2006;5(10):835–44.
64. Keating GM. Sorafenib: a review in hepatocellular carcinoma. Target Oncol. 2017;12(2):243–53.
65. Liu Q, Dai Y. Sorafenib combined with transarterial chemoembolization prolongs survival of patients with advanced hepatocellular carcinoma. J BUON. 2020;25(2):945–51.
66. Yamamoto Y, Matsui J, Matsushima T, Obaishi H, Miyazaki K, Nakamura K, et al. Lenvatinib, an angiogenesis inhibitor targeting VEGFR/FGFR, shows broad antitumor activity in human tumor xenograft models associated with microvessel density and pericyte coverage. Vasc Cell. 2014;6:18.
67. Tohyama O, Matsui J, Kodama K, Hata-Sugi N, Kimura T, Okamoto K, et al. Antitumor activity of lenvatinib (e7080): an angiogenesis inhibitor that targets multiple receptor tyrosine kinases in preclinical human thyroid cancer models. J Thyroid Res. 2014;2014:638747.
68. Okamoto K, Ikemori-Kawada M, Jestel A, von König K, Funahashi Y, Matsushima T, et al. Distinct binding mode of multikinase inhibitor lenvatinib revealed by biochemical characterization. ACS Med Chem Lett. 2015;6(1):89–94.
69. Finn RS, Ikeda M, Zhu AX, Sung MW, Baron AD, Kudo M, et al. Phase Ib study of Lenvatinib plus Pembrolizumab in patients with unresectable hepatocellular carcinoma. J Clin Oncol. 2020;38(26):2960–70.
70. Finn RS, Ryoo B-Y, Merle P, Kudo M, Bouattour M, Lim H-Y, et al. Results of KEYNOTE-240: phase 3 study of pembrolizumab (Pembro) vs best supportive care (BSC) for second line therapy in advanced hepatocellular carcinoma (HCC). J Clin Oncol. 2019;37(15_suppl):4004.
71. Maruta S, Koshima Y, Tsuchiya T, Tamura R, Takahashi M, Ohshima T, et al. Combination therapy of Lenvatinib and hepatic arterial infusion chemotherapy using cisplatin with Lipiodol and 5-Fluorouracil: a potential breakthrough therapy for unresectable advanced hepatocellular carcinoma. Cureus. 2024;16(8):e66185.
72. Suzuki K, Iwai H, Utsunomiya K, Kono Y, Kobayashi Y, Van Bui D, et al. Combination therapy with lenvatinib and radiation significantly inhibits thyroid cancer growth by uptake of tyrosine kinase inhibitor. Exp Cell Res. 2021;398(1):112390.
73. Zhang XH, Cao MQ, Li XX, Zhang T. Apatinib as an alternative therapy for advanced hepatocellular carcinoma. World J Hepatol. 2020;12(10):766–74.
74. Gu H, Li J, You N, Wu K, Wang Z, Wang L, et al. Efficacy and safety of apatinib combined with transarterial chemoembolization (TACE) in treating patients with recurrent hepatocellular carcinoma. Ann Transl Med. 2020;8(24):1677.
75. Lin J-X, Xu Y-C, Lin W, Xue F-Q, Ye J-X, Zang W-D, et al. Effectiveness and safety of Apatinib plus chemotherapy as neoadjuvant treatment for locally advanced gastric cancer: a nonrandomized controlled trial. JAMA Netw Open. 2021;4(7):e2116240-e.
76. Qiu H, Ke S, Cai G, Wu Y, Wang J, Shi W, et al. An exploratory clinical trial of apatinib combined with intensity-modulated radiation therapy for patients with unresectable hepatocellular carcinoma. Cancer Med. 2023;12(1):213–22.

77. Wood JM, Bold G, Buchdunger E, Cozens R, Ferrari S, Frei J, et al. PTK787/ZK 222584, a novel and potent inhibitor of vascular endothelial growth factor receptor tyrosine kinases, impairs vascular endothelial growth factor-induced responses and tumor growth after oral administration. Cancer Res. 2000;60(8):2178–89.
78. Jaeger-Lansky A, Cejka D, Ying L, Preusser M, Hoeflmayer D, Fuereder T, et al. Effects of vatalanib on tumor growth can be potentiated by mTOR blockade in vivo. Cancer Biol Ther. 2010;9(11):919–27.
79. Batchelor T, Eichler AF, Plotkin SR, Drappatz J, Wen P, Sorensen AG, et al. Phase I trial of vatalanib (PTK787) in combination with standard radiation and temozolomide in patients with newly diagnosed glioblastoma. J Clin Oncol. 2009;27(15_suppl):2035.
80. Eskens FA, de Jonge MJ, Bhargava P, Isoe T, Cotreau MM, Esteves B, et al. Biologic and clinical activity of tivozanib (AV-951, KRN-951), a selective inhibitor of VEGF receptor-1,-2, and-3 tyrosine kinases, in a 4-week-on, 2-week-off schedule in patients with advanced solid tumors. Clin Cancer Res. 2011;17(22):7156–63.
81. Nakamura K, Taguchi E, Miura T, Yamamoto A, Takahashi K, Bichat F, et al. KRN951, a highly potent inhibitor of vascular endothelial growth factor receptor tyrosine kinases, has antitumor activities and affects functional vascular properties. Cancer Res. 2006;66(18):9134–42.
82. Bhargava P, Robinson MO. Development of second-generation VEGFR tyrosine kinase inhibitors: current status. Curr Oncol Rep. 2011;13:103–11.
83. Mayer EL, Scheulen M, Beckman J, Richly H, Duarte A, Cotreau M, et al. A phase I dose-escalation study of the VEGFR inhibitor tivozanib hydrochloride with weekly paclitaxel in metastatic breast cancer. Breast Cancer Res Treat. 2013;140:331–9.
84. Motzer RJ, Nosov D, Eisen T, Bondarenko I, Lesovoy V, Lipatov O, et al. Tivozanib versus sorafenib as initial targeted therapy for patients with metastatic renal cell carcinoma: results from a phase III trial. J Clin Oncol. 2013;31(30):3791–9.
85. Heckman CA, Holopainen T, Wirzenius M, Keskitalo S, Jeltsch M, Ylä-Herttuala S, et al. The tyrosine kinase inhibitor cediranib blocks ligand-induced vascular endothelial growth factor receptor-3 activity and lymphangiogenesis. Cancer Res. 2008;68(12):4754–62.
86. Miller KD, Miller M, Mehrotra S, Agarwal B, Mock BH, Zheng Q-H, et al. A physiologic imaging pilot study of breast cancer treated with AZD2171. Clin Cancer Res. 2006;12(1):281–8.
87. Sahade M, Caparelli F, Hoff PM. Cediranib: a VEGF receptor tyrosine kinase inhibitor. Future Oncol. 2012;8(7):775–81.
88. Batchelor TT, Mulholland P, Neyns B, Nabors LB, Campone M, Wick A, et al. Phase III randomized trial comparing the efficacy of cediranib as monotherapy, and in combination with lomustine, versus lomustine alone in patients with recurrent glioblastoma. J Clin Oncol. 2013;31(26):3212–8.
89. Schmoll H-J, Cunningham D, Sobrero A, Karapetis CS, Rougier P, Koski SL, et al. Cediranib with mFOLFOX6 versus bevacizumab with mFOLFOX6 as first-line treatment for patients with advanced colorectal cancer: a double-blind, randomized phase III study (HORIZON III). J Clin Oncol. 2012;30(29):3588–95.
90. Chi AS, Gerstner ER, Eichler AF, Chea HK, Drappatz J, Wen PY, et al. Phase Ib study of cediranib in combination with daily temozolomide and radiation in patients with newly diagnosed glioblastoma. J Clin Oncol. 2009;27(15_suppl)):e13010-e.
91. Batchelor TT, Won M, Chakravarti A, Hadjipanayis CG, Shi W, Ashby LS, et al. NRG/RTOG 0837: randomized, phase II, double-blind, placebo-controlled trial of chemoradiation with or without cediranib in newly diagnosed glioblastoma. Neurooncol Adv. 2023;5(1):vdad116.
92. Ono M, Kuwano M. Molecular mechanisms of epidermal growth factor receptor (EGFR) activation and response to gefitinib and other EGFR-targeting drugs. Clin Cancer Res. 2006;12(24):7242–51.
93. Morimoto K, Yamada T, Furuya N, Tanaka H, Yoshimura A, Oba T, et al. Exploration of clinical biomarkers for guiding treatment selection between chemotherapy and combination therapy with Atezolizumab, bevacizumab, carboplatin, and paclitaxel in EGFR-mutant NSCLC patients after EGFR-TKI therapy: the SPIRAL-STEP study. Lung Cancer. 2025;201:108447.

94. Robinson KW, Sandler AB. EGFR tyrosine kinase inhibitors: difference in efficacy and resistance. Curr Oncol Rep. 2013;15:396–404.
95. Dienstmann R, De Dosso S, Felip E, Tabernero J. Drug development to overcome resistance to EGFR inhibitors in lung and colorectal cancer. Mol Oncol. 2012;6(1):15–26.
96. De La Vega M, Diaz-Canton E, Alvarez RH. Novel targeted agents for the treatment of advanced breast cancer. Future Med Chem. 2012;4(7):893–914.
97. Burtness B. Her signaling in pancreatic cancer. Expert Opin Biol Ther. 2007;7(6):823–9.
98. Ioannou N, Seddon AM, Dalgleish A, Mackintosh D, Modjtahedi H. Expression pattern and targeting of HER family members and IGF-IR in pancreatic cancer. Front Biosci (Landmark Ed). 2012;17(7):2698–724.
99. Herbst RS. Review of epidermal growth factor receptor biology. Int J Radiat Oncol Biol Phys. 2004;59(2):S21–S6.
100. O'Donnell P, Ferguson J, Shyu J, Current R, Rehage T, Tsai J, et al. Analytic performance studies and clinical reproducibility of a real-time PCR assay for the detection of epidermal growth factor receptor gene mutations in formalin-fixed paraffin-embedded tissue specimens of non-small cell lung cancer. BMC Cancer. 2013;13:1–10.
101. Moore MJ, Goldstein D, Hamm J, Figer A, Hecht JR, Gallinger S, et al. Erlotinib plus gemcitabine compared with gemcitabine alone in patients with advanced pancreatic cancer: a phase III trial of the National Cancer Institute of Canada Clinical Trials Group. J Clin Oncol. 2007;25(15):1960–6.
102. Tan F, Shen X, Wang D, Xie G, Zhang X, Ding L, et al. Icotinib (BPI-2009H), a novel EGFR tyrosine kinase inhibitor, displays potent efficacy in preclinical studies. Lung Cancer. 2012;76(2):177–82.
103. Han B, Jin B, Chu T, Niu Y, Dong Y, Xu J, et al. Combination of chemotherapy and gefitinib as first-line treatment for patients with advanced lung adenocarcinoma and sensitive EGFR mutations: a randomized controlled trial. Int J Cancer. 2017;141(6):1249–56.
104. Fu Z, Yang X, Wang W, Deng L, Zhang T, Bi N, et al. Radiotherapy combined with gefitinib for patients with locally advanced non-small cell lung cancer who are unfit for surgery or concurrent chemoradiotherapy: a phase II clinical trial. Radiat Oncol. 2020;15(1):155.
105. Komaki R, Allen PK, Wei X, Blumenschein GR, Tang X, Lee JJ, et al. Adding erlotinib to chemoradiation improves overall survival but not progression-free survival in stage III non-small cell lung cancer. Int J Radiat Oncol Biol Phys. 2015;92(2):317–24.
106. Solca F, Dahl G, Zoephel A, Bader G, Sanderson M, Klein C, et al. Target binding properties and cellular activity of afatinib (BIBW 2992), an irreversible ErbB family blocker. J Pharmacol Exp Ther. 2012;343(2):342–50.
107. Subramaniam DS, Hwang J. BIBW 2992 in non-small cell lung cancer. Expert Opin Investig Drugs. 2011;20(3):415–22.
108. Tsou H-R, Overbeek-Klumpers EG, Hallett WA, Reich MF, Floyd MB, Johnson BD, et al. Optimization of 6,7-disubstituted-4-(arylamino)quinoline-3-carbonitriles as orally active, irreversible inhibitors of human epidermal growth factor receptor-2 kinase activity. J Med Chem. 2005;48(4):1107–31.
109. Schuler M, Yang JC, Park K, Kim JH, Bennouna J, Chen YM, et al. Afatinib beyond progression in patients with non-small-cell lung cancer following chemotherapy, erlotinib/gefitinib and afatinib: phase III randomized LUX-lung 5 trial. Ann Oncol. 2016;27(3):417–23.
110. Howe LR, Brown PH. Targeting the HER/EGFR/ErbB family to prevent breast cancer. Cancer Prev Res. 2011;4(8):1149–57.
111. Xia W, Mullin RJ, Keith BR, Liu L-H, Ma H, Rusnak DW, et al. Anti-tumor activity of GW572016: a dual tyrosine kinase inhibitor blocks EGF activation of EGFR/erbB2 and downstream Erk1/2 and AKT pathways. Oncogene. 2002;21(41):6255–63.
112. Rusnak DW, Lackey K, Affleck K, Wood ER, Alligood KJ, Rhodes N, et al. The effects of the novel, reversible epidermal growth factor receptor/ErbB-2 tyrosine kinase inhibitor, GW2016, on the growth of human normal and tumor-derived cell lines in vitro and in vivo. Mol Cancer Ther. 2001;1(2):85–94.

113. Cho WC, Roukos DH. Trastuzumab emtansine for advanced HER2-positive breast cancer and beyond: genome landscape-based targets. Expert Rev Anticancer Ther. 2013;13(1):5–8.
114. Gonzales AJ, Hook KE, Althaus IW, Ellis PA, Trachet E, Delaney AM, et al. Antitumor activity and pharmacokinetic properties of PF-00299804, a second-generation irreversible pan-erbB receptor tyrosine kinase inhibitor. Mol Cancer Ther. 2008;7(7):1880–9.
115. Heinrich MC, Griffith D, McKinley A, Patterson J, Presnell A, Ramachandran A, et al. Crenolanib inhibits the drug-resistant PDGFRA D842V mutation associated with imatinib-resistant gastrointestinal stromal tumors. Clin Cancer Res. 2012;18(16):4375–84.
116. Andrae J, Gallini R, Betsholtz C. Role of platelet-derived growth factors in physiology and medicine. Genes Dev. 2008;22(10):1276–312.
117. Roberts WG, Whalen PM, Soderstrom E, Moraski G, Lyssikatos JP, Wang H-F, et al. Antiangiogenic and antitumor activity of a selective PDGFR tyrosine kinase inhibitor, CP-673,451. Cancer Res. 2005;65(3):957–66.
118. Ehnman M, Missiaglia E, Folestad E, Selfe J, Strell C, Thway K, et al. Distinct effects of ligand-induced PDGFRα and PDGFRβ signaling in the human rhabdomyosarcoma tumor cell and stroma cell compartments. Cancer Res. 2013;73(7):2139–49.
119. Lewis NL, Lewis LD, Eder JP, Reddy NJ, Guo F, Pierce KJ, et al. Phase I study of the safety, tolerability, and pharmacokinetics of oral CP-868,596, a highly specific platelet-derived growth factor receptor tyrosine kinase inhibitor in patients with advanced cancers. J Clin Oncol. 2009;27(31):5262–9.
120. Wang ES, Stone RM, Tallman MS, Walter RB, Eckardt JR, Collins R. Crenolanib, a type I FLT3 TKI, can be safely combined with Cytarabine and Anthracycline induction chemotherapy and results in high response rates in patients with newly diagnosed FLT3 mutant acute myeloid leukemia (AML). Blood. 2016;128(22):1071.
121. Liang G, Liu Z, Wu J, Cai Y, Li X. Anticancer molecules targeting fibroblast growth factor receptors. Trends Pharmacol Sci. 2012;33(10):531–41.
122. Dimitroff CJ, Klohs W, Sharma A, Pera P, Driscoll D, Veith J, et al. Anti-angiogenic activity of selected receptor tyrosine kinase inhibitors, PD166285 and PD173074: implications for combination treatment with photodynamic therapy. Investig New Drugs. 1999;17:121–35.
123. Bansal R, Magge S, Winkler S. Specific inhibitor of FGF receptor signaling: FGF-2-mediated effects on proliferation, differentiation, and MAPK activation are inhibited by PD173074 in oligodendrocyte-lineage cells. J Neurosci Res. 2003;74(4):486–93.
124. Anreddy N, Patel A, Sodani K, Kathawala RJ, Chen EP, Wurpel JND, et al. PD173074, a selective FGFR inhibitor, reverses MRP7 (ABCC10)-mediated MDR. Acta Pharm Sin B. 2014;4(3):202–7.
125. Byron SA, Loch DC, Pollock PM. Fibroblast growth factor receptor inhibition synergizes with paclitaxel and doxorubicin in endometrial cancer cells. Int J Gynecol Cancer. 2012;22(9):1517–26.

Therapeutic Potential of Combining Receptor Tyrosine Kinase Inhibitor with Immune Checkpoint Blockade

Alicia D'Souza and Ryuhjin Ahn

1 Introduction

Receptor tyrosine kinases (RTKs) are central regulators of cellular signaling, orchestrating key physiological processes such as proliferation, differentiation, migration, and survival in physiology and pathology. RTKs, through ligand binding or transactivation, initiate complex downstream signaling cascades. The dysregulation of RTKs is a defining feature of cancer, from tumor initiation and progression to therapy resistance. Given their role as upstream signaling nodes, RTKs have been targeted by various therapeutic strategies in cancers with aberrant oncogenic signaling. Despite their clinical success, however, resistance to RTK-targeted therapies remains a critical challenge.

Advances in cancer immunotherapy have transformed treatment paradigms, demonstrating durable clinical responses in diverse cancer types. However, primary and acquired resistance to immunotherapy driven by tumor-intrinsic and tumor-extrinsic mechanisms remains a significant hurdle to long-term efficacy. Mounting

Alicia D'Souza and Ryuhjin Ahn contributed equally with all other contributors.

A. D'Souza
MIT-Harvard Health Sciences and Technology, Cambridge, MA, USA

David H. Koch Institute for Integrative Cancer Research, Massachusetts Institute of Technology (MIT), Cambridge, MA, USA
e-mail: adsouza@mit.edu

R. Ahn (✉)
David H. Koch Institute for Integrative Cancer Research, Massachusetts Institute of Technology (MIT), Cambridge, MA, USA

Department of Biological Engineering, MIT, Cambridge, MA, USA
e-mail: ahnr@mit.edu

A. Sabarwal et al. (eds.), *Receptor Tyrosine Kinases in Cancer*, Cancer Drug Discovery and Development, https://doi.org/10.1007/978-3-031-93894-8_10

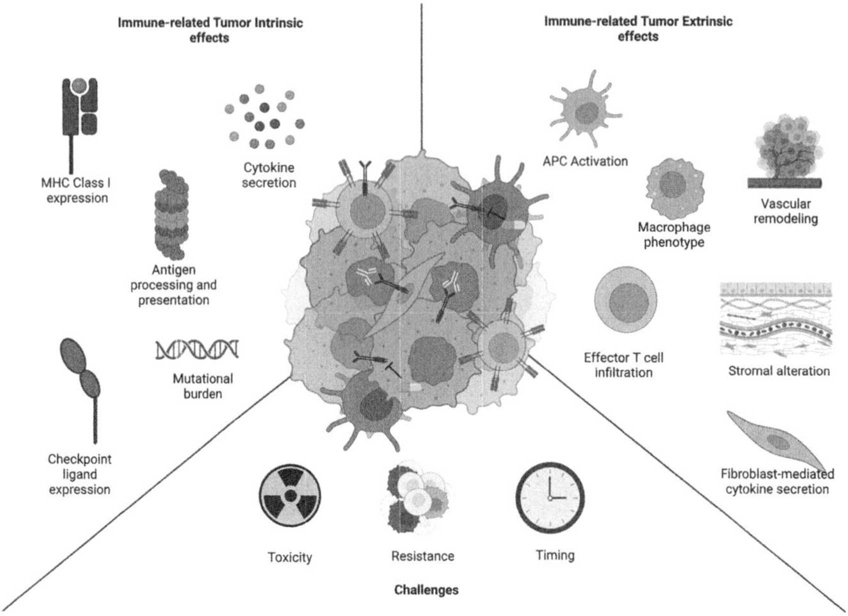

Fig. 1 Effects of RTK inhibition on tumor microenvironment and therapeutics challenges

evidence has underscored the interplay between RTK-mediated signaling and immune evasion and immunosuppression, establishing a strong rationale for combining anti-RTK therapy with immunotherapy. The inhibition of RTKs not only disrupts oncogenic signaling but also can enhance tumor immunogenicity, normalize the tumor vasculature, and restore antitumor immune responses, thereby priming the TME for immunotherapy and enabling greater therapeutic synergy (Fig. 1).

Here, we summarize the immunomodulatory roles of RTKs in both tumor and host cells. We examine RTK-targeted therapies that have been investigated in combination with various cancer immunotherapy modalities, exploring the rationale behind these combinations, their clinical trial outcomes, evidence of synergy, the molecular mechanisms driving their cooperative effects, and, where understood, the potential causes of therapeutic failures.

2 T Cells and Immune Checkpoints

Our immune system maintains a delicate balance between activation and tolerance, ensuring host defense while preventing autoimmunity [1]. Central to this regulation are T cells, which can recognize peptide antigens presented on major histocompatibility complex (MHC) molecules via their T cell receptor (TCR). This antigen recognition initiates signal transduction cascades that drive T cell activation, proliferation, and differentiation. Upon activation, CD8+ cytotoxic T lymphocytes

(CTLs) directly kill infected or malignant cells by releasing perforin and granzymes, while CD4+ helper T cells orchestrate immune responses by secreting cytokines that modulate both adaptive and innate immunity. The specificity of the TCR ensures precise immune targeting, while co-stimulatory and inhibitory signals fine-tune T cell responses to maintain immune homeostasis and prevent autoimmunity [2]. Key inhibitory signals are mediated by immune checkpoint (IC) receptors, including programmed death-1 (PD-1), cytotoxic T-lymphocyte-associated antigen 4 (CTLA-4), lymphocyte activation gene-3 (LAG-3), T cell immunoglobulin and mucin-domain containing-3 (TIM-3), V-domain Ig suppressor of T cell activation (VISTA), and T cell immunoreceptor with Ig and ITIM domains (TIGIT) [3, 4]. For instance, PD-1 is expressed on activated T cells, and its ligands PD-L1 and PD-L2 inhibit T cell responses by engaging SHP2 (Src homology region 2-containing protein tyrosine phosphatase-2), which then dephosphorylates signaling molecules downstream of the TCR and CD28 co-stimulatory pathway, leading to decreased cytokine production, reduced proliferation, impaired effector function in T cells, and T cell exhaustion due to chronic antigen exposure [5]. In cancer, tumor cells often upregulate the expression of IC ligands to suppress T cell-driven antitumor immunity and promote an immunosuppressive TME [1]. Combating the ability of tumors to activate the ICs on T cells using IC inhibitors (ICIs) has transformed oncology across diverse cancer types [6]. Thus far, FDA-approved ICIs include anti-CTLA-4 antibodies (ipilimumab, tremelimumab), anti-PD-1 antibodies (nivolumab, pembrolizumab, cemiplimab, dostarlimab), anti-PD-L1 antibodies (atezolizumab, avelumab, durvalumab), and anti-LAG3 antibody (relatlimab) [6]. These immunotherapies can generate durable and adaptable responses in some patients, leading to long-term remissions through the establishment of immunological memory. This capability sets immunotherapy apart from anti-RTK therapies, which primarily provide transient effects [6].

3 Resistance Mechanisms to Immune Checkpoint Inhibitors

Despite the clinical efficacy of ICIs, many patients experience primary, adaptive, and acquired resistance to ICIs due to multifactorial tumor-intrinsic and tumor-extrinsic mechanisms (reviewed in [7]). Tumor-intrinsic resistance arises from genetic, epigenetic, translational, and posttranslational regulations [8–12]. Tumors that lack the expression of ICs do not respond to ICIs, and they can upregulate alternative ICs in response to targeted ICs [13]. In addition to ICs, tumors avoid immune recognition through the downregulation, mutation, or alternative splicing of tumor antigens and antigen presentation and processing (APP) machinery, including the expression of major histocompatibility complex (MHC) [14, 15]. Tumor cells also secrete cytokines that can promote immunosuppression, aberrant angiogenesis, and ECM remodeling, all of which sustain poor antitumor immune response [16]. High mutation rates in tumor cells can also generate immune-evasive, apoptosis-resistant clones, while neoantigen evolution and phenotypic plasticity, such as epithelial-to-mesenchymal transition, further enable immune escape [17, 18]. Additionally, the silencing of immune-related genes via DNA methylation or histone modifications and genomic instability exacerbate tumor evolution under immune pressure [19–22]. Together, these tumor-intrinsic mechanisms prevent ICI and cancer immunotherapy responsiveness overall.

Tumor-extrinsic resistance factors from the local TME and systemic influences comprise immunosuppressive immune cells such as regulatory T cells (Tregs), myeloid-derived suppressor cells (MDSCs), tumor-associated macrophages (TAMs), and tumor-associated neutrophils. Immunosuppressive cytokines secreted by both host cells, along with systemic immunosuppressants during cancer treatment, further contribute to immune evasion [16, 23–25]. Physical barriers, including a dense extracellular matrix (ECM) produced by cancer-associated fibroblasts (CAFs), restrict immune cell infiltration [26, 27]. A hypoxic TME created by tumor and stroma-driven neovascularization further compounds these effects by impairing immune cell infiltration and activity [28]. Host genetic variability, including polymorphisms in the immune signaling pathways [29], and systemic variables, such as gut microbiome [30], also influence immune response. Poor antitumor effector T cell function is also driven by chronic antigen exposure and immune tolerance mechanisms (e.g., anergy), leading to the removal of tumor-specific effector cells [1]. Systemic toxicities, such as cytokine release syndrome and immune-related adverse events, complicate cancer immunotherapy [31–33], necessitating new combination strategies that mitigate these effects. Unlike tumor-intrinsic factors, tumor-extrinsic factors (e.g., endothelial cells driving angiogenesis, myeloid populations) present compelling therapeutic opportunities due to their genetic stability, commonality across cancers, and their critical roles in supporting immune evasion and immunotherapy resistance. When combined with therapies targeting tumor-intrinsic factors, this strategy can offer a comprehensive approach to overcoming tumor heterogeneity and adaptive resistance by dismantling the collective tumor-supportive networks. As we will discuss in Sect. 4, combining RTK-targeted therapies with ICIs has been proposed as a strategy to overcome both tumor-intrinsic and tumor-extrinsic resistance mechanisms against ICIs and cancer immunotherapy overall.

3.1 Determinants of Immune Checkpoint Inhibitor Response

Given the variability in ICI response across patients, extensive research has focused on identifying the biomarkers that predict treatment efficacy (reviewed in [34, 35]). These biomarkers not only inform patient selection but also provide insights into the mechanisms of ICI. Understanding these determinants is critical for refining immunotherapy strategies and identifying rational combination, particularly with therapies that can enhance these predictive markers. As we will discuss in Sect. 5, several ICI predictive markers are enhanced by RTK targeted therapies, suggesting their potential synergy with ICIs.

The most robust predictor of response to ICI is the presence of an active antitumor immune response [36], consistent with the role of ICIs in enhancing endogenous immune activity. Given the central role of the PD-L1/PD-1 axis in cancer immunotherapy, most established biomarkers have been defined within the context of PD-L1/PD-1 inhibition, underscoring the need for expanded biomarker studies for other ICIs. Deficiencies in DNA repair, including mismatch repair deficiency (dMMR) and microsatellite instability-high (MSI-H), and tumor mutational burden (TMB) have been associated with ICI efficacy across multiple cancer types [35, 37]. However, the composition and clonality of tumor mutations may be more predictive of ICI response than the overall TMB [38]. Additional features linked to improved ICI response include high tumor-infiltrating lymphocyte (TIL) density and its association with tertiary lymphoid structures (TLS), increased tumor neoantigen burden, and the activation of IFN-γ signaling pathways [39–42].

On the other hand, the predictive significance of PD-L1 expression on tumor cells and immune cells of the TME is dependent on the ICI, treatment context, and cancer type, requiring careful consideration of its importance (reviewed in [35, 43]). For example, in metastatic NSCLC, response rates are the highest in patients with high tumor PD-L1 expression [44]. In metastatic triple negative breast cancer (TNBC), PD-L1 expression in tissue predicts response and 2-year survival in metastatic TNBC patients receiving PD-1/PD-L1 inhibitors, with stronger predictive value in first-line treatment and with chemotherapy [45]. In esophageal squamous cell carcinoma, tumor-specific PD-L1 expression is positively correlated with pembrolizumab responsiveness [46]. While useful in these contexts, prognostic values in other cancer types such as melanoma remain inconclusive [43], highlighting the need to discover additional biomarkers for patient selection.

4 Leveraging Anti-RTK Therapy to Overcome Immunotherapy Resistance

Many of the tumor-intrinsic and tumor-extrinsic resistance mechanisms to cancer immunotherapy are regulated by oncogenic and physiologic RTK activity within tumor, immune, and stromal compartments as we will discuss below [47–50]. Moreover, accumulating evidence demonstrates that RTK inhibition induces changes aligning with biomarkers of ICI responsiveness. Consequently, combining immunotherapy with RTK-targeted therapies, either sequentially or concomitantly, has emerged as a promising approach to overcoming resistance to immunotherapy. Several combination regimens of ICIs and RTK-targeted therapies have already received approval for cancer treatment in the USA, with anti-VEGF/VEGFR therapies demonstrating the greatest clinical success to date (Table 1). As of February 2025, more than 600 clinical trials have been initiated or completed or are ongoing (based on ClinicalTrials.gov) to explore the efficacy of combining RTK-targeted therapies with immunotherapies, reflecting the high interest in such strategies [51].

Table 1 RTK-targeted therapy and immunotherapy combinations approved by the FDA as of February 2025

Indications	Immunotherapy	Targeted therapy	Year
Metastatic non-squamous NSCLC [52]	Atezolizumab	Bevacizumab	2018
aRCC (not MSI-high or dMMR) [53, 54]	Pembrolizumab	Lenvatinib	2019
aRCC [55]	Avelumab	Axitinib	2019
aRCC [56, 57]	Nivolumab	Cabozantinib	2021
aEC that is not MSI-H/dMMR, those who have disease progression following prior systemic therapy [58, 59]	Pembrolizumab	Lenvatinib	2019, 2021
Unresectable or metastatic HCC [60]	Atezolizumab	Bevacizumab	2020
Locally advanced unresectable or metastatic HER2-positive gastric or GEJ cancer [61, 62]	Pembrolizumab	Trastuzumab	2021
Cervical cancer [63]	Pembrolizumab	Bevacizumab	2021

aNSCLC advanced NSCLC, *aRCC* advanced renal cell carcinoma, *EC* endometrial carcinoma, *MSI* microsatellite instability, *dMMR* mismatch repair deficient, *HCC* hepatocellular carcinoma, *GEJ* gastroesophageal junction

5 Types of RTK-Targeted Therapies

The accessibility of RTKs on the cell surface, along with their high frequency of oncogenic mutation and activation, has positioned them as one of the most actively targeted protein families for precision oncology. Decades of RTK-targeted therapy development—spanning small-molecule inhibitors (SMIs), monoclonal antibodies (mAbs), and antibody-drug conjugates (ADCs)—have primed these strategies for integration with immunotherapy, creating new opportunities to enhance antitumor immune responses. Since the landmark FDA approvals of trastuzumab, imatinib, and gefitinib in the late 1990s and early 2000s [48, 64], 43 SMIs that target RTKs have gained FDA approval [65]. ADCs, such as T-DM1 (ado-trastuzumab emtansine) and trastuzumab deruxtecan, combine monoclonal antibodies with cytotoxic payloads to enhance tumor-specific delivery [66, 67]. Bispecific antibodies, like amivantamab (Rybrevant) approved by FDA in 2021 [68], bind two antigens or epitopes simultaneously to bridge two RTKs or RTKs with receptors on immune effector cells [69, 70]. In addition to these approved targeted therapies, multiple novel anti-RTK therapies are in development [71–76], collectively expanding the arsenal against RTK-driven cancers and combination possibilities.

Highly specific anti-RTK therapies are designed to inhibit a single RTK (e.g., HER2 mAb trastuzumab, RET inhibitor selpercatinib). On the other hand, RTKs such as VEGFR1–3, FGFR1–4, PDGFRa, RET, and c-Kit have relatively conserved ATP-binding pockets and have significant signaling cross-talk with overlapping biological functions, prompting the development of multi-kinase inhibitors (MKIs) that target them concurrently [77]. MKIs such as lenvatinib and nintedanib have demonstrated significant efficacy in combination with ICIs across various cancers (Table 1). Despite the improved efficacy observed however, the combination treatments remain non-curative. Here we summarize the tumor and TME-regulatory roles of key cancer-associated RTKs (VEGFR, EGFR, HER2, HER3, c-Met, FGFR, PDGFR, c-Kit, ALK, RET, CSF1R, and Trk, FLT3, ROS1) targeted by existing anti-RTK therapies (Table 2). We examine preclinical evidence supporting their combination with ICIs and highlight clinical studies evaluating their efficacy and the known factors that have contributed to their success and failure.

Table 2 Biological roles of RTKs and the effects of their inhibition

RTK	Baseline	Inhibition
EGFR	↑PD-1, PD-L1, and CTLA-4 in lung cancer [81–84]	↓Treg presence, ↑CD8+ T cell recruitment, ↑CXCL10 expression [85]. ↑TNF signaling [200]
HER2 and HER3	↓APP, ↓MHC I, ↑PD-L1 expression [94, 95]	↑Antitumor immunity, ↑sensitivity to PD-1 inhibition [100]. ↑CD8+ T cell activation, NK cell frequency, and APP expression [101]
c-MET	↑PD-L1 expression [110–112]. ↑Immunosuppressive macrophages, MDSC expansion, and angiogenesis [117, 119]	↑ICD [113]
ALK	↑Tregs, neutrophils, and exhausted CD8+ T cells [28, 133]. ↓MHC-I expression, ↑PD-L1 expression [134, 135]	↑ICD [113], ↑T cell infiltration, ↑MHC I expression [134, 136]
ROS1	↑PD-L1 expression [173]	
RET	RET alterations are associated with ↑immunotherapy response, ↑cytokine production, ↑immune infiltration [149, 151, 201, 202]	↑MHC-I expression in PTC cells [134]
c-Kit	↑Mast cell survival, ↑MDSC expansion, ↑Th2 polarization, ↓Th1 differentiation, ↓DC function [203–205]	↓Type I IFN production, ↓MHC class I expression [206]
CSF1R	↑Macrophage polarization [207]	↓TAMs, reprograms the TME to a pro-inflammatory state, ↑immune cell recruitment and activation [208–210]
FLT3	Required for DC development [160, 161]	
VEGFR	Signaling correlates with ↑PD-L1/PD-1 expression [165, 168]	↓PD-L1 expression on endothelial cells and macrophages ↑CD8+ T cell activation [170, 172, 173]
FGFR	↑Immunosuppression via STAT3/PD-L1 and macrophage polarization [177]	↑CD8+ T cell infiltration, ↓MDSCs [178–180]
PDGFR	↑ECM remodeling, ↑immune evasion [2, 186, 187]	↓Tumor-associated macrophage survival, ↑CD8+ T cell infiltration [189, 192, 193]

5.1 EGFR

Epidermal growth factor receptor (EGFR) regulates proliferation, differentiation, and survival in normal and malignant cells by activating numerous pathways (e.g., PI3K/AKT/mTOR, MAPK) [78]. Oncogenic amplification or mutations aberrantly sustain these pathways to drive tumor progression and therapy resistance [79]. Beyond tumor-intrinsic roles, EGFR activation has also been implicated in multiple immune evasion mechanisms [80]. In lung cancer, it can upregulate PD-1, PD-L1, and CTLA-4 [81–84], and its inhibition reduces Treg presence and enhances CD8+ T cell recruitment via IRF-1-mediated CXCL10 induction in tumor cells [85]. Consistently, combinatorially targeting PD-1 and EGFR has been shown to reduce lung tumor burden in preclinical models [85]. EGFR inhibition in some NSCLC cell lines using osimertinib also leads to increased IFN-induced transcriptional programs, suggesting EGFR actively suppresses IFN-inducible pathways [86].

In clinical trials, the combinations of EGFR inhibitors with ICIs have unfortunately been limited by hepatic toxicity and interstitial pneumonitis [87–89]. EGFR inhibitor-induced epithelial and hepatic stress primes tissues for damage, which is then exacerbated by ICI-induced inflammation. These observations highlight the challenge of combining EGFR-targeted therapies with immunotherapy. Combining other immunotherapy approaches with EGFR inhibitors has also not yielded success. For instance, in NSCLC patients treated with a DC vaccine in combination with erlotinib, the combination group did not have improved overall survival (OS) compared to the erlotinib alone group [90]. Overcoming toxicities of combination regimens will require precision dosing, next-generation agents that spare wild-type EGFR, and immune toxicity management (e.g., biomarker-driven monitoring, selective immunosuppression).

5.2 HER2 and HER3

HER2 and HER3 are ErbB RTK family members with complementary signaling roles [91]. HER2 regulates cell growth and differentiation but lacks a direct ligand [92], instead serving as a preferred dimerization partner for other ErbB receptors, particularly HER3. When dimerized with HER2, HER3 activates downstream signaling, particularly the PI3K/AKT pathway, driving cellular proliferation and survival in cancer [93]. HER2/HER3 signaling also facilitates immune evasion by downregulating APP machinery and MHC I expression, and upregulating PD-L1 expression, altogether reducing tumor recognition by CTLs [94, 95]. Conversely, disrupting HER2/HER3 signaling has been shown to enhance tumor immunogenicity [96–99]. For example, HER3-targeting ADC enhanced antitumor immunity by promoting tumor cell lysis, immune cell infiltration, and alarmin release, ultimately sensitizing tumors to PD-1 inhibition [100]. HER2 inhibitor tucatinib also enhanced CD8+ T cell activation, NK cell frequency, and APP expression while reducing immunosuppressive myeloid cells, synergizing with PD-1 inhibition in trastuzumab-resistant HER2-positive breast cancer (BC) [101]. These preclinical findings showed HER2/HER3-targeted therapy could enhance tumor immunogenicity, thus potentially making tumors more susceptible to immunotherapy.

Consistent with the preclinical studies, neoadjuvant trastuzumab plus chemotherapy increased TILs in HER2-positive BC patients [102]. Accordingly, multiple immunotherapies have been tested and in development for combination with HER2-targeted therapy in BC [103]. Thus far, clinical outcomes have been mixed. In early HER2-positive BC (EBC) setting, the chemotherapy backbone has been shown to influence the outcome. In the phase III IMpassion050 trial, adding atezolizumab to neoadjuvant HER2-targeted therapy pertuzumab-trastuzumab along with chemotherapy doxorubicin, cyclophosphamide, and paclitaxel did not increase pCR [104]. On the other hand, the ABCSG-52 phase II trial showed that adding atezolizumab to trastuzumab, pertuzumab, and epirubicin in a chemotherapy-de-escalation strategy could achieve a pathological complete response (pCR) rate in EBC [105].

In advanced BC, the efficacy of HER2-ICI combination treatment has been modest, with data suggesting that a pre-existing immune response is critical. PANACEA trial found that pembrolizumab plus trastuzumab had clinical benefit in HER2-positive, trastuzumab-resistant BC but only in PD-L1-positive patients [106]. Phase II KATE2 study also showed that adding atezolizumab to T-DM1 improves progression-free survival (PFS) and OS compared to T-DM1 alone in pretreated patients with HER2-positive, PD-L1-positive locally advanced or metastatic BC but not in those without PD-L1 positivity [107], prompting the ongoing phase III KATE3 study of this specific patient subpopulations [108]. Similarly, in HER2-positive gastric or GEJ adenocarcinoma, patients with PD-L1 expression showed improved PFS when pembrolizumab was added to the first-line trastuzumab and chemotherapy, leading to the FDA approval of this regimen [61, 62]. These studies underscore the urgent need for strategies that invigorate the TME in advanced tumors that are PD-L1 negative to improve synergy with ICIs.

5.3 c-MET

Cellular mesenchymal-epithelial transition factor (c-MET) is frequently dysregulated in tumor cells through overexpression, amplification, mutations, or autocrine signaling, promoting tumorigenesis, metastasis, and therapy resistance [109]. c-MET influences immune responses by modulating PD-L1 expression, though its effects vary across cancer types [110–112]. In glioma, c-MET-driven STAT4 activation enhances PD-L1 expression [112], while in HCC, this occurs via the MAPK/NF-κB pathway [110]. Conversely, in NSCLC, c-MET inhibition by tivantinib stabilizes PD-L1, potentially aiding immune evasion [111]. These findings highlight the complex role of c-MET in immune regulation and underscore the need for cancer-specific mechanistic studies. Beyond PD-L1 modulation, high-dose crizotinib induced immunogenic cell death (ICD) in NSCLC preclinical models regardless of canonical crizotinib target expression, enhancing T cell infiltration and sensitizing tumors to ICI when combined with cisplatin [113]. Thus, a sequential treatment strategy integrating chemotherapy, targeted therapy, and ICI could enhance antitumor immunity.

c-MET signaling can also promote immunosuppressive TME by polarizing macrophages toward pro-regenerative states [114–116], expanding MDSC by activating their STAT3 pathway [117], and enhancing angiogenesis through endothelial activation [118–124]. CAFs and TAMs further reinforce c-MET signaling by secreting the c-MET ligand HGF, driving resistance to erlotinib and sorafenib [125, 126]. The therapeutic potential for combining c-MET inhibitor with immunotherapy for uveal melanoma was also reviewed recently [127]. These studies position c-MET/HGF signaling as a key contributor of pro-tumorigenic tumor-host cross-talk and adaptive resistance mechanism, emphasizing the need to target it in both the TME and tumor cells.

Highly selective c-MET inhibitors (capmatinib, tepotinib) and MKIs (crizotinib, cabozantinib) have been combined with ICIs in various clinical trials with mixed outcomes. In the CheckMate 9ER trial, nivolumab plus cabozantinib significantly improved PFS and OS compared to sunitinib in previously untreated advanced RCC [56, 57], leading to its FDA approval. For NSCLC, however, tolerability concerns have significantly limited success [128]. For example, sequential ICI-crizotinib treatment increased hepatotoxicity risk [129], while capmatinib and pembrolizumab combination in treatment-naive advanced NSCLC with high tumor PD-L1 expression showed poor tolerability and no activity [130]. On the other hand, capmatinib and spartalizumab combination in treatment-naive NSCLC patients with *MET* exon 14 skipping mutation showed modest antitumor activity but poor tolerability [131]. Although capmatinib monotherapy had a manageable toxicity profile [132], its combination with ICIs exacerbated immune-related toxicities, particularly hepatotoxicity and pneumonitis. These toxicities may stem from heightened inflammation driven by the suppression of c-MET-mediated regenerative, tissue-protective, and immunosuppressive signaling in normal cells, rendering liver and lung tissues more vulnerable to damage. Thus, there is urgent need for optimized regimens and more precise immunotherapy targeting tumor-specific immunity.

5.4 ALK

In cancer, anaplastic lymphoma kinase (ALK) activation not only drives oncogenesis through regulating survival and proliferation but also contributes to immune evasion. For example, ALK-positive LUADs are characterized by increased infiltration of Tregs, neutrophils, and exhausted CD8+ T cells compared to ALK-negative tumors [28, 133]. ALK-positive tumors also exhibit lower MHC class I and increased PD-L1 expression, which altogether may contribute to tumor immune evasion [134, 135]. Consistent with this observation, neoadjuvant ALK inhibition in NSCLC patients led to increased MHC class I expression, antigen presentation, and immune cell infiltration within the TME, suggestive of restored tumor immunogenicity [134, 136]. Despite these studies indicating that ALK-rearranged tumors may benefit from combining ALK inhibitors with immunotherapy, in the JAVELIN Lung 101 phase Ib/II trial of ALK inhibitors (lorlatinib and crizotinib) in combination with avelumab in NSCLC, patients failed to show increased antitumor activity [137], highlighting the need for further refinement in combination strategies.

5.5 *ROS1*

ROS1 is an orphan RTK with an unclear physiological role [138]. However, ROS1 mutations occur in several cancer types [139]. MKIs targeting ROS1 (crizotinib and entrectinib) have been approved for patients with advanced ROS1 fusion-positive NSCLC [140]. ROS activation is associated with increased PD-L1 expression due to signaling through the MEK/ERK pathway, suggesting it may contribute to immune evasion [141]. Patients treated with crizotinib often develop ROS1 mutations that increase ROS1 activation, and it remains to be explored whether combination therapy approaches could abrogate these resistance mechanisms [142].

5.6 *Trk*

Tropomyosin receptor kinases (Trk), encoded by neurotrophic RTK (NTRK) genes, are normally expressed in the nervous system and non-neuronal tissues including the immune system [143–145]. The aberrant overexpression of constitutively active Trk fusion proteins has been detected in up to 1% of all solid tumors [138]. Larotrectinib and entrectinib, the SMIs targeting Trk, have been approved for tumors harboring NTRK gene fusions in 2018 and 2019, marking the first FDA-approved tumor-agnostic RTK treatment [138]. A case report has detailed a patient with Trk fusion-positive colon cancer who, after progressing on both single-agent and combination immunotherapies, was treated with larotrectinib, leading to a significant and durable response [146]. However, comprehensive clinical studies investigating the combination of these Trk inhibitors with immunotherapies are yet to be explored.

5.7 *RET*

Rearranged during transfection (*RET*) gene fusions are found in 1–2% of NSCLC and 30% of papillary thyroid carcinoma (PTC), while *RET* point mutations are associated with medullary thyroid cancer and multiple endocrine neoplasia type 2 (MEN2) [134, 147–149]. Aberrant RET signaling activates Ras/MAPK, PI3K/

AKT, and PLCg pathways, promoting cell survival, differentiation, and migration of tumor cells [147]. Highly selective inhibitors (selpercatinib, pralsetinib) as well as MKIs (cabozantinib, lenvatinib, vandetanib) against RET have gained FDA approval [134, 150]. Interestingly, pralsetinib was shown to increase MHC-I expression in a PTC cell line [134], suggesting RET inhibitors may increase tumor immunogenicity. RET activation has also been linked to increased production of pro-inflammatory cytokines and immune infiltration [151]. In the clinic, however, the combination of targeted therapy and ICIs for RET-driven cancers has been discouraged. In thyroid cancer with *RET* mutations, treatment has primarily focused on targeting driver mutations with highly selective KIs, given the low TMB, low PD-L1, and limited responses to immunotherapy [149, 152, 153]. In NSCLC also, patients harboring *RET* rearrangements have been shown to derive limited clinical benefit from ICIs compared to chemotherapy, supporting the preferential use of selective RET-targeted therapies or chemotherapy until reliable biomarkers predicting immunotherapy responsiveness are identified [154].

5.8 FLT3

Mutations in FMS-like tyrosine kinase 3 (FLT3) which lead to its constitutive activation are found in approximately 30% of AML cases [155]. Because many FLT3-mutated AML cells exhibit oncogenic addiction to this pathway, three SMIs (midostaurin, gilteritinib, and quizartinib) have received FDA approval for the treatment of FLT3-positive AML [156, 157]. Since FLT3 is expressed on immune cells such as DCs and hematopoietic stem cells, and it is required for DC development in peripheral lymphoid tissue, its inhibition stands to have immune-altering effects beyond the AML cells [158–161].

5.9 VEGFR, FGFR, and PDGFR

Tumor angiogenesis is rapid and disorganized, leading to aberrant vasculature that fosters hypoxia, suppresses immunity, and impairs drug delivery [51, 162–165]. RTKs such as vascular endothelial growth factor (VEGF) receptor, fibroblast growth factor receptor (FGFR), and platelet-derived growth factor receptor (PDGFR) have been shown to regulate endothelial proliferation, pericyte recruitment, and vessel maturation, contributing to the neovascularization in cancer [166]. Thus, these RTKs have been extensively targeted—primarily using MKIs—to normalize vasculature, boost immune infiltration, and enhance treatment efficacy [163, 164, 166–169]. Here, we summarize known immunomodulatory roles of VEGFR, FGFR, and PDGFR signaling and the clinical trials (Sect. 5.9.4) that tested the combination of immunotherapy and MKIs against these RTKs.

5.9.1 VEGFR

VEGFR is broadly expressed across various tumor types and immune cells (e.g., macrophages, fibroblasts, DCs, and T cells) in addition to endothelial cells [170]. In response to hypoxia and pro-inflammatory signals, VEGF can also be secreted by tumor cells, immune cells (e.g., macrophages, neutrophils, T cells), and CAFs [171]. Thus, anti-VEGFR and anti-VEGF therapy simultaneously affect multiple cell types. Multiple studies have shown positive correlation or causal relationship between the activation of VEGF/VEGFR signaling and PD-L1/PD-1 expression, pointing to their concerted role in immunosuppression. VEGF and PD-L1/PD-L2 expression, along with the microvessel density, significantly correlate in patients with Hodgkin lymphoma [172], while the presence of PD-L1-positive tumor-associated vascular endothelial cells strongly correlates with the infiltration of immunosuppressive Treg in kidney and colon cancer [173]. The use of VEGFR inhibitors anlotinib and axitinib downregulates PD-L1 on vascular endothelial cells in vitro [173] and baicalin-polarized M2 macrophages in vitro [174], respectively. The direct immunomodulatory effects of anti-VEGF treatment include the induction of peripheral DC population in patients and increased allo-stimulatory capacity and proliferation of DC in vitro [175].

5.9.2 FGFR

Aberrant FGFR signaling drives tumor growth, angiogenesis, and therapy resistance by promoting uncontrolled proliferation, survival, and migration [176]. In cancer, FGFR is expressed in epithelial, mesenchymal, and endothelial cells, where its activation in both tumor and stromal cells modulates immune responses and shapes antitumor immunity from multiple angles [177]. FGFR inhibition with an SMI, FIIN4, enhanced antitumor immunity in metastatic BC by increasing CD8+ T cell infiltration and reducing MDSCs, delaying pulmonary metastasis growth [178]. While FIIN4 initially impaired ICI efficacy by inhibiting T cell receptor signaling, a sequential dosing strategy restored immune-mediated tumor killing, highlighting FGFR and ICI combinatorial targeting as a potential approach to improve ICI response [178]. FGFR signaling also drives tumor-intrinsic immunosuppression. FGFR can signal through the STAT3 pathway which increases PD-L1 expression, and this signaling pathway is upregulated upon EGFR TKI treatment, pointing to the role of FGFR in TKI resistance [179, 180]. FGFR signaling also contributes to an immunosuppressive TME where FGFR/FGF2 signaling in macrophages polarized them to be immunosuppressive, and FGF2 depletion led to prolonged survival of mice and an increase in $iNOS^{+}/CD206^{+}$ TAM ratio [181]. Consistent with these studies, FGFR inhibition with erdafitinib led to increased T cell infiltration and prolonged survival in combination with anti-PD1 therapy in TNBC preclinical mouse models [182]. The equipotent FGFR/VEGFR inhibitor ODM-203 also suppresses tumor growth, angiogenesis, and metastasis in various FGFR-dependent cancer cells in part by downregulating PD-1/PD-L1 expression and enhancing

CD8+ T cell activation [183]. The combination of anti-PD-1 and lenvatinib, a VEGFR/FGFR inhibitor, demonstrated superior antitumor activity and prolonged survival in RCC preclinical models compared to anti-PD-1 plus axitinib, a VEGFR TKI [184]. These were associated with reduced TAMs, increased IFNγ-positive CD8+ T cells, and restored IFNγ-stimulated JAK/STAT signaling, demonstrating the therapeutic benefits of combination therapy [184].

5.9.3 PDGFR

PDGFR, predominantly expressed in mesenchymal cells (e.g., fibroblasts, pericytes), plays an important role in cellular processes like proliferation, migration, and differentiation by binding to PDGF [185]. PDGF can be secreted by macrophages and tumor cells, which in turn modulate PDGFR-mediated stromal and immune cell functions in the TME [186, 187]. For instance, PDGFR activation in fibroblasts remodels the ECM, increasing tissue stiffness and promoting tumor growth [188], while in CAFs, it facilitates the recruitment of immunosuppressive polymorphonuclear MDSCs [189]. PDGF/PDGFR signaling in DCs inhibits their maturation and increases the expression of CLEC-2 on DCs, which helps the differentiation of naive T cells into Tregs [190]. In endothelial cells, PDGF/PDGFR signaling elevates interstitial pressure, impairing drug uptake, while its inhibition reserves this effect [191]. Consistent with these observations, preclinical evidence shows PDGFR inhibition reduces TME immunosuppression. PDGF-C knockdown led to the increased apoptosis of TAMs, indicating PDGF may promote their survival in TNBC [192]. Using an aptamer against PDGFRβ in combination with anti-PD-L1 also led to tumor growth inhibition, increased intra-tumoral CD8+ T cells, and fewer Treg cells in breast cancer mouse models [193]. Importantly, stromal reprogramming via PDGFRα/β blockade reversed immune suppression, enhanced lymphocyte infiltration, and synergized with anti-PD-1 therapy to improve immunotherapy efficacy in a fibrotic gastric cancer model [189]. These findings point to the broad immunosuppressive effects of PDGFR signaling in the TME, with preclinical evidence suggesting its inhibition may synergize with immunotherapy.

5.9.4 Clinical Trials

Consistent with the pro-tumorigenic, immunosuppressive roles of VEGFR, FGFR, PDGFR, and RET in the tumor and TME, significant clinical benefits of combining MKIs targeting these RTKs with ICIs have led to combination regimens receiving the FDA approval (Table 1). These results are consistent with preclinical studies where anti-angiogenic MKI nintedanib showed vascular stabilization, increased MHC I presentation, and tumor cell apoptosis [194]. Beyond these combinations, multiple ongoing trials have shown promise. For example, combining pembrolizumab with bevacizumab significantly improved the response rate in patients with platinum-resistant recurrent or metastatic nasopharyngeal carcinoma compared to pembrolizumab alone, warranting further validation in phase III trials [195]. In a phase Ib trial of PDGFR/VEGFR inhibitor vorolanib, its combination with pembrolizumab or nivolumab showed manageable toxicities and promising efficacy in patients with advanced solid tumors [196]. These results show clear benefits of combining MKIs with immunotherapy over existing therapies.

Despite these successes, the treatments remain non-curative. One of the resistance mechanisms proposed is endothelial cell anergy, exemplified by their reduced expression of endothelial adhesion molecules [28]. In addition, tumors refractory to bevacizumab have been shown to recruit CD11b + Gr + myeloid cells, highlighting the need for combinatorial strategies where the innate immune cell recruitment or activity needs to be targeted [197]. In contrast, exceptional response to VEGFR and ICI combinations in advanced RCCs has been attributed to clonal neoantigen-driven T cell responses or B cell-mediated immunity with TLS [198]. Critically, targeting TIGIT, an IC expressed on T cells and NK cells, with tiragolumab demonstrated tolerability and promising antitumor activity in combination with atezolizumab and bevacizumab in recent early-phase trial in advanced HCC [199], reinforcing the existence of additional ICs and immunosuppressive mechanisms that need to be targeted for patients who are resistant to existing ICIs.

5.10 c-Kit

c-Kit mutations drive tumor growth in multiple cancers, most prominently in gastrointestinal stromal tumors (GIST; ~80–90%) and less frequently in melanoma (~3%) [211, 212]. Targeted inhibitors, including the highly selective avapritinib and several MKIs (e.g., imatinib, ripretinib, regorafenib, tivozanib, nilotinib, sunitinib), effectively suppress c-Kit-driven malignancies but can also impair immune function by disrupting hematopoiesis, antigen presentation, and T cell responses [212–217], potentially limiting cancer immunosurveillance and cancer immunotherapy efficacy. These immunomodulatory effects, linked to the role of c-Kit in mast cells, DCs, and T cells, underscore the complexity of c-Kit inhibition in shaping the TME [205, 218–220]. While the immunoregulatory impact of avapritinib remains unclear, tivozanib-mediated survival benefits in HCC have been associated with reduced immune suppression by Tregs and MDSCs [221], suggesting the potential benefit of targeting c-Kit in the immunosuppressive immune cells. In GIST, imatinib suppresses type I IFN signaling and MHC class I expression [206] but also enhances CD8+ T cell activity and reduces Tregs via IDO inhibition [222], and synergizes with PD-1/PD-L1 blockade [223], together providing a mechanistic rationale for combining c-Kit inhibitors with ICIs. Yet, clinical trials evaluating the combination of c-Kit inhibition and immunotherapies, especially ICIs, have yielded mixed results in GIST, underscoring the need for improved biomarkers and patient stratification [212, 224]. The mixed results have been partially attributed to relatively low PD-L1/PD-1 expression in tumors and possible immunosuppressive effects of long-term c-Kit inhibition in advanced GIST [212]. The ongoing area of interest is the combination of c-Kit inhibitors with immunotherapy for *KIT*-mutated advanced melanoma [224–226]. Taken together, c-Kit inhibition results in both pro- and anti-tumor immune responses and likely necessitates careful regimen design with consideration for the addition of immune-stimulatory agents (e.g., STING agonists, cytokine therapy) to counteract the immunosuppressive effects of c-Kit inhibition [206, 227].

5.11 CSF1R

Unlike other oncogenic RTKs discussed, colony-stimulating factor 1 receptor (CSF1R) has been primarily targeted as an anti-myeloid therapy due to its predominant function in tumor-associated monocytes and TAMs [207], which support angiogenesis, suppress cytotoxic immune responses, and enhance metastatic potential in cancer [228, 229]. Anti-CSF1R therapies in preclinical models deplete TAMs through apoptosis and simultaneously reprogram the TME to a pro-inflammatory state [208–210]. This shift fosters enhanced recruitment and activation of effector immune cells, reinvigorating antitumor immunity. Despite the extensive clinical trials testing these agents in cancers that have high TAM infiltration [207, 230, 231], CSF1R therapies have failed to show survival benefit in patients. To date, pexidartinib, which targets CSF1R, c-Kit, and FLT3, remains the only FDA-approved therapy for benign tenosynovial giant cell tumor characterized by aberrant autocrine CSF1/CSF1R signaling caused by CSF1 gene translocation [232]. Although pexidartinib in combination with ICI in advanced solid tumors showed manageable toxicities, it failed to show survival benefits, likely due to pexidartinib also inhibiting FLT3 signaling, which is critical for DC differentiation [233]. A phase II trial of cabiralizumab (anti-CSF1R antibody) in combination with nivolumab in advanced pancreatic cancer also failed to meet its endpoint [234]. In a phase I study, patients treated with emactuzumab (anti-CSF1R antibody) had fewer M2 macrophages in the TME. However, this did not increase T cell infiltration nor lead to the reduction of tumor, indicating that other resistance mechanisms likely exist and need to be targeted to boost antitumor immune responses [235]. Further studies are needed to evaluate the efficacy of anti-CSF1R therapy in combination with ICI.

6 Challenges and Future Directions

Despite significant improvements in patient survival with combination therapies, they are generally not curative. The mechanisms underlying their success and failure in altering patient survival remain poorly understood, partly due to the lack of comprehensive longitudinal analyses. Thus, a systematic biomarker analysis and detailed evaluation of both effective and ineffective combinations will be essential for refining therapeutic approaches and improving patient stratification. Additionally, the precise interplay between combined modalities and the contribution of key variables such as race, treatment schedules, circadian rhythm, dosing, and sex to combination therapy efficacy is undefined [236–243]. Mechanistic understanding of KIs will also further help with effective treatment strategy [152]. Furthermore, identifying and targeting new adaptive and innate ICs, especially upregulated in response to existing ICIs, will expand the combination treatment options. Innovative approaches to identify optimal combination therapy strategies are also needed.

Kinase-driven signaling networks are highly dynamic and context-dependent, and cannot be inferred from genomic, transcriptomic, or whole proteomic data. Improvements in mass spectrometry-driven phosphoproteomics technologies [244–247] have enabled systems-level mapping of phosphorylation events, providing deeper insights into intracellular signaling networks and their regulation in disease [248–250]. Systems-level phosphoproteomics capable of analyzing specific cell types or targeted subpopulations in patients and preclinical animal models will advance our understanding of the on- and off-target effects of kinase perturbations and the efficacy of next-generation combination therapies.

7 Conclusion

The combination of anti-RTK therapies with immunotherapy represents a synergistic and complementary approach to overcoming the limitations of monotherapies in cancer treatment. We have outlined various combination strategies that have been tested and examined their rationale, clinical trial outcomes, mechanisms underlying their success and failures where known, as well as potential opportunities for optimization for improved clinical benefit. Anti-RTK therapies, in addition to disrupting tumor-intrinsic pro-survival and proliferation pathways, mitigate systemic immunosuppression and enhance tumor cell immunogenicity by restoring cytokine balance, enhancing antigen presentation, stabilizing tumor-associated vasculature, and reducing structural barriers within the TME. Collectively, these effects facilitate immune cell infiltration, antitumor immune responses, and therapeutic delivery, ultimately creating a supportive milieu for tumor eradication by immunotherapy. The unique capability of immunotherapy to induce durable and adaptable responses may complement the transient but targeted effects of anti-RTK therapies. Despite substantial progress in integrating these therapeutic modalities, resistance and toxicities remain as hurdles. Improved understanding of the factors that affect success and failure of combination strategies and the development of novel anti-RTK therapies and immunotherapy modalities will guide future combination strategies.

Acknowledgments We are grateful to Dr. Forest White for the critical reading of the chapter. The figures were created using BioRender.com.

References

1. Pardoll DM. The blockade of immune checkpoints in cancer immunotherapy. Nat Rev Cancer. 2012;12(4):252–64.
2. Zou W, Wolchok JD, Chen L. PD-L1 (B7-H1) and PD-1 pathway blockade for cancer therapy: mechanisms, response biomarkers and combinations. Sci Transl Med. 2016;8(328):328rv4.
3. Waldman AD, Fritz JM, Lenardo MJ. A guide to cancer immunotherapy: from T cell basic science to clinical practice. Nat Rev Immunol. 2020;20(11):651–68.
4. Anderson AC, Joller N, Kuchroo VK. Lag-3, Tim-3, and TIGIT co-inhibitory receptors with specialized functions in immune regulation. Immunity. 2016;44(5):989–1004.
5. Topalian SL, Drake CG, Pardoll DM. Immune checkpoint blockade: A common denominator approach to cancer therapy. Cancer Cell. 2015;27(4):450–61.
6. Gougis P, Jochum F, Abbar B, Dumas E, Bihan K, Lebrun-Vignes B, et al. Clinical spectrum and evolution of immune-checkpoint inhibitors toxicities over a decade—a worldwide perspective. eClinicalMedicine [Internet]. 2024 Apr 1 [cited 2025 Jan 27];70. Available from: https://www.thelancet.com/journals/eclinm/article/PIIS2589-5370(24)00115-9/fulltext
7. Sharma P, Hu-Lieskovan S, Wargo JA, Ribas A. Primary, adaptive, and acquired resistance to cancer immunotherapy. Cell. 2017;168(4):707–23.
8. Piper M, Kluger H, Ruppin E, Hu-Lieskovan S. Immune resistance mechanisms and the road to personalized immunotherapy. Am Soc Clin Oncol Educ Book. 2023;43:e390290.
9. Hwang EI, Sayour EJ, Flores CT, Grant G, Wechsler-Reya R, Hoang-Minh LB, et al. The current landscape of immunotherapy for pediatric brain tumors. Nat Cancer. 2022;3(1):11–24.
10. Braun DA, Bakouny Z, Hirsch L, Flippot R, Van Allen EM, Wu CJ, et al. Beyond conventional immune-checkpoint inhibition—novel immunotherapies for renal cell carcinoma. Nat Rev Clin Oncol. 2021;18(4):199–214.
11. Wang Y, Safi M, Hirsch FR, Lu S, Peters S, Govindan R, et al. Immunotherapy for advanced-stage squamous cell lung cancer: the state of the art and outstanding questions. Nat Rev Clin Oncol. 2025;6:1–15.
12. Huang AC, Zappasodi R. A decade of checkpoint blockade immunotherapy in melanoma: understanding the molecular basis for immune sensitivity and resistance. Nat Immunol. 2022;23(5):660–70.
13. Cai L, Li Y, Tan J, Xu L, Li Y. Targeting LAG-3, TIM-3, and TIGIT for cancer immunotherapy. J Hematol Oncol. 2023;16:101.
14. Dhatchinamoorthy K, Colbert JD, Rock KL. Cancer immune evasion through loss of MHC Class I antigen presentation. Front Immunol [Internet]. 2021 Mar 9 [cited 2024 Dec 5];12. Available from: https://www.frontiersin.org/journals/immunology/articles/10.3389/fimmu.2021.636568/full
15. Hazini A, Fisher K, Seymour L. Deregulation of HLA-I in cancer and its central importance for immunotherapy. J Immunother Cancer. 2021;9(8):e002899.
16. Propper DJ, Balkwill FR. Harnessing cytokines and chemokines for cancer therapy. Nat Rev Clin Oncol. 2022;19(4):237–53.
17. Vitale I, Shema E, Loi S, Galluzzi L. Intratumoral heterogeneity in cancer progression and response to immunotherapy. Nat Med. 2021;27(2):212–24.
18. Messmer MN, Snyder AG, Oberst A. Comparing the effects of different cell death programs in tumor progression and immunotherapy. Cell Death Differ. 2019;26(1):115–29.
19. Wang N, Ma T, Yu B. Targeting epigenetic regulators to overcome drug resistance in cancers. Sig Transduct Target Ther. 2023;8(1):1–24.
20. Karin M, Shalapour S. Regulation of antitumor immunity by inflammation-induced epigenetic alterations. Cell Mol Immunol. 2022;19(1):59–66.
21. Laisné M, Lupien M, Vallot C. Epigenomic heterogeneity as a source of tumour evolution. Nat Rev Cancer. 2025;25(1):7–26.

22. Roulois D, Loo Yau H, Singhania R, Wang Y, Danesh A, Shen SY, et al. DNA-Demethylating agents target colorectal cancer cells by inducing viral mimicry by endogenous transcripts. Cell. 2015;162(5):961–73.
23. Kureshi CT, Dougan SK. Cytokines in cancer. Cancer Cell. 2025;43(1):15–35.
24. Habashy KJ, Mansour R, Moussalem C, Sawaya R, Massaad MJ. Challenges in glioblastoma immunotherapy: mechanisms of resistance and therapeutic approaches to overcome them. Br J Cancer. 2022;127(6):976–87.
25. Reardon DA, Brandes AA, Omuro A, Mulholland P, Lim M, Wick A, et al. Effect of Nivolumab vs Bevacizumab in patients with recurrent glioblastoma: the CheckMate 143 phase 3 randomized clinical trial. JAMA Oncol. 2020;6(7):1003–10.
26. Wu B, Zhang B, Li B, Wu H, Jiang M. Cold and hot tumors: from molecular mechanisms to targeted therapy. Sig Transduct Target Ther. 2024;9(1):1–65.
27. Schmied L, Höglund P, Meinke S. Platelet-mediated protection of cancer cells from immune surveillance—possible implications for cancer immunotherapy. Front Immunol. 2021;12:640578.
28. Huinen ZR, Huijbers EJM, van Beijnum JR, Nowak-Sliwinska P, Griffioen AW. Anti-angiogenic agents—overcoming tumour endothelial cell anergy and improving immunotherapy outcomes. Nat Rev Clin Oncol. 2021;18(8):527–40.
29. Wagner M, Jasek M, Karabon L. Immune checkpoint molecules—inherited variations as markers for cancer risk. Front Immunol. 2021;11:606721.
30. Lenehan JG, Maleki Vareki S. An adjunct microbiome therapy to enhance immunotherapy efficacy in melanoma patients. Genes Immun. 2024;25(1):87–8.
31. Sullivan RJ, Weber JS. Immune-related toxicities of checkpoint inhibitors: mechanisms and mitigation strategies. Nat Rev Drug Discov. 2022;21(7):495–508.
32. Ramos-Casals M, Brahmer JR, Callahan MK, Flores-Chávez A, Keegan N, Khamashta MA, et al. Immune-related adverse events of checkpoint inhibitors. Nat Rev Dis Primers. 2020;6(1):1–21.
33. Morris EC, Neelapu SS, Giavridis T, Sadelain M. Cytokine release syndrome and associated neurotoxicity in cancer immunotherapy. Nat Rev Immunol. 2022;22(2):85–96.
34. Yamaguchi H, Hsu JM, Sun L, Wang SC, Hung MC. Advances and prospects of biomarkers for immune checkpoint inhibitors. CR Med [Internet]. 2024 Jul 16 [cited 2025 Feb 11];5(7). Available from: https://www.cell.com/cell-reports-medicine/abstract/S2666-3791(24)00325-2
35. Holder AM, Dedeilia A, Sierra-Davidson K, Cohen S, Liu D, Parikh A, et al. Defining clinically useful biomarkers of immune checkpoint inhibitors in solid tumours. Nat Rev Cancer. 2024;24(7):498–512.
36. Goodman RS, Jung S, Balko JM, Johnson DB. Biomarkers of immune checkpoint inhibitor response and toxicity: challenges and opportunities. Immunol Rev. 2023;318(1):157–66.
37. Yarchoan M, Hopkins A, Jaffee EM. Tumor mutational burden and response rate to PD-1 inhibition. N Engl J Med. 2017;377(25):2500–1.
38. Westcott PMK, Muyas F, Hauck H, Smith OC, Sacks NJ, Ely ZA, et al. Mismatch repair deficiency is not sufficient to elicit tumor immunogenicity. Nat Genet. 2023;55(10):1686–95.
39. Wang M, Zhai X, Li J, Guan J, Xu S, Li Y, et al. The role of cytokines in predicting the response and adverse events related to immune checkpoint inhibitors. Front Immunol. 2021;12:670391.
40. Wang P, Chen Y, Wang C. Beyond tumor mutation burden: tumor Neoantigen burden as a biomarker for immunotherapy and other types of therapy. Front Oncol. 2021;11:672677.
41. Bevins NJ, Okamura R, Montesion M, Adashek JJ, Goodman AM, Kurzrock R. Tumor infiltrating lymphocyte expression of PD-1 predicts response to anti-PD-1/PD-L1 immunotherapy. J Immunother Precis Oncol. 2022;5(4):90–7.
42. Presti D, Dall'Olio FG, Besse B, Ribeiro JM, Di Meglio A, Soldato D. Tumor infiltrating lymphocytes (TILs) as a predictive biomarker of response to checkpoint blockers in solid tumors: A systematic review. Crit Rev Oncol Hematol. 2022;177:103773.

43. Doroshow DB, Bhalla S, Beasley MB, Sholl LM, Kerr KM, Gnjatic S, et al. PD-L1 as a biomarker of response to immune-checkpoint inhibitors. Nat Rev Clin Oncol. 2021;18(6):345–62.
44. Gandhi L, Rodríguez-Abreu D, Gadgeel S, Esteban E, Felip E, De Angelis F, et al. Pembrolizumab plus chemotherapy in metastatic non-Small-cell lung cancer. N Engl J Med. 2018;378(22):2078–92.
45. Khan M, Du K, Ai M, Wang B, Lin J, Ren A, et al. PD-L1 expression as biomarker of efficacy of PD-1/PD-L1 checkpoint inhibitors in metastatic triple negative breast cancer: A systematic review and meta-analysis. Front Immunol. 2023;14:1060308.
46. Shah MA, Kojima T, Hochhauser D, Enzinger P, Raimbourg J, Hollebecque A, et al. Efficacy and safety of Pembrolizumab for heavily pretreated patients with advanced, metastatic adenocarcinoma or squamous cell carcinoma of the esophagus: the phase 2 KEYNOTE-180 study. JAMA Oncol. 2019;5(4):546–50.
47. Petroni G, Buqué A, Coussens LM, Galluzzi L. Targeting oncogene and non-oncogene addiction to inflame the tumour microenvironment. Nat Rev Drug Discov. 2022;21(6):440–62.
48. Attwood MM, Fabbro D, Sokolov AV, Knapp S, Schiöth HB. Trends in kinase drug discovery: targets, indications and inhibitor design. Nat Rev Drug Discov. 2021;20(11):839–61.
49. Ferguson FM, Gray NS. Kinase inhibitors: the road ahead. Nat Rev Drug Discov. 2018;17(5):353–77.
50. Tomuleasa C, Tigu AB, Munteanu R, Moldovan CS, Kegyes D, Onaciu A, et al. Therapeutic advances of targeting receptor tyrosine kinases in cancer. Sig Transduct Target Ther. 2024;9(1):1–51.
51. Petrazzuolo A, Maiuri MC, Zitvogel L, Kroemer G, Kepp O. Trial watch: combination of tyrosine kinase inhibitors (TKIs) and immunotherapy. Onco Targets Ther. 2022;11(1):2077898.
52. Socinski MA, Nishio M, Jotte RM, Cappuzzo F, Orlandi F, Stroyakovskiy D, et al. IMpower150 final overall survival analyses for Atezolizumab plus bevacizumab and chemotherapy in first-line metastatic nonsquamous NSCLC. J Thorac Oncol. 2021;16(11):1909–24.
53. Arora S, Balasubramaniam S, Zhang W, Zhang L, Sridhara R, Spillman D, et al. FDA approval summary: Pembrolizumab plus Lenvatinib for endometrial carcinoma, a collaborative international review under project Orbis. Clin Cancer Res. 2020;26(19):5062–7.
54. Motzer R, Alekseev B, Rha SY, Porta C, Eto M, Powles T, et al. Lenvatinib plus Pembrolizumab or Everolimus for advanced renal cell carcinoma. N Engl J Med. 2021;384(14):1289–300.
55. Motzer RJ, Penkov K, Haanen J, Rini B, Albiges L, Campbell MT, et al. Avelumab plus Axitinib versus Sunitinib for advanced renal-cell carcinoma. N Engl J Med. 2019;380(12):1103–15.
56. Choueiri TK, Powles T, Burotto M, Escudier B, Bourlon MT, Zurawski B, et al. Nivolumab plus Cabozantinib versus Sunitinib for advanced renal-cell carcinoma. N Engl J Med. 2021;384(9):829–41.
57. Powles T, Burotto M, Escudier B, Apolo AB, Bourlon MT, Shah AY, et al. Nivolumab plus cabozantinib versus sunitinib for first-line treatment of advanced renal cell carcinoma: extended follow-up from the phase III randomised CheckMate 9ER trial. ESMO Open. 2024;9(5):102994.
58. FDA grants regular approval to pembrolizumab and lenvatinib for advanced endometrial carcinoma. FDA [Internet]. 2024 Aug 9 [cited 2025 Mar 4]; Available from: https://www.fda.gov/drugs/resources-information-approved-drugs/fda-grants-regular-approval-pembrolizumab-and-lenvatinib-advanced-endometrial-carcinoma
59. Makker V, Colombo N, Herráez AC, Santin AD, Colomba E, Miller DS, et al. Lenvatinib plus Pembrolizumab for advanced endometrial cancer. N Engl J Med. 2022;386(5):437–48.
60. Casak SJ, Donoghue M, Fashoyin-Aje L, Jiang X, Rodriguez L, Shen YL, et al. FDA approval summary: Atezolizumab plus bevacizumab for the treatment of patients with advanced unresectable or metastatic hepatocellular carcinoma. Clin Cancer Res. 2021;27(7):1836–41.
61. Janjigian YY, Kawazoe A, Bai Y, Xu J, Lonardi S, Metges JP, et al. Pembrolizumab plus trastuzumab and chemotherapy for HER2-positive gastric or gastro-oesophageal junction adenocarcinoma: interim analyses from the phase 3 KEYNOTE-811 randomised placebo-controlled trial. Lancet. 2023;402(10418):2197–208.

62. Janjigian YY, Kawazoe A, Yañez P, Li N, Lonardi S, Kolesnik O, et al. The KEYNOTE-811 trial of dual PD-1 and HER2 blockade in HER2-positive gastric cancer. Nature. 2021;600(7890):727–30.
63. Colombo N, Dubot C, Lorusso D, Caceres MV, Hasegawa K, Shapira-Frommer R, et al. Pembrolizumab for persistent, recurrent, or metastatic cervical cancer. N Engl J Med. 2021;385(20):1856–67.
64. Druker BJ. Imatinib and the dawn of precision cancer therapy. Nat Med. 2025;23:1–2.
65. Roskoski R. Properties of FDA-approved small molecule protein kinase inhibitors: a 2024 update. Pharmacol Res. 2024;200:107059.
66. von Minckwitz G, Huang CS, Mano MS, Loibl S, Mamounas EP, Untch M, et al. Trastuzumab Emtansine for residual invasive HER2-positive breast cancer. N Engl J Med. 2019;380(7):617–28.
67. Narayan P, Osgood CL, Singh H, Chiu HJ, Ricks TK, Chow ECY, et al. FDA approval summary: fam-trastuzumab deruxtecan-nxki for the treatment of unresectable or metastatic HER2-positive breast cancer. Clin Cancer Res. 2021;27(16):4478–85.
68. Chon K, Larkins E, Chatterjee S, Mishra-Kalyani PS, Aungst S, Wearne E, et al. FDA approval summary: Amivantamab for the treatment of patients with non-small cell lung cancer with EGFR exon 20 insertion mutations. Clin Cancer Res. 2023;29(17):3262–6.
69. Klein C, Brinkmann U, Reichert JM, Kontermann RE. The present and future of bispecific antibodies for cancer therapy. Nat Rev Drug Discov. 2024;23(4):301–19.
70. Goebeler ME, Stuhler G, Bargou R. Bispecific and multispecific antibodies in oncology: opportunities and challenges. Nat Rev Clin Oncol. 2024;21(7):539–60.
71. Bennasroune A, Fickova M, Gardin A, Dirrig-Grosch S, Aunis D, Crémel G, et al. Transmembrane peptides as inhibitors of ErbB receptor signaling. Mol Biol Cell. 2004;15(7):3464–74.
72. Rosenbaum MI, Clemmensen LS, Bredt DS, Bettler B, Strømgaard K. Targeting receptor complexes: a new dimension in drug discovery. Nat Rev Drug Discov. 2020;19(12):884–901.
73. Cordani N, Nova D, Sala L, Abbate MI, Colonese F, Cortinovis DL, et al. Proteolysis targeting chimera agents (PROTACs): new Hope for overcoming the resistance mechanisms in oncogene-addicted non-Small cell lung cancer. Int J Mol Sci. 2024;25(20):11214.
74. Zhou J, Rossi J. Aptamers as targeted therapeutics: current potential and challenges. Nat Rev Drug Discov. 2017;16(3):181–202.
75. Kim KH, Kim H. Progress of antibody-based inhibitors of the HGF–cMET axis in cancer therapy. Exp Mol Med. 2017;49(3):e307–7.
76. Yu F, Cai M, Shao L, Zhang J. Targeting Protein Kinases Degradation by PROTACs. Front Chem [Internet]. 2021 Jun 30 [cited 2025 Jan 23];9. Available from: https://www.frontiersin.org/journals/chemistry/articles/10.3389/fchem.2021.679120/full
77. Chae YK, Ranganath K, Hammerman PS, Vaklavas C, Mohindra N, Kalyan A, et al. Inhibition of the fibroblast growth factor receptor (FGFR) pathway: the current landscape and barriers to clinical application. Oncotarget. 2016;8(9):16052–74.
78. Yarden Y. The EGFR family and its ligands in human cancer: signalling mechanisms and therapeutic opportunities. Eur J Cancer. 2001;37:3–8.
79. Liu H, Zhang B, Sun Z. Spectrum of EGFR aberrations and potential clinical implications: insights from integrative pan-cancer analysis. Cancer Commun (Lond). 2020;40(1):43–59.
80. Kumagai S, Koyama S, Nishikawa H. Antitumour immunity regulated by aberrant ERBB family signalling. Nat Rev Cancer. 2021;21(3):181–97.
81. van Weverwijk A, de Visser KE. Mechanisms driving the immunoregulatory function of cancer cells. Nat Rev Cancer. 2023;23(4):193–215.
82. Akbay EA, Koyama S, Carretero J, Altabef A, Tchaicha JH, Christensen CL, et al. Activation of the PD-1 pathway contributes to immune escape in EGFR-driven lung tumors. Cancer Discov. 2013;3(12):1355–63.
83. Chen N, Fang W, Zhan J, Hong S, Tang Y, Kang S, et al. Upregulation of PD-L1 by EGFR activation mediates the immune escape in EGFR-driven NSCLC: implication for optional

immune targeted therapy for NSCLC patients with EGFR mutation. J Thorac Oncol. 2015;10(6):910–23.
84. Tang Y, Fang W, Zhang Y, Hong S, Kang S, Yan Y, et al. The association between PD-L1 and EGFR status and the prognostic value of PD-L1 in advanced non-small cell lung cancer patients treated with EGFR-TKIs. Oncotarget. 2015;6(16):14209–19.
85. Sugiyama E, Togashi Y, Takeuchi Y, Shinya S, Tada Y, Kataoka K, et al. Blockade of EGFR improves responsiveness to PD-1 blockade in EGFR-mutated non–small cell lung cancer. Science Immunology. 2020;5(43):eaav3937.
86. Gurule NJ, McCoach CE, Hinz TK, Merrick DT, Van Bokhoven A, Kim J, et al. A tyrosine kinase inhibitor-induced interferon response positively associates with clinical response in EGFR-mutant lung cancer. npj Precis Onc. 2021;5(1):1–11.
87. Rizvi NA, Chow LQM, Borghaei H, Shen Y, Harbison C, Alaparthy S, et al. Safety and response with nivolumab (anti-PD-1; BMS-936558, ONO-4538) plus erlotinib in patients (pts) with epidermal growth factor receptor mutant (EGFR MT) advanced NSCLC. J Clin Oncol [Internet] 2014 May 20 [cited 2024 Dec 5]; Available from: https://ascopubs.org/doi/10.1200/jco.2014.32.15_suppl.8022
88. Oshima Y, Tanimoto T, Yuji K, Tojo A. EGFR–TKI-associated interstitial pneumonitis in Nivolumab-treated patients with non–small cell lung cancer. JAMA Oncol. 2018;4(8):1112–5.
89. Spigel DR, Reynolds C, Waterhouse D, Garon EB, Chandler J, Babu S, et al. Phase 1/2 study of the safety and tolerability of nivolumab plus crizotinib for the first-line treatment of anaplastic lymphoma kinase translocation—positive advanced non–small cell lung cancer (CheckMate 370). J Thorac Oncol. 2018;13(5):682–8.
90. Shi SB, Tang XY, Tian J, Chang CX, Li P, Qi JL. Efficacy of erlotinib plus dendritic cells and cytokine-induced killer cells in maintenance therapy of advanced non–small cell lung cancer. J Immunother. 2014;37(4):250.
91. Gandullo-Sánchez L, Ocaña A, Pandiella A. HER3 in cancer: from the bench to the bedside. J Exp Clin Cancer Res. 2022;41(1):310.
92. Mishra R, Patel H, Alanazi S, Yuan L, Garrett JT. HER3 signaling and targeted therapy in cancer. Oncol Rev. 2018;12(1):355.
93. Gala K, Chandarlapaty S. Molecular pathways: HER3 targeted therapy. Clin Cancer Res. 2014;20(6):1410–6.
94. Herrmann F, Lehr HA, Drexler I, Sutter G, Hengstler J, Wollscheid U, et al. HER-2/neu-mediated regulation of components of the MHC class I antigen-processing pathway. Cancer Res. 2004;64(1):215–20.
95. Triulzi T, Forte L, Regondi V, Di Modica M, Ghirelli C, Carcangiu ML, et al. HER2 signaling regulates the tumor immune microenvironment and trastuzumab efficacy. Onco Targets Ther. 2019;8(1):e1512942.
96. Mimura K, Ando T, Poschke I, Mougiakakos D, Johansson CC, Ichikawa J, et al. T cell recognition of HLA-A2 restricted tumor antigens is impaired by the oncogene HER2. Int J Cancer. 2011;128(2):390–401.
97. Okita R, Mougiakakos D, Ando T, Mao Y, Sarhan D, Wennerberg E, et al. HER2/HER3 signaling regulates NK cell-mediated cytotoxicity via MHC class I chain-related molecule A and B expression in human breast cancer cell lines. J Immunol. 2012;188(5):2136–45.
98. Inoue M, Mimura K, Izawa S, Shiraishi K, Inoue A, Shiba S, et al. Expression of MHC Class I on breast cancer cells correlates inversely with HER2 expression. Onco Targets Ther. 2012;1(7):1104–10.
99. Wu S, Zhang Q, Zhang F, Meng F, Liu S, Zhou R, et al. HER2 recruits AKT1 to disrupt STING signalling and suppress antiviral defence and antitumour immunity. Nat Cell Biol. 2019;21(8):1027–40.
100. Haratani K, Yonesaka K, Takamura S, Maenishi O, Kato R, Takegawa N, et al. U3-1402 sensitizes HER3-expressing tumors to PD-1 blockade by immune activation. J Clin Invest. 130(1):374–88.

101. Li R, Sant S, Brown E, Caramia F, Nikolic B, Clarke K, et al. Tucatinib promotes immune activation and synergizes with programmed cell death–1 and programmed cell death–ligand 1 inhibition in HER2-positive breast cancer. JNCI J Natl Cancer Inst. 2023;115(7):805–14.
102. Shang M, Chi Y, Zhang J, Chang J, Yang H, Yin S, et al. The therapeutic effectiveness of neoadjuvant Trastuzumab plus chemotherapy for HER2-positive breast cancer can be predicted by tumor-infiltrating lymphocytes and PD-L1 expression. Front Oncol. 2022;11:706606.
103. Swain SM, Shastry M, Hamilton E. Targeting HER2-positive breast cancer: advances and future directions. Nat Rev Drug Discov. 2023;22(2):101–26.
104. Huober J, Barrios CH, Niikura N, Jarząb M, Chang YC, Huggins-Puhalla SL, et al. Atezolizumab with neoadjuvant anti–human epidermal growth factor receptor 2 therapy and chemotherapy in human epidermal growth factor receptor 2–positive early breast cancer: primary results of the randomized Phase III IMpassion050 Trial. JCO. 2022;40(25):2946–56.
105. Rinnerthaler G, Egle D, Bartsch R, Schmitt CA, Petzer A, Balic M, et al. Neoadjuvant atezolizumab in combination with dual HER2 blockade plus epirubicin in women with early HER2-positive breast cancer: the randomized phase 2 ABCSG-52/ATHENE trial. Nat Cancer. 2025;16:1–10.
106. Loi S, Giobbie-Hurder A, Gombos A, Bachelot T, Hui R, Curigliano G, et al. Pembrolizumab plus trastuzumab in trastuzumab-resistant, advanced, HER2-positive breast cancer (PANACEA): a single-arm, multicentre, phase 1b–2 trial. Lancet Oncol. 2019;20(3):371–82.
107. Emens LA, Esteva FJ, Beresford M, Saura C, Laurentiis MD, Kim SB, et al. Trastuzumab emtansine plus atezolizumab versus trastuzumab emtansine plus placebo in previously treated, HER2-positive advanced breast cancer (KATE2): a phase 2, multicentre, randomised, double-blind trial. Lancet Oncol. 2020;21(10):1283–95.
108. Loi S, Schneeweiss A, Song E, Harries M, Laurentiis MD, Li Y, et al. 329TiP KATE3: A phase III study of trastuzumab emtansine (T-DM1) in combination with atezolizumab or placebo in patients with previously treated HER2-positive and PD-L1–positive locally advanced or metastatic breast cancer. Ann Oncol. 2021;32:S509.
109. Zhang Y, Xia M, Jin K, Wang S, Wei H, Fan C, et al. Function of the c-Met receptor tyrosine kinase in carcinogenesis and associated therapeutic opportunities. Mol Cancer. 2018;17(1):45.
110. Xu R, Liu X, Li A, Song L, Liang J, Gao J, et al. c-Met up-regulates the expression of PD-L1 through MAPK/NF-κBp65 pathway. J Mol Med. 2022;100(4):585–98.
111. Sun X, Li CW, Wang WJ, Chen MK, Li H, Lai YJ, et al. Inhibition of c-MET upregulates PD-L1 expression in lung adenocarcinoma. Am J Cancer Res. 2020;10(2):564–71.
112. Wang QW, Sun LH, Zhang Y, Wang Z, Zhao Z, Wang ZL, et al. MET overexpression contributes to STAT4-PD-L1 signaling activation associated with tumor-associated, macrophages-mediated immunosuppression in primary glioblastomas. J Immunother Cancer. 2021;9(10):e002451.
113. Liu P, Zhao L, Pol J, Levesque S, Petrazzuolo A, Pfirschke C, et al. Crizotinib-induced immunogenic cell death in non-small cell lung cancer. Nat Commun. 2019;10(1):1486.
114. Nishikoba N, Kumagai K, Kanmura S, Nakamura Y, Ono M, Eguchi H, et al. HGF-MET signaling shifts M1 macrophages toward an M2-like phenotype through PI3K-mediated induction of arginase-1 expression. Front Immunol [Internet]. 2020 Sep 2 [cited 2025 Feb 2];11. Available from: https://www.frontiersin.org/journals/immunology/articles/10.3389/fimmu.2020.02135/full
115. Choi W, Lee J, Lee J, Lee SH, Kim S. Hepatocyte growth factor regulates macrophage transition to the M2 phenotype and promotes murine skeletal muscle regeneration. Front Physiol [Internet] 2019 Jul 25 [cited 2025 Feb 2];10. Available from: https://www.frontiersin.org/journals/physiology/articles/10.3389/fphys.2019.00914/full
116. Giannoni P, Pietra G, Travaini G, Quarto R, Shyti G, Benelli R, et al. Chronic lymphocytic leukemia nurse-like cells express hepatocyte growth factor receptor (c-MET) and indoleamine 2,3-dioxygenase and display features of immunosuppressive type 2 skewed macrophages. Haematologica. 2014;99(6):1078–87.

117. Yen BL, Yen ML, Hsu PJ, Liu KJ, Wang CJ, Bai CH, et al. Multipotent human mesenchymal stromal cells mediate expansion of myeloid-derived suppressor cells via hepatocyte growth factor/c-met and STAT3. Stem Cell Reports. 2013;1(2):139–51.
118. Ding S, Merkulova-Rainon T, Han ZC, Tobelem G. HGF receptor up-regulation contributes to the angiogenic phenotype of human endothelial cells and promotes angiogenesis in vitro. Blood. 2003;101(12):4816–22.
119. Finisguerra V, Prenen H, Mazzone M. Preclinical and clinical evaluation of MET functions in cancer cells and in the tumor stroma. Oncogene. 2016;35(42):5457–67.
120. Ilangumaran S, Villalobos-Hernandez A, Bobbala D, Ramanathan S. The hepatocyte growth factor (HGF)–MET receptor tyrosine kinase signaling pathway: diverse roles in modulating immune cell functions. Cytokine. 2016;82:125–39.
121. Papaccio F, Della Corte CM, Viscardi G, Di Liello R, Esposito G, Sparano F, et al. HGF/MET and the immune system: relevance for cancer immunotherapy. Int J Mol Sci. 2018;19(11):3595.
122. Gordin M, Tesio M, Cohen S, Gore Y, Lantner F, Leng L, et al. C-met and its ligand HGF regulate mature B cell survival in a pathway induced by CD74. J Immunol. 2010;185(4):2020–31.
123. Jabbarzadeh Kaboli P, Roozitalab G, Farghadani R, Eskandarian Z, Zerrouqi A. c-MET and the immunological landscape of cancer: novel therapeutic strategies for enhanced anti-tumor immunity. Front Immunol [Internet]. 2024 Nov 27 [cited 2025 Feb 2];15. Available from: https://www.frontiersin.org/journals/immunology/articles/10.3389/fimmu.2024.1498391/full
124. Xia Y, Huang C, Zhong M, Zhong H, Ruan R, Xiong J, et al. Targeting HGF/c-MET signaling to regulate the tumor microenvironment: implications for counteracting tumor immune evasion. Cell Commun Signal. 2025;23(1):46.
125. Mueller KL, Madden JM, Zoratti GL, Kuperwasser C, List K, Boerner JL. Fibroblast-secreted hepatocyte growth factor mediates epidermal growth factor receptor tyrosine kinase inhibitor resistance in triple-negative breast cancers through paracrine activation of Met. Breast Cancer Res. 2012;14(4):R104.
126. Dong Q, Du Y, Li H, Liu C, Wei Y, Chen MK, et al. EGFR and c-MET cooperate to enhance resistance to PARP inhibitors in hepatocellular carcinoma. Cancer Res. 2019;79(4):819–29.
127. Machiraju D, Hassel JC. Targeting the cMET pathway to enhance immunotherapeutic approaches for mUM patients. Front Oncol. 2023;12:1068029.
128. Felip E, Metro G, Tan DSW, Wolf J, Mark M, Boyer M, et al. Capmatinib plus nivolumab in pretreated patients with *EGFR* wild-type advanced non–small cell lung cancer. Lung Cancer. 2024;192:107820.
129. Lin JJ, Chin E, Yeap BY, Ferris LA, Kamesan V, Lennes IT, et al. Increased hepatotoxicity associated with sequential immune checkpoint inhibitor and crizotinib therapy in patients with non–small cell lung cancer. J Thorac Oncol. 2019;14(1):135–40.
130. Mok TSK, Cortinovis DL, Majem M, Johnson ML, Ingriani F, Zhao X, et al. Efficacy and safety of capmatinib plus pembrolizumab in treatment (tx)-naïve patients with advanced non–small cell lung cancer (NSCLC) with high tumor PD-L1 expression: Results of a randomized, open-label, multicenter, phase 2 study.
131. Wolf J, Heist R, Kim TM, Nishio M, Dooms C, Kanthala RR, et al. 994P efficacy and safety of capmatinib plus spartalizumab in treatment-naïve patients with advanced NSCLC harboring MET exon 14 skipping mutation. Ann Oncol. 2022;33:S1007–8.
132. Wolf J, Hochmair M, Han JY, Reguart N, Souquet PJ, Smit EF, et al. Capmatinib in MET exon 14-mutated non-small-cell lung cancer: final results from the open-label, phase 2 GEOMETRY mono-1 trial. Lancet Oncol. 2024;25(10):1357–70.
133. Budczies J, Kirchner M, Kluck K, Kazdal D, Glade J, Allgäuer M, et al. Deciphering the immunosuppressive tumor microenvironment in ALK- and EGFR-positive lung adenocarcinoma. Cancer Immunol Immunother. 2022;71(2):251–65.

134. Oh CY, Klatt MG, Bourne C, Dao T, Dacek MM, Brea EJ, et al. ALK and RET inhibitors promote HLA class I antigen presentation and unmask new antigens within the tumor immunopeptidome. Cancer Immunol Res. 2019;7(12):1984–97.
135. Chu YY, Chen MK, Wei Y, Lee HH, Xia W, Wang YN, et al. Targeting the ALK–CDK9-Tyr19 kinase cascade sensitizes ovarian and breast tumors to PARP inhibition via destabilization of the P-TEFb complex. Nat Cancer. 2022;3(10):1211–27.
136. Zheng N, Zhang Y, Zeng Y, Ma Q, Zhang R, Zhao Q, et al. Pathological response and tumor immune microenvironment remodeling upon neoadjuvant ALK-TKI treatment in ALK-rearranged non-Small cell lung cancer. Targ Oncol. 2023;18(4):625–36.
137. Solomon BJ, Dagogo-Jack I, Lee SH, Boyer MJ, Ramalingam SS, Carcereny E, et al. Avelumab in combination with Lorlatinib or Crizotinib in patients with previously treated advanced NSCLC: phase 1b/2 results from the JAVELIN lung 101 trial. JTO Clin Res Report. 2024;5(7):100685.
138. Drilon A. TRK inhibitors in TRK fusion-positive cancers. Ann Oncol. 2019;30:viii23–30.
139. Drilon A, Jenkins C, Iyer S, Schoenfeld A, Keddy C, Davare MA. ROS1-dependent cancers—biology, diagnostics and therapeutics. Nat Rev Clin Oncol. 2021;18(1):35–55.
140. Shaw AT, Ou SHI, Bang YJ, Camidge DR, Solomon BJ, Salgia R, et al. Crizotinib in ROS1-rearranged non–small-cell lung cancer. N Engl J Med. 2014;371(21):1963–71.
141. Liu Z, Zhao K, Wei S, Liu C, Zhou J, Gou Q, et al. ROS1-fusion protein induces PD-L1 expression via MEK-ERK activation in non-small cell lung cancer. Onco Targets Ther. 2020;9(1):1758003.
142. Lin JJ, Choudhury NJ, Yoda S, Zhu VW, Johnson TW, Sakhtemani R, et al. Spectrum of mechanisms of resistance to Crizotinib and Lorlatinib in ROS1 fusion-positive lung cancer. Clin Cancer Res. 2021;27(10):2899–909.
143. Cheong TC, Jang A, Wang Q, Leonardi GC, Ricciuti B, Alessi JV, et al. Mechanistic patterns and clinical implications of oncogenic tyrosine kinase fusions in human cancers. Nat Commun. 2024;15(1):5110.
144. Drilon A, Laetsch TW, Kummar S, DuBois SG, Lassen UN, Demetri GD, et al. Efficacy of larotrectinib in TRK fusion–positive cancers in adults and children. N Engl J Med 2018;378(8):731–739.
145. Coppola V, Barrick CA, Southon EA, Celeste A, Wang K, Chen B, et al. Ablation of TrkA function in the immune system causes B cell abnormalities. Development. 2004;131(20):5185–95.
146. Kasi PM, Afghan MK, Bellizzi AM, Chan CH. Larotrectinib in mismatch-repair-deficient TRK fusion-positive metastatic colon cancer after progression on immunotherapy. Cureus. 14(7):e26648.
147. Phay JE, Shah MH. Targeting RET receptor tyrosine kinase activation in cancer. Clin Cancer Res. 2010;16(24):5936–41.
148. Regua AT, Najjar M, Lo HW. RET signaling pathway and RET inhibitors in human cancer. Front Oncol [Internet] 2022 Jul 25 [cited 2025 Feb 2];12. Available from: https://www.frontiersin.org/journals/oncology/articles/10.3389/fonc.2022.932353/full
149. Castellone MD, Melillo RM. RET-mediated modulation of tumor microenvironment and immune response in multiple endocrine neoplasia type 2 (MEN2). Endocr Relat Cancer. 2018;25(2):T105–19.
150. Clark L, Fisher G, Brook S, Patel S, Arkenau HT. Selective RET inhibitors (SRIs) in cancer: a journey from multi-kinase inhibitors to the next generation of SRIs. Cancers (Basel). 2023;16(1):31.
151. Mulligan LM. RET revisited: expanding the oncogenic portfolio. Nat Rev Cancer. 2014;14(3):173–86.
152. Gild ML, Tsang VHM, Clifton-Bligh RJ, Robinson BG. Multikinase inhibitors in thyroid cancer: timing of targeted therapy. Nat Rev Endocrinol. 2021;17(4):225–34.

153. Bongiovanni M, Rebecchini C, Saglietti C, Bulliard JL, Marino L, Leval L de, et al. Very low expression of PD-L1 in medullary thyroid carcinoma. 2017 Jun 1 [cited 2025 Feb 24]; Available from: https://erc.bioscientifica.com/view/journals/erc/24/6/L35.xml
154. Lu C, Dong XR, Zhao J, Zhang XC, Chen HJ, Zhou Q, et al. Association of genetic and immuno-characteristics with clinical outcomes in patients with RET-rearranged non-small cell lung cancer: a retrospective multicenter study. J Hematol Oncol. 2020;13(1):37.
155. Stone RM, Mandrekar SJ, Sanford BL, Laumann K, Geyer S, Bloomfield CD, et al. Midostaurin plus chemotherapy for acute myeloid leukemia with a FLT3 mutation. N Engl J Med. 2017;377(5):454–64.
156. Perl AE, Martinelli G, Cortes JE, Neubauer A, Berman E, Paolini S, et al. Gilteritinib or chemotherapy for relapsed or refractory FLT3-mutated AML. N Engl J Med. 2019;381(18):1728–40.
157. Cortes JE, Khaled S, Martinelli G, Perl AE, Ganguly S, Russell N, et al. Quizartinib versus salvage chemotherapy in relapsed or refractory FLT3-ITD acute myeloid leukaemia (QuANTUM-R): a multicentre, randomised, controlled, open-label, phase 3 trial. Lancet Oncol. 2019;20(7):984–97.
158. Gilliland DG, Griffin JD. Role of FLT3 in leukemia. Curr Opin Hematol. 2002;9(4):274.
159. Small D. FLT3 mutations: biology and treatment. Hematology. 2006;2006(1):178–84.
160. Régnier P, Vetillard M, Bansard A, Pierre E, Li X, Cagnard N, et al. FLT3L-dependent dendritic cells control tumor immunity by modulating Treg and NK cell homeostasis. Cell Report Med. 2023;4(12):101256.
161. Waskow C, Liu K, Darrasse-Jèze G, Guermonprez P, Ginhoux F, Merad M, et al. The receptor tyrosine kinase Flt3 is required for dendritic cell development in peripheral lymphoid tissues. Nat Immunol. 2008;9(6):676–83.
162. De Palma M, Hanahan D. Milestones in tumor vascularization and its therapeutic targeting. Nat Cancer. 2024;5(6):827–43.
163. Guelfi S, Hodivala-Dilke K, Bergers G. Targeting the tumour vasculature: from vessel destruction to promotion. Nat Rev Cancer. 2024;24(10):655–75.
164. Jain RK. Normalizing tumor vasculature with anti-angiogenic therapy: a new paradigm for combination therapy. Nat Med. 2001;7(9):987–9.
165. Mirchandani AS, Sanchez-Garcia MA, Walmsley SR. How oxygenation shapes immune responses: emerging roles for physioxia and pathological hypoxia. Nat Rev Immunol. 2024;30:1–17.
166. Liu ZL, Chen HH, Zheng LL, Sun LP, Shi L. Angiogenic signaling pathways and anti-angiogenic therapy for cancer. Signal Transduct Target Ther. 2023;8(1):198.
167. Bourhis M, Palle J, Galy-Fauroux I, Terme M. Direct and Indirect Modulation of T Cells by VEGF-A Counteracted by Anti-Angiogenic Treatment. Front Immunol [Internet]. 2021 Mar 29 [cited 2025 Feb 5];12. Available from: https://www.frontiersin.org/journals/immunology/articles/10.3389/fimmu.2021.616837/full
168. Park JA, Espinosa-Cotton M, Guo H, Fen, Monette S, Cheung NKV. Targeting tumor vasculature to improve antitumor activity of T cells armed ex vivo with T cell engaging bispecific antibody. J Immunother Cancer. 2023;11(3):e006680.
169. Dammeijer F, Dumoulin DW, Aerts JGJV. Anti–vascular endothelial growth factor/programmed cell death protein 1 bispecific antibodies: using nunchucks to fight an old adversary. J Thorac Oncol. 2024;19(3):366–9.
170. Kim DK, Jeong J, Lee DS, Hyeon DY, Park GW, Jeon S, et al. PD-L1-directed PlGF/VEGF blockade synergizes with chemotherapy by targeting CD141+ cancer-associated fibroblasts in pancreatic cancer. Nat Commun. 2022;13(1):6292.
171. Wang FT, Sun W, Zhang JT, Fan YZ. Cancer-associated fibroblast regulation of tumor neo-angiogenesis as a therapeutic target in cancer. Oncol Lett. 2019;17(3):3055–65.
172. Koh YW, Han JH, Yoon DH, Suh C, Huh J. PD-L1 expression correlates with VEGF and microvessel density in patients with uniformly treated classical Hodgkin lymphoma. Ann Hematol. 2017;96(11):1883–90.

173. Liu S, Qin T, Liu Z, Wang J, Jia Y, Feng Y, et al. Anlotinib alters tumor immune microenvironment by downregulating PD-L1 expression on vascular endothelial cells. Cell Death Dis. 2020;11(5):1–16.
174. Lai YS, Wahyuningtyas R, Aui SP, Chang KT. Autocrine VEGF signalling on M2 macrophages regulates PD-L1 expression for immunomodulation of T cells. J Cell Mol Med. 2019;23(2):1257–67.
175. Osada T, Chong G, Tansik R, Hong T, Spector N, Kumar R, et al. The effect of anti-VEGF therapy on immature myeloid cell and dendritic cells in cancer patients. Cancer Immunol Immunother. 2008;57(8):1115–24.
176. Zhou Y, Wu C, Lu G, Hu Z, Chen Q, Du X. FGF/FGFR signaling pathway involved resistance in various cancer types. J Cancer. 2020;11(8):2000–7.
177. Ruan R, Li L, Li X, Huang C, Zhang Z, Zhong H, et al. Unleashing the potential of combining FGFR inhibitor and immune checkpoint blockade for FGF/FGFR signaling in tumor microenvironment. Mol Cancer. 2023;22(1):60.
178. Akhand SS, Liu Z, Purdy SC, Abdullah A, Lin H, Cresswell GM, et al. Pharmacological inhibition of FGFR modulates the metastatic immune microenvironment and promotes response to immune checkpoint blockade. Cancer Immunol Res. 2020;8(12):1542–53.
179. Li P, Huang T, Zou Q, Liu D, Wang Y, Tan X, et al. FGFR2 promotes expression of PD-L1 in colorectal cancer via the JAK/STAT3 signaling pathway. J Immunol. 2019;202(10):3065–75.
180. Song X, Tang W, Peng H, Qi X, Li J. FGFR leads to sustained activation of STAT3 to mediate resistance to EGFR-TKIs treatment. Investig New Drugs. 2021;39(5):1201–12.
181. Im JH, Buzzelli JN, Jones K, Franchini F, Gordon-Weeks A, Markelc B, et al. FGF2 alters macrophage polarization, tumour immunity and growth and can be targeted during radiotherapy. Nat Commun. 2020;11(1):4064.
182. Wu Y, Yi Z, Li J, Wei Y, Feng R, Liu J, et al. FGFR blockade boosts T cell infiltration into triple-negative breast cancer by regulating cancer-associated fibroblasts. Theranostics. 2022;12(10):4564–80.
183. Holmström TH, Moilanen AM, Ikonen T, Björkman ML, Linnanen T, Wohlfahrt G, et al. ODM-203, a selective inhibitor of FGFR and VEGFR, shows strong antitumor activity, and induces antitumor immunity. Mol Cancer Ther. 2019;18(1):28–38.
184. Adachi Y, Kamiyama H, Ichikawa K, Fukushima S, Ozawa Y, Yamaguchi S, et al. Inhibition of FGFR reactivates IFNγ signaling in tumor cells to enhance the combined antitumor activity of Lenvatinib with anti-PD-1 antibodies. Cancer Res. 2022;82(2):292–306.
185. Zou X, Tang XY, Qu ZY, Sun ZW, Ji CF, Li YJ, et al. Targeting the PDGF/PDGFR signaling pathway for cancer therapy: a review. Int J Biol Macromol. 2022;202:539–57.
186. Pandey P, Khan F, Upadhyay TK, Seungjoon M, Park MN, Kim B. New insights about the PDGF/PDGFR signaling pathway as a promising target to develop cancer therapeutic strategies. Biomed Pharmacother. 2023;161:114491.
187. Strell C, Rodríguez-Tomàs E, Östman A. Functional and clinical roles of stromal PDGF receptors in tumor biology. Cancer Metastasis Rev. 2024;43(4):1593–609.
188. Hammer AM, Sizemore GM, Shukla VC, Avendano A, Sizemore ST, Chang JJ, et al. Stromal PDGFR-α activation enhances matrix stiffness, impedes mammary ductal development, and accelerates tumor growth. Neoplasia. 2017;19(6):496–508.
189. Akiyama T, Yasuda T, Uchihara T, Yasuda-Yoshihara N, Tan BJY, Yonemura A, et al. Stromal reprogramming through dual PDGFRα/β blockade boosts the efficacy of anti–PD-1 immunotherapy in fibrotic tumors. Cancer Res. 2023;83(5):753–70.
190. Agrawal S, Ganguly S, Hajian P, Cao JN, Agrawal A. PDGF upregulates CLEC-2 to induce T regulatory cells. Oncotarget. 2015;6(30):28621–32.
191. Pietras K, Ostman A, Sjöquist M, Buchdunger E, Reed RK, Heldin CH, et al. Inhibition of platelet-derived growth factor receptors reduces interstitial hypertension and increases transcapillary transport in tumors. Cancer Res. 2001;61(7):2929–34.
192. Son D, Na YR, Hwang ES, Seok SH. Platelet-derived growth factor-C (PDGF-C) induces anti-apoptotic effects on macrophages through Akt and bad phosphorylation. J Biol Chem. 2014;289(9):6225–35.

193. Camorani S, Passariello M, Agnello L, Esposito S, Collina F, Cantile M, et al. Aptamer targeted therapy potentiates immune checkpoint blockade in triple-negative breast cancer. J Exp Clin Cancer Res. 2020;39(1):180.
194. Tu J, Xu H, Ma L, Li C, Qin W, Chen X, et al. Nintedanib enhances the efficacy of PD-L1 blockade by upregulating MHC-I and PD-L1 expression in tumor cells. Theranostics. 2022;12(2):747–66.
195. Chong WQ, Low JL, Tay JK, Le TBU, Goh GSQ, Sooi K, et al. Pembrolizumab with or without bevacizumab in platinum-resistant recurrent or metastatic nasopharyngeal carcinoma: a randomised, open-label, phase 2 trial. The Lancet Oncology [Internet] 2025 Jan 15 [cited 2025 Jan 18];0(0). Available from: https://www.thelancet.com/journals/lanonc/article/PIIS1470-2045(24)00677-6/abstract?rss=yes
196. Bagegni NA, Park H, Kraft K, O-Toole M, Gao F, Waqar SN, et al. Phase 1b trial of anti-VEGF/PDGFR vorolanib combined with immune checkpoint inhibitors in patients with advanced solid tumors. Cancer Chemother Pharmacol. 2022;89(4):487–97.
197. Shojaei F, Wu X, Malik AK, Zhong C, Baldwin ME, Schanz S, et al. Tumor refractoriness to anti-VEGF treatment is mediated by CD11b+Gr1+ myeloid cells. Nat Biotechnol. 2007;25(8):911–20.
198. Jammihal T, Saliby RM, Labaki C, Soulati H, Gallegos J, Peris A, et al. Immunogenomic determinants of exceptional response to immune checkpoint inhibition in renal cell carcinoma. Nat Cancer. 2025;9:1–13.
199. Finn RS, Ryoo BY, Hsu CH, Li D, Burgoyne AM, Cotter C, et al. Tiragolumab in combination with atezolizumab and bevacizumab in patients with unresectable, locally advanced or metastatic hepatocellular carcinoma (MORPHEUS-liver): a randomised, open-label, phase 1b–2, study. Lancet Oncol. 2025;26(2):214–26.
200. Gong K, Guo G, Gerber DE, Gao B, Peyton M, Huang C, et al. TNF-driven adaptive response mediates resistance to EGFR inhibition in lung cancer. J Clin Invest. 2018;128(6):2500–18.
201. Long JY, Li RZ, Wang DX, Liu H, Tian J, Ding ZN, et al. Comprehensive molecular analysis identifies *RET* alterations association with response of ICIs in multi-immunotherapy cohorts. Int Immunopharmacol. 2024;126:111281.
202. Ward LS. Immune response in thyroid cancer: widening the boundaries. Scientifica (Cairo). 2014;2014:125450.
203. Stahl M, Gedrich R, Peck R, LaVallee T, Eder JP. Targeting KIT on innate immune cells to enhance the antitumor activity of checkpoint inhibitors. Immunotherapy. 2016;8(7):767–74.
204. Yang Z, Zhang B, Li D, Lv M, Huang C, Shen GX, et al. Mast cells mobilize myeloid-derived suppressor cells and Treg cells in tumor microenvironment via IL-17 pathway in murine hepatocarcinoma model. PLoS One. 2010;5(1):e8922.
205. Yang B, Yang Q, Huang Q, Yan H, Sun T, Tong H. Silencing c-Kit expression in human DCs suppresses Th2, Th17 response but enhances Th1 response. Am J Transl Res. 2015;7(9):1499–509.
206. Liu M, Etherington MS, Hanna A, Medina BD, Vitiello GA, Bowler TG, et al. Oncogenic KIT modulates type I IFN–mediated antitumor immunity in GIST. Cancer Immunol Res. 2021;9(5):542–53.
207. Cannarile MA, Weisser M, Jacob W, Jegg AM, Ries CH, Rüttinger D. Colony-stimulating factor 1 receptor (CSF1R) inhibitors in cancer therapy. J Immun Cancer. 2017;5(1):53.
208. Chen K, Li X, Dong S, Guo Y, Luo Z, Zhuang SM, et al. Modulating tumor-associated macrophages through CSF1R inhibition: a potential therapeutic strategy for HNSCC. J Transl Med. 2025;23(1):27.
209. O'Brien SA, Orf J, Skrzypczynska KM, Tan H, Kim J, DeVoss J, et al. Activity of tumor-associated macrophage depletion by CSF1R blockade is highly dependent on the tumor model and timing of treatment. Cancer Immunol Immunother. 2021;70(8):2401–10.
210. Chen TW, Hung WZ, Chiang SF, Chen WTL, Ke TW, Liang JA, et al. Dual inhibition of TGFβ signaling and CSF1/CSF1R reprograms tumor-infiltrating macrophages and improves response to chemotherapy *via* suppressing PD-L1. Cancer Lett. 2022;543:215795.

211. Pham D, Daniel M, Guhan S, Tsao H. KIT and melanoma: biological insights and clinical implications. Yonsei Med J. 2020;61(7):562–71.
212. Li B, Chen H, Yang S, Chen F, Xu L, Li Y, et al. Advances in immunology and immunotherapy for mesenchymal gastrointestinal cancers. Mol Cancer. 2023;22(1):71.
213. Kumar V, Doros L, Thompson M, Mushti SL, Charlab R, Spehalski EI, et al. FDA approval summary: Ripretinib for advanced gastrointestinal stromal tumor (GIST). Clin Cancer Res. 2023;29(11):2020–4.
214. Galanis A, Levis M. Inhibition of c-Kit by tyrosine kinase inhibitors. Haematologica. 2015;100(3):e77–9.
215. Mumprecht S, Matter M, Pavelic V, Ochsenbein AF. Imatinib mesylate selectively impairs expansion of memory cytotoxic T cells without affecting the control of primary viral infections. Blood. 2006;108(10):3406–13.
216. Seggewiss R, Loré K, Greiner E, Magnusson MK, Price DA, Douek DC, et al. Imatinib inhibits T-cell receptor-mediated T-cell proliferation and activation in a dose-dependent manner. Blood. 2005;105(6):2473–9.
217. Zitvogel L, Rusakiewicz S, Routy B, Ayyoub M, Kroemer G. Immunological off-target effects of imatinib. Nat Rev Clin Oncol. 2016;13(7):431–46.
218. Lennartsson J, Rönnstrand L. Stem cell factor receptor/c-Kit: from basic science to clinical implications. Physiol Rev. 2012;92(4):1619–49.
219. Ray P, Krishnamoorthy N, Oriss TB, Ray A. Signaling of c-kit in dendritic cells influences adaptive immunity. Ann N Y Acad Sci. 2010;1183(1):104–22.
220. Tsai M, Valent P, Galli SJ. KIT as a master regulator of the mast cell lineage. J Allergy Clin Immunol. 2022;149(6):1845–54.
221. Kalathil SG, Wang K, Hutson A, Iyer R, Thanavala Y. Tivozanib mediated inhibition of c-Kit/SCF signaling on Tregs and MDSCs and reversal of tumor induced immune suppression correlates with survival of HCC patients. Onco Targets Ther. 2020;9(1):1824863.
222. Balachandran VP, Cavnar MJ, Zeng S, Bamboat ZM, Ocuin LM, Obaid H, et al. Imatinib potentiates antitumor T cell responses in gastrointestinal stromal tumor through the inhibition of Ido. Nat Med. 2011;17(9):1094–100.
223. Seifert AM, Zeng S, Zhang JQ, Kim TS, Cohen NA, Beckman MJ, et al. PD-1/PD-L1 blockade enhances T-cell activity and antitumor efficacy of imatinib in gastrointestinal stromal tumors. Clin Cancer Res. 2017;23(2):454–65.
224. Reilley MJ, Bailey A, Subbiah V, Janku F, Naing A, Falchook G, et al. Phase I clinical trial of combination imatinib and ipilimumab in patients with advanced malignancies. J Immunother Cancer. 2017;5(1):35.
225. Abdou Y, Kapoor A, Hamad L, Ernstoff MS. Combination of pembrolizumab and imatinib in a patient with double KIT mutant melanoma. Medicine (Baltimore). 2019;98(44):e17769.
226. Abdellateif MS, Bayoumi AK, Mohammed MA. c-Kit receptors as a therapeutic target in cancer: current insights. Onco Targets Ther. 2023;16:785–99.
227. Wolf D, Tilg H, Rumpold H, Gastl G, Wolf AM. The kinase inhibitor imatinib—an immunosuppressive drug? http://www.eurekaselect.com [Internet]. [cited 2025 Feb 23]. Available from: https://www.eurekaselect.com/article/4109
228. Pan Y, Yu Y, Wang X, Zhang T. Tumor-associated macrophages in tumor immunity. Front Immunol [Internet]. 2020 Dec 3 [cited 2025 Feb 3];11. Available from: https://www.frontiersin.org/journals/immunology/articles/10.3389/fimmu.2020.583084/full
229. Noy R, Pollard JW. Tumor-associated macrophages: from mechanisms to therapy. Immunity. 2014;41(1):49–61.
230. Lin CC. Clinical development of Colony-stimulating factor 1 receptor (CSF1R) inhibitors. J Immun Precision Oncol. 2021;4(2):105–14.
231. Ho WJ, Jaffee EM. Macrophage-targeting by CSF1/1R blockade in pancreatic cancers. Cancer Res. 2021;81(24):6071–3.
232. Spierenburg G, van der Heijden L, van Langevelde K, Szuhai K, Bovée JVGM, van de Sande MAJ, et al. Tenosynovial giant cell tumors (TGCT): molecular biology, drug targets and non-surgical pharmacological approaches. Expert Opin Ther Targets. 2022;26(4):333–45.

233. Voissière A, Gomez-Roca C, Chabaud S, Rodriguez C, Nkodia A, Berthet J, et al. The CSF-1R inhibitor pexidartinib affects FLT3-dependent DC differentiation and may antagonize durvalumab effect in patients with advanced cancers. Sci Transl Med. 2024;16(731):eadd1834.
234. Wang-Gillam A, O'Reilly EM, Bendell JC, Wainberg ZA, Borazanci EH, Bahary N, et al. A randomized phase II study of cabiralizumab (cabira) + nivolumab (nivo) ± chemotherapy (chemo) in advanced pancreatic ductal adenocarcinoma (PDAC). JCO. 2019;37(4_suppl):TPS465–8.
235. Gomez-Roca CA, Italiano A, Le Tourneau C, Cassier PA, Toulmonde M, D'Angelo SP, et al. Phase I study of emactuzumab single agent or in combination with paclitaxel in patients with advanced/metastatic solid tumors reveals depletion of immunosuppressive M2-like macrophages. Ann Oncol. 2019;30(8):1381–92.
236. Bornes L, van Winden LJ, Geurts VCM, de Bruijn B, Azarang L, Lanfermeijer M, et al. The oestrous cycle stage affects mammary tumour sensitivity to chemotherapy. Nature. 2025;637(8044):195–204.
237. Brahmer JR, Dahlberg SE, Gray RJ, Schiller JH, Perry MC, Sandler A, et al. Sex differences in outcome with bevacizumab therapy: analysis of patients with advanced-stage non-small cell lung cancer treated with or without bevacizumab in combination with paclitaxel and carboplatin in the eastern cooperative oncology group trial 4599. J Thorac Oncol. 2011;6(1):103–8.
238. Gerritse SL, Janssen JBE, Labots M, de Vries R, Rudek M, Carducci M, et al. High-dose administration of tyrosine kinase inhibitors to improve clinical benefit: a systematic review. Cancer Treat Rev. 2021;97:102171.
239. Zemek RM, Anagnostou V, Pires da Silva I, Long GV, Lesterhuis WJ. Exploiting temporal aspects of cancer immunotherapy. Nat Rev Cancer. 2024;24(7):480–97.
240. Lamia KA. Stressing the importance of circadian time in treatment responses. Cancer Cell. 2025;43(1):13–4.
241. Damato AR, Luo J, Katumba RGN, Talcott GR, Rubin JB, Herzog ED, et al. Temozolomide chronotherapy in patients with glioblastoma: a retrospective single-institute study. Neuro-Oncol Adv. 2021;3(1):vdab041.
242. Liang J, Hong J, Tang X, Qiu X, Zhu K, Zhou L, et al. Sex difference in response to non-small cell lung cancer immunotherapy: an updated meta-analysis. Ann Med. 2022;54(1):2605–15.
243. Lee J, Nicosia M, Hong ES, Silver DJ, Li C, Bayik D, et al. Sex-biased T-cell exhaustion drives differential immune responses in glioblastoma. Cancer Discov. 2023;13(9):2090–105.
244. Ficarro SB, McCleland ML, Stukenberg PT, Burke DJ, Ross MM, Shabanowitz J, et al. Phosphoproteome analysis by mass spectrometry and its application to Saccharomyces cerevisiae. Nat Biotechnol. 2002;20(3):301–5.
245. Rush J, Moritz A, Lee KA, Guo A, Goss VL, Spek EJ, et al. Immunoaffinity profiling of tyrosine phosphorylation in cancer cells. Nat Biotechnol. 2005;23(1):94–101.
246. Zhang Y, Wolf-Yadlin A, Ross PL, Pappin DJ, Rush J, Lauffenburger DA, et al. Time-resolved mass spectrometry of tyrosine phosphorylation sites in the epidermal growth factor receptor signaling network reveals dynamic modules. Mol Cell Proteomics. 2005;4(9):1240–50.
247. Del Rosario AM, White FM. Quantifying oncogenic phosphotyrosine signaling networks through systems biology. Curr Opin Genet Dev. 2010;20(1):23–30.
248. Kohale IN, Burgenske DM, Mladek AC, Bakken KK, Kuang J, Boughey JC, et al. Quantitative analysis of tyrosine phosphorylation from FFPE tissues reveals patient-specific signaling networks. Cancer Res. 2021;81(14):3930–41.
249. Friedrich C, Schallenberg S, Kirchner M, Ziehm M, Niquet S, Haji M, et al. Comprehensive micro-scaled proteome and phosphoproteome characterization of archived retrospective cancer repositories. Nat Commun. 2021;12(1):3576.
250. Bica V, Venafra V, Massacci G, Graziosi S, Gualdi S, Minnella G, et al. Phosphoproteomics reveals novel BCR::ABL1-independent mechanisms of resistance in chronic myeloid leukemia [Internet]. bioRxiv; 2025 [cited 2025 Feb 3]. p. 2025.01.08.631869. Available from: https://www.biorxiv.org/content/10.1101/2025.01.08.631869v1

Natural Compounds as Inhibitors of Receptor Tyrosine Kinases in Cancer

Satyabrata Nayak and Rana P. Singh

Abbreviations

EGFR	Epidermal growth factor receptor
EMT	Epithelial to mesenchymal transition
ER	Endoplasmic reticulum
FDA	Food and Drug Administration
HCC	Hepatocellular carcinoma
HNSCC	Head and neck squamous cell carcinoma
IGF-1R	Insulin-like growth factor 1 receptor
NSCLC	Non-small cell lung cancer
OSCC	Human oral squamous cell carcinoma
ROS	Reactive oxygen species
RTK	Receptor tyrosine kinases
TKI	Tyrosine kinase inhibitor
TNBC	Triple-negative breast cancer
VEGF	Vascular endothelial growth factor

1 Introduction

Cancer encompasses diseases distinguished by unrestricted cell division, ability to replicate indefinitely, and resistance to the natural course of apoptosis. Tumor formation occurs as cancerous cells multiply and grow into an abnormal mass. The

S. Nayak
Cancer Biology Laboratory, School of Life Sciences, Jawaharlal Nehru University, New Delhi, India

R. P. Singh (✉)
Cancer Biology Laboratory, School of Life Sciences, Jawaharlal Nehru University, New Delhi, India

Department of Pharmaceutical Sciences, Skaggs School of Pharmacy and Pharmaceutical Sciences, University of Colorado, Anschutz Medical Campus, Aurora, CO, USA

Gautam Buddha University, Greater Noida, UP, India
e-mail: rana_singh@mail.jnu.ac.in

A. Sabarwal et al. (eds.), *Receptor Tyrosine Kinases in Cancer*, Cancer Drug Discovery and Development, https://doi.org/10.1007/978-3-031-93894-8_11

onset of cancer is mainly due to mutations in proto-oncogenes, which encourage cell growth, and tumor suppressor genes, which typically inhibit cell proliferation and induce apoptosis. Tumor progression involves changes in DNA repair genes which are often influenced by mutations that manage DNA damage and repair. These mutations facilitate unchecked cell division, bypassing growth regulation, evading the immune system, resisting cell death, and accumulating further genetic errors [1]. According to the GLOBOCAN 2022 report, there were 20 million new cases of cancer globally.

Cellular communications and functions mostly depend on membrane proteins, known as receptor tyrosine kinases (RTKs). Through the phosphorylation of intracellular protein tyrosine residues, these proteins promote signaling and control a range of cellular processes, such as growth, differentiation, and metabolism. Tyrosine kinase inhibitors (TKIs) are specifically designed molecules which were developed to inhibit the abnormal activation of RTKs that are often linked to the progression of cancer. Overexpression, mutations, or some changes in RTKs generally result in the development of TKI resistance. During treatment, this resistance makes cancer seriously difficult to manage. Currently, research suggests that phytochemicals or natural products extracted from plants as modulators of RTK activity can be used to address these serious problems. By blocking RTK signaling pathways, these phytochemicals may also enhance the effectiveness of established cancer therapies when used in combination. Phytochemicals may be utilized as adjuvants to overcome TKI resistance and decrease the side effects of cancer treatments, providing a better path to treat cancer [2].

2 Phytochemicals

Chemoprevention is the process of using synthetic products, natural products, and biological agents to prevent, suppress, or reverse the early stages of carcinogenesis. It also stops premalignant tumor cells from spreading to other locations. The advances from the ongoing research on cancer biology have revealed several molecular targets for the treatment of tumor and are also enhancing our knowledge and strategies for the management of cancer [3]. Vegetables, seeds, fruits, and everyday used foods are the main sources of various phytochemicals with anticancer effects. Curcumin from turmeric, capsaicin from chili peppers, lycopene from tomato extract, resveratrol from grape seed extract, sulforaphane from broccoli, and allicin from common garlic are a few of these phytochemicals that have cancer chemopreventive properties [4]. The primary resources for preventing cancer are thought to be natural plant compounds [5]. Bioactive natural products are becoming the primary means of producing semisynthetic compounds from natural resources, such as marine and microbiological ecosystems, for improved pharmaceutical applications as interest in natural products grows [6–8]. Natural products can be chemically altered to create derivatives based on phytochemicals with improved pharmacological qualities, making them extremely significant substances. There are numerous instances of natural agents that have the potential to combat medication resistance. The multidrug resistance (MDR) phenotype is one such example. Drug efflux pumps, which belong to the ATP-binding cassette transporter family, generate resistance to a variety of anticancer drugs, including specific small compounds. The use of natural agents and medicinal herbs can lessen the side effects of both targeted therapy and non-targeted classical cytotoxic therapy [9].

Because natural products are a rich source of active therapeutic agents, including those with anticancer characteristics, their application in cancer control has emerged as a critical component of cancer management (Fig. 1). Particularly, phytochemicals exhibit potential as novel options for the development of antiangiogenic therapies. An epidemiological study suggests that a diet high in flavonoids may be linked to a decreased risk of getting a number of cancer forms [10]. A wealth of medicinal compounds can be found in natural products, especially when it comes to preventing or treating cancer. Many plant-based foods contain flavonoids, a class of polyphenolic compounds that have been shown in both animal and human studies to have a number of positive health effects, including antiviral, antimicrobial, antimutagenic, antineoplastic, anti-inflammatory, anti-allergic, antithrombotic, and cytoprotective qualities across a range of cell types [11]. The anticancer efficacy and associated mechanisms of selected phytochemicals are discussed below (Table 1).

2.1 *Curcumin*

Curcumin, a naturally occurring polyphenolic chemical with anti-inflammatory, anticancer, and antioxidant qualities, is extracted from the rhizome of the plant, *Curcuma longa* Linn. Recent clinical trials have demonstrated its therapeutic and chemopreventive potential for a range of cancer types. Its minimal toxicity even at

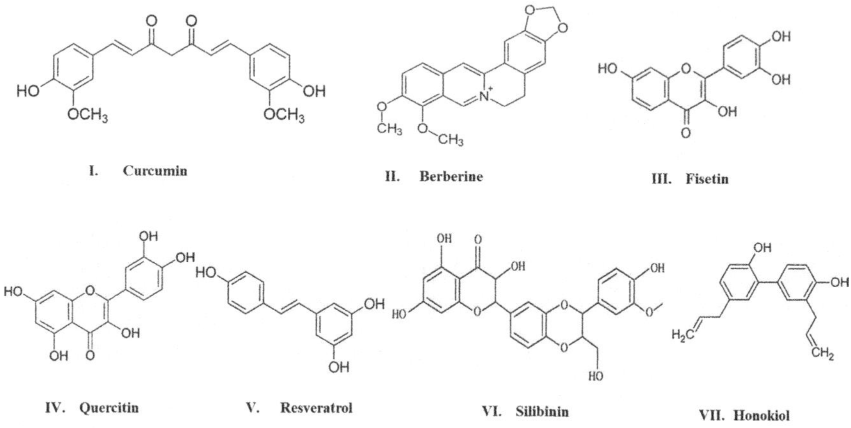

Fig. 1 Chemical structure of phytochemicals

Table 1 Phytochemicals targeting RTKs in different cancers

S. no.	Phytochemicals	Cancer	Targeted RTKs	References
1	Curcumin	Lung cancer	AXL↓, EGFR↓, IGF-1R↓, FGFR↓, VEGFR↓	[13, 15–17]
		Breast cancer	EGFR↓, IGF-1R↓,VEGFR↓, AXL↓	[12, 19–24]
		Prostate cancer	EGFR↓	[25, 26]
		HNSCC	EGFR↓	[27, 28]
		Colorectal cancer	EGFR↓, EGFR and IGF-1R↓, VEGFR↓, insulin receptor and IGF-1R↓	[29–32]
		Ovarian cancer	EGFR↓	[33]
		Hepatocellular carcinoma	EGFR↓	[34]
		Melanoma	EGFR↓	[35]
2	Berberine	Lung cancer	EGFR↓, C-MET↓	[36–38]
		Breast cancer	EGFR↓, VEGFR2↓	[39–41, 43, 44]
		Prostate cancer	EGFR↓	[42]
		Glioblastoma	EGFR↓, VEGFR2↓	[46, 48]
		Ovarian cancer	EGFR and ERBB2↓	[122]
		Colorectal cancer	EGFR↓	[49, 50]
3	Fisetin	Lung cancer	EGFR↓, AXL↓	[52, 53]
		HNSCC	EGFR↓, C-MET↓	[54, 55]
		Colorectal cancer	EGFR↓	[51]
4	Quercetin	Lung cancer	EGFRT790M mutation↓, EGFR C797S mutation↓, and AXL↓	[56, 57]
		Breast cancer	IGF-1R↓, EGFR↓, P-EGFR↓, VEGFR-2↓	[59, 62–65]
		Prostate cancer	EGFR↓, VEGF-R2↓, ERBB-2 and ERBB-3, C-MET↓	[10, 11, 66–68]
		HNSCC	EGFR↓	[71]
		Colorectal cancer	EGFR↓, ERBB2 and ERBB3↓	[58, 60, 61]
		Ovarian cancer	EGFR↓	[72]
		Hepatocellular carcinoma	EGFR↓	[73]
		Melanoma	C-MET↓, AXL↓	[69, 70]
5	Resveratrol	Lung cancer	EGFR↓, NGFR↓	[75, 76, 78]
		Breast cancer	EGFR↓	[79]
		Prostate cancer	EGFR↓, C-MET ↓	[80, 81]
		HNSCC	EGFR↓	[82]
		Colorectal cancer	EGFR↓, IGF-1R↓	[74, 83]
		Ovarian cancer	IGF-1R↓	[84]
		Hepatocellular carcinoma	C-MET↓	[77]

(continued)

Table 1 (continued)

S. no.	Phytochemicals	Cancer	Targeted RTKs	References
6	Silibinin	Lung cancer	EGFR↓, TIE2↓	[85, 86, 88]
		Breast cancer	EGFR↓, ERBB2↓, IGF-1R↓	[89, 90, 92]
		Prostate cancer	EGFR↓, ERBB1↓	[91, 93, 94]
		Basal cell carcinoma	EGFR↓	[97, 98]
		Renal cell carcinoma	EGFR↓	[99]
7	Honokiol	Lung cancer	EGFR↓	[102]
		Breast cancer	EGFR↓, ERBB2↓	[100, 104]
		HNSCC	EGFR↓	[101, 105, 106]
		Hepatocellular carcinoma	EGFR↓	[103]
		Glioblastoma	EGFR↓	[107]

effective levels is demonstrated by research showing that human oral doses up to 12 g/day are safe. It affects several signaling pathways linked to programmed cell death, cancer formation, and cell survival. Curcumin slows down the progression of certain cancers, such as advanced pancreatic cancer, squamous cancer of the head and neck, colon cancer, breast cancer, and cervical cancer [12, 13]. Curcumin effectively inhibits the expression of Axl molecule, which results in antiproliferative effects in both chemoresistant and parental NSCLC cells. This is achieved by reducing XIAP and inducing p21, which activates apoptosis-related pathways and inhibits cell cycle progression [13]. It suppresses the growth of lung and pancreatic cancer cells by downregulating the expression of EGFR and COX-2 which results in decreased cell survival, increased apoptosis, and decreased Erk1/2 activity, especially in cell lines that express COX-2 [14]. Treatment with curcumin reduces the migration and proliferation of non-small cell lung cancer (NSCLC) cells, by decreasing EMT and blocking AP-1 protein via the TLR4/MyD88 and EGFR pathways [15].

In patients with NSCLC, curcumin taken at a low dose in conjunction with therapies that target EGFR, IGF-1R, PI3K, FGFR, or NF-κB may work as an adjuvant treatment [16]. It has demonstrated antiangiogenic effects that were controlled by the HIF1α/mTOR/VEGF/VEGFR pathway and may inhibit neuropeptide-driven angiogenesis in lung cancer [17]. It suppresses tumorigenesis-related phosphorylation of RTK-PI3K-AKT/MEK-ERK1/2 signaling network and decreases the mRNA levels of important targets in breast cancer cells [18]. Through their synergistic effects on Bcl-2 downregulation, Bax upregulation, and inhibition of EGFR signaling, curcumin and paclitaxel increase cell death in breast cancer cells, pointing to a potential approach for less toxic and more effective treatment [12].

Curcumin affects vital pathways involved in cell proliferation and survival by downregulating IGF-1R expression and interfering with the IGF-1 signaling cascade, which reduces the growth of breast cancer and triggers cell death [19]. It reduces breast cancer cells, including TNBC, proliferation, and invasion while decreasing STAT3 and EGFR expression and signaling, impacting on PI3K-AKT and JAK-STAT or pERK1/2 [20, 21]. It provides promising potential against TNBC by decreasing tumor growth and angiogenesis by lowering VEGFR levels [22]. It suppresses invasion and migration of cells by inhibiting epithelial to mesenchymal transition (EMT) and modulating EMT-related genes like N-cadherin, E-cadherin, Axl, slug, and CD24 and regulating miR-34a in breast cancer cells [23, 24].

Curcumin works against prostate cancer cell lines by decreasing size of spheroid and cell viability and stimulating apoptosis by targeting EGFR/AKT/NF-κB signaling pathways [25, 26]. It targets EGFR-STAT3 signaling and EGFR-AKT-C/EBPβ regulatory axis and downregulates miR-31 expression for its anticancer activity against OSCC [27, 28]. It reduces EGFR expression by lowering the trans-activation activity of Egr-1, thereby preventing the growth of human colon cancer cells [29]. Curcumin is also shown to inhibit IGF-1R, VEGFR, and CXCR4 signaling pathways in colon tumors [30, 31]. It reduces the cell proliferation and migration and chemoresistance in colorectal cancer cells, by reducing insulin receptor, IGF-1R, and MYC expression [32]. By reducing RCP's stimulation of β1 integrin stabilization and inhibiting the EGFR and β1 integrin/ILK/FAK signaling pathways, it decreases ovarian cancer cell invasion [33]. In hepatocellular carcinoma (HCC), it targets EGFR, and VEGFR-2, and activates mitochondrial apoptosis, inhibits invasion and metastasis, and reduces angiogenesis [34]. Further, by inhibiting EGFR, activating reactive oxygen species and mitochondrial signaling pathways, curcumin decreases cell viability in vemurafenib-resistant melanoma cells [35].

Overall, curcumin inhibits RTK signaling pathways, including IGF-1R, EGFR, and VEGFR, to reduce cancer cell proliferation, migration, and survival, enhancing its anticancer effects across various cancer types.

2.2 *Berberine*

Berberine, an isoquinoline alkaloid mainly derived from the plant *Coptis chinensis*, provides a thorough understanding of herbal medicine for its antimicrobial, anti-inflammatory, and metabolic benefits. At the moment, researchers are looking into how it could be beneficial for the treatment of cancer and other disorders like polycystic ovarian syndrome, obesity, coronary artery disease (CAD), and diabetes. Its anticancer properties have been shown in studies on different tumor types, including leukemia, lung, breast, and liver cancers. It causes cell cycle arrest and apoptosis, prevents migration and invasion of cancer cells, and modulates the tumor microenvironment [36]. It has various pharmacological properties including anti-inflammatory, anti-arrhythmic, and anticancer effects.

In lung cancer studies, berberine exhibits a strong binding affinity to the EGFR protein and downregulates its signaling causing apoptosis and autophagic cell death [36, 37]. The mutation in c-Met, a proto-oncogene receptor tyrosine kinase, is linked with the growth and survival of c-Met-amplified NSCLC which is shown to be inhibited by berberine leading to preferential apoptosis in lung cancer cells [38]. EGFR inhibition by berberine leads to the downregulation of MAPK/ERK1/2/p38 and PI3K-AKT signaling cascades and induces apoptosis and lowers IL-8 levels in breast cancer [39–41]. It is also reported to induce G1 phase arrest and apoptosis in prostate cancer cells by reducing the expression and activity of EGFR [42].

By inhibiting NF-κB and elevating AMPK activity, berberine causes the activation of signaling pathways that are initiated by EGFR or doxorubicin-induced ROS. These factors then impact the transport and efflux of small-molecule inhibitors as well as expression of NGAL in cancer cells [43]. Berberine has a strong inhibitory effect on ephrin-B2-expressing HEK293 cells than on normal HEK293 cells. It reduces levels of ephrin-B2- and the PDZ-binding proteins that are linked to it and also lowers phosphorylation of VEGFR2 and downstream signaling proteins, ERK1/2 and AKT, which in turn reduces the expression of MMP2 and MMP9 [44]. In human glioblastoma cells, berberine induces G1 phase arrest and apoptosis via the mitochondrial and caspase pathways and induces autophagy via the AMPK/mTOR/ULK1 and the EGFR/MEK/ERK1/2 pathways [45–47]. Antiangiogenic effects of berberine are mediated via the inhibition of VEGFR2 phosphorylation and ERK activation in glioblastoma xenograft [48]. The EGFR signaling inhibition is also known to cause cellular senescence in glioblastoma cells [46]. In colon cancer, berberine-caused inhibition of EGFR activation suppresses tumor xenograft growth and tumorigenesis in $APC^{min/+}$ mice [49]. This is also linked to the reduction of pro-inflammatory cytokines, IL-6 and TNF-α, in colonic macrophages, which inhibits the development of colitis-associated colorectal cancer [50].

Berberine inhibits RTKs, such as EGFR, VEGFR, and c-Met, leading to downregulation of key signaling pathways, including MAPK/ERK, PI3K-AKT, and NF-κB; induction of apoptosis, autophagy, and cell cycle arrest; and reduction of tumor growth in various cancers, including lung, breast, prostate, glioblastoma, and colon cancer.

2.3 *Fisetin*

Fisetin (3,7,3′,4′-tetrahydroxyflavone), a natural flavonoid found in various fruits and vegetables, such as cucumbers, apples, onions, persimmons, and strawberries, has anti-inflammatory, antioxidant, and antiproliferative properties across different cancer types [51]. It has notable anti-invasive and anti-migratory effects on various cancer types by suppressing EMT processes by downregulating key signaling pathways, like MAPK and NF-κB. It reduces EGFR and other downstream signaling proteins involved in the EMT process, hence decreasing migration and invasion in NSCLC [52]. In lung cancer, by inhibiting the abnormally regulated AKT and MAPK signaling pathways and reducing Axl, it restores the sensitivity in cancer cells that are resistant to therapy [53]. In HNSCC, by suppressing the EGFR-AKT/ERK1/2-STAT3 signaling pathway, fisetin causes apoptosis and G2/M phase arrest and increases activation of JNK and caspases [54]. It also targets the c-Met/Src signaling and decreases c-Met, Src, and ADAM9 protein expression in oral cancer, thereby inducing apoptosis and inhibiting proliferation [55]. Inhibition of EGFR and NF-κB activation and disruption of Wnt signaling by fisetin cause apoptosis and halt growth of COX-2-overexpressing HT29 colon cancer cells [51].

Fisetin inhibits cancer cell migration, invasion, and proliferation by targeting RTKs (EGFR, c-Met) and key signaling pathways, such as MAPK, AKT, NF-κB, and Wnt, inducing apoptosis and cell cycle arrest, and restoring therapy sensitivity in lung, head and neck, oral, and colon cancer (Fig. 2), which highlight the effects of various phytochemicals on cancer and RTKs.

2.4 *Quercetin*

Quercetin (3,3′,4′,5,7-pentahydroxyflavone) and its glycosides, the naturally produced polyphenolic flavonoids, are present in a variety of vegetables and fruits, like apples, grapes, red onions, leafy greens, tea, broccoli, capers, red grapes, parsley, and a variety of berries. In animal experiments, it has been found to improve longevity and prevent premature aging, among other health advantages. It has antioxidant potential and capacity to block enzymes that cause cancer, impact signaling pathways, and interact with other proteins and cell receptors. Apart from its antiviral, anti-inflammatory, anti-fibrotic, and antioxidant properties, quercetin also exhibits encouraging promise in preventing several forms of cancer [10, 56]. By impacting several biological targets, it exhibits significant anticancer effects including induction of apoptosis; decreased cell proliferation in lung, prostate, and colon cancers; and inhibition of metastasis in breast cancer [57]. Furthermore, it is shown to inhibit a variety of protein kinases, including RTK [58, 59].

In colorectal cancer, quercetin significantly inhibits growth and induces apoptosis concomitant with the decrease in EGF receptor level and signaling pathway [58]. It suppresses cell growth and increases cell death in colon cancer HT29 and SW480 cell lines by reducing ErbB2 and ErbB3 expression and decreasing phosphorylated AKT levels. These effects are dose-dependent and are accompanied by the activation of apoptotic markers like cleaved caspase-3 and a characteristic DNA laddering pattern [60]. It highly elevates the cytotoxic effects on colorectal cancer cells when encapsulated in nanoliposomes, by increasing cell death and reducing the EGFR gene expression [61].

In lung cancer, quercetin inhibits glucose-6-phosphate dehydrogenase (G6PD), reducing NADPH levels and promoting the degradation of the EGFRT790M mutation, thereby delaying TKI resistance [56]. It inhibits Axl, triggers apoptosis, and slows tumor growth in non-small cell lung cancer cells with the EGFR-C797S mutation [57].

In breast cancer, in addition to improving epithelial phenotype indicators and preventing tumor growth in xenograft model, it also suppresses invasiveness, reverses EMT, and lowers IGF-1R activation in TNBC cells [59]. By inhibiting

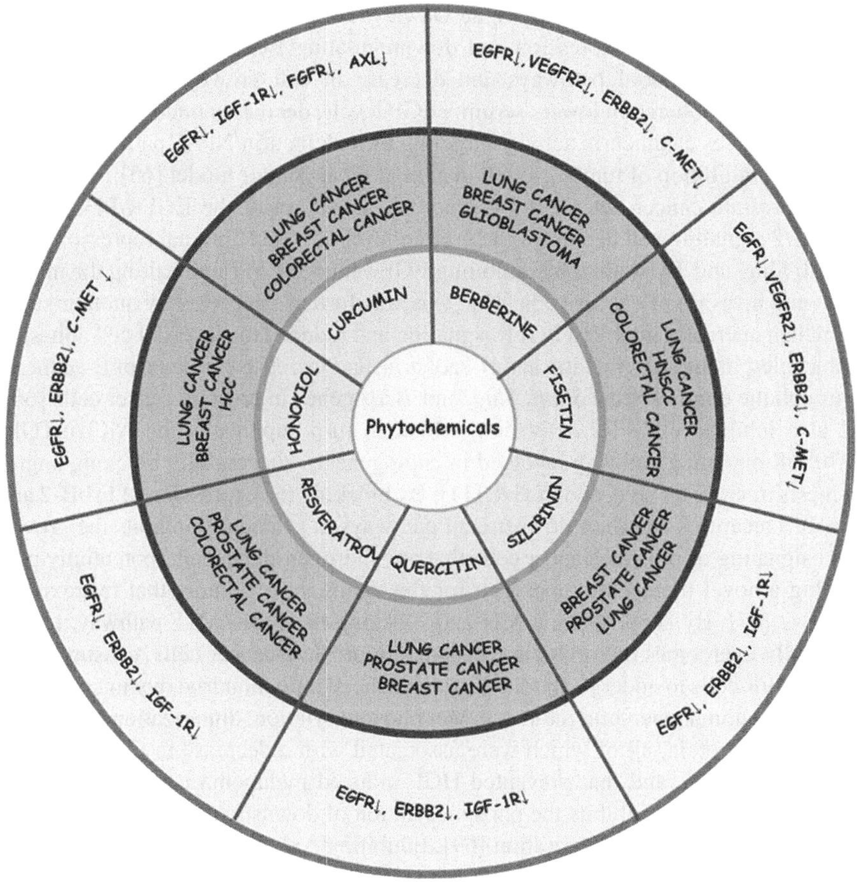

Fig. 2 Effects of phytochemicals on RTKs and associated cancers. The center and second lane are dedicated to several phytochemicals known for their anticancer potential, such as curcumin, resveratrol, quercetin, berberine, silibinin, honokiol, and fisetin. Surrounding this central node, the second layer specifies the individual cancers that each phytochemical influences, such as lung, breast, colorectal, and prostate cancers. The outermost layer focuses on the RTKs that are affected by these phytochemicals, such as EGFR (epidermal growth factor receptor), VEGFR (vascular endothelial growth factor receptor), PDGFR (platelet-derived growth factor receptor), and IGF-1R (insulin-like growth factor 1 receptor). This diagram provides a comprehensive picture of interrelated roles of phytochemicals in cancer regulation, highlighting their therapeutic potential in cancer treatments

EGFR/ERK pathway, it exerts antiproliferative and pro-apoptotic actions that increase the chemosensitivity in MCF-7 breast cancer cells [62]. By significantly lowering the invasiveness and viability of breast cancer cells, increasing E-cadherin, decreasing vimentin and N-cadherin, and blocking angiogenic signaling pathways, quercetin gold nanoparticles (AuNPs-Qu-5) reduce tumor volume and prolong the survival in a DMBA-induced Sprague Dawley rat model of breast cancer [63]. By increasing miR-146a expression and downregulating EGFR, increased apoptosis through mitochondrial pathways and decrease in cell growth and invasion are reported [64]. Quercetin lowers serum VEGF levels, decreases microvessel density, and suppresses calcineurin activity targeting the calcineurin/NFAT pathway leading to strong inhibition of tumor growth in animal breast tumor model [65].

In prostate cancer, quercetin prevents the activation of the EGFR/PI3K/AKT/ERK1/2 signaling and decreases the protein level of transcriptional repressors, like Snail, Slug, and Twist, thereby inhibiting EGF-induced EMT and halting the migration and invasion of cancer cells [10]. This was further supported by another study in which quercetin inhibited EGFR signaling and reduced the levels of cell adhesion molecules, including vimentin and N-cadherin, and increased the level of E-cadherin through the control of the *Snail*, *Slug*, and *Twist* genes in prostate cancer cells [66]. It also inhibits VEGFR2 activation, which in turn suppresses the AKT/mTOR/P70S6K signaling pathway involved in angiogenesis, successfully blocking angiogenesis in vitro, ex vivo, and in vivo [11]. By blocking the expression of ErbB-2 and ErbB-3 receptors and their downstream pathways, it efficiently inhibits the growth and signaling of prostate cancer cells that are androgen-dependent, potentially providing a novel therapeutic approach for the treatment of tumors that overexpress ErbB-2 [67]. By suppressing c-Met and blocking the PI3K/AKT pathway, it successfully overcomes doxorubicin resistance in prostate cancer cells, causing resistant PC3/R cells to undergo apoptosis and enhanced mitochondrial dysfunction [68].

In melanoma, quercetin reduces c-Met phosphorylation, dimerization, and levels of its total protein, all of which were associated with a decrease in fatty acid synthase expression, and that prevented HGF-induced melanoma cell migration and invasion. Further, it inhibits the phosphorylation of downstream targets, like Gab1, FAK, and PAK, and cell migration [69]. Inhibiting Axl expression while changing PIM-1 levels, it lowers the viability of melanoma cells. It also decreased HIF-1 alpha expression and demonstrated higher efficacy than kinase inhibitors [70].

In other cancers, such as head and neck cancer, it inhibits the activation of MMP-9 and MMP-2 and reduces migration, invasion, and colony formation in EGFR-overexpressing HNSCC cells [71]. It highly elevates the cytotoxic effects on colorectal cancer cells when encapsulated in nanoliposomes, increasing cell death and reducing the EGFR gene expression [61]. By suppressing the Ras/Raf and PI3K/AKT pathways, reducing EGFR and N-cadherin expression, decreasing MMP-2 and MMP-9 activity and gene expression, and modifying tight junction proteins, it lowers the proliferation, survival, migration, and adhesion of metastatic ovarian cancer PA-1 cells [72]. In hepatocellular carcinoma, quercetin suppresses EGFR/AKT signaling pathway to partially overcome sorafenib resistance (SOR) in Huh7R cells, offering a great approach to treating acquired resistance in HCC [73].

Overall, quercetin inhibits cancer cell proliferation, migration, and invasion by targeting RTKs like EGFR, VEGFR, and c-Met, disrupting key signaling pathways including PI3K/AKT, MAPK, and NF-κB, and promotes apoptosis across various cancers.

2.5 Resveratrol

Resveratrol (3,5,4′-trihydroxystilbene, or RSV), a naturally occurring polyphenolic molecule, is found in over 70 plant species, including grapes (*Vitis vinifera)*, knotweed (*Polygonum cuspidatum)*, peanuts (*Arachis hypogaea*), blueberries (*Vaccinium cyanococcus*), and mulberries (*Morus alba*). Resveratrol has been investigated for a number of malignancies, including those of the breast, blood, kidney, liver, bladder, thyroid, prostate, lung, and skin. Resveratrol has been shown in several in vitro and in vivo investigations to function as a chemopreventive agent through anti-inflammatory properties, cell growth suppression, and apoptosis induction [74–76]. At the molecular level, resveratrol, with its anticancer effects, inhibits many RTKs and other protein kinases, including EGFR, AKT, and ERK1/2 [77].

In lung cancer, resveratrol decreases cell viability and proliferation, elevates hydrogen peroxide production and apoptotic markers, and decreases the levels of pEGFR, pAKT, c-FLIP, and NF-κB [75]. It targets glucose metabolism system, specifically via suppressing PI3K/AKT and HK2 signaling pathways, both of which are frequently dysregulated in NSCLC patients due to PIK3CA amplification or EGFR mutations [78]. The NGFR-AMPK-mTOR signaling pathway that controls the apoptosis and autophagy is also targeted by resveratrol in A549 lung cancer cells [76]. Through a mechanism that involves increased production of ROS and enhanced DLC1 expression, it causes cancer cell senescence. Resveratrol inhibits the DYRK1A-EGFR pathway, causing DNA damage and an increase in γH2AX phosphorylation while downregulating the DNA repair proteins. This ultimately promotes senescence in breast and liver cancer cells and has been shown to be effective in reducing tumor volume in a chicken embryo xenograft model [79].

Additionally, resveratrol causes transcriptional regulation of PTEN by the androgen receptor, highlighting a significant connection between nuclear receptors and the PI3K/AKT signaling pathway. It suppresses cellular proliferation by the processes that are both AR-dependent and AR-independent in prostate cancer. It increases PTEN expression by inhibiting AR activity while simultaneously binding to EGFR and inhibiting its phosphorylation, thereby decreasing AKT phosphorylation [80]. By inhibiting EMT process and interfering with HGF-c-Met interaction between the stroma and epithelium, it prevents migration of epithelial prostate cancer cells [81]. Resveratrol inhibits the invasion of oral squamous cell carcinoma (OSCC) cells by interfering with the RCP-induced signaling pathways, including β1 integrin, EGFR, and β-catenin, which together result in elevated Zeb1 expression and MT1-MMP upregulation [82].

Resveratrol and curcumin work together to inhibit tumor growth, reduce cell proliferation, promote apoptosis, and downregulate NF-κB and RTKs, like EGFR and IGF-1R, leading to higher efficacy in the colon cancer prevention and treatment [74, 83]. By downregulating IGF-1R and its downstream AKT/Wnt pathways, it significantly inhibits the proliferation of colon cancer cells and promotes apoptosis despite IGF-1 exposure. It activates the p53 pathway and suppresses the growth of ovarian cancer cells by interfering with the interactions between the ERα and

IGF-1R pathways [84]. By lowering c-Met expression, reducing anchorage-dependent and anchorage-independent growth, suppressing HGF-induced cell invasion, and inhibiting the HGF-c-Met signaling pathway, resveratrol significantly decreases the growth of human hepatocellular carcinoma [77].

Moreover, resveratrol inhibits cancer cell growth, migration, and invasion by targeting RTK, such as EGFR, IGF-1R, and c-Met, and modulates key signaling pathways like PI3K/AKT, NF-κB, and AMPK-mTOR, promoting apoptosis, DNA damage, and senescence.

2.6 Silibinin

Silibinin, a naturally occurring flavonoid derived from the plant *Silybum marianum* (milk thistle plant), has long been used to treat liver diseases and has recently gained notice for its strong antitumor properties against various cancers, including those of the breast, lung, skin, kidney, colon, and prostate. Many researchers have looked into its underlying anticancer mechanisms involving different cellular signaling pathways, including RTK [85].

In lung cancer, silibinin binds to the EGFR to inhibit phosphorylated EGFR levels and suppresses PI3K/AKT and JAK/STAT signaling pathways and lowers the expression of PD-L1 and VEGF. This effectively reduces the proliferation of NSCLC cells by inducing cell cycle arrest and apoptosis and also inhibits cancer cell migration, invasion, and angiogenesis [85]. A novel mechanism in NSCLC was revealed by its effect on EGFR/LOX pathway, which is regulated by EGFR through the PI3K/AKT, SAPK/JNK, and MEK/ERK pathways [86]. By suppressing microvessel density and inhibiting the development of new blood vessels in tumors, it exhibits antitumor activity for both pre-existing and urethane-induced lung adenocarcinomas. This is explained by a decrease in tumor-associated macrophage infiltration, a decrease in angiogenic cytokine levels, an increase in TIMP-1 and TIMP-2 levels, and the suppression of STAT3, HIF-1α, and NF-κB activation. Additionally, it raises Tie-2 and Ang-2 levels without altering VEGF levels [87, 88].

In breast cancer, with expression levels of EGFR, MYC, and TXNIP showing metabolic aggressiveness, silibinin decreases glucose utilization and important metabolic pathways in TNBC, preventing cell proliferation and also increasing the sensitivity to paclitaxel [89]. Targeting apoptotic factors, such as BCL2L11, PTEN, and caspase-9, and downregulation of expression of miR-20b, it has shown antiproliferative effects in breast cancer models [90].

In prostate cancer, silibinin targets several pathways, including DNA repair signaling through inhibiting the nuclear translocation of EGFR, and functions as a strong radiosensitizer in human prostate cancer cells while offering considerable protection to healthy tissues from radiation-induced harm [91]. It causes MCF-7 cells to undergo considerable apoptosis by upregulating the expression of Bax and FasL which shows potent antitumor effects by inhibiting IGF-1R signaling pathway in prostate cancer cells [92]. Silibinin inhibits androgen-independent prostate cancer PC-3 cell proliferation by increasing IGFBP-3 accumulation and mRNA expression while also reducing insulin receptor substrate 1 tyrosine phosphorylation, suggesting its antiproliferative effects via modulation of the IGF-1 receptor signaling pathway [93]. By upregulating IGFBP-3, it sequesters IGF-1 and suppresses the ligand-induced activation of IGF-1R-mediated signaling pathway in prostate cancer cells [94]. Dietary silibinin significantly suppresses the growth of advanced human prostate carcinoma (DU145) xenografts in nude mice, with its anticancer effects linked to increased plasma levels of insulin-like growth factor-binding protein-3 (IGFBP-3), supporting its potential as a therapeutic agent for prostate cancer [95].

Silibinin protects against UVB-induced skin damage in A431 cells by enhancing autophagy through the suppression of the IGF-1R signaling pathway, promoting cell survival and reducing apoptosis [96]. Silibinin effectively inhibits growth and induces apoptosis in hedgehog inhibitor-resistant basal cell carcinoma (BCC) cells by modulating key signaling pathways, including EGFR, AKT, ERK1/2, and Gli-1 [97]. Silibinin and its oxidation product 2,3-dehydrosilybin (DHS) effectively inhibit growth, induce apoptosis, and suppress key signaling pathways, including EGFR, ERK1/2, AKT, STAT3, NF-κB, and AP-1, in BCC cells [98]. Silibinin inhibits EGFR-induced migration and invasion of renal cell carcinoma (RCC) cells by blocking EGFR signaling and reducing MMP-9 expression and activity [99].

Overall, silibinin inhibits cancer progression by targeting receptor tyrosine kinases such as EGFR and IGF-1R and blocking downstream signaling pathways such as PI3K/AKT, JAK/STAT, and VEGF, reducing cell proliferation, migration, invasion, and angiogenesis in various cancers.

2.7 *Honokiol*

Honokiol is a biphenolic compound obtained from the root and bark of the plant, *Magnolia officinalis*, which has an ancient history of use in traditional medicine for its antioxidant, anti-inflammatory, muscle relaxation, anti-allergic, and antibacterial properties, with a well-established safety profile. Recently, it has gained interest as a potential anticancer agent. Preclinical studies show that it has anti-inflammatory, antioxidant, and antithrombotic activities. Notably, its anticancer effects such as induction of apoptosis and suppression of proliferation in various cancer types, including leukemia, multiple myeloma, lung cancer cells, and colorectal cancer, are reported. Recent studies have looked into honokiol's diverse biological and pharmacological effects, including its anti-inflammatory, antioxidant, and anticancer activities [100–102].

In lung cancer, honokiol has significant pro-apoptotic and antiproliferative effects on EGFR-overexpressing cells and NNK-induced animal model of lung cancer, where it suppresses total as well as phosphorylated levels of EGFR, AKT, and STAT3 [102]. It effectively inhibits cell growth and induces cell death in lung adenocarcinoma, with its action dependent on the Lyn and EGFR signaling pathways, showing greater efficacy in the EGFR mutant PC-9 cell line than in A549 cells, and this antitumor effect is diminished after Lyn knockdown [103].

In breast cancer, honokiol inhibits the growth of various human breast cancer cell lines, including drug-resistant ones, leading to G1 phase cell cycle arrest and caspase-dependent apoptosis. Its effect was influenced by the levels of HER-2, and additionally, it downregulated the PI3K/AKT/mTOR pathway [100]. Another study shows that honokiol inhibited growth in ER-negative breast cancer cells by processes involving the activation of apoptosis and cell cycle arrest in the G0/G1 phase by modulating signaling pathway, specifically, the c-Src/EGFR and AKT/mTOR cascades [104].

In head and neck cancer, it reduces EGFR and mTOR levels and induces apoptosis, hence inhibiting cell survival in HNSCC cells [101]. Other studies also show that it inhibited the growth of head and neck cancer and induces cell death, via inhibiting EGFR signaling and AKT, MAPK, and STAT3 downstream pathways [105]. It has been shown to affect EGFR signalling via STAT3 in head and neck cancer, potentially increasing the efficacy of EGFR-targeted treatments [106]. Honokiol exhibits the capacity to inhibit the glioma stem cell proliferation by inducing apoptosis through inhibiting EGFR-mediated JAK-STAT3 signaling pathway, promoting E-cadherin expression to suppress EMT to impede the progression of gliomas and glioblastomas [107].

Overall, honokiol targets receptor tyrosine kinases by inhibiting their downstream signaling pathways, such as AKT, STAT3, and mTOR, to reduce cell proliferation, migration, and survival in lung, breast, and head and neck cancers and gliomas, making it a promising RTK-targeted cancer therapeutic.

3 RTK Inhibitors in Combination with Phytochemicals

One kind of targeted medication that prevents the activity of oncogenic tyrosine kinases (TKs) is called a tyrosine kinase inhibitor (TKI). Their modes of action may involve interacting with ATP binding kinase activation domains reducing the phosphorylation activity. This impairs tumor cell DNA repair, promotes apoptosis, and inhibits cell division during the G1 phase. Most of these inhibitors are safe and can be given orally. In 2019, the FDA authorized 48 inhibitors for protein kinases, and 25 of those inhibit receptor tyrosine kinases. A few well-known EGFR inhibitors have been approved for the treatment of a variety of malignancies of epithelial origin. These include erlotinib, gefitinib, and lapatinib, along with monoclonal antibodies including panitumumab, cetuximab, and trastuzumab. Lenvatinib acts as VEGFR inhibitor. An increasing amount of research has been conducted on combinatorial effects of phytochemicals with different multi-kinase inhibitors, including regorafenib, sorafenib, and sunitinib. These comprise breast cancer, hepatocellular carcinoma, squamous cell carcinoma of the head and neck, advanced lung cancer, and melanoma (Table 2). However, problems with drug resistance and side effects frequently limit the benefits of these therapies for many individuals with cancer [2].

Silibinin when combined with tyrphostin AG1024, an inhibitor of IGF-1R autophosphorylation, enhances growth inhibition and apoptosis in breast cancer cells. Additionally, it activates apoptotic pathways independently of IGF-1R signaling, suggesting that blocking IGF-1R could further increase the antitumor efficacy of silibinin [92]. When combined with EGFR inhibitors, lapatinib and AG1478, silibinin efficiently inhibits the activation of EGFR/ERK1/2 and lowers the EGF-induced CD44 and MMP-9 expression in breast cancer cells [108]. The silibinin combination with EGFR inhibitors, such as AG1478 and gefitinib, blocks the phosphorylation of STAT3, and EGF-induced fibronectin expression in TNBC cells, suggesting that it may be used as an adjuvant therapeutic agent against TNBC [109]. Berberine combination with lapatinib increases AKT signaling in breast cancer cells and promotes chemoresistance; however, the combination also downregulated genes linked to proliferation, such as DNMT1 and CDK6 [110]. In addition to downregulating EGFR/MEK/ERK phosphorylation, berberine when combined with EGFR inhibitors, neratinib and afatinib, lowers IL-8 expression and enhances inhibition of cell invasiveness in breast cancer [40].

In hepatocellular carcinoma (HCC), by dramatically reducing the EGFR-PI3K-AKT signaling in resistant cell, the combination of curcumin with lenvatinib successfully reverses lenvatinib resistance in hepatocellular carcinoma [111]. Silibinin elevates the antiproliferative effects of gefitinib and sorafenib on hepatocellular carcinoma cells by the inhibition of EGFR-mediated AKT signaling pathway [112].

In erlotinib-resistant oral squamous cell carcinoma cells, quercetin elevates the effectiveness of erlotinib by re-sensitizing the cells to erlotinib treatment by decreasing EMT-related proteins, inhibiting the PKM2-mediated glycolytic pathway, and increasing apoptosis via the EGFR signaling pathway [113]. Honokiol has the potential to act as a synergistic phytochemical in overcoming resistance to EGFR-targeted therapies because it increases growth inhibition in human HNSCC cell lines when treated in combination with the EGFR inhibitor, cetuximab, by inducing apoptosis and downregulating key signaling pathways like MAPK and AKT [105]. The combination of honokiol and osimertinib effectively reduces colony formation and growth, increases apoptosis, and decreases survival in a variety of osimertinib-resistant cell lines. It also significantly reduces growth in osimertinib-resistant xenograft tumors with mutations, such as C797S, 19del, and T790M, and primarily through increased degradation of Mcl-1. A synthetic derivative of honokiol exerts comparable but higher effects [114].

In lung cancer, berberine reduces the survival of NSCLC cells, and when combined with osimertinib, a c-Met inhibitor, it selectively targets and decreases the survival of c-Met-amplified osimertinib-resistant EGFR mutant NSCLC cells by inducing Bim-dependent apoptosis and also stops the growth of these tumors in vivo [36, 38]. However, more studies are needed to address the acquired resistance to osimertinib and other EGFR tyrosine kinase inhibitors, with berberine and its derivatives. The combination of icotinib and berberine (BBR) exerts a synergistic inhibitory effect on EGFR-resistant NSCLC, indicating that it could be useful to overcome icotinib resistance [36]. Berberine and erlotinib work together to suppress cell growth, resulting in much larger decreases in pEGFR and pAKT levels along with

Table 2 Phytochemicals and RTKI combination studies in various cancers

S. no.	Cancer	Phytochemicals + RTK inhibitors	In vitro	In vivo	Mode of action	References
1	Breast cancer	Berberine + lapatinib	MD-MB-231, MCF-7	–	Downregulates CDK6 and DNMT1 and upregulates AKT by upregulating alternative growth factors such as PlGF and FGFs	[110]
		Berberine + neratinib and afatinib	MCF7, Hs578T, and MDA-MB231	–	Downregulates the expression of IL-8 by blocking the EGFR/MEK/ERK	[40]
		Silibinin + AG1478 and lapatinib	SKBR3 and BT474	–	Reduces EGFR ligand-induced CD44 expression	[108]
		Silibinin + AG1478 and gefitinib	T47D, ZR75–1, MDA468, and BT20	–	Reduces EGF-induced fibronectin expression in TNBC cells by inhibiting STAT3 phosphorylation	[109]
2	HCC	Curcumin + lenvatinib	Huh-7, SNU449, PLC-PRF-5, SNU398, and Sk-Hep-1	–	Inhibits the EGFR-PI3K-AKT signaling pathway	[111]
		Silibinin + either sorafenib or gefitinib	Huh-BAT, Hep3B, Huh7, HepG2, HepG2, PLC/PRF5, SNU449, SNU387, SNU398, SNU475, and SNU761	–	Suppresses EGFR-mediated AKT signaling	[112]
3	HNSCC	Honokiol + cetuximab	Cal-33, 1483 cells, 686LN and 686LNR30 cells	Xenograft	Reduces MAPK and AKT	[105]
		Quercetin + erlotinib	HSC-3, OECM-1, and SAS	Xenograft	Downregulating EMT-related proteins and promoting apoptosis through the EGFR signaling pathway	[113]
		Honokiol + osimertinib	PC-9/AR, PC-9/GR/AR, PC-9/3 M, and HCC827/AR	Xenograft	Increased degradation of Mcl-1	[114]

4	Lung cancer	Honokiol + gefitinib	HCC827, H1975, SPC-A1, H460, H441, H226, H522, H1650, H1993	Xenograft	Inhibits EGFR signaling by promoting degradation of HSP90	[118]
		Honokiol + cetuximab	H226	Patient-derived xenografts	Downregulating key signaling pathways, including MAPK and AKT, while also reducing DRP1 phosphorylation and reactive oxygen species levels	[119]
		Silibinin + erlotinib	PC9	Xenograft	Repressing mesenchymal markers such as SNAIL, ZEB, and N-cadherin, effectively inducing a mesenchymal-to-epithelial transition (MET)	[121]
		Silibinin + gefitinib and erlotinib	A549, H460, H1299, H1975,HCC827, and 293 T	Xenograft	Inhibiting EGFR family receptor dimerization	[120]
		Curcumin + gefitinib	A549, H1299, H1650, H1975, PC-9	Xenograft	Promoting EGFR ubiquitination and apoptosis through modulation of p38 activation	[116]
		Berberine + icotinib	H460, H1299, HCC827, BEAS-2B	Xenograft	Inhibiting EGFR protein expression	[36]
		Berberine + osimertinib	HCC827, pc-9	Xenograft	Enhancing apoptosis through increased Bim levels and decreased Mcl-1 while effectively suppressing the growth of MET	[38]
		Berberine + erlotinib	A431, NCI-H1666, NCI-H441, and NCI-H1781	Xenograft	Reductions in pEGFR and pAKT levels, as well as decreased expression of cyclin D and Bcl-2	[115]
		Fisetin + erlotinib	HCC827	–	Inhibiting aberrant activation of the MAPK and AKT signaling pathways through AXL suppression	[53]
		Fisetin + erlotinib	A549 and H1299	–	Upregulating epithelial markers, like E-cadherin and ZO-1, downregulating mesenchymal markers, such as vimentin and N-cadherin, andreducing stem cell signature markers CD44 and CD133	[52]
		Quercetin + erlotinib	A549 and NCI-H460	–	Reducing the expression of P-glycoprotein and nuclear EGFR	[117]
5	Melanoma	Curcumin + gefitinib	A375.S2	–	Activating ROS and mitochondrial signaling and enhancing the effects of EGFR inhibition	[35]

lower expression of cyclin D and Bcl-2 than with either treatment alone. This combination also effectively inhibits tumor growth in nude mice [115].

Curcumin increases the ubiquitination EGFR and cell death and increases the antitumor effects of gefitinib in NSCLC concomitant with the reduction of gefitinib-induced gastrointestinal side effects, indicating the potential of curcumin in adjuvant therapy [116]. By preventing abnormal activation of the MAPK and AKT signaling pathways through Axl suppression, fisetin increases the susceptibility of erlotinib-resistant lung adenocarcinoma cells to erlotinib, suggesting that it may be used as a therapeutic agent to reverse the acquired erlotinib resistance [53]. Fisetin decreases the migration as well as invasion of NSCLC by attenuating EMT and increasing epithelial markers like ZO-1 and E-cadherin and also lowers the stem cell signature markers CD44 and CD133. It also enhances the cytotoxic effects of erlotinib, suggesting its potential use against lung cancer metastases [52]. The combination of erlotinib and quercetin in solid lipid nanoparticles (EQNPs) significantly enhances the anticancer effects against erlotinib-resistant A549 and NCI H460 cells by increasing drug uptake, inducing apoptosis, and reducing the expression of P-glycoprotein and nuclear epidermal growth factor receptor, suggesting EQNPs as a promising targeted therapy for non-small cell lung cancer with improved efficacy and reduced side effects [117]. In melanoma, by promoting ROS and mitochondrial signaling pathways, curcumin significantly reduces the cell viability and triggers apoptosis in vemurafenib-resistant melanoma cells. When paired with gefitinib, it enhances the effects of EGFR suppression, indicating that curcumin may be a useful treatment option for drug resistance in melanoma [35].

Honokiol and erlotinib combination demonstrated enhanced anticancer activity in tumorigenic 1170 bronchial cells and EGFR mutant cell lines, H1650 and H1975, suggesting the potential use of this combination to treat pulmonary cancers associated with EGFR abnormalities or resistance [102]. In NSCLC xenograft models, liposomal honokiol causes cell death and antitumor effects. It reduces EGFR signaling by promoting the lysosomal breakdown of HSP90, which results in ER stress and autophagy. Autophagy inhibition increases the anticancer efficacy of liposomal honokiol, which is authorized for first-in-human clinical trials in NSCLC [118]. In cetuximab-resistant NSCLC and patient-derived xenografts, honokiol enhances the antiproliferative effects of cetuximab which is mainly due to alteration in mitochondrial activity and suppression of the HER family signaling pathways [119].

Silibinin induces mesenchymal-to-epithelial transition (MET) in resistant cells and inhibits their migratory phenotype by restoring high miR-200c levels and reversing the EMT-related microRNA signature and suppressing SNAIL, ZEB, and N-cadherin. Further, by specifically suppressing the dimerization of EGFR family receptors and subsequently EGFR signaling, silibinin enhances the effectiveness of gefitinib and erlotinib in overcoming T790M-induced resistance in lung cancer. This study also showed a significant reduction of tumor growth in erlotinib-resistant PC-9 xenografts. This study also showed a significant reduction of tumor growth in erlotinib-resistant PC-9 xenografts [120]. Another study has also reported its tumor-suppressing effect when combined with erlotinib showing nearly 50% reduction in tumor volume in erlotinib-refractory NSCLC xenografts. These studies suggest that silibinin has combinatorial therapeutic potential against NSCLC [121].

Overall, the combination of phytochemicals with tyrosine kinase inhibitors (TKIs) demonstrates significant potential in overcoming drug resistance and enhancing antitumor efficacy across various cancers, including breast cancer, hepatocellular carcinoma, non-small cell lung cancer (NSCLC), and melanoma.

4 Conclusions

One intriguing approach to improving cancer treatments is the investigation of natural agents as receptor tyrosine kinase (RTK) inhibitors. As this area of study develops, a number of potential viewpoints become apparent. Firstly, investigating new natural compounds or products with RTK-inhibitory properties can take a lead by the incorporation of high-throughput screening techniques. Technological progresses, like genomics and CRISPR technique, may help us to understand the molecular mechanisms by which these agents function, paving the way to more specialized and efficient applications. Secondly, there is a chance that phytochemicals in combination with TKIs that are currently available will create synergistic efficacy for the treatment. These combinations may improve therapeutic efficacy and also resolve resistance concerns. Improving the application of these natural compounds in clinical trials will need an understanding of their pharmacokinetics and bioavailability. Furthermore, personalized medicine may be beneficial in customizing therapies according to each patient's tumor characteristics. Clinicians might select the most appropriate natural product(s) for cancer treatment, improving their efficacy while decreasing adverse effects, by examining the precise RTK mutations in a patient's malignancy. Lastly, increased awareness and funding for research on natural products will be essential. Collaborations between researchers, clinicians, and the pharmaceutical industry can facilitate the translation of laboratory findings into clinical applications, ultimately leading to more effective and accessible cancer treatments.

Overall, natural agents and their derivatives have a strong potential as RTK inhibitors in cancer treatment. To fully utilize these natural chemicals and enhance the therapeutic responses in cancer patients, further research for their precise mechanisms and safety and interdisciplinary cooperation will be essential. As highlighted in Fig. 3, natural phytochemicals can suppress RTKs and impact key downstream signaling pathways, such as PI3K/AKT, ERK, JAK/STAT, mTOR, JNK, and NF-κB, which are crucial in cancer progression. This insight into the molecular actions of natural agents will be fundamental in enhancing their therapeutic efficacy and addressing challenges like drug resistance in cancer therapy.

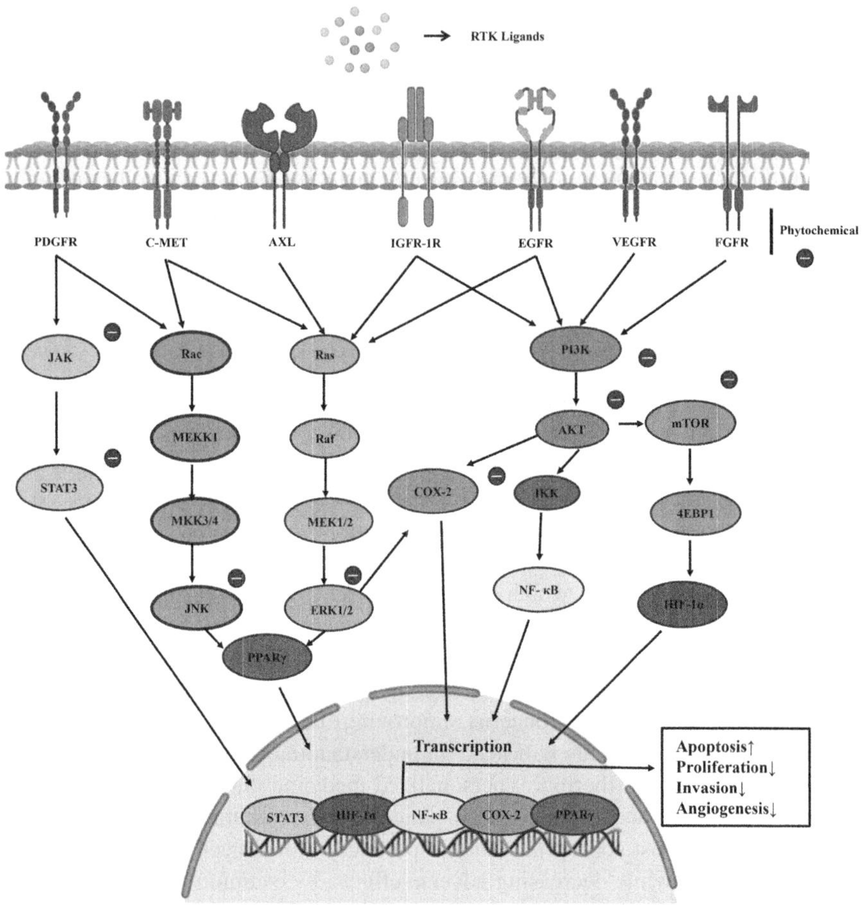

Fig. 3 An overview of the phytochemical suppression of RTKs and downstream components, PI3K/Akt, ERK, JAK/STAT, mTOR, JNK, and NF-κB. RTK receptor tyrosine kinase; PDGFR platelet-derived growth factor receptor; c-MET mesenchymal epithelial transition factor; IGF-1R insulin-like growth factor 1 receptor; VEGF vascular endothelial growth factor; VEGFR vascular endothelial growth factor receptor; FGFR fibroblast growth factor receptor; JAK Janus kinase; STAT signal transducer and activator of transcription; Rac substrate of the Ras-related C3 botulinum toxin; MEKK1/MAP3K mitogen-activated protein kinase kinase kinase 1; MKK3/4 mitogen-activated protein kinase kinase 3/4; JNK c-Jun N-terminal kinases; PPARγ peroxisome proliferator-activated receptor γ; Ras rat sarcoma virus protein; Raf, rapidly accelerated fibrosarcoma protein; ERK1/2 extracellular signal-regulated kinase 1/2; MEK1/2 mitogen-activated protein kinase kinase 1/2; COX-2 cyclooxygenase-2; PI3K phosphoinositide 3-kinase; Akt/PKB transformation of Ak strain/protein kinase B; IKK inhibitor of nuclear factor kappa-B kinase; mTOR mammalian target of rapamycin; NF-κB nuclear factor kappa-light-chain-enhancer of activated B cells; 4EBP1 eukaryotic translation initiation factor 4E (eIF4E)-binding protein 1; HIF-α hypoxia-inducible factor-alpha

Acknowledgment The figures are created with BioRender.com

References

1. Pérez-Herrero E, Fernández-Medarde A. Advanced targeted therapies in cancer: drug nanocarriers, the future of chemotherapy. Eur J Pharm Biopharm. 2015;93:52–79.
2. Sudhesh Dev S, Zainal Abidin SA, Farghadani R, Othman I, Naidu R. Receptor tyrosine kinases and their signaling pathways as therapeutic targets of curcumin in cancer. Front Pharmacol. 2021;12:772510.
3. Ranjan A, Ramachandran S, Gupta N, Kaushik I, Wright S, Srivastava S, et al. Role of phytochemicals in cancer prevention. Int J Mol Sci. 2019;20(20).
4. Surh YJ. Cancer chemoprevention with dietary phytochemicals. Nat Rev Cancer. 2003;3(10):768–80.
5. Newman DJ, Cragg GM. Natural products as sources of new drugs over the last 25 years. J Nat Prod. 2007;70(3):461–77.
6. Efferth T. Molecular pharmacology and pharmacogenomics of artemisinin and its derivatives in cancer cells. Curr Drug Targets. 2006;7(4):407–21.
7. Kuete V, Efferth T. Cameroonian medicinal plants: pharmacology and derived natural products. Front Pharmacol. 2010;1:123.
8. Kuete V, Efferth T. Pharmacogenomics of Cameroonian traditional herbal medicine for cancer therapy. J Ethnopharmacol. 2011;137(1):752–66.
9. Gillet JP, Efferth T, Remacle J. Chemotherapy-induced resistance by ATP-binding cassette transporter genes. Biochim Biophys Acta. 2007;1775(2):237–62.
10. Bhat FA, Sharmila G, Balakrishnan S, Arunkumar R, Elumalai P, Suganya S, et al. Quercetin reverses EGF-induced epithelial to mesenchymal transition and invasiveness in prostate cancer (PC-3) cell line via EGFR/PI3K/AKT pathway. J Nutr Biochem. 2014;25(11):1132–9.
11. Pratheeshkumar P, Budhraja A, Son YO, Wang X, Zhang Z, Ding S, et al. Quercetin inhibits angiogenesis mediated human prostate tumor growth by targeting VEGFR- 2 regulated AKT/mTOR/P70S6K signaling pathways. PLoS One. 2012;7(10):e47516.
12. Zhan Y, Chen Y, Liu R, Zhang H, Zhang Y. Potentiation of paclitaxel activity by curcumin in human breast cancer cell by modulating apoptosis and inhibiting EGFR signaling. Arch Pharm Res. 2014;37(8):1086–95.
13. Kim KC, Baek SH, Lee C. Curcumin-induced downregulation of Axl receptor tyrosine kinase inhibits cell proliferation and circumvents chemoresistance in non-small lung cancer cells. Int J Oncol. 2015;47(6):2296–303.
14. Lev-Ari S, Starr A, Vexler A, Karaush V, Loew V, Greif J, et al. Inhibition of pancreatic and lung adenocarcinoma cell survival by curcumin is associated with increased apoptosis, down-regulation of COX-2 and EGFR and inhibition of Erk1/2 activity. Anticancer Res. 2006;26(6B):4423–30.
15. Zhang L, Tao X, Fu Q, Ge C, Li R, Li Z, et al. Curcumin inhibits cell proliferation and migration in NSCLC through a synergistic effect on the TLR4/MyD88 and EGFR pathways. Oncol Rep. 2019;42(5):1843–55.
16. Lin HP, Kuo LK, Chuu CP. Combined treatment of curcumin and small molecule inhibitors suppresses proliferation of A549 and H1299 human non-small-cell lung cancer cells. Phytother Res. 2012;26(1):122–6.
17. Fan S, Xu Y, Li X, Tie L, Pan Y, Li X. Opposite angiogenic outcome of curcumin against ischemia and Lewis lung cancer models: in silico, in vitro and in vivo studies. Biochim Biophys Acta. 2014 Sep;1842(9):1742–54.

18. Wang R, Yu H, Chen P, Yuan T, Zhang J. Integrated transcriptome and molecular docking to identify the hub superimposed attenuation targets of curcumin in breast cancer cells. Int J Mol Sci. 2023;24(15)
19. Xia Y, Jin L, Zhang B, Xue H, Li Q, Xu Y. The potentiation of curcumin on insulin-like growth factor-1 action in MCF-7 human breast carcinoma cells. Life Sci. 2007;80(23):2161–9.
20. Sun XD, Liu XE, Huang DS. Curcumin induces apoptosis of triple-negative breast cancer cells by inhibition of EGFR expression. Mol Med Rep. 2012;6(6):1267–70.
21. Deng Z, Chen G, Shi Y, Lin Y, Ou J, Zhu H, et al. Curcumin and its nano-formulations: defining triple-negative breast cancer targets through network pharmacology, molecular docking, and experimental verification. Front Pharmacol. 2022;13:920514.
22. Somers-Edgar TJ, Scandlyn MJ, Stuart EC, Le Nedelec MJ, Valentine SP, Rosengren RJ. The combination of epigallocatechin gallate and curcumin suppresses ER alpha-breast cancer cell growth in vitro and in vivo. Int J Cancer. 2008;122(9):1966–71.
23. Gallardo M, Calaf GM. Curcumin inhibits invasive capabilities through epithelial mesenchymal transition in breast cancer cell lines. Int J Oncol. 2016;49(3):1019–27.
24. Gallardo M, Kemmerling U, Aguayo F, Bleak TC, Muñoz JP, Calaf GM. Curcumin rescues breast cells from epithelial-mesenchymal transition and invasion induced by anti-miR-34a. Int J Oncol. 2020;56(2):480–93.
25. Kim JH, Xu C, Keum YS, Reddy B, Conney A, Kong ANT. Inhibition of EGFR signaling in human prostate cancer PC-3 cells by combination treatment with beta-phenylethyl isothiocyanate and curcumin. Carcinogenesis. 2006;27(3):475–82.
26. Boccellino M, Ambrosio P, Ballini A, De Vito D, Scacco S, Cantore S, et al. The role of curcumin in prostate cancer cells and derived spheroids. Cancers (Basel). 2022;14(14)
27. Lu WC, Kao SY, Yang CC, Tu HF, Wu CH, Chang KW, et al. EGF up-regulates miR-31 through the C/EBPβ signal cascade in oral carcinoma. PLoS One. 2014;9(9):e108049.
28. Zhen L, Fan D, Yi X, Cao X, Chen D, Wang L. Curcumin inhibits oral squamous cell carcinoma proliferation and invasion via EGFR signaling pathways. Int J Clin Exp Pathol. 2014;7(10):6438–46.
29. Chen A, Xu J, Johnson AC. Curcumin inhibits human colon cancer cell growth by suppressing gene expression of epidermal growth factor receptor through reducing the activity of the transcription factor Egr-1. Oncogene. 2006;25(2):278–87.
30. Kunnumakkara AB, Diagaradjane P, Anand P, Harikumar KB, Deorukhkar A, Gelovani J, et al. Curcumin sensitizes human colorectal cancer to capecitabine by modulation of cyclin D1, COX-2, MMP-9, VEGF and CXCR4 expression in an orthotopic mouse model. Int J Cancer. 2009;125(9):2187–97.
31. Nautiyal J, Banerjee S, Kanwar SS, Yu Y, Patel BB, Sarkar FH, et al. Curcumin enhances dasatinib-induced inhibition of growth and transformation of colon cancer cells. Int J Cancer. 2011;128(4):951–61.
32. Hosseini SA, Zand H, Cheraghpour M. The influence of curcumin on the downregulation of MYC, insulin and IGF-1 receptors: a possible mechanism underlying the anti-growth and anti-migration in chemoresistant colorectal cancer cells. Medicina (Kaunas). 2019;55(4):90.
33. Choe SR, Kim YN, Park CG, Cho KH, Cho DY, Lee HY. RCP induces FAK phosphorylation and ovarian cancer cell invasion with inhibition by curcumin. Exp Mol Med. 2018;50(4):1–10.
34. Zhang HH, Zhang Y, Cheng YN, Gong FL, Cao ZQ, Yu LG, et al. Metformin in combination with curcumin inhibits the growth, metastasis, and angiogenesis of hepatocellular carcinoma in vitro and in vivo. Mol Carcinog. 2018;57(1):44–56.
35. Chiu YJ, Yang JS, Tsai FJ, Chiu HY, Juan YN, Lo YH, et al. Curcumin suppresses cell proliferation and triggers apoptosis in vemurafenib-resistant melanoma cells by downregulating the EGFR signaling pathway. Environ Toxicol. 2022;37(4):868–79.
36. Chen P, Dai CH, Shi ZH, Wang Y, Wu JN, Chen K, et al. Synergistic inhibitory effect of berberine and icotinib on non-small cell lung cancer cells via inducing autophagic cell death and apoptosis. Apoptosis. 2021;26(11–12):639–56.

37. Ayaz Z, Zainab B, Rashid U, Darwish NM, Gatasheh MK, Abbasi AM. In silico screening of synthetic and natural compounds to inhibit the binding capacity of heavy metal compounds against EGFR protein of lung cancer. Biomed Res Int. 2022;2022:2941962.
38. Chen Z, Vallega KA, Chen H, Zhou J, Ramalingam SS, Sun SY. The natural product berberine synergizes with osimertinib preferentially against MET-amplified osimertinib-resistant lung cancer via direct MET inhibition. Pharmacol Res. 2022;175:105998.
39. Refaat A, Abdelhamed S, Saiki I, Sakurai H. Inhibition of p38 mitogen-activated protein kinase potentiates the apoptotic effect of berberine/tumor necrosis factor-related apoptosis-inducing ligand combination therapy. Oncol Lett. 2015;10(3):1907–11.
40. Kim S, You D, Jeong Y, Yu J, Kim SW, Nam SJ, et al. Berberine down-regulates IL-8 expression through inhibition of the EGFR/MEK/ERK pathway in triple-negative breast cancer cells. Phytomedicine. 2018;50:43–9.
41. Jabbarzadeh Kaboli P, Leong MPY, Ismail P, Ling KH. Antitumor effects of berberine against EGFR, ERK1/2, P38 and AKT in MDA-MB231 and MCF-7 breast cancer cells using molecular modelling and in vitro study. Pharmacol Rep. 2019;71(1):13–23.
42. Huang ZH, Zheng HF, Wang WL, Wang Y, Zhong LF, Wu JL, et al. Berberine targets epidermal growth factor receptor signaling to suppress prostate cancer proliferation in vitro. Mol Med Rep. 2015;11(3):2125–8.
43. Chappell WH, Abrams SL, Franklin RA, LaHair MM, Montalto G, Cervello M, et al. Ectopic NGAL expression can alter sensitivity of breast cancer cells to EGFR, Bcl-2, CaM-K inhibitors and the plant natural product berberine. Cell Cycle. 2012;11(23):4447–61.
44. Ma W, Zhu M, Zhang D, Yang L, Yang T, Li X, et al. Berberine inhibits the proliferation and migration of breast cancer ZR-75-30 cells by targeting Ephrin-B2. Phytomedicine. 2017 Feb;25:45–51.
45. Eom KS, Hong JM, Youn MJ, So HS, Park R, Kim JM, et al. Berberine induces G1 arrest and apoptosis in human glioblastoma T98G cells through mitochondrial/caspases pathway. Biol Pharm Bull. 2008;31(4):558–62.
46. Liu Q, Xu X, Zhao M, Wei Z, Li X, Zhang X, et al. Berberine induces senescence of human glioblastoma cells by downregulating the EGFR-MEK-ERK signaling pathway. Mol Cancer Ther. 2015;14(2):355–63.
47. Wang J, Qi Q, Feng Z, Zhang X, Huang B, Chen A, et al. Berberine induces autophagy in glioblastoma by targeting the AMPK/mTOR/ULK1-pathway. Oncotarget. 2016;7(41):66944–58.
48. Jin F, Xie T, Huang X, Zhao X. Berberine inhibits angiogenesis in glioblastoma xenografts by targeting the VEGFR2/ERK pathway. Pharm Biol. 2018;56(1):665–71.
49. Wang L, Cao H, Lu N, Liu L, Wang B, Hu T, et al. Berberine inhibits proliferation and down-regulates epidermal growth factor receptor through activation of Cbl in colon tumor cells. PLoS One. 2013;8(2):e56666.
50. Li D, Zhang Y, Liu K, Zhao Y, Xu B, Xu L, et al. Berberine inhibits colitis-associated tumorigenesis via suppressing inflammatory responses and the consequent EGFR signaling-involved tumor cell growth. Lab Investig. 2017;97(11):1343–53.
51. Suh Y, Afaq F, Johnson JJ, Mukhtar H. A plant flavonoid fisetin induces apoptosis in colon cancer cells by inhibition of COX2 and Wnt/EGFR/NF-kappaB-signaling pathways. Carcinogenesis. 2009;30(2):300–7.
52. Tabasum S, Singh RP. Fisetin suppresses migration, invasion and stem-cell-like phenotype of human non-small cell lung carcinoma cells via attenuation of epithelial to mesenchymal transition. Chem Biol Interact. 2019;303:14–21.
53. Zhang L, Huang Y, Zhuo W, Zhu Y, Zhu B, Chen Z. Fisetin, a dietary phytochemical, overcomes Erlotinib-resistance of lung adenocarcinoma cells through inhibition of MAPK and AKT pathways. Am J Transl Res. 2016;8(11):4857–68.
54. Yadav M, Kandhari K, Mathan SV, Ali M, Singh RP. Fisetin induces G2/M phase arrest and caspase-mediated cleavage of p21(Cip1) and p27(Kip1) leading to apoptosis and tumor growth inhibition in HNSCC. Mol Carcinog. 2024;63(9):1697–711.

55. Li YS, Qin XJ, Dai W. Fisetin suppresses malignant proliferation in human oral squamous cell carcinoma through inhibition of Met/Src signaling pathways. Am J Transl Res. 2017;9(12):5678–83.
56. Ge Z, Xu M, Ge Y, Huang G, Chen D, Ye X, et al. Inhibiting G6PD by quercetin promotes degradation of EGFR T790M mutation. Cell Rep. 2023;42(11):113417.
57. Huang KY, Wang TH, Chen CC, Leu YL, Li HJ, Jhong CL, et al. Growth suppression in lung cancer cells harboring EGFR-C797S mutation by quercetin. Biomol Ther. 2021;11(9)
58. Richter M, Ebermann R, Marian B. Quercetin-induced apoptosis in colorectal tumor cells: possible role of EGF receptor signaling. Nutr Cancer. 1999;34(1):88–99.
59. Chen WJ, Tsai JH, Hsu LS, Lin CL, Hong HM, Pan MH. Quercetin blocks the aggressive phenotype of triple-negative breast cancer by inhibiting IGF1/IGF1R-mediated EMT program. J Food Drug Anal. 2021;29(1):98–112.
60. Kim WK, Bang MH, Kim ES, Kang NE, Jung KC, Cho HJ, et al. Quercetin decreases the expression of ErbB2 and ErbB3 proteins in HT-29 human colon cancer cells. J Nutr Biochem. 2005;16(3):155–62.
61. Keshavarz F, Dorfaki M, Bardania H, Khosravani F, Nazari P, Ghalamfarsa G. Quercetin-loaded liposomes effectively induced apoptosis and decreased the epidermal growth factor receptor expression in colorectal cancer cells: an in vitro study. Iran J Med Sci. 2023;48(3):321–8.
62. Yang Y, Yan J, Huang J, Wu X, Yuan Y, Yuan Y, et al. Exploring the mechanism by which quercetin re-sensitizes breast cancer to paclitaxel: network pharmacology, molecular docking, and experimental verification. Naunyn Schmiedeberg's Arch Pharmacol. 2023;396(11):3045–59.
63. Balakrishnan S, Bhat FA, Raja Singh P, Mukherjee S, Elumalai P, Das S, et al. Gold nanoparticle-conjugated quercetin inhibits epithelial-mesenchymal transition, angiogenesis and invasiveness via EGFR/VEGFR-2-mediated pathway in breast cancer. Cell Prolif. 2016;49(6):678–97.
64. Tao SF, He HF, Chen Q. Quercetin inhibits proliferation and invasion acts by up-regulating miR-146a in human breast cancer cells. Mol Cell Biochem. 2015;402(1–2):93–100.
65. Zhao X, Wang Q, Yang S, Chen C, Li X, Liu J, et al. Quercetin inhibits angiogenesis by targeting calcineurin in the xenograft model of human breast cancer. Eur J Pharmacol. 2016;781:60–8.
66. Firdous AB, Sharmila G, Balakrishnan S, RajaSingh P, Suganya S, Srinivasan N, et al. Quercetin, a natural dietary flavonoid, acts as a chemopreventive agent against prostate cancer in an in vivo model by inhibiting the EGFR signaling pathway. Food Funct. 2014;5(10):2632–45.
67. Huynh H, Nguyen TTT, Chan E, Tran E. Inhibition of ErbB-2 and ErbB-3 expression by quercetin prevents transforming growth factor alpha (TGF-alpha)- and epidermal growth factor (EGF)-induced human PC-3 prostate cancer cell proliferation. Int J Oncol. 2003;23(3):821–9.
68. Shu Y, Xie B, Liang Z, Chen J. Quercetin reverses the doxorubicin resistance of prostate cancer cells by downregulating the expression of c-met. Oncol Lett. 2018;15(2):2252–8.
69. Cao HH, Cheng CY, Su T, Fu XQ, Guo H, Li T, et al. Quercetin inhibits HGF/c-Met signaling and HGF-stimulated melanoma cell migration and invasion. Mol Cancer. 2015 May;14:103.
70. Rocha-Brito KJP, Clerici SP, Cordeiro HG, Scotá Ferreira AP, Barreto Fonseca EM, Gonçalves PR, et al. Quercetin increases mitochondrial proteins (VDAC and SDH) and downmodulates AXL and PIM-1 tyrosine kinase receptors in NRAS melanoma cells. Biol Chem. 2022;403(3):293–303.
71. Chan CY, Lien CH, Lee MF, Huang CY. Quercetin suppresses cellular migration and invasion in human head and neck squamous cell carcinoma (HNSCC). Biomedicine. 2016;6(3):15.
72. Dhanaraj T, Mohan M, Arunakaran J. Quercetin attenuates metastatic ability of human metastatic ovarian cancer cells via modulating multiple signaling molecules involved in cell survival, proliferation, migration and adhesion. Arch Biochem Biophys. 2021;701:108795.

73. Zhang Z, Wu H, Zhang Y, Shen C, Zhou F. Dietary antioxidant quercetin overcomes the acquired resistance of Sorafenib in Sorafenib-resistant hepatocellular carcinoma cells through epidermal growth factor receptor signaling inactivation. Naunyn Schmiedeberg's Arch Pharmacol. 2024;397(1):559–74.
74. Majumdar APN, Banerjee S, Nautiyal J, Patel BB, Patel V, Du J, et al. Curcumin synergizes with resveratrol to inhibit colon cancer. Nutr Cancer. 2009;61(4):544–53.
75. Wright C, Iyer AKV, Yakisich JS, Azad N. Anti-tumorigenic effects of resveratrol in lung cancer cells through modulation of c-FLIP. Curr Cancer Drug Targets. 2017;17(7):669–80.
76. Li J, Fan Y, Zhang Y, Liu Y, Yu Y, Ma M. Resveratrol induces autophagy and apoptosis in non-small-cell lung cancer cells by activating the NGFR-AMPK-mTOR pathway. Nutrients. 2022;14(12)
77. Gao F, Deng G, Liu W, Zhou K, Li M. Resveratrol suppresses human hepatocellular carcinoma via targeting HGF-c-Met signaling pathway. Oncol Rep. 2017;37(2):1203–11.
78. Li W, Ma X, Li N, Liu H, Dong Q, Zhang J, et al. Resveratrol inhibits Hexokinases II mediated glycolysis in non-small cell lung cancer via targeting AKT signaling pathway. Exp Cell Res. 2016 Dec;349(2):320–7.
79. Ma F, Ma Y, Liu K, Gao J, Li S, Sun X, et al. Resveratrol induces DNA damage-mediated cancer cell senescence through the DLC1-DYRK1A-EGFR axis. Food Funct. 2023;14(3):1484–97.
80. Wang Y, Romigh T, He X, Orloff MS, Silverman RH, Heston WD, et al. Resveratrol regulates the PTEN/AKT pathway through androgen receptor-dependent and -independent mechanisms in prostate cancer cell lines. Hum Mol Genet. 2010;19(22):4319–29.
81. Hsieh TC, Wu JM. Resveratrol suppresses prostate cancer epithelial cell scatter/invasion by targeting inhibition of hepatocyte growth factor (HGF) secretion by prostate stromal cells and upregulation of E-cadherin by prostate cancer epithelial cells. Int J Mol Sci. 2020;21(5)
82. Kim JY, Cho KH, Jeong BY, Park CG, Lee HY. Zeb1 for RCP-induced oral cancer cell invasion and its suppression by resveratrol. Exp Mol Med. 2020;52(7):1152–63.
83. Vanamala J, Reddivari L, Radhakrishnan S, Tarver C. Resveratrol suppresses IGF-1 induced human colon cancer cell proliferation and elevates apoptosis via suppression of IGF-1R/Wnt and activation of p53 signaling pathways. BMC Cancer. 2010;10:238.
84. Kang NH, Hwang KA, Lee HR, Choi DW, Choi KC. Resveratrol regulates the cell viability promoted by 17β-estradiol or bisphenol A via down-regulation of the cross-talk between estrogen receptor α and insulin growth factor-1 receptor in BG-1 ovarian cancer cells. Food Chem Toxicol. 2013;59:373–9.
85. Rugamba A, Kang DY, Sp N, Jo ES, Lee JM, Bae SW, et al. Silibinin regulates tumor progression and Tumorsphere formation by suppressing PD-L1 expression in non-small cell lung cancer (NSCLC) cells. Cells. 2021;10(7)
86. Hou X, Du H, Quan X, Shi L, Zhang Q, Wu Y, et al. Silibinin inhibits NSCLC metastasis by targeting the EGFR/LOX pathway. Front Pharmacol. 2018;9:21.
87. Singh RP, Deep G, Chittezhath M, Kaur M, Dwyer-Nield LD, Malkinson AM, et al. Effect of silibinin on the growth and progression of primary lung tumors in mice. J Natl Cancer Inst. 2006;98(12):846–55.
88. Tyagi A, Singh RP, Ramasamy K, Raina K, Redente EF, Dwyer-Nield LD, et al. Growth inhibition and regression of lung tumors by silibinin: modulation of angiogenesis by macrophage-associated cytokines and nuclear factor-kappaB and signal transducers and activators of transcription 3. Cancer Prev Res (Phila). 2009;2(1):74–83.
89. Iqbal MA, Chattopadhyay S, Siddiqui FA, Ur Rehman A, Siddiqui S, Prakasam G, et al. Silibinin induces metabolic crisis in triple-negative breast cancer cells by modulating EGFR-MYC-TXNIP axis: potential therapeutic implications. FEBS J. 2021;288(2):471–85.
90. Ranapour S, Motamed N. Effect of Silibinin on the expression of Mir-20b, Bcl2L11, and Erbb2 in breast cancer cell lines. Mol Biotechnol. 2023;65(12):1979–90.
91. Nambiar DK, Rajamani P, Deep G, Jain AK, Agarwal R, Singh RP. Silibinin preferentially Radiosensitizes prostate cancer by inhibiting DNA repair signaling. Mol Cancer Ther. 2015;14(12):2722–34.

92. Wang HJ, Tashiro S, ichi, Onodera S, Ikejima T. Inhibition of insulin-like growth factor 1 receptor signaling enhanced silibinin-induced activation of death receptor and mitochondrial apoptotic pathways in human breast cancer MCF-7 cells. J Pharmacol Sci. 2008;107(3):260–9.
93. Zi X, Zhang J, Agarwal R, Pollak M. Silibinin up-regulates insulin-like growth factor-binding protein 3 expression and inhibits proliferation of androgen-independent prostate cancer cells. Cancer Res. 2000;60(20):5617–20.
94. Sharma Y, Agarwal C, Singh AK, Agarwal R. Inhibitory effect of silibinin on ligand binding to erbB1 and associated mitogenic signaling, growth, and DNA synthesis in advanced human prostate carcinoma cells. Mol Carcinog. 2001;30(4):224–36.
95. Singh RP, Dhanalakshmi S, Tyagi AK, Chan DCF, Agarwal C, Agarwal R. Dietary feeding of silibinin inhibits advance human prostate carcinoma growth in athymic nude mice and increases plasma insulin-like growth factor-binding protein-3 levels. Cancer Res. 2002;62(11):3063–9.
96. Liu W, Otkur W, Li L, Wang Q, He H, Ye Y, et al. Autophagy induced by silibinin protects human epidermoid carcinoma A431 cells from UVB-induced apoptosis. J Photochem Photobiol B. 2013;123:23–31.
97. Dheeraj A, Rigby CM, O'Bryant CL, Agarwal C, Singh RP, Deep G, et al. Silibinin treatment inhibits the growth of hedgehog inhibitor-resistant basal cell carcinoma cells via targeting EGFR-MAPK-AKT and hedgehog signaling. Photochem Photobiol. 2017;93(4):999–1007.
98. Tilley C, Deep G, Agarwal C, Wempe MF, Biedermann D, Valentová K, et al. Silibinin and its 2,3-dehydro-derivative inhibit basal cell carcinoma growth via suppression of mitogenic signaling and transcription factors activation. Mol Carcinog. 2016;55(1):3–14.
99. Liang L, Li L, Zeng J, Gao Y, Chen YL, Wang ZQ, et al. Inhibitory effect of silibinin on EGFR signal-induced renal cell carcinoma progression via suppression of the EGFR/MMP-9 signaling pathway. Oncol Rep. 2012;28(3):999–1005.
100. Liu H, Zang C, Emde A, Planas-Silva MD, Rosche M, Kühnl A, et al. Anti-tumor effect of honokiol alone and in combination with other anti-cancer agents in breast cancer. Eur J Pharmacol. 2008;591(1–3):43–51.
101. Singh T, Gupta NA, Xu S, Prasad R, Velu SE, Katiyar SK. Honokiol inhibits the growth of head and neck squamous cell carcinoma by targeting epidermal growth factor receptor. Oncotarget. 2015;6(25):21268–82.
102. Song JM, Anandharaj A, Upadhyaya P, Kirtane AR, Kim JH, Hong KH, et al. Honokiol suppresses lung tumorigenesis by targeting EGFR and its downstream effectors. Oncotarget. 2016;7(36):57752–69.
103. Dai X, Li RZ, Jiang ZB, Wei CL, Luo LX, Yao XJ, et al. Honokiol inhibits proliferation, invasion and induces apoptosis through targeting Lyn kinase in human lung adenocarcinoma cells. Front Pharmacol. 2018;9:558.
104. Park EJ, Min HY, Chung HJ, Hong JY, Kang YJ, Hung TM, et al. Down-regulation of c-Src/EGFR-mediated signaling activation is involved in the honokiol-induced cell cycle arrest and apoptosis in MDA-MB-231 human breast cancer cells. Cancer Lett. 2009;277(2):133–40.
105. Leeman-Neill RJ, Cai Q, Joyce SC, Thomas SM, Bhola NE, Neill DB, et al. Honokiol inhibits epidermal growth factor receptor signaling and enhances the antitumor effects of epidermal growth factor receptor inhibitors. Clin Cancer Re. 2010;16(9):2571–9.
106. Hale AJ, Smith CA, Sutherland LC, Stoneman VE, Longthorne VL, Culhane AC, et al. Apoptosis: molecular regulation of cell death. Eur J Biochem. 1996;236(1):1–26.
107. Fan Y, Xue W, Schachner M, Zhao W. Honokiol eliminates glioma/glioblastoma stem cell-like cells via JAK-STAT3 signaling and inhibits tumor progression by targeting epidermal growth factor receptor. Cancers (Basel). 2018;11(1)
108. Kim S, Han J, Kim JS, Kim JH, Choe JH, Yang JH, et al. Silibinin suppresses EGFR ligand-induced CD44 expression through inhibition of EGFR activity in breast cancer cells. Anticancer Res. 2011;31(11):3767–73.

109. Kim S, Jeon M, Lee J, Han J, Oh SJ, Jung T, et al. Induction of fibronectin in response to epidermal growth factor is suppressed by silibinin through the inhibition of STAT3 in triple negative breast cancer cells. Oncol Rep. 2014;32(5):2230–6.
110. Jabbarzadeh Kaboli P, Luo S, Chen Y, Jomhori M, Imani S, Xiang S, et al. Pharmacotranscriptomic profiling of resistant triple-negative breast cancer cells treated with lapatinib and berberine shows upregulation of PI3K/AKT signaling under cytotoxic stress. Gene. 2022;816:146171.
111. Miyazaki K, Morine Y, Xu C, Nakasu C, Wada Y, Teraoku H, et al. Curcumin-mediated resistance to Lenvatinib via EGFR signaling pathway in hepatocellular carcinoma. Cells. 2023;12(4)
112. Gu HR, Park SC, Choi SJ, Lee JC, Kim YC, Han CJ, et al. Combined treatment with silibinin and either sorafenib or gefitinib enhances their growth-inhibiting effects in hepatocellular carcinoma cells. Clin Mol Hepatol. 2015;21(1):49–59.
113. Chan CY, Hong SC, Chang CM, Chen YH, Liao PC, Huang CY. Oral squamous cell carcinoma cells with acquired resistance to Erlotinib are sensitive to anti-cancer effect of quercetin via Pyruvate Kinase M2 (PKM2). Cells. 2023;12(1)
114. Zang H, Qian G, Arbiser J, Owonikoko TK, Ramalingam SS, Fan S, et al. Overcoming acquired resistance of EGFR-mutant NSCLC cells to the third generation EGFR inhibitor, osimertinib, with the natural product honokiol. Mol Oncol. 2020;14(4):882–95.
115. Cuan X, Yang X, Zhu W, Zhao Y, Luo R, Huang Y, et al. Antitumor effects of erlotinib in combination with berberine in A431 cells. BMC Pharmacol Toxicol. 2023;24(1):29.
116. Lee JY, Lee YM, Chang GC, Yu SL, Hsieh WY, Chen JJW, et al. Curcumin induces EGFR degradation in lung adenocarcinoma and modulates p38 activation in intestine: the versatile adjuvant for gefitinib therapy. PLoS One. 2011;6(8):e23756.
117. Ganthala PD, Alavala S, Chella N, Andugulapati SB, Bathini NB, Sistla R. Co-encapsulated nanoparticles of Erlotinib and Quercetin for targeting lung cancer through nuclear EGFR and PI3K/AKT inhibition. Colloids Surf B Biointerfaces. 2022;211:112305.
118. Yang J, Wu W, Wen J, Ye H, Luo H, Bai P, et al. Liposomal honokiol induced lysosomal degradation of Hsp90 client proteins and protective autophagy in both gefitinib-sensitive and gefitinib-resistant NSCLC cells. Biomaterials. 2017;141:188–98.
119. Pearson HE, Iida M, Orbuch RA, McDaniel NK, Nickel KP, Kimple RJ, et al. Overcoming resistance to cetuximab with honokiol, a small-molecule polyphenol. Mol Cancer Ther. 2018;17(1):204–14.
120. Rho JK, Choi YJ, Jeon BS, Choi SJ, Cheon GJ, Woo SK, et al. Combined treatment with silibinin and epidermal growth factor receptor tyrosine kinase inhibitors overcomes drug resistance caused by T790M mutation. Mol Cancer Ther. 2010;9(12):3233–43.
121. Cufí S, Bonavia R, Vazquez-Martin A, Oliveras-Ferraros C, Corominas-Faja B, Cuyàs E, et al. Silibinin suppresses EMT-driven erlotinib resistance by reversing the high miR-21/low miR-200c signature in vivo. Sci Rep. 2013;3:2459.
122. Chuang TC, Wu K, Lin YY, Kuo HP, Kao MC, Wang V, et al. Dual down-regulation of EGFR and ErbB2 by berberine contributes to suppression of migration and invasion of human ovarian cancer cells. Environ Toxicol. 2021;36(5):737–47.

Future Directions for Receptor Tyrosine Kinases in Cancer Treatment

Akansha Sharma, Debashish Chakraborty, and Monika Yadav

1 Introduction

Receptor tyrosine kinases (RTKs) belong to a class of cell surface receptors that play critical roles in cellular activities comprising cell growth, proliferation, differentiation, metabolism, survival, and communication between cells and their surroundings. As of now, over 20 distinct classes of RTKs have been thoroughly investigated on basis of their extracellular domain structure as well as the ability to interact with diverse ligands [1, 2]. When a specific ligand binds, receptor dimerization occurs, and the activated kinase causes autophosphorylation of the tyrosine on the receptor's intracellular domain. The phosphorylated tyrosine serves as docking points for substantial protein complexes, activating a range of signaling pathways.

Essentially, RTKs lie at the heart of complex interconnected signaling networks and play a substantial role in maintaining cellular homeostasis in healthy cells; any

Akansha Sharma and Debashish Chakraborty contributed equally with all other contributors.

A. Sharma
Department of Microbiology and Immunology, Thomas Jefferson University,
Philadelphia, PA, USA
e-mail: Akansha.Sharma@jefferson.edu

D. Chakraborty
School of Biological Sciences, National Institute of Science Education and Research
(NISER) Bhubaneswar, An OCC of Homi Bhabha National Institute, Khurda, Odisha, India
e-mail: debashish.chakraborty@niser.ac.in

M. Yadav (✉)
Laboratory of Nanotechnology and Chemical Biology, Regional Centre for Biotechnology,
Faridabad, India
e-mail: monikayadav.mkb@rcb.res.in

A. Sabarwal et al. (eds.), *Receptor Tyrosine Kinases in Cancer*, Cancer Drug Discovery and Development, https://doi.org/10.1007/978-3-031-93894-8_12

problem pertaining to their regulation or activation can alter a cell's ability to grow, survive, differentiate, and metabolize. A prominent consequence of RTK dysregulation is cancer and its associated phenotypes [3]. For instance, RTK class I members of the epidermal growth factor receptor (EGFR)/ErbB family are central to a variety of cellular functions, like homeostasis regulation [4]. However, the inadequate level of the EGFR/ErbBs led to the production of neurodegenerative disorders and embryonic fatality, while excessive levels are connected with the high grade of malignancies [4, 5]. Similarly, the c-MET receptor regulates cell growth and motility by binding to the hepatocyte growth factor (HGF). Yet, in malignancies, c-MET can become abnormally active due to mutation, amplification, or protein overexpression, resulting in uncontrolled cell proliferation and dissemination [6]. Vascular endothelial growth factors (VEGFs) are crucial for the formation of blood vessels, and vascular permeability, gain or loss, has been related to ischemic strokes, diabetes, and cancer [7]. The RTK AXL belongs to the TAM (TYRO3-AXL-MERTK) family and, when interacting with its ligand GAS6, regulates a variety of critical biological processes such as proliferation, survival, motility, and immune response. Despite not being implicated as an oncogene, AXL has been found to be overexpressed in a number of solid and hematologic cancers [8]. Correspondingly, alterations in other RTKs resulting from point mutations, gain or loss of function, amplification, deletions, and insertions caused the signaling network to malfunction and oncogenesis to occur [3].

RTK's critical function in the development, spread, metastasis, and resistance to treatment of cancer made them a promising biomarker that can be used to assess the spread and progression of tumors as well as to diagnose and predict outcomes in cancer patients [9–11]. As a result, multiple RTK-targeting drugs such as tyrosine kinase inhibitors (TKIs), monoclonal antibodies (mAbs), and antibody-drug conjugates (ADCs) have been employed in oncology as first- or second-line therapy against different cancers. However, their therapeutic value is restricted by the emergence of resistance or side effects [12].

It has been observed that during TKI therapy, tumor cells eventually become resistant to the drugs; nevertheless, acquired drug resistance does not take long to develop [13]. Moreover, a small percentage of tumor cells also exhibit primary resistance to TKIs [14]. Tumor development and resistance to TKIs are further influenced by the TME, EMT, cell death resistance, crosstalk with other RTKs, tumor metabolism, bypass signaling, epigenetic modification, and aberrant metabolism of TKIs [15–17]. Furthermore, selection pressure and clonal evolution might also result in the emergence of novel resistance mechanisms against TKIs. In cancer patients, TKI resistance ultimately leads to recurrences, progressions, treatment failures, low compliance, and mortality. Consequently, it is practical to repeat molecular analyses via active monitoring using serial noninvasive liquid biopsy to identify emerging resistance mechanisms as soon as possible, along with treatment response [18, 19]. Aside from that, there are still a few significant issues with RTKs that need more detailed future research.

The therapeutic value of TKIs is restricted by their inherent limitations, which involve on-target side effects by excessively inhibiting the intended tyrosine kinase activity in normal tissues to cause undesirable effects involving hypertension, skin rashes, hypothyroidism, and proteinuria and/or off-target side effects from suppressing multiple kinases and unexpected toxicities, particularly in the case of multi-targeted TKIs [20, 21]. In this regard, TKI-based nano-delivery technologies that target tumor cells precisely may offer solutions [22, 23]. Furthermore, innovative methods of anticancer drug screening, such as single-cell drug sensitivity testing, the potential use of conditionally reprogrammed primary tumor cells, and using ImmunoPET as tool for assessing the efficacy of treatment, while investigating the potential of chimeric antigen receptor (CAR) T cell and proteolysis targeting chimera (PROTAC) modality, will aid in the personalized determination of an optimal drug choice for cancer patients [24–30]. Therapeutic strategies employing multiple combinations of immunotherapies, TKIs, monoclonal antibodies, and cytotoxic medications have demonstrated enhanced clinical benefits and are currently being investigated for designing promising treatment regimens to target multiple cancer hallmarks simultaneously while offering long-term protection against disease relapse [31, 32]. In addition, the development of artificial intelligence (AI) and its innovative tools yields state-of-the-art algorithms and expedites therapeutic prospects [33, 34]. Overall, TKI resistance involves a complex set of networks and needs further clinical exploration to delineate the molecular mechanisms and pinpoint a viable target using cutting-edge precision medicine approaches and tools.

The present chapter provides a comprehensive vision for the future of four major RTKs (EGFR, AXL, c-MET, and VEGFR) targeted in cancer therapies by discussing research challenges related to these RTKs, focusing on direction of forthcoming research in identifying new RTK targets, and addressing drug development challenges with technological innovations, precision approaches, and relevant clinical trials.

2 Future Perspectives of RTKs

2.1 EGFR

RTKs, particularly the EGFR, have played critical roles in developing targeted cancer therapeutics. It is also one of the most extensively studied RTKs due to its role in promoting oncogenic signaling in various cancers, especially in NSCLC, in which 40–80% of individuals exhibit upregulation of EGFR due to epigenetic abnormalities and variations in the copy number of the gene [35–38]. Despite the effectiveness of EGFR TKIs like erlotinib and gefitinib, higher toxicity and drug resistance have reduced their long-term efficacy [39, 40]. The main goal of current and future research is to understand and resolve these problems by coming up with new ways to overcome resistance, make treatments more specific, and improve patients' long-term health through combination therapies, molecular profiling, and personalized medicine.

Secondary mutations like T790M, which render first- or second-generation inhibitors ineffective within 9–14 months of treatment, are the main obstacle to EGFR-related treatments [41]. Third-generation EGFR inhibitor, osimertinib, is the current gold standard for first-line treatment in advanced NSCLC, as it effectively and specifically targets both EGFR-sensitizing and T790M mutations and provides better clinical results with fewer adverse events related to wild-type EGFR inhibition than first-generation TKIs [42, 43]. Furthermore, osimertinib may be able to penetrate the blood-brain barrier and exhibit favorable effectiveness in patients with metastases to the central nervous system (CNS) [44, 45]. Other third-generation EGFR TKIs, including furmonertinib, lazertinib, and nazartinib, have also shown reasonable results as well as adequate efficacy in advanced NSCLC with a positive EGFR T790M mutation [37, 46, 47]. Though the therapeutic algorithm has undergone a spectacular transformation, yet the third-generation EGFR TKIs' long-term effects are often compromised by acquired resistance mutations, among which the EGFR exon 20 C797S mutation is the most frequent cause [48]. It has been identified that all EGFR-mutant patients are known to develop resistance to EGFR TKI therapy following first-, second-, or third-generation EGFR TKI. Current research is focused on developing fourth-generation EGFR TKIs to combat osimertinib resistance caused by EGFR-related mutation processes [49]. For example, EAI045 shields against EGFR T790M and C797S mutations [50], TQB3804 inhibits the EGFR triple mutants D746–750/T790M/C797S and L858R/T790M/C797S [51], and LS-106 targets 19del/T790M/C797S and L858R/T790M/C797S [52]. Even while fourth-generation EGFR TKIs exhibit encouraging outcomes, their development is still ongoing. Again, the question is what comes next? The problem of drug resistance would reappear with targeted therapy for lung cancer since the mechanism underlying resistance is still unclear. We are unsure if the three-generation EGFR TKIs would have an effect equivalent to that of "bacterial multiple drug-resistant" on the front line, rendering the subsequent therapy unresponsive. Therefore, the relevant studies should constantly track the tumor's response to further treatment in order to identify additional undiscovered mutations. Additional investigation of molecular resistance mechanisms and intratumor heterogeneity is widely acknowledged as essential for expanding biological understanding, enhancing therapeutic viewpoint in clinical settings, and helping to create novel treatment approaches.

Combination therapy of cetuximab and brigatinib has also shown some initial efficacy in lung adenocarcinoma patients with EGFR/T790M/cis-C797S triple mutations [53]. In the ADAURA trial, the effectiveness of osimertinib is investigated in an adjuvant context (NCT02511106), where in patients with fully resected, mutated-EGFR, stage IB to IIIA lung cancer, adjuvant osimertinib significantly improved overall survival [54]. Clinical trials investigating the combination of immune checkpoint inhibitors (ICIs) with anti-angiogenic therapy and/or chemotherapy also demonstrated prolonged progression-free survival (PFS) and overall survival (OS) in lung cancer patients harboring EGFR mutations [55–57]. However, to further validate the effectiveness of this approach, additional prospective clinical investigations are still required. Furthermore, innovative dual- or multi-target drugs that can target EGFR in conjunction with other pathways (e.g., EGFR-MET or EGFR-HER2 dual inhibitors) are being developed to address the compensatory mechanisms that allow cancer cells to circumvent EGFR inhibition [39]. EGFR dual- or multi-target drugs have great prospects because of their decreased likelihood of drug resistance, improved efficacy, lower dosage, and less side effects compared to single-target drugs [58–60]. Recently, multiple-low-dosage therapy is also identified as an efficient method for treating EGFR inhibitor-resistant NSCLC tumors [61]. A study by Du et al. uncovered the role of novel EGFR-PROTACs in breaking down several EGFR-mutated proteins to overcome EGFR TKI resistance [62]. Nonetheless, the research warrants more thorough investigation. Further, phase I clinical trials demonstrated that patients with EGFR-positive advanced relapsed NSCLC and metastatic pancreatic cancer respond safely and effectively to CAR T-EGFR cell therapy [63, 64]. Subsequent research will need to evaluate the effectiveness of CAR T-EGFR therapy in additional patients or perhaps at an optimum/higher dosage.

Precision medicine based custom-tailored treatments to an individual with distinct molecular profile and holds the key to the future of EGFR-targeted therapy. Advancement in NGS and liquid biopsies has enabled real-time monitoring of tumor genomic evolution, allowing for prompt therapy adjustments to address emergent resistance mechanisms [65, 66]. Molecular profiling can help identify individuals who may benefit from combination therapy that includes EGFR inhibitors and ICIs as well as other targeted drugs, allowing for improved clinical outcomes [67].

Advancements in AI technology enable cutting-edge algorithms and expedite the therapeutic potential. It can decrease hurdles and expenses throughout the discovery, optimization, and development phases [68]. AI has now increasingly been integrated into cancer treatment, particularly in optimizing EGFR-targeted therapies [69]. AI-based techniques can be implemented to accurately discover certain critical alterations from histopathological images, particularly those that are clinically relevant and function as response indicators for targeted treatments (e.g., activating mutations in EGFR) [69–71]. AI also has the potential for improving patients' treatment plans by suggesting effective cure approaches. For instance, Wang et al. developed a multitasking AI system based on deep learning, radiomics modules, and a combination of deep learning, radiomics modules, and clinical features, trained and optimized with computed tomography (CT) images which showed promising findings for determining gene status of programmed death-ligand 1 (PD-L1) and EGFR and further evaluating the genotypes [72]. Correspondingly, Zhou et al. also employed an AI algorithm to examine the relationship between EGFR gene mutation and characteristics seen in CT images of lung adenocarcinoma and identified substantial variations in the EGFR gene mutation rate and gender, smoking status, form of pulmonary nodules, and cavity, hair prick, thymus depression signs, and malignancy with vascular convergence sign categories of lung adenocarcinoma patients, while the EGFR gene mutation rate and the patient's tumor size, lobulated sign, pleural and pericardial effusion, and pleural thickening did not significantly differ from one another [73]. It's interesting to note that this interpretable, high-throughput, AI-based model combined with CT semantic features may be a noninvasive biomarker which helps determine the presence of EGFR mutations with a high degree of accuracy [74]. Altogether these studies emphasize the potential of AI in tailoring personalized treatment strategies, which may improve patient outcomes and reduce toxic side effects. Also, the role of AI in diagnosing and predicting treatment outcomes and the correlation between imaging features and EGFR gene mutation status offer a reference for personalized medicine in cancer management. This system could potentially revolutionize decision-making in oncology by providing fast, evidence-based treatment recommendations [71].

Overall, the future of EGFR-targeted therapy in cancer treatment lies in the integration of novel inhibitors, combination therapies, and personalized medicine approaches. By overcoming resistance mechanisms and addressing tumor heterogeneity, these advancements offer the potential for more effective and durable treatments. Continued research, use of AI-based high-throughput algorithms, and clinical trials are essential to validate these strategies and bring them into mainstream oncology practice, ultimately improving clinical outcomes for patients with EGFR-driven cancers.

2.2 *AXL*

AXL, a receptor tyrosine kinase, has appeared as a promising anticancer target. The potential to discover AXL's translational ability as a biomarker for evaluating treatment response is demonstrated by its signaling network, which is associated with the cancer cell growth, development, EMT, drug resistance, and immune modulation [75]. Over the past decades, numerous AXL-targeted therapeutic interventions have been developed including TKIs, mAbs, ADCs, and soluble receptors, while multiple clinical trials are still in progress to evaluate the potential clinical advantages of targeting AXL [76]. Nonetheless, there remain a number of significant unsolved questions with AXL.

It has been observed that unlike other tailored drugs like EGFR inhibitors, DNA alterations related to AXL inhibitor responsiveness are uncommon and may not always be an accurate predictor; thus, it will be critical to discover whether AXL's elevated expression, related to poor prognosis, serves as an effective biomarker to recognize individuals who might benefit primarily from AXL-targeted interventions. As of yet, there are few reliable response predictors for AXL-targeted therapies; for instance, tumor AXL expression may be a useful indicator for bemcentinib, the first and most researched AXL-specific TKI. In a phase II trial of bemcentinib plus pembrolizumab in patients with advanced NSCLC who had received prior treatment, AXL positivity (determined by IHC scoring method) correlated with response irrespective of PD-L1 expression. In AXL-positive cases, which totalled 50% of the research group, the response rate was more than 30%, while it was below 15% for AXL-negative individuals (NCT03184571) [77]. Additional, investigation is required to identify the epigenetic modifications influencing AXL expression.

Until now, different trials have used different methods to define and evaluate AXL expression. Therefore, coordinated attempts to acquire and assess biospecimens will be necessary to get deeper insights into this prospective biomarker; yet, in many advanced solid tumors, such actions may present difficult real-world problems. In this regard, personalized cancer therapy decision-making is anticipated to benefit from noninvasive diagnostics employing circulating tumor cells (CTCs). Nevertheless, due to EMT, the epithelial markers often utilized for CTC detection might not be sufficient to identify AXL-expressing CTCs. Ikeda et al.'s [78] study utilizing microcavity array technology for identifying AXL-expressing CTCs in advanced NSCLC patients by employing the mesenchymal marker revealed that in peripheral blood collected from 20 advanced NSCLC patients, single CTCs expressing AXL were more common in vimentin-positive than cytokeratin-positive CTCs. Thus, reliable outcomes desire the use of an approach that integrates mesenchymal markers alongside CTC detection.

Another challenge is that AXL's modulation of the TME induces immune escape and immune checkpoint blockade resistance, with survival benefits only being shown in a limited proportion of patients [79]. Consequently, to gain a better understanding of tumor microenvironment-induced anti-tumor responses, it is crucial to define the molecular relevance of AXL signaling within distinct populations of stromal cells. Future studies should also concentrate on a range of parameters impacting the metabolism of TMEs, including hypoxia, pH dysregulation, nutritional deprivation, and mechanical and/or oxidative stress, as they may affect the expression of AXL and/or the cell type that expresses it. In addition to immunological suppression, AXL is essential for EMT and stemness; for instance, Asiedu et al. [80] have discovered the role of AXL in stimulating EMT in healthy and immortalized human mammary epithelial cells via a positive feedback loop system to regulate self-renewal of breast cancer stem cells and chemoresistance. This implies that suppressing AXL could shed light on how these mechanisms affect chemosensitivity and resistance as well as their function in metastatic illness. Studies exploring the possibility of blocking AXL with potential single- or multi-target agents to overcome resistance to traditional treatments are presently in progress.

Furthermore, owing to the intricate relationships between RTKs and activation of alternative signaling pathways, tumor cells can circumvent targeted suppression of certain receptor pathways leading to drug resistance [81]. As such, targeted drugs are usually believed to have cytostatic effects extending PFS and OS, rather than curative ones, yet some individuals have shown substantial responses. Patients may get targeted drugs in a sequential manner to overcome resistance to initial therapy, depending upon their level of treatment tolerance. It may be more effective to simultaneously inhibit several RTKs using a multi-TKI rather than focusing on one pathway either singly or consecutively [82]. Though multi-targeted AXL TKIs have numerous benefits, they also have drawbacks, like unforeseen toxicities and drug resistance that significantly limit the further clinical application of these drugs [83, 84]. For instance, long-term sunitinib therapy activates AXL and MET signaling, which is linked to worse patient outcomes in cases of renal cell carcinoma (RCC) [85]. In order to overcome drug resistance and substantially improve therapeutic efficacy, nanocarrier-based multi-targeted TKI delivery systems have garnered a lot of attention lately. A study by Q. Yang et al. (2017) demonstrated that copper oxide nanoparticles have the potential to induce endoplasmic reticulum (ER) stress-directed apoptosis via controlling copper trafficking and downregulating AXL, MET, AKT, and ERK expression in sunitinib resistance RCC cells to restore sunitinib responsiveness [86]. Collectively, investigating the cooperation and interaction between AXL and other RTKs is essential for pinpointing the specific downstream signaling pathways which encourage cancer cell viability and developing plausible single, combination, and nanocarrier-based therapies to improve anti-tumor immunity and individualized therapy.

Furthermore, the fact that patients with AXL-driven cancers will inevitably relapse highlights the necessity of employing high-throughput single-cell sequencing techniques, examining the potential of CAR T cells and AXL degraders, developing combination treatments, and an understanding of resistance mechanisms in order to create novel treatment plans that can effectively manage possible toxicities and suppress drug resistance, resulting in more profound and prolonged therapeutic responses. High-throughput techniques such as single-cell transcriptomic analysis determine drug-tolerant states, discover distinct cell subpopulations indicative of the functional heterogeneity of drug-tolerant cells, and select optimal drugs and drug combinations that are effective in targeting persister cell subpopulations [87]. Correspondingly, cytometry by time-of-flight (CyTOF), a proteomic profiling technique, provides imperative insights about intracellular and surface markers that respond to drug treatments at a single cell level and serves as an effective tool for the rational development of AXL-targeting methods. Taverna et al. [88] discovered combined inhibition of AXL and JAK1 in lung cancer as a unique treatment approach by employing CyTOF technique. In general, the use of high-throughput single-cell profiling in clinical settings may facilitate the discovery of novel therapeutic targets, make interpretation of data from early-phase clinical trials easier, and enable the development of predictive biomarkers to stratify patients.

Recently, an anti-AXL CAR T cell therapy, a novel cancer immunotherapy, has emerged as a viable therapeutic alternative [89, 90]. However, the use of CAR T cell therapy is shown to be less successful in treating solid tumors due to the unfavorable side effects incurred [91]. Consequently, a great deal of research is being carried out to modify the CAR construct and use different kinds of immune cells, such as AXL-targeting CAR-NK cells as a platform for enhancing the safety and efficacy of CAR-based therapies in AXL-positive advanced solid tumors (NCT05410717) [91]. Meanwhile, there exists an urgent need to engineer CAR constructs in order to generate "smart" tools that can assist and enhance current CAR therapies. PROTAC is another promising tool for targeting and degrading "undruggable" proteins via the ubiquitin-proteasome system to overcome cancer cell drug resistance [92]. Shi et al. [93] employed a structure-based drug discovery approach to design and synthesize 20 AXL degraders, of which compounds 20 and 22 demonstrated significant anti-tumor activity by specifically degrading AXL and triggering methuosis (an over synthesis of macropinosomes resulting in non-apoptotic cell death). In triple-negative breast cancer (TNBC) cells, AXL PROTAC degrader compound 6n is identified to deplete AXL with a 5 nM half-maximal degradation concentration (DC50) value while substantially enhancing effectiveness against the AXL signaling induction, cell proliferation, migration, and invasion, contrasting to the related kinase inhibitor [94]. This finding suggests that rather than directly suppressing kinase activity, PROTAC-mediated AXL degradation could deactivate the signal completely, preventing or delaying the compensation feedback activation of other kinases involved to overcome acquired drug resistance development.

Up to now, numerous clinical trials assessing drugs used in targeted therapy, immunotherapy, and combination therapy have been carried out, with favorable outcomes [95–97]. Trial results will provide insights into the precise role of targeted therapy in the management of different types of cancer, prediction of possible secondary effects, as well as the functions of immune-based monotherapies, combination therapies, and targeted therapy-immunotherapy combos. Finally, both preclinical and clinical findings demonstrated that therapeutic targeting of AXL can significantly improve the efficacy of other standard therapies. However, the knowledge of improved molecular biomarkers using high-throughput techniques and development of combination medicines may offer a successful cure. In the future, precise selection of patients employing predictive biomarkers will be critical for theranostics and personalized cancer treatment in order to maximize limited resources and reduce vulnerability. Future improvements in patient survival are expected if resistance is tackled, the appropriate inhibitor or combination of inhibitors is administered, and side effects are minimized.

2.3 MET

A member of RTK family, c-MET undergoes dimerization, autophosphorylation, and subsequent activation in response to its ligand HGF. c-MET-mediated signaling is essential for both healthy development and the advancement of numerous types of human cancer. c-MET activation is also known to trigger multiple pathways such as PI3K/Akt, SRC, MAPK, and STAT to mediate a range of biological processes, involving growth, invasion, motility, and angiogenesis. Gene amplification or autocrine signaling, mutational activation of c-MET, and overexpression of HGF or c-MET can all result in uncontrolled c-MET signaling [98, 99]. Considering its substantial impact on cancer, the c-MET receptor can be targeted with a number of therapeutics including TKIs, mAbs, and CAR T cell therapies. Nevertheless, it has been observed that the majority of these small molecule TKIs have an early therapeutic response but do not progress to phase III of clinical trials [100, 101].

Presently, cabozantinib and crizotinib are the only two MET-TKIs approved by US Food and Drug Administration (FDA) and the European Medicines Agency (EMA) for treatment of multiple cancers [101]. Tepotinib, capmatinib, and savolitinib are recently approved small molecule inhibitors by FDA for the treatment of advanced NSCLC patients with MET exon 14 skipping mutations [102, 103]. Despite the therapeutic impact, risks associated with off-target side effects and toxicity remain a matter of concern while using these inhibitors. Moreover, activation of MET signaling is also identified to induce acquired drug resistance against the TKIs of EGFR, HER2, and VEGF receptors to promote disease advancement, suggesting the incorporation of combination regime targeting MET with other RTKs would be advantageous [104]. Likewise, targeted therapy approaches against HGF such as rilotumumab (AMG-102) and ficlatuzumab (AV-299) also failed to produce any favorable outcomes and have been associated with side effects and poor response [105, 106]. The inefficiencies of mAbs lead to the development of engineered proteins such as designed ankyrin repeat protein (DARPin), possessing different domains to target different oncogenic proteins simultaneously. For instance, MP0250, a DARPin® molecule that neutralizes VEGF and HGF, exhibits strong anti-tumor activity as a single agent and also improves anti-programmed death-1 (anti-PD-1) efficacy suggesting the possibility of immune-therapeutic molecule combination therapy as well as the investigation of DARPin® molecules in conjunction with other conventional drugs [107]. Future research and analysis of additional crucial crosstalk pathways involving the c-MET signaling will also be necessary as they may influence patient survival and result in more significant gains in the effectiveness of new anticancer treatments.

Immune responses are pivotal in defining the pathway for cancer progression. Interestingly, immune response modulation is another function of c-MET signaling. It has been demonstrated that in a contact hypersensitivity mouse model, HGF/c-MET signaling enhances immunity by causing skin-resident dendritic cells (DCs) to migrate to the draining lymph node [108]. Contrastingly, the HGF/c-MET axis has also been shown in multiple studies to have immunosuppressive effects, including enhancing T cell tolerance and impairing DC activities [109, 110]. Cancer immunotherapy is limited by reactive neutrophil responses reliant on the c-MET signaling [111]. In such situation, tailoring targeted therapy to protect neutrophils with the application of co-treatment with c-MET inhibitors may enhance responses to cancer immunotherapy.

PD-L1 is a negative regulator of immune checkpoint and functions by thwarting immune responses against cancer. Certain malignancies exhibit a correlation between PD-L1 and MET expression, while others do not. For instance, combined MET and PD-L1 inhibition enhances the effectiveness of immunotherapy for pancreatic cancer [112]. PD-L1 expression and MET gene amplification are also found to be positively correlated in individuals with advanced NSCLC [113]. Comparative analysis of matched biopsies and surgical resection samples, on the other hand, shows no association between MET and PD-L1 expression in NSCLC [114]. This emphasizes the value of tailored therapies based on each individual's unique expression profile

CAR T cell immunotherapy has shown exciting results in targeting c-MET expressing cancer cells in preclinical models [115]. However, associated toxicities stemming from the hyperactive immune response limit the benefits of CAR immunotherapy. Such toxicities could be addressed by reducing the affinity of the antigen-binding domain such that its optimal activation will only occur in presence of higher levels of the antigen. Chiriaco et al. developed a CAR-T immunotherapy for malignancies overexpressing MET and those that are ineligible for anti-MET-targeted treatment with limited toxicities [116]. In another study, Qin et al. developed MET-specific CARs that target hepatocellular carcinoma (HCC) with MET overexpression using CD28ζ or 4-1BBζ as co-stimulation domains, independent of signaling pathway activity [117]. MET-targeting ADCs are another effective tool akin to CAR T cells that could specifically target MET and deliver a cytotoxic compound to kill tumor cells [118]. Such ADCs could be beneficial in settings where MET blockers are ineffective. However, off-target toxicity is still remaining a concern in the application of ADC. Moreover, risk of immunogenicity in both CAR T cells and ADC-based therapies demands for taking a more nuanced approach in their design. Aptamer-based drugs are also being developed to target RTKs in cancer. Recently a group developed a chimera aptamer combining two modules targeting protein tyrosine phosphatase receptor type F (PTPRF) and MET/EGFR. PTPRF is a phosphatase that dephosphorylates MET rendering it inactivated during its oncogenic activation [119]. Owing to the small size of aptamers, the risk associated with its immunogenicity is minimal.

The role of upstream events in regulating MET activity must be taken into account while designing targeted therapies against MET. Glycosylation of MET is important for its conformation, maturation, and function. Inhibition of glycosylation by tunicamycin was found to compromise the MET activity by promoting its degradation, preventing its localization to the membrane, and inhibiting its phosphorylation [120]. Proteasomal degradation of MET is achieved through recruitment of c-Cbl, an E3 ubiquitin ligase which binds to phosphorylated Y1003. Loss of this site in TPR-MET fusion oncoprotein is identified to abolish the proteasomal regulation of MET and promotes oncogenesis [121]. Interestingly, MET inhibitors (SU11274 and PHA665752) are identified to promote MET accumulation via reducing the phosphorylation at Y1003 and Cbl interaction, and MET ubiquitination ultimately results in decreased MET lysosomal degradation [122]. These findings might imply that anti-MET clinical studies should be carefully considered. FasL is another molecule that interacts with MET and triggers its activation. Therefore, small molecules could be developed to disrupt such interactions and inhibit MET activation. Additionally, caspases can cleave MET and generate a pro-apoptotic fragment. Strategies could be developed to harness this activity to inactivate MET as well as induce apoptosis in the cancer cells [120].

Immunohistochemistry (IHC) and fluorescence in situ hybridization (FISH) analysis remain the gold standard in identification of gene overexpression and gene amplification, respectively, in any clinical trial. However, these techniques have poor coverage due to the limitations in combining multiple probes to identify multiple genes simultaneously. Moreover, it is not possible to get a complete picture of the gene signatures owing to tumor heterogeneity. Poor identification of biomarkers is often the reason behind failure of clinical trials. Liquid biopsy is a minimally invasive technique that samples cell-free RNA, ctDNA, and CTCs derived from tumors from the blood. These noninvasive approaches can be combined with more sensitive sequencing techniques such as next-generation sequencing (NGS) and single-cell RNA-seq analysis to get a realistic view of biomarkers in specific tumors. In this regard, ctDNA could be employed as meaningful tool for NGS molecular profiling from advanced NSCLC to detect specific rearrangements and MET exon 14 mutations which are difficult to be identified using traditional methods as result of low sensitivity [123].

Another challenge in the case of c-MET is to determine which genetically specified responsive patient subsets may benefit from c-MET inhibition, as this will allow future clinical research to employ suitable patient selection criteria. In this context, a commercially available proteomic test called VeriStrat® could be employed to determine which patients are most or least likely to benefit from TKI treatment [124]. When assessing patients participating in clinical trials, evaluating biomarkers is also crucial as in NSCLC high MET levels are identified to be associated with receptor phosphorylation but not with the amount of soluble form in the serum [125]. With advancement in computational methods and advent of AI, designing druggable novel c-MET inhibitors is also feasible in lesser time. Further, growing evidence suggests the replacement of "compound-to-trial" approach of traditional drug development with a "biology-to-trial" strategy, beginning with the discovery of the basic mechanisms underlying cancer targets. This may then serve as the basis for initial drug discovery and follow-up clinical studies [126]. Growing usage of precision medicine to continually monitor each patient's responses against a treatment regimen in a timely manner is also key to design a highly potent and less toxic anticancer therapy.

2.4 VEGFR

An essential component of angiogenesis, the VEGF/vascular endothelial growth factor receptor (VEGFR) axis has been identified as a major contributor to tumor vascularization [127]. As a result, various cancer treatment approaches including mAbs and small molecules that target VEGF and its corresponding receptors, VEGFR-1, VEGFR-2, and VEGFR-3, have been developed [128]. As a recombinant humanized mAb targeting VEGF, bevacizumab has demonstrated positive clinical outcomes [129]. There are, however, a number of issues with anti-VEGF mAbs, including their immunogenicity, instability, and expensiveness [130]. Moreover, the clinical applicability of anti-VEGF mAbs is severely impeded by the significant adverse effects attributed to the suppression of physiological angiogenesis process—one of the most frequent negative consequences of anti-angiogenic treatments [131]. However, the concern has been resolved with the development of a number of oncotherapy tactics involving VEGFR targeting for ultimately blocking tumor angiogenesis.

The FDA has so far approved 72 drugs aimed at various protein kinases, the majority of which are multi-kinase inhibitors [132]. Thirteen of them directly block the activity of VEGFR, alongside related receptor RTKs, notably PDGFRs, FGFRs,

TGF-βRs, and others [133]. The multi-target VEGFR TKIs, sunitinib, pazopanib, sorafenib, and cabozantinib, are well established in their application against RCC and HCC [134]. However, the clinical therapeutic impact of these VEGFR inhibitors is compromised by their frequent severe cytotoxicity, resistance, and indicational restrictions [134, 135]. The third-generation VEGFR TKI, vatalanib, thus aims to increase target inhibition and simultaneously enhance target specificity while sparing physiological functions [136]. Even though vatalanib seems to be a safer and more bearable alternative to conventional VEGFR TKIs, the efficacy improvement has not yet demonstrated superiority, and there is still a concerning indication about the therapeutic resistance and off-target toxicity [21].

Stimulation of alternative pro-angiogenic networks, attraction of both distant and local stromal cells, and diverse tumor vascularization mechanisms (such as hypoxia) are the three distinct resistance mechanisms majorly observed in the case of VEGFR inhibitors [137]. Additionally, because of the kinase domain similarity between VEGFR and other receptors, these inhibitors also inhibit targets like PDGFR, c-KIT, and FLT3, potentially leading to off-target effects. VEGFR inhibitors have also been studied for a number of clinical adverse effects, including hypertension, arterial thrombosis, hypothyroidism, proteinuria, and leukoencephalopathy syndrome. Lastly, growing research has demonstrated that a number of VEGFR inhibitors have not consistently demonstrated significant overall efficacy in clinical settings [138, 139]. In this regard, the combination of VEGFR inhibitors with additional inhibitors (immunotherapeutic drugs, epigenetic agents, and other RTK inhibitors) appeared as effective clinical strategies to address these limitations by efficiently targeting tumor cells and the tumor microenvironment. For instance, in two different phase III trials, VEGFR inhibitor axitinib shows encouraging clinical outcomes in advanced RCC when combined with ICIs pembrolizumab and avelumab (PD-1 and PD-L1 inhibitors, respectively) in comparison to sunitinib alone treatment [140, 141]. These combinations aim to enhance anti-tumor immune responses while simultaneously inhibiting tumor angiogenesis [142]. Nonetheless, the intricate dosages and time frame, questionable pharmacokinetic or pharmacodynamic profile, and possible side effects necessitate thorough investigation of these combinations [143, 144].

Dual-target or multi-target drugs are identified as an emerging alternative approach to combination treatment because of their improved pharmacokinetic profiles, reduced incidence of undesirable drug-drug interactions, enhanced patient compliance, low chances of target-based resistance development, and assured safety [145, 146]. Isakova et al. demonstrated potential advantages of employing multivalent fusion protein SRH-DR5-B-iRGD for reducing tumor growth and angiogenesis by dual targeting of VEGFR2 and DR5 pathways [147]. Sulfatinib, a VEGFR-1/2/3, FGFR1, and CSF1R multi-target inhibitor, has also demonstrated therapeutic effectiveness in various phase II (NCT02267967, NCT02614495) and phase III (NCT02588170) studies in patients with solid malignancies, and additional trials are still being conducted (NCT04579757, NCT02549937). Drug repurposing, computational methods such as high-throughput virtual drug screening, and pharmacophore-based combinations are different strategies currently employed for the discovery of dual- or multi-target agents [148, 149]. Bispecific antibodies such as CTX-009, which targets VEGF-A and delta-like ligand 4 concurrently, are among the other treatments in the early phases of research and are presently undergoing evaluation in further trials (NCT03292783, NCT04492033).

Co-targeting treatment using a novel humanized bispecific ANG2/VEGF mAb, vanucizumab, displays an acceptable tolerability and safety in first-in-human phase I trial of advanced solid cancers [150]. Other novel strategies for addressing VEGF/VEGFR signaling are being developed. These include a decoy receptor fusion protein, conbercept, which interacts with VEGFA, VEGFB, and PIGF, as well as VEGF-Grab that targets VEGF-A and PIGF. Both of these have demonstrated encouraging outcomes in preventing the formation of tumors in colon cancer models [151, 152]. PROTACs aimed at VEGFR have also shown potential as anticancer drugs [153, 154]. Given the fact that hypoxia diminishes the sensitivity to VEGFR therapies, hypoxia-responsive nanoparticles leading to a significantly higher drug release rate under hypoxic settings in contrast to normoxic conditions appear to be a shrewd technique to overcome this issue [155].

In spite of the advancement in the field of VEGF/VEGFR inhibitors, challenges remain [156]. Although a number of VEGF/VEGFR inhibitors have shown beneficial results in a variety of tumor types such as NSCLC, RCC, and HCC, we still have little knowledge about how to use them optimally. As there are currently no established biomarkers in clinical practice for identifying which individuals are most likely to gain advantages from such drugs, it is frequently unclear whether the advantages of treatment stem from effects on the immune microenvironment, tumor angiogenesis, cancer cells, or a combination of these. In this regard, advances in next-generation genomic profiling, employed routinely for various tumors, along with the increasing use of transcriptomic profiling offer the possibility to investigate the relationship between particular oncogene (EGFR, KRAS, etc.), tumor suppressor (p53) drivers, and transcriptionally defined phenotypes (mesenchymal or inflamed tumors) and their correlation with clinical outcomes to utilize this knowledge for the identification of prognostic indicators and development of more plausible combination therapies. Additionally, inhibitors of VEGF/VEGFR have the potential to join the trend of targeted treatments and immunotherapies, which are usually administered in a customized, biomarker-driven strategy.

Nevertheless, the pairing of an anti-angiogenic medication with an anti-tumor drug is the basis for many cancer treatments, which enables the drug to reach the tumor location by temporarily restoring normal blood vessels [157]. These effects, however, have a time limit, vary according to the cancer types, and indicate a distinct "normalization window." Thus, new therapeutic prospects may need to be developed to overcome these constraints and treat the tumor with pro-angiogenic (as opposed to anti-angiogenic) therapies, which could fortify a network of concise blood vessels that aid in drug delivery [158]. While VEGFR TKIs are predominantly used in cancer therapy, their role in treating other diseases characterized by pathological angiogenesis (e.g., age-related macular degeneration, diabetic retinopathy) is also being explored. Future research could expand VEGFR inhibitor applications to conditions such as chronic inflammation, fibrosis, and cardiovascular diseases where abnormal vascular growth or endothelial dysfunction plays a role [21].

2.5 Other Receptor Tyrosine Kinases in Cancer

2.5.1 Platelet-Derived Growth Factor Receptors

RTK family consists of numerous members and earlier sections have focused on the major members. Herein, we briefly cover the future direction of other significant members of the RTK family in cancer therapy [159]. Platelet-derived growth factor receptors (PDGFRs) akin to many RTKs are important for tissue development and homeostasis. PDGFRs bind to platelet-derived growth factors (PDGFs) released by fibroblasts, endothelial cells, and macrophages [160]. Aberrant signaling due to elevated levels of PDGF as well as PDGFR has been observed in multiple cancers such as pancreatic, prostate, and ovarian cancer [161]. As a result, several therapeutic drugs such as TKIs, antibodies, and their combinations which block ligands or receptors have been developed [162]. Neutralizing antibodies have also been tested against PDGFs to prevent its binding with PDGFR. PDGF antibody MOR8457 binds PDGF-BB and hinders its interaction with PDGFRβ [163].

Aptamers are known for their ability to show target specificity. Multitude of aptamers targeting PDGF/PDGFR in cancer has been developed. PDGFRβ aptamer holds immense potential as an anti-tumor agent, inhibiting receptor signaling, tumor growth, and targeted imaging as well as metastasis in glioblastoma and breast cancer [164, 165]. A novel PDGFB-CGO Platelet-derived growth factor B Platelet-derived growth factor B (PDGFB)-conjugated copper-gadolinium oxide nanocluster) exhibited anti-tumor activity and inhibited angiogenesis and cell migration in cancer cells via disrupting intracellular copper balance [166]. Additionally, PDGFB-CGO's systemic administration greatly improved the tumor's MRI signal [166] and offers a viable approach for developing cutting-edge biomaterials for successful theranostic cancer applications. However, crosstalk of PDGF/PDGFR with various other signaling pathways continues to be a significant obstacle for effective therapy. Consequently, it is crucial to build combinatorial treatment regimens for different tumors employing an expanded mechanistic understanding of PDGFR and the signaling network of various RTKs. For instance, anlotinib, a TKI that can target multiple RTKs including PDGFR, has demonstrated promising effectiveness and safety both alone and in combination with bevacizumab (mAb against VEGF-A) in recurrent high-grade glioma in a phase II study [167]. The functional and clinical significance of stromal PDGF receptors in cancer biology should also be the subject of future research in light of their relevance in the mutual communication between stroma and tumor cells [160].

2.5.2 Fibroblast Growth Factor Receptors

Fibroblast growth factor receptors (FGFRs) are well-known players in signaling pathways related to cellular survival, proliferation, migration, and differentiation [168]. Malfunctioning of FGFR signaling resulting from amplification, mutation, fusions, and overexpression is connected to several human malignancies, making it an attractive therapeutic target [169]. Current guidelines from the European Society for Medical Oncology and the National Comprehensive Cancer Network recommend the use of RT-PCR, FISH, and NGS methods for detection of abnormalities in bladder cancer [170, 171] and cholangiocarcinoma [172–174]. The FDA approved the use of erdafitinib for the treatment of FGFR3-mutated bladder cancer and infigratinib, pemigatinib, and futibatinib for treating cholangiocarcinoma having FGFR2 fusion [169]. The other FGFR therapies such as AZD4547 and RLY-4008 are presently being studied in clinics [175, 176].

Though FGFRs emerge as viable target, some questions still need to be considered and explored. For instance, in breast cancer FGFR1 amplification is often accompanied with co-amplification of neighboring genes, making FGFR1-targeted treatments less effective as additional amplified genes may trigger cancer growth on their own [177]. The abnormal signaling of FGFR1 in breast cancers is identified to increase the risk of cancer recurrence in ER+ breast cancer patients, rendering them resistant to endocrine interventions and CDK4/6 inhibitors [178–180]. Therefore, the clinical prognosis of these patients could therefore be improved by attempting suitable clinical studies. The methods employed to pinpoint the precise population subset that responds best to FGFR-TKI also have room for improvement.

Recently, in FORT-1 phase II/III investigation, patients with uroepithelial carcinoma overexpressing FGFR1–3 mRNA and receiving rogaratinib showed benefits similar to those of chemotherapy in the FORT-1 phase II/III study. Nonetheless, the ORR rose to 52 percent in the subgroup with FGFR3 DNA level alterations, as opposed to 27 percent in the chemotherapy group [181]. The study posed a crucial query regarding the discrepancy, wondering whether FGFR3 DNA alterations are a more accurate indicator than overexpressed mRNA. Further, overexpression of FGFR might also be a result of epigenetic alteration or downstream action of some other driver gene [182]. Therefore, identification of FGFR gene aberrations holds significant importance while selecting patients for clinical trials.

In the future, AI-based algorithms could be deployed to screen for FGFR mutations within cancer. Loeffler et al. created a deep learning-based model to predict FGFR3 mutations within H&E slides from patients of bladder cancer which could aid in preselecting patients for additional molecular screening [183]. However, bigger cohort is required to validate these approaches. Drug resistance development in individuals with cancer during treatment is an inevitable subject. Although there are several factors that contribute to drug resistance, the most common ones include FGFR gatekeeper mutations, bypass signaling activation, molecular-brake and secondary mutations in FGFR, and TP53 mutations [184, 185]. In the setting of therapeutic resistance, more-selective therapeutic agents and combination therapy showed improved clinical outcomes [185–187]. For instance, the phase II clinical trial exploring the combination of bemarituzumab, a mAb against FGFR2b, with mFOLFOX6 resulted in longer median PFS and OS than mFOLFOX6 alone treatment in FGFR2b-positive metastatic gastric cancer [188, 189]. Phase III confirmatory trials (NCT05052801, NCT05111626) are still in progress. Due to limited efficacy, ADCs (aprutumab ixadotin) targeting FGFR in phase I clinical trials were terminated [190]. Repurposing these ADCs and understanding the role of FGFR in TME could be a viable option for achieving better efficacy and efficiency.

2.5.3 Rearranged During Transfection Receptor

Rearranged during transfection (RET) RTK plays an important role during the developmental stage of the fetus. Fusion and mutations are major form of aberrations associated with the dysregulation of RET signaling in cancer. Mutations could be found in medullary thyroid cancer (MTC), while fusions are described in papillary thyroid carcinoma (PTC) and lung adenocarcinoma [191–193]. Treatment against RET includes existing selective TKI such as cabozantinib and lenvatinib [194, 195]. New-generation RET-specific inhibitors such as pralsetinib and selpercatinib are effective against $RET^{V804L/M}$ gatekeeper mutants as well as metastatic NSCLC in central nervous system [196]. Selpercatinib is also being used as a neoadjuvant therapy agent in a clinical trial (NCT04759911). Major drawbacks of these inhibitors are toxicities and resistance development (Hamidi and Hu, 2024). Clinical trials are now underway for next-generation RET inhibitors that can prevent steric clash caused by solvent front mutations and combinatorial approaches to prevent bypass changes like amplification of EGFR and MET signaling [197]. In future identification of novel RET mutations, using NGS technique could further help in refining the targets under treatment.

2.5.4 Receptor Tyrosine Kinase-Like Orphan Receptor 1

Similar to RET, receptor tyrosine kinase-like orphan receptor 1 (ROR1) is essentially important for embryonic development and is absent in adult cells. Nonetheless, ROR1 is overexpressed in a variety of solid tumors and hematological malignancies and has numerous pro-survival roles in cancer, including promoting cell proliferation, EMT, stemness, and drug resistance [198, 199]. It has been identified that ROR1 regulates the noncanonical Wnt signaling pathway and triggers downstream carcinogenic cascades such as the PI3K/Akt and RAS-MAPK pathways [200]. Research has demonstrated that ROR1 inhibition efficiently targets a variety of ROR1 overexpressing cancers [201, 202]. In mantle cell lymphoma (MCL) cells, a ROR1 inhibitor (KAN0441571C) demonstrated substantial cell death alone and exhibited synergistic apoptotic effect as well as blocked the phosphorylation of ROR1-associated signaling molecules when used in combination with inhibitors of BCL-2, PI3Kδ, mTOR, and BTK [203]. These studies indicate that ROR1 targeting by small molecule inhibitors needs to be further explored, especially in combination with other targeted drugs as a novel treatment strategy for different cancer types. ADCs were also observed to be efficient for the elimination of ROR1 expressing cells. For instance, VLS-101 has been identified as a possible candidate whose binding to ROR1 causes monomethyl auristatin E (MMAE) to be delivered and internalized by the cell rapidly, causing tumor cell death [204].

Endotoxin-based targeted therapies have also been exploited where mAbs are conjugated with bacterial endotoxins to achieve anti-tumor responses. For example, pseudomonas exotoxin A (PE38)-conjugate developed by Baskar et al. have shown activity against cells derived from patients with CLL and MCL [205]. Bispecific T cell engagers (BiTEs) such as NVG-11 are currently being explored for its effectiveness in phase I clinical trial against relapsed or refractory ROR1-positive malignancies (NCT04763083). Even after numerous clinical trials, CAR T therapies for ROR1 have not shown encouraging outcomes. A fully human TGFβRIIDN-armored ROR1-CAR-1 T cell was recently developed by Tran et al., and it demonstrated strong anti-ROR1-positive tumor activity and reversed TGFβ's immunosuppressive effects [206]. In the context of TGFβ1-rich TME, this strategy of pairing the CAR-1 construct along with the TGFβRIIDN protective armor could be investigated further for addressing additional solid tumor antigens and disease types. Focus should be on designing novel small molecule inhibitors and improving the CAR T cell therapy against ROR1. Vaccination is another strategy to activate the immune response against ROR1 as shown by Wu et al. in ROR1-expressing ovarian cancer stem cells [207].

2.5.5 Discoidin Domain-Containing Receptors

Discoidin domain-containing receptors (DDRs) are relatively newly discovered members that play a key role in extracellular matrix (ECM) remodeling. DDRs mainly interact with the collagen and are involved in organ homeostasis. Any disruption in their functioning affects the ECM, altering cell motility, proliferation, and differentiation [208]. DDRs are regarded as a crucial component of immunotherapy because of their function in several solid-state cancers [209]. Recent studies indicate DDRs function as a reliable tumor biomarker for a number of tumor types which include lung, ovarian, pancreatic, and breast cancer, glioma, hepatocellular carcinoma, and melanoma [210, 211]. However, understanding of DDRs' signaling and their interacting partners in many circumstances still needs to be expanded, though.

Likewise, for determining novel therapeutic possibilities for DDR-driven malignancies, machine learning techniques could eventually be utilized with large next-generation imaging and sequencing techniques. For instance, machine learning approach such as generative tensorial reinforcement learning (GENTRL) has shown promise in screening new inhibitors against DDRs but needs to be explored further for better outcomes [212]. Additionally, the fact that DDR plays a kinase-independent function in tumor maintenance implies that kinase inhibitors could not be the sole tactic used to counteract the serious threat posed by an elevated level of DDRs.

Supporting the implementation of combination therapy, such as antibody-drug conjugates, kinase inhibitors, nanobodies, and additional cytotoxic medicines, to reduce the fatal consequences of cancers driven by DDRs. Recently, a delivery method based on nanomedicine demonstrated effectiveness in administering a combination of siRNA and the chemotherapeutic drug SN-38 to target DDR1 in colorectal cancer [213], suggesting exploration of RNA interference-based treatment strategies in DDR-overexpressing cancer.

2.5.6 Tropomyosin Receptor Kinase Receptor

NTRK genes encode tropomyosin receptor kinase (TRK) family of RTK [214]. Expressed in neuronal tissues, these kinases are important for developing and functioning of the nervous system [215, 216]. TRK can influence cell proliferation and death majorly via regulation of PI3K/Akt, RAS/MAPKs, and PLCγ pathways [217]. Chromosomal rearrangement and fusion of the NTRK genes to support constitutive activation of kinase are reported majorly in thyroid, colorectal, and lung cancers [218–221]. Mutations as well as protein overexpression of the NTRK have also been observed in other cancers [222]. Gene fusion detection is majorly accomplished through techniques like FISH, RT-PCR, IHC, NGS, and RNA sequencing in cancers with high NTRK expression [214]. Although each test has unique features, drawbacks, and advantages, sample variety and the test's sensitivity and specificity usually influence the choice of test. Although liquid biopsies are a substitute to DNA NGS, RNA NGS has not yet been commonly accessible in liquid biopsies and is relatively costly. Currently, targeting NTRK in cancer has involved the use of numerous TKIs including entrectinib, larotrectinib, belizatinib, Cpd5n, AZ23, PHA-E429, GNF-5837, milciclib, cabozantinib, altiratinib, and sitravatinib. Larotrectinib and entrectinib, two of these anticancer drugs, were approved by the FDA based on biomarkers instead of tumor agnostic properties [223, 224]. This alters the conventional treatment approach for the primary tumor site and demonstrates the advancements in employing biomarkers to direct drug development and improve drug administration. However, like other targeted therapies, acquired drug resistance restricts the effectiveness of TRK inhibitors.

Owing to the resistance of other kinases, dual-blocking TRKs and other kinases have been shown to be an alternate strategy for drug-resistant conditions. Another intriguing treatment approach to address drug resistance is the combination of TRK inhibitors with chemo- or immunotherapy; nevertheless, it is important to note that additional side effects should be avoided. The PROTAC technology has also been used to generate TRK drugs [225], and it appeared to offer an innovative approach to develop novel targeted TRK therapeutics with a similar degradation mechanism to PROTAC. Antibody-based treatments, however, have not yet been investigated. Ultimately, a thorough grasp of how TRKs operate in cancer therapy, more precise methods for detecting NTRK fusions, and the application of computer-aided drug design to identify novel TRK inhibitors are all essential for enhancing clinical care.

3 Conclusion

RTKs remain promising targets for anticancer therapies. With improvement in X-ray free electron lasers and cryo-EM techniques along with AI-based algorithms, it is possible to determine detailed structure of different domains of RTKs. This information would aid in the repurposing of existing inventory of small molecule inhibitors and antibody-based agents to target RTKs. Interconnection of RTK signaling pathways makes it difficult to achieve desired clinical benefits. For instance, IGF-1R is a widely studied RTK and a plethora of targeted therapies exists against the molecule [226]. But not a single agent has successfully made it beyond clinical trials. The reason behind the failure is partly triggered by compensatory signaling pathways being activated, which encourages drug resistance. Investigating RTK inhibitors in conjunction with other therapeutic approaches thus necessitates a thorough grasp of alternate signaling or bypass mechanisms. Additionally, many promising RTK inhibitors have failed in clinical settings as the impact of components from TME is often not considered. Traditional approaches target only the tumor cells, overlooking key TME components such as immune cells, fibroblasts, and extracellular matrix, which may explain why RTK inhibitors have poor efficacy or increased cytotoxicity. A prudent approach would be to study the effect of the inhibitors on the various TME counterparts. In this context, patient-derived organoid-based screening for RTK inhibitors provides an innovative platform for identifying the most effective drug combination. With gradual advancement in organoid technology, the stromal cell inclusion can improve the ability to reproduce the tumor microenvironment, enabling a thorough assessment of RTK inhibitor efficacy and resistance in preclinical settings.

Finally, the success of clinical trials depends upon the use of a biomarker-based precision medicine approach, the identification of patient-specific RTK aberrations, the use of high-throughput techniques for drug discovery, and the simultaneous employment of novel drug delivery strategies for anticipating and monitoring therapy responses. Robust detection techniques with minimum artifacts can also be employed for the identification of RTK alteration during patient selection.

Acknowledgments D.C. is supported by the Department of Atomic Energy (DAE), India. M.Y. acknowledges the M K Bhan—Young Researcher Fellowship Programme (YRFP), Department of Biotechnology (DBT), Ministry of Science & Technology, Government of India.

Conflict of Interest Statement The authors declare no conflict of interest.

References

1. Lemmon MA, Schlessinger J. Cell signaling by receptor tyrosine kinases. Cell. 2010;141(7):1117–34.
2. Sengupta P, Das R, Majumder P, Mukhopadhyay D. Connecting the ends: signaling via receptor tyrosine kinases and cytoskeletal degradation in neurodegeneration. Explor Neurosci. 2024;3(1):1–26.

3. Saraon P, Pathmanathan S, Snider J, Lyakisheva A, Wong V, Stagljar I. Receptor tyrosine kinases and cancer: oncogenic mechanisms and therapeutic approaches. Oncogene. 2021;40(24):4079–93.
4. Wang Z. ErbB receptors and cancer. Methods Mol Biol Clifton NJ. 2017;1652:3–35.
5. Wieduwilt MJ, Moasser MM. The epidermal growth factor receptor family: biology driving targeted therapeutics. Cell Mol Life Sci. 2008;65(10):1566–84.
6. Organ SL, Tsao MS. An overview of the c-MET signaling pathway. Ther Adv Med Oncol. 2011;3(1 Suppl):S7–19.
7. Apte RS, Chen DS, Ferrara N. VEGF in signaling and disease: beyond discovery and development. Cell. 2019;176(6):1248–64.
8. Gay CM, Balaji K, Byers LA. Giving AXL the axe: targeting AXL in human malignancy. Br J Cancer. 2017;116(4):415–23.
9. Goss G, Tsai CM, Shepherd FA, Bazhenova L, Lee JS, Chang GC, et al. Osimertinib for pretreated EGFR Thr790Met-positive advanced non-small-cell lung cancer (AURA2): a multicentre, open-label, single-arm, phase 2 study. Lancet Oncol. 2016;17(12):1643–52.
10. Greig SL. Osimertinib: first global approval. Drugs. 2016;76(2):263–73.
11. Yang JCH, Ahn MJ, Kim DW, Ramalingam SS, Sequist LV, Su WC, et al. Osimertinib in pretreated T790M-positive advanced non-small-cell lung cancer: AURA study phase II extension component. J Clin Oncol. 2017;35(12):1288–96.
12. Pottier C, Fresnais M, Gilon M, Jérusalem G, Longuespée R, Sounni NE. Tyrosine kinase inhibitors in cancer: breakthrough and challenges of targeted therapy. Cancers. 2020;12(3):731.
13. Leonetti A, Assaraf YG, Veltsista PD, El Hassouni B, Tiseo M, Giovannetti E. MicroRNAs as a drug resistance mechanism to targeted therapies in EGFR-mutated NSCLC: current implications and future directions. Drug Resist Updates. 2019;42:1–11.
14. Ng KP, Hillmer AM, Chuah CTH, Juan WC, Ko TK, Teo ASM, et al. A common BIM deletion polymorphism mediates intrinsic resistance and inferior responses to tyrosine kinase inhibitors in cancer. Nat Med. 2012;18(4):521–8.
15. Aziz MH, Ahmad A. Epigenetic basis of cancer drug resistance. Cancer Drug Resist. 2020;3(1):113–6.
16. Dela Cruz MACP, Medina PMB. Epithelial–mesenchymal transition (EMT) and its role in acquired epidermal growth factor receptor-tyrosine kinase inhibitor (EGFR-TKI) chemoresistance in non-small cell lung cancer (NSCLC). Cancer Pathog Ther [Internet]. 2024 [cited 2024 Aug 12]; Available from: https://www.sciencedirect.com/science/article/pii/S2949713224000521.
17. Yang Y, Li S, Wang Y, Zhao Y, Li Q. Protein tyrosine kinase inhibitor resistance in malignant tumors: molecular mechanisms and future perspective. Signal Transduct Target Ther. 2022;7(1):1–36.
18. Alix-Panabières C, Marchetti D, Lang JE. Liquid biopsy: from concept to clinical application. Sci Rep. 2023;13(1):21685.
19. Zheng MM, Li YS, Tu HY, Jiang BY, Yang JJ, Zhou Q, et al. Genotyping of cerebrospinal fluid associated with Osimertinib response and resistance for leptomeningeal metastases in EGFR-mutated NSCLC. J Thorac Oncol. 2021;16(2):250–8.
20. Shah DR, Shah RR, Morganroth J. Tyrosine kinase inhibitors: their on-target toxicities as potential indicators of efficacy. Drug Saf. 2013;36(6):413–26.
21. Shyam Sunder S, Sharma UC, Pokharel S. Adverse effects of tyrosine kinase inhibitors in cancer therapy: pathophysiology, mechanisms and clinical management. Signal Transduct Target Ther. 2023;8(1):1–27.
22. Paranthaman S, Goravinahalli Shivananjegowda M, Mahadev M, Moin A, Hagalavadi Nanjappa S, Nanjaiyah ND, et al. Nanodelivery systems targeting epidermal growth factor receptors for glioma management. Pharmaceutics. 2020;12(12):1198.
23. Xu W, Ye C, Qing X, Liu S, Lv X, Wang W, et al. Multi-target tyrosine kinase inhibitor nanoparticle delivery systems for cancer therapy. Mater Today Bio. 2022;16:100358.

24. Burslem GM, Smith BE, Lai AC, Jaime-Figueroa S, McQuaid DC, Bondeson DP, et al. The advantages of targeted protein degradation over inhibition: a RTK case study. Cell Chem Biol. 2018;25(1):67–77.e3.
25. Linguanti F, Abenavoli EM, Calabretta R, Berti V, Lopci E. ImmunoPET targeting receptor tyrosine kinase: clinical applications. Cancers. 2023;15(24):5886.
26. Manafi-Farid R, Ataeinia B, Ranjbar S, Jamshidi Araghi Z, Moradi MM, Pirich C, et al. ImmunoPET: antibody-based PET imaging in solid tumors. Front Med [Internet]. 2022 [cited 2024 Aug 12];9. Available from: https://www.frontiersin.org/journals/medicine/articles/10.3389/fmed.2022.916693/full.
27. Sakemura R, Yang N, Cox MJ, Sinha S, Hefazi M, Hansen MJ, et al. Axl-RTK inhibition modulates T cell functions and synergizes with chimeric antigen receptor T cell therapy in B cell malignancies. Biol Blood Marrow Transplant. 2019;25(3, Supplement):S165.
28. Yang C, Yang C, Yarden Y, To KKW, Fu L. The prospects of tumor chemosensitivity testing at the single-cell level. Drug Resist Updates. 2021;54:100741.
29. Zhai J, Liu Y, Ji W, Huang X, Wang P, Li Y, et al. Drug screening on digital microfluidics for cancer precision medicine. Nat Commun. 2024;15(1):4363.
30. Zhong M, Fu L. Culture and application of conditionally reprogrammed primary tumor cells. Gastroenterol Rep. 2020;8(3):224–33.
31. Rager T, Eckburg A, Patel M, Qiu R, Gantiwala S, Dovalovsky K, et al. Treatment of metastatic melanoma with a combination of immunotherapies and molecularly targeted therapies. Cancers. 2022;14(15):3779.
32. Van Cutsem E, Eng C, Nowara E, Świeboda-Sadlej A, Tebbutt NC, Mitchell E, et al. Randomized phase Ib/II trial of Rilotumumab or Ganitumab with Panitumumab versus Panitumumab alone in patients with wild-type KRAS metastatic colorectal cancer. Clin Cancer Res. 2014;20(16):4240–50.
33. Chouaïd C, Gendarme S, Auliac JB. Artificial intelligence to finally enable precision medicine for the management of resected non-small-cell lung cancer. Ann Oncol. 2023;34(7):565–6.
34. Zhang Z, Wei X. Artificial intelligence-assisted selection and efficacy prediction of antineoplastic strategies for precision cancer therapy. Semin Cancer Biol. 2023;(90):57–72.
35. Nan X, Xie C, Yu X, Liu J. EGFR TKI as first-line treatment for patients with advanced EGFR mutation-positive non-small-cell lung cancer. Oncotarget. 2017;8(43):75712–26.
36. Rehmani HS, Issaeva N. EGFR in head and neck squamous cell carcinoma: exploring possibilities of novel drug combinations. Ann Transl Med. 2020;8(13):813.
37. Tan DSW, Kim SW, Ponce Aix S, Sequist LV, Smit EF, Yang JCH, et al. Nazartinib for treatment-naive EGFR-mutant non-small cell lung cancer: results of a phase 2, single-arm, open-label study. Eur J Cancer Oxf Engl. 2022;1990(172):276–86.
38. Ebrahimi N, Fardi E, Ghaderi H, Palizdar S, Khorram R, Vafadar R, et al. Receptor tyrosine kinase inhibitors in cancer. Cell Mol Life Sci. 2023;80(4):104.
39. Kumar R, Goel H, Solanki R, Rawat L, Tabasum S, Tanwar P, et al. Recent developments in receptor tyrosine kinase inhibitors: a promising mainstay in targeted cancer therapy. Med Drug Discov. 2024;23:100195.
40. Dickerson H, Diab A, Al MO. Epidermal growth factor receptor tyrosine kinase inhibitors in cancer: current use and future prospects. Int J Mol Sci. 2024;25(18):10008.
41. Singh S, Sadhukhan S, Sonawane A. 20 years since the approval of first EGFR-TKI, gefitinib: insight and foresight. Biochim Biophys Acta Rev Cancer. 2023;1878(6):188967.
42. Soria JC, Ohe Y, Vansteenkiste J, Reungwetwattana T, Chewaskulyong B, Lee KH, et al. Osimertinib in untreated EGFR-mutated advanced non–small-cell lung cancer. N Engl J Med. 2018;378(2):113–25.
43. Scott LJ. Osimertinib as first-line therapy in advanced NSCLC: a profile of its use. Drugs Ther Perspect. 2018;34(8):351–7.
44. Cross DAE, Ashton SE, Ghiorghiu S, Eberlein C, Nebhan CA, Spitzler PJ, et al. AZD9291, an irreversible EGFR TKI, overcomes T790M-mediated resistance to EGFR inhibitors in lung cancer. Cancer Discov. 2014;4(9):1046–61.

45. Mok TS, Wu YL, Ahn MJ, Garassino MC, Kim HR, Ramalingam SS, et al. Osimertinib or platinum–pemetrexed in EGFR T790M–positive lung cancer. N Engl J Med. 2017;376(7):629–40.
46. Cho BC, Ahn MJ, Kang JH, Soo RA, Reungwetwattana T, Yang JCH, et al. Lazertinib versus Gefitinib as first-line treatment in patients with EGFR-mutated advanced non–small-cell lung cancer: results from LASER301. J Clin Oncol. 2023;41(26):4208–17.
47. Ding J, Ding X, Zeng J, Liu X. Furmonertinib for EGFR-mutant advanced non-small cell lung cancer: a glittering diamond in the rough of EGFR-TKI. Front Pharmacol. 2024;(15):1357913.
48. Le T, Gerber DE. Newer-generation EGFR inhibitors in lung cancer: how are they best used? Cancers. 2019;11(3):366.
49. Corvaja C, Passaro A, Attili I, Aliaga PT, Spitaleri G, Signore ED, et al. Advancements in fourth-generation EGFR TKIs in EGFR-mutant NSCLC: bridging biological insights and therapeutic development. Cancer Treat Rev [Internet]. 2024 [cited 2024 Dec 16];130. Available from: https://www.cancertreatmentreviews.com/article/S0305-7372(24)00152-X/fulltext.
50. Wang S, Song Y, Liu D. EAI045: the fourth-generation EGFR inhibitor overcoming T790M and C797S resistance. Cancer Lett. 2017;385:51–4.
51. Liu X, Zhang X, Yang L, Tian X, Dong T, Ding CZ, et al. Abstract 1320: preclinical evaluation of TQB3804, a potent EGFR C797S inhibitor. Cancer Res. 2019;79(13_Supplement):1320.
52. Liu Y, Lai M, Li S, Wang Y, Feng F, Zhang T, et al. LS-106, a novel EGFR inhibitor targeting C797S, exhibits antitumor activities both in vitro and in vivo. Cancer Sci. 2022;113(2):709–20.
53. Wang Y, Yang N, Zhang Y, Li L, Han R, Zhu M, et al. Effective treatment of lung adenocarcinoma harboring *EGFR*-activating mutation, T790M, and *cis*-C797S triple mutations by Brigatinib and Cetuximab combination therapy. J Thorac Oncol. 2020;15(8):1369–75.
54. Tsuboi M, Herbst RS, John T, Kato T, Majem M, Grohé C, et al. Overall survival with osimertinib in resected EGFR-mutated NSCLC. N Engl J Med. 2023;389(2):137–47.
55. Si J, Hao Y, Wei J, Xiang J, Xu C, Shen Q, et al. Clinical outcomes of immune checkpoint inhibitors to treat non-small cell lung cancer patients harboring epidermal growth factor receptor mutations. BMC Pulm Med. 2023;23(1):158.
56. Ding J, Ding X, Leng Z. Immunotherapy-based therapy as a promising treatment for EGFR-mutant advanced non-small cell lung cancer patients after EGFR-TKI resistance. Expert Rev Anticancer Ther. 2023;23(2):187–98.
57. Li X, Huang H, Sun Y, Jiang Q, Yu Y. Immunotherapy strategies for EGFR-mutated advanced NSCLC after EGFR tyrosine-kinase inhibitors failure. Front Oncol. 2023;13:1265236.
58. Doostmohammadi A, Jooya H, Ghorbanian K, Gohari S, Dadashpour M. Potentials and future perspectives of multi-target drugs in cancer treatment: the next generation anti-cancer agents. Cell Commun Signal. 2024;22(1):228.
59. Al-Wahaibi LH, El-Sheref EM, Tawfeek HN, Abou-Zied HA, Rabea SM, Bräse S, et al. Design, synthesis, and biological evaluation of novel quinoline-based EGFR/HER-2 dual-target inhibitors as potential anti-tumor agents. RSC Adv. 2024;14(45):32978–91.
60. Tan L, Zhang J, Wang Y, Wang X, Wang Y, Zhang Z, et al. Development of dual inhibitors targeting epidermal growth factor receptor in cancer therapy. J Med Chem [Internet]. 2022 [cited 2024 Dec 16];65(7). Available from: https://pubmed.ncbi.nlm.nih.gov/35311289/.
61. Fernandes Neto JM, Nadal E, Bosdriesz E, Ooft SN, Farre L, McLean C, et al. Multiple low dose therapy as an effective strategy to treat EGFR inhibitor-resistant NSCLC tumours. Nat Commun. 2020;11(1):1–9.
62. Du Y, Shi S, Shu C, He Y, Xu W, Wu D, et al. Discovery of novel EGFR-PROTACs capable of degradation of multiple EGFR-mutated proteins. Eur J Med Chem. 2024;272:116489.
63. Liu Y, Guo Y, Wu Z, Feng K, Tong C, Wang Y, et al. Anti-EGFR chimeric antigen receptor-modified T cells in metastatic pancreatic carcinoma: a phase I clinical trial. Cytotherapy. 2020;22(10):573–80.
64. Zhang Y, Zhang Z, Ding Y, Fang Y, Wang P, Chu W, et al. Phase I clinical trial of EGFR-specific CAR-T cells generated by the piggyBac transposon system in advanced relapsed/refractory non-small cell lung cancer patients. J Cancer Res Clin Oncol. 2021;147(12):3725–34.

65. Knebel FH, Bettoni F, Shimada AK, Cruz M, Alessi JV, Negrão MV, et al. Sequential liquid biopsies reveal dynamic alterations of EGFR driver mutations and indicate EGFR amplification as a new mechanism of resistance to osimertinib in NSCLC. Lung Cancer Amst Neth. 2017;108:238–41.
66. Chen M, Zhao H. Next-generation sequencing in liquid biopsy: cancer screening and early detection. Hum Genomics. 2019;13(1):34.
67. Tomuleasa C, Tigu AB, Munteanu R, Moldovan CS, Kegyes D, Onaciu A, et al. Therapeutic advances of targeting receptor tyrosine kinases in cancer. Signal Transduct Target Ther. 2024;9(1):1–51.
68. Li J, Gong C, Zhou H, Liu J, Xia X, Ha W, et al. Kinase inhibitors and kinase-targeted cancer therapies: recent advances and future perspectives. Int J Mol Sci. 2024;25(10):5489.
69. Nguyen HS, Ho DKN, Nguyen NN, Tran HM, Tam KW, Le NQK. Predicting EGFR Mutation Status in Non–Small Cell Lung Cancer Using Artificial Intelligence: A Systematic Review and Meta-Analysis. Acad Radiol. 2024;31(2):660–83.
70. Coudray N, Ocampo PS, Sakellaropoulos T, Narula N, Snuderl M, Fenyö D, et al. Classification and mutation prediction from non–small cell lung cancer histopathology images using deep learning. Nat Med. 2018;24(10):1559–67.
71. Bhinder B, Gilvary C, Madhukar NS, Elemento O. Artificial intelligence in cancer research and precision medicine. Cancer Discov. 2021;11(4):900–15.
72. Wang C, Ma J, Shao J, Zhang S, Liu Z, Yu Y, et al. Predicting EGFR and PD-L1 status in NSCLC patients using multitask AI system based on CT images. Front Immunol [Internet]. 2022 [cited 2024 Dec 17];13. Available from: https://www.frontiersin.org/journals/immunology/articles/10.3389/fimmu.2022.813072/full
73. Zhou G, Xu S, Liu X, Ge J, He Q, Cao W, et al. Relationship between the image characteristics of artificial intelligence and EGFR gene mutation in lung adenocarcinoma. Front Genet. 2022;13:1090180.
74. Mahajan A, Kania V, Agarwal U, Ashtekar R, Shukla S, Patil VM, et al. Deep-learning-based predictive imaging biomarker model for EGFR mutation status in non-small cell lung cancer from CT imaging. Cancers. 2024;16(6):1130.
75. Tang Y, Zang H, Wen Q, Fan S. AXL in cancer: a modulator of drug resistance and therapeutic target. J Exp Clin Cancer Res. 2023;42(1):148.
76. Zhu C, Wei Y, Wei X. AXL receptor tyrosine kinase as a promising anti-cancer approach: functions, molecular mechanisms and clinical applications. Mol Cancer. 2019;18(1):153.
77. Felip E, Brunsvig P, Helland Å, Viñolas N, Aix S, Carcereny E, et al. MA03.06 efficacy results of selective AXL inhibitor Bemcentinib with Pembrolizumab following chemotherapy in patients with NSCLC. J Thorac Oncol. 2019;14(10):S258–9.
78. Ikeda M, Koh Y, Teraoka S, Sato K, Kanai K, Hayata A, et al. Detection of AXL expression in circulating tumor cells of lung cancer patients using an automated microcavity array system. Cancer Med. 2020;9(6):2122–33.
79. Engelsen AST, Lotsberg ML, Abou Khouzam R, Thiery JP, Lorens JB, Chouaib S, et al. Dissecting the role of AXL in cancer immune escape and resistance to immune checkpoint inhibition. Front Immunol [Internet]. 2022 [cited 2023 Jun 6];13. Available from: https://www.frontiersin.org/articles/10.3389/fimmu.2022.869676.
80. Asiedu MK, Beauchamp-Perez FD, Ingle JN, Behrens MD, Radisky DC, Knutson KL. AXL induces epithelial-to-mesenchymal transition and regulates the function of breast cancer stem cells. Oncogene. 2014;33(10):1316–24.
81. Zhang Y, Xia M, Jin K, Wang S, Wei H, Fan C, et al. Function of the c-met receptor tyrosine kinase in carcinogenesis and associated therapeutic opportunities. Mol Cancer. 2018;17(1):45.
82. Santoni M, Iacovelli R, Colonna V, Klinz S, Mauri G, Nuti M. Antitumor effects of the multi-target tyrosine kinase inhibitor cabozantinib: a comprehensive review of the preclinical evidence. Expert Rev Anticancer Ther. 2021;21(9):1029–54.

83. Hellerstedt BA, Vogelzang NJ, Kluger HM, Yasenchak CA, Aftab DT, Ramies DA, et al. Results of a phase II placebo-controlled randomized discontinuation trial of Cabozantinib in patients with non-small-cell lung carcinoma. Clin Lung Cancer. 2019;20(2):74–81.e1.
84. Powles T, Motzer RJ, Escudier B, Pal S, Kollmannsberger C, Pikiel J, et al. Outcomes based on prior therapy in the phase 3 METEOR trial of cabozantinib versus everolimus in advanced renal cell carcinoma. Br J Cancer. 2018;119(6):663–9.
85. Zhou L, Liu XD, Sun M, Zhang X, German P, Bai S, et al. Targeting MET and AXL overcomes resistance to sunitinib therapy in renal cell carcinoma. Oncogene. 2016;35(21):2687–97.
86. Yang Q, Wang Y, Yang Q, Gao Y, Duan X, Fu Q, et al. Cuprous oxide nanoparticles trigger ER stress-induced apoptosis by regulating copper trafficking and overcoming resistance to sunitinib therapy in renal cancer. Biomaterials. 2017;146:72–85.
87. Aissa AF, Islam ABMMK, Ariss MM, Go CC, Rader AE, Conrardy RD, et al. Single-cell transcriptional changes associated with drug tolerance and response to combination therapies in cancer. Nat Commun. 2021;12(1):1628.
88. Taverna JA, Hung CN, DeArmond DT, Chen M, Lin CL, Osmulski PA, et al. Single-cell proteomic profiling identifies combined AXL and JAK1 inhibition as a novel therapeutic strategy for lung cancer. Cancer Res. 2020;80(7):1551–63.
89. Cho JH, Okuma A, Al-Rubaye D, Intisar E, Junghans RP, Wong WW. Engineering Axl specific CAR and SynNotch receptor for cancer therapy. Sci Rep. 2018;8(1):3846.
90. Wei J, Sun H, Zhang A, Wu X, Li Y, Liu J, et al. A novel AXL chimeric antigen receptor endows T cells with anti-tumor effects against triple negative breast cancers. Cell Immunol. 2018;331:49–58.
91. Włodarczyk M, Pyrzynska B. CAR-NK as a rapidly developed and efficient immunotherapeutic strategy against cancer. Cancers. 2023;15(1):117.
92. Liu Z, Hu M, Yang Y, Du C, Zhou H, Liu C, et al. An overview of PROTACs: a promising drug discovery paradigm. Mol Biomed. 2022;3:46.
93. Shi W, Feng Z, Chi F, Zhou J, Qiu Q, Jiang Y, et al. Structure-based discovery of receptor tyrosine kinase AXL degraders with excellent anti-tumor activity by selectively degrading AXL and inducing methuosis. Eur J Med Chem. 2022;234:114253.
94. He R, Song Z, Bai Y, He S, Huang J, Wang Y, et al. Discovery of AXL degraders with improved potencies in triple-negative breast cancer (TNBC) cells. J Med Chem. 2023;66(3):1873–91.
95. Fan L, Liu X, Jin X, Yang Y, Chen L, Hu X, et al. The safety, tolerability, and preliminary antitumor activity of sitravatinib plus tislelizumab in patients with locally recurrent or metastatic triple-negative breast cancer. J Clin Oncol. 2022;40(16_suppl):1070.
96. Hirohashi T, Beck JT, McDermott JD, Abbruzzese JL, Doi T, Ingram K, et al. Trials in progress: TAM kinase inhibitor PF-07265807 and sasanlimab plus axitinib in patients with advanced or metastatic renal cell carcinoma—a phase 1, open-label, pharmacokinetic, safety and tolerability study. J Clin Oncol. 2023;41(6_suppl):TPS743.
97. Powles T, Burotto M, Escudier B, Apolo AB, Bourlon MT, Shah AY, et al. Nivolumab plus cabozantinib versus sunitinib for first-line treatment of advanced renal cell carcinoma: extended follow-up from the phase III randomised CheckMate 9ER trial. ESMO Open. 2024;9(5):102994.
98. Graveel CR, Tolbert D, Woude GFV. MET: a critical player in tumorigenesis and therapeutic target. Cold Spring Harb Perspect Biol. 2013;5(7):a009209.
99. Recondo G, Che J, Jänne PA, Awad MM. Targeting MET dysregulation in cancer. Cancer Discov. 2020;10(7):922–34.
100. Hughes VS, Siemann DW. Have clinical trials properly assessed c-met inhibitors? Trends Cancer. 2018;4(2):94–7.
101. Puccini A, Marín-Ramos NI, Bergamo F, Schirripa M, Lonardi S, Lenz HJ, et al. Safety and tolerability of c-MET inhibitors in cancer. Drug Saf. 2019;42(2):211.
102. Markham A. Savolitinib: First Approval. Drugs. 2021;81(14):1665–70.

103. Mathieu LN, Larkins E, Akinboro O, Roy P, Amatya AK, Fiero MH, et al. FDA approval summary: Capmatinib and Tepotinib for the treatment of metastatic NSCLC harboring MET exon 14 skipping mutations or alterations. Clin Cancer Res. 2022;28(2):249–54.
104. Acker F, Klein A, Rasokat A, Eisert A, Kron A, Christopoulos P, et al. Multicenter real-world analysis of combined MET and EGFR inhibition in patients with non-small cell lung cancer and acquired MET amplification or polysomy after EGFR inhibition. Clin Lung Cancer [Internet]. 2024 [cited 2024 Nov 20]; Available from: https://www.sciencedirect.com/science/article/pii/S1525730424001517.
105. Patnaik A, Weiss GJ, Papadopoulos KP, Hofmeister CC, Tibes R, Tolcher A, et al. Phase I ficlatuzumab monotherapy or with erlotinib for refractory advanced solid tumours and multiple myeloma. Br J Cancer. 2014;111(2):272–80.
106. Tarhini AA, Rafique I, Floros T, Tran P, Gooding WE, Villaruz LC, et al. Phase 1/2 study of rilotumumab (AMG 102), a hepatocyte growth factor inhibitor, and erlotinib in patients with advanced non-small cell lung cancer. Cancer. 2017;123(15):2936–44.
107. Fiedler U, Ekawardhani S, Cornelius A, Gilboy P, Bakker TR, Dolado I, et al. MP0250, a VEGF and HGF neutralizing DARPin® molecule shows high anti-tumor efficacy in mouse xenograft and patient-derived tumor models. Oncotarget. 2017;8(58):98371–83.
108. Baek JH, Birchmeier C, Zenke M, Hieronymus T. The HGF receptor/Met tyrosine kinase is a key regulator of dendritic cell migration in skin immunity. J Immunol Baltim Md 1950. 2012;189(4):1699–707.
109. Okunishi K, Dohi M, Nakagome K, Tanaka R, Mizuno S, Matsumoto K, et al. A novel role of hepatocyte growth factor as an immune regulator through suppressing dendritic cell function. J Immunol Baltim Md 1950. 2005;175(7):4745–53.
110. Benkhoucha M, Santiago-Raber ML, Schneiter G, Chofflon M, Funakoshi H, Nakamura T, et al. Hepatocyte growth factor inhibits CNS autoimmunity by inducing tolerogenic dendritic cells and CD25+Foxp3+ regulatory T cells. Proc Natl Acad Sci. 2010;107(14):6424–9.
111. Glodde N, Bald T, Boorn-Konijnenberg D van den, Nakamura K, O'Donnell JS, Szczepanski S, et al. Reactive neutrophil responses dependent on the receptor tyrosine kinase c-MET limit cancer immunotherapy. Immunity 2017;47(4):789–802.e9.
112. Li E, Huang X, Zhang G, Liang T. Combinational blockade of MET and PD-L1 improves pancreatic cancer immunotherapeutic efficacy. J Exp Clin Cancer Res. 2021;40(1):279.
113. Domènech M, Marmol AMM, Mate JL, Estival A, Moran T, Cucurull M, et al. Correlation between PD-L1 expression and MET gene amplification in patients with advanced non-small cell lung cancer and no other actionable oncogenic driver. Oncotarget. 2021;12(18):1802.
114. Ilié M, Hofman V, Bontoux C, Goffinet S, Benzaquen J, Heeke S, et al. Lack of correlation between MET and PD-L1 expression in non-small cell lung cancer revealed by comparative study of matched biopsies and surgical resection samples. Lung Cancer. 2023;181:107230.
115. Thayaparan T, Petrovic RM, Achkova DY, Zabinski T, Davies DM, Klampatsa A, et al. CAR T-cell immunotherapy of MET-expressing malignant mesothelioma. Onco Targets Ther. 2017;6(12):e1363137.
116. Chiriaco C, Donini C, Cortese M, Ughetto S, Modica C, Martinelli I, et al. Efficacy of CAR-T immunotherapy in MET overexpressing tumors not eligible for anti-MET targeted therapy. J Exp Clin Cancer Res. 2022;41(1):309.
117. Qin A, Qin Y, Lee J, Musket A, Ying M, Krenciute G, et al. Tyrosine kinase signaling-independent MET-targeting with CAR-T cells. J Transl Med. 2023;21(1):682.
118. DaSilva JO, Yang K, Surriga O, Nittoli T, Kunz A, Franklin MC, et al. A biparatopic antibody-drug conjugate to treat MET-expressing cancers, including those that are unresponsive to MET pathway blockade. Mol Cancer Ther. 2021;20(10):1966–76.
119. Wu S, Shang Y, Yan Y, Zhou A, Bing T, Zhao Z, et al. Aptamer-based enforced phosphatase-recruiting chimeras inhibit receptor tyrosine kinase signal transduction. J Am Chem Soc. 2024;146(32):22445–54.
120. Zhang J, Babic A. Regulation of the MET oncogene: molecular mechanisms. Carcinogenesis. 2016;37(4):345–55.

121. Peschard P, Fournier TM, Lamorte L, Naujokas MA, Band H, Langdon WY, et al. Mutation of the c-Cbl TKB domain binding site on the met receptor tyrosine kinase converts it into a transforming protein. Mol Cell. 2001;8(5):995–1004.
122. Leiser D, Pochon B, Blank-Liss W, Francica P, Glück AA, Aebersold DM, et al. Targeting of the MET receptor tyrosine kinase by small molecule inhibitors leads to MET accumulation by impairing the receptor downregulation. FEBS Lett. 2014;588(5):653–8.
123. Verzè M, Boscolo Bragadin A, Minari R, Pasello G, Perrone F, Scattolin D, et al. NGS detection of gene rearrangements and METexon14 mutations in liquid biopsy of advanced NSCLC patients: a study of two Italian centers. J Liq Biopsy. 2024;4:100143.
124. Lee SM, Upadhyay S, Lewanski C, Falk S, Skailes G, Woll PJ, et al. The clinical role of VeriStrat testing in patients with advanced non–small cell lung cancer considered unfit for first-line platinum-based chemotherapy. Eur J Cancer. 2019;120:86–96.
125. Copin MC, Lesaffre M, Berbon M, Doublet L, Leroy C, Tresch E, et al. High-MET status in non-small cell lung tumors correlates with receptor phosphorylation but not with the serum level of soluble form. Lung Cancer Amst Neth. 2016;101:59–67.
126. Yap TA, de Bono JS. Targeting the HGF/c-Met axis: state of play. Mol Cancer Ther. 2010;9(5):1077–9.
127. Mabeta P, Steenkamp V. The VEGF/VEGFR Axis revisited: implications for cancer therapy. Int J Mol Sci. 2022;23(24):15585.
128. Meadows KL, Hurwitz HI. Anti-VEGF therapies in the clinic. Cold Spring Harb Perspect Med. 2012;2(10):a006577.
129. Garcia J, Hurwitz HI, Sandler AB, Miles D, Coleman RL, Deurloo R, et al. Bevacizumab (Avastin®) in cancer treatment: a review of 15 years of clinical experience and future outlook. Cancer Treat Rev. 2020;86:102017.
130. Xu M, Fan R, Fan X, Shao Y, Li X. Progress and challenges of anti-VEGF agents and their sustained-release strategies for retinal angiogenesis. Drug Des Devel Ther. 2022;16:3241–62.
131. Żak K, Satora M, Skrabalak I, Tarkowski R, Ostrowska-Leśko M, Bobiński M. The potential influence of residual or recurrent disease on bevacizumab treatment efficacy in ovarian cancer: current evidence and future perspectives. Cancers. 2024;16(5):1063.
132. Roskoski R. Properties of FDA-approved small molecule protein kinase inhibitors: a 2024 update. Pharmacol Res. 2024;200:107059.
133. Wang L, Liu WQ, Broussy S, Han B, Fang H. Recent advances of anti-angiogenic inhibitors targeting VEGF/VEGFR axis. Front Pharmacol [Internet]. 2024 [cited 2024 Dec 1];14. Available from: https://www.frontiersin.org/journals/pharmacology/articles/10.3389/fphar.2023.1307860/full.
134. Yamaoka T, Kusumoto S, Ando K, Ohba M, Ohmori T. Receptor tyrosine kinase-targeted cancer therapy. Int J Mol Sci. 2018;19(11):3491.
135. van Geel RMJM, Beijnen JH, Schellens JHM. Concise drug review: pazopanib and axitinib. Oncologist. 2012;17(8):1081–9.
136. Roskoski R. Vascular endothelial growth factor (VEGF) and VEGF receptor inhibitors in the treatment of renal cell carcinomas. Pharmacol Res. 2017;120:116–32.
137. Thakur A, Rana M, Mishra A, Kaur C, Pan CH, Nepali K. Recent advances and future directions on small molecule VEGFR inhibitors in oncological conditions. Eur J Med Chem. 2024;272:116472.
138. Liang W, Zheng Y, Zhang J, Sun X. Multiscale modeling reveals angiogenesis-induced drug resistance in brain tumors and predicts a synergistic drug combination targeting EGFR and VEGFR pathways. BMC Bioinform. 2019;20(7):203.
139. Khanna P, Soh HJ, Chen CH, Saxena R, Amin S, Naughton M, et al. ACE2 abrogates tumor resistance to VEGFR inhibitors suggesting angiotensin-(1-7) as a therapy for clear cell renal cell carcinoma. Sci Transl Med. 2021;13(577):eabc0170.
140. Motzer RJ, Penkov K, Haanen J, Rini B, Albiges L, Campbell MT, et al. Avelumab plus axitinib versus sunitinib for advanced renal-cell carcinoma. N Engl J Med. 2019;380(12):1103–15.

141. Powles T, Plimack ER, Soulières D, Waddell T, Stus V, Gafanov R, et al. Pembrolizumab plus axitinib versus sunitinib monotherapy as first-line treatment of advanced renal cell carcinoma (KEYNOTE-426): extended follow-up from a randomised, open-label, phase 3 trial. Lancet Oncol. 2020;21(12):1563–73.
142. Lee WS, Yang H, Chon HJ, Kim C. Combination of anti-angiogenic therapy and immune checkpoint blockade normalizes vascular-immune crosstalk to potentiate cancer immunity. Exp Mol Med. 2020;52(9):1475–85.
143. Choueiri TK, Motzer RJ. Systemic therapy for metastatic renal-cell carcinoma. N Engl J Med. 2017;376(4):354–66.
144. Moriguchi M, Kataoka S, Itoh Y. Evolution of systemic treatment for hepatocellular carcinoma: changing treatment strategies and concepts. Cancers. 2024;16(13):2387.
145. Meric-Bernstam F, Larkin J, Tabernero J, Bonini C. Enhancing anti-tumour efficacy with immunotherapy combinations. Lancet Lond Engl. 2021;397(10278):1010–22.
146. Liu Y, Li Y, Wang Y, Lin C, Zhang D, Chen J, et al. Recent progress on vascular endothelial growth factor receptor inhibitors with dual targeting capabilities for tumor therapy. J Hematol Oncol. 2022;15(1):89.
147. Isakova AA, Artykov AA, Plotnikova EA, Trunova GV, Khokhlova VA, Pankratov AA, et al. Dual targeting of DR5 and VEGFR2 molecular pathways by multivalent fusion protein significantly suppresses tumor growth and angiogenesis. Int J Biol Macromol. 2024;255:128096.
148. Wang X, Song K, Li L, Chen L. Structure-based drug design strategies and challenges. Curr Top Med Chem. 2018;18(12):998–1006.
149. Nayarisseri A, Abdalla M, Joshi I, Yadav M, Bhrdwaj A, Chopra I, et al. Potential inhibitors of VEGFR1, VEGFR2, and VEGFR3 developed through Deep Learning for the treatment of Cervical Cancer. Sci Rep. 2024;14(1):13251.
150. Hidalgo M, Martinez-Garcia M, Le Tourneau C, Massard C, Garralda E, Boni V, et al. First-in-human phase I study of single-agent Vanucizumab, a first-in-class bispecific anti-Angiopoietin-2/anti-VEGF-A antibody, in adult patients with advanced solid tumors. Clin Cancer Res. 2018;24(7):1536–45.
151. Wang Q, Li T, Wu Z, Wu Q, Ke X, Luo D, et al. Novel VEGF decoy receptor fusion protein conbercept targeting multiple VEGF isoforms provide remarkable anti-angiogenesis effect in vivo. PLoS One. 2013;8(8):e70544.
152. Lee JE, Kim C, Yang H, Park I, Oh N, Hua S, et al. Novel glycosylated VEGF decoy receptor fusion protein, VEGF-Grab, efficiently suppresses tumor angiogenesis and progression. Mol Cancer Ther. 2015;14(2):470–9.
153. Wang XR, Wang S, Mu HX, Xu KY, Wang XT, Shi JT, et al. Discovery of novel VEGFR-2-PROTAC degraders based on the localization of lysine residues via recruiting VHL for the treatment of gastric cancer. Eur J Med Chem. 2022;244:114821.
154. Feng Z, Xie D, Qiu F, Huang J, Wang Z, Liang C. Development of hybrid aptamers-engineered PROTACs for degrading VEGF165 in both tumor- and vascular endothelial cells. Eur J Med Chem. 2025;281:117027.
155. Xia Y, Duan S, Han C, Jing C, Xiao Z, Li C. Hypoxia-responsive nanomaterials for tumor imaging and therapy. Front Oncol. 2022;12:1089446.
156. Patel SA, Nilsson MB, Le X, Cascone T, Jain RK, Heymach JV. Molecular mechanisms and future implications of VEGF/VEGFR in cancer therapy. Clin Cancer Res. 2023;29(1):30–9.
157. Jain RK. Normalizing tumor microenvironment to treat cancer: bench to bedside to biomarkers. J Clin Oncol. 2013;31(17):2205–18.
158. Lupo G, Caporarello N, Olivieri M, Cristaldi M, Motta C, Bramanti V, et al. Anti-angiogenic therapy in cancer: Downsides and new pivots for precision medicine. Front Pharmacol [Internet]. 2017 [cited 2024 Dec 7];7. Available from: https://www.frontiersin.org/journals/pharmacology/articles/10.3389/fphar.2016.00519/full
159. Ségaliny AI, Tellez-Gabriel M, Heymann MF, Heymann D. Receptor tyrosine kinases: characterisation, mechanism of action and therapeutic interests for bone cancers. J Bone Oncol. 2015;4(1):1–12.

160. Strell C, Rodríguez-Tomàs E, Östman A. Functional and clinical roles of stromal PDGF receptors in tumor biology. Cancer Metastasis Rev. 2024;43(4):1593–609.
161. Pandey P, Khan F, Upadhyay TK, Seungjoon M, Park MN, Kim B. New insights about the PDGF/PDGFR signaling pathway as a promising target to develop cancer therapeutic strategies. Biomed Pharmacother. 2023;161:114491.
162. Zou X, Tang XY, Qu ZY, Sun ZW, Ji CF, Li YJ, et al. Targeting the PDGF/PDGFR signaling pathway for cancer therapy: a review. Int J Biol Macromol. 2022;202:539–57.
163. Kuai J, Mosyak L, Brooks J, Cain M, Carven GJ, Ogawa S, et al. Characterization of binding mode of action of a blocking anti-platelet-derived growth factor (PDGF)-B monoclonal antibody, MOR8457, reveals conformational flexibility and avidity needed for PDGF-BB to bind PDGF receptor-β. Biochemistry. 2015;54(10):1918–29.
164. Camorani S, Esposito CL, Rienzo A, Catuogno S, Iaboni M, Condorelli G, et al. Inhibition of receptor signaling and of glioblastoma-derived tumor growth by a novel PDGFRβ aptamer. Mol Ther J Am Soc Gene Ther. 2014;22(4):828–41.
165. Camorani S, Hill BS, Collina F, Gargiulo S, Napolitano M, Cantile M, et al. Targeted imaging and inhibition of triple-negative breast cancer metastases by a PDGFRβ aptamer. Theranostics. 2018;8(18):5178–99.
166. Yan M, Qi X, Xie W, Xiao J, Gan Y, Wang P, et al. PDGF receptor-β-targeted copper-gadolinium-oxide self-assembled nanoclusters suppress tumor growth and metastasis via copper overload-mediated apoptosis and anti-angiogenesis. Chem Eng J. 2023;457:141227.
167. Zhao S, Zhang M, Zhang Q, Wu J, Dai H. Anlotinib alone or in combination with bevacizumab in the treatment of recurrent high-grade glioma: a prospective single-arm, open-label phase II trial. BMC Cancer. 2024;24(1):6.
168. Xie Y, Su N, Yang J, Tan Q, Huang S, Jin M, et al. FGF/FGFR signaling in health and disease. Signal Transduct Target Ther. 2020;5(1):1–38.
169. Zhang P, Yue L, Leng Q, Chang C, Gan C, Ye T, et al. Targeting FGFR for cancer therapy. J Hematol Oncol. 2024;17(1):39.
170. Miyake M, Sugano K, Kawashima K, Ichikawa H, Hirabayashi K, Kodama T, et al. Sensitive detection of FGFR3 mutations in bladder cancer and urine sediments by peptide nucleic acid-mediated real-time PCR clamping. Biochem Biophys Res Commun. 2007;362(4):865–71.
171. Powles T, Bellmunt J, Comperat E, De Santis M, Huddart R, Loriot Y, et al. Bladder cancer: ESMO Clinical Practice Guideline for diagnosis, treatment and follow-up. Ann Oncol. 2022;33(3):244–58.
172. Arai Y, Totoki Y, Hosoda F, Shirota T, Hama N, Nakamura H, et al. Fibroblast growth factor receptor 2 tyrosine kinase fusions define a unique molecular subtype of cholangiocarcinoma. Hepatol Baltim Md. 2014;59(4):1427–34.
173. Benson AB, D'Angelica MI, Abrams T, Abbott DE, Ahmed A, Anaya DA, et al. NCCN guidelines® insights: biliary tract cancers, version 2.2023. J Natl Compr Cancer Netw. 2023;21(7):694–704.
174. Chen LT, Vogel A, Hsu C, Chen MH, Fang W, Pangarsa EA, et al. Pan-Asian adapted ESMO Clinical Practice Guidelines for the diagnosis, treatment and follow-up of patients with biliary tract cancer. ESMO Open [Internet]. 2024 [cited 2024 Nov 24];9(8). Available from: https://www.esmoopen.com/article/S2059-7029(24)01416-9/fulltext.
175. Coombes RC, Badman PD, Lozano-Kuehne JP, Liu X, Macpherson IR, Zubairi I, et al. Results of the phase IIa RADICAL trial of the FGFR inhibitor AZD4547 in endocrine resistant breast cancer. Nat Commun. 2022;13(1):3246.
176. Subbiah V, Sahai V, Maglic D, Bruderek K, Touré BB, Zhao S, et al. RLY-4008, the first highly selective FGFR2 inhibitor with activity across FGFR2 alterations and resistance mutations. Cancer Discov. 2023;13(9):2012–31.
177. Condorelli R, Mosele F, Verret B, Bachelot T, Bedard PL, Cortes J, et al. Genomic alterations in breast cancer: level of evidence for actionability according to ESMO Scale for Clinical Actionability of Molecular Targets (ESCAT). Ann Oncol. 2019;30(3):365–73.

178. Turner N, Pearson A, Sharpe R, Lambros M, Geyer F, Lopez-Garcia MA, et al. FGFR1 amplification drives endocrine therapy resistance and is a therapeutic target in breast cancer. Cancer Res. 2010;70(5):2085–94.
179. Formisano L, Lu Y, Servetto A, Hanker AB, Jansen VM, Bauer JA, et al. Aberrant FGFR signaling mediates resistance to CDK4/6 inhibitors in ER+ breast cancer. Nat Commun. 2019;10(1):1373.
180. Mouron S, Manso L, Caleiras E, Rodriguez-Peralto JL, Rueda OM, Caldas C, et al. FGFR1 amplification or overexpression and hormonal resistance in luminal breast cancer: rationale for a triple blockade of ER, CDK4/6, and FGFR1. Breast Cancer Res. 2021;23(1):21.
181. Sternberg CN, Petrylak DP, Bellmunt J, Nishiyama H, Necchi A, Gurney H, et al. FORT-1: phase II/III study of Rogaratinib versus chemotherapy in patients with locally advanced or metastatic urothelial carcinoma selected based on FGFR1/3 mRNA expression. J Clin Oncol Off J Am Soc Clin Oncol. 2023;41(3):629–39.
182. DeSalvo J, Ban Y, Li L, Sun X, Jiang Z, Kerr DA, et al. ETV4 and ETV5 drive synovial sarcoma through cell cycle and DUX4 embryonic pathway control. J Clin Invest. 2021;131(13):e141908, 141908.
183. Loeffler CML, Ortiz Bruechle N, Jung M, Seillier L, Rose M, Laleh NG, et al. Artificial intelligence-based detection of FGFR3 mutational status directly from routine histology in bladder cancer: a possible preselection for molecular testing? Eur Urol Focus. 2022;8(2):472–9.
184. Yue S, Li Y, Chen X, Wang J, Li M, Chen Y, et al. FGFR-TKI resistance in cancer: current status and perspectives. J Hematol Oncol. 2021;14(1):23.
185. Katoh M, Loriot Y, Brandi G, Tavolari S, Wainberg ZA, Katoh M. FGFR-targeted therapeutics: clinical activity, mechanisms of resistance and new directions. Nat Rev Clin Oncol. 2024;21(4):312–29.
186. Jones RH, Anthoney DA, Jones RJ, Crabb SJ, Hussain SA, Birtle AJ, et al. FIESTA: a phase Ib and pharmacokinetic trial of AZD4547 in combination with gemcitabine and cisplatin. J Clin Oncol. 2016;34(15_suppl):4521.
187. Ma Q, Liao R, Wang Y. 1015P The efficacy and mechanism of pan-FGFR inhibitor (AZD4547) combined with immunoagonists or immunosuppressants in FGFR-positive tumors. Ann Oncol. 2024;35:S687–8.
188. Wainberg ZA, Enzinger PC, Kang YK, Qin S, Yamaguchi K, Kim IH, et al. Bemarituzumab in patients with FGFR2b-selected gastric or gastro-oesophageal junction adenocarcinoma (FIGHT): a randomised, double-blind, placebo-controlled, phase 2 study. Lancet Oncol. 2022;23(11):1430–40.
189. Wainberg ZA, Kang YK, Lee KW, Qin S, Yamaguchi K, Kim IH, et al. Bemarituzumab as first-line treatment for locally advanced or metastatic gastric/gastroesophageal junction adenocarcinoma: final analysis of the randomized phase 2 FIGHT trial. Gastric Cancer. 2024;27(3):558–70.
190. Kim SB, Meric-Bernstam F, Kalyan A, Babich A, Liu R, Tanigawa T, et al. First-in-human phase I study of Aprutumab Ixadotin, a fibroblast growth factor receptor 2 antibody-drug conjugate (BAY 1187982) in patients with advanced cancer. Target Oncol. 2019;14(5):591–601.
191. Grieco M, Santoro M, Berlingieri MT, Melillo RM, Donghi R, Bongarzone I, et al. PTC is a novel rearranged form of the ret proto-oncogene and is frequently detected in vivo in human thyroid papillary carcinomas. Cell. 1990;60(4):557–63.
192. Ju YS, Lee WC, Shin JY, Lee S, Bleazard T, Won JK, et al. A transforming KIF5B and RET gene fusion in lung adenocarcinoma revealed from whole-genome and transcriptome sequencing. Genome Res. 2012;22(3):436–45.
193. Li AY, McCusker MG, Russo A, Scilla KA, Gittens A, Arensmeyer K, et al. RET fusions in solid tumors. Cancer Treat Rev. 2019;81:101911.
194. Drilon A, Rekhtman N, Arcila M, Wang L, Ni A, Albano M, et al. Cabozantinib in patients with advanced RET-rearranged non-small-cell lung cancer: an open-label, single-centre, phase 2, single-arm trial. Lancet Oncol. 2016;17(12):1653–60.

195. Hida T, Velcheti V, Reckamp KL, Nokihara H, Sachdev P, Kubota T, et al. A phase 2 study of lenvatinib in patients with RET fusion-positive lung adenocarcinoma. Lung Cancer Amst Neth. 2019;138:124–30.
196. Subbiah V, Velcheti V, Tuch BB, Ebata K, Busaidy NL, Cabanillas ME, et al. Selective RET kinase inhibition for patients with RET-altered cancers. Ann Oncol. 2018;29(8):1869–76.
197. Shabbir A, Kojadinovic A, Shafiq T, Mundi PS. Targeting RET alterations in cancer: recent progress and future directions. Crit Rev Oncol Hematol. 2023;181:103882.
198. Wang J, Li Z, Zhao Q. Receptor tyrosine kinase-like orphan receptor serves as a potential target in cancer immunotherapy. J Leukoc Biol. 2024;117:qiae141.
199. Tigu AB, Munteanu R, Moldovan C, Rares D, Kegyes D, Tomai R, et al. Therapeutic advances in the targeting of ROR1 in hematological cancers. Cell Death Discov. 2024;10(1):1–13.
200. Menck K, Heinrichs S, Baden C, Bleckmann A. The WNT/ROR pathway in cancer: from signaling to therapeutic intervention. Cells. 2021;10(1):142.
201. Liu Y, Yang H, Chen T, Luo Y, Xu Z, Li Y, et al. Silencing of receptor tyrosine kinase ROR1 inhibits tumor-cell proliferation via PI3K/AKT/mTOR signaling pathway in lung adenocarcinoma. PLoS One. 2015;10(5):e0127092.
202. Daneshmanesh AH, Hojjat-Farsangi M, Ghaderi A, Moshfegh A, Hansson L, Schultz J, et al. A receptor tyrosine kinase ROR1 inhibitor (KAN0439834) induced significant apoptosis of pancreatic cells which was enhanced by erlotinib and ibrutinib. PLoS One. 2018;13(6):e0198038.
203. Ghaderi A, Zhong W, Okhovat MA, Aschan J, Svensson A, Sander B, et al. A ROR1 small molecule inhibitor (KAN0441571C) induced significant apoptosis of mantle cell lymphoma (MCL) cells. Pharmaceutics. 2022;14(10):2238.
204. Vaisitti T, Arruga F, Vitale N, Lee TT, Ko M, Chadburn A, et al. ROR1 targeting with the antibody-drug conjugate VLS-101 is effective in Richter syndrome patient-derived xenograft mouse models. Blood. 2021;137(24):3365–77.
205. Baskar S, Wiestner A, Wilson WH, Pastan I, Rader C. Targeting malignant B cells with an immunotoxin against ROR1. MAbs. 2012;4(3):349–61.
206. Tran TM, Chand Thakuri BK, Nurmukhambetova S, Lee JJ, Hu P, Tran NQ, et al. Armored TGFβRIIDN ROR1-CAR T cells reject solid tumors and resist suppression by constitutively-expressed and treatment-induced TGFβ1. J Immunother Cancer. 2024;12(4):e008261.
207. Wu D, Yu X, Wang J, Hui X, Zhang Y, Cai Y, et al. Ovarian cancer stem cells with high ROR1 expression serve as a new prophylactic vaccine for ovarian cancer. J Immunol Res. 2019;2019:9394615.
208. Gao Y, Zhou J, Li J. Discoidin domain receptors orchestrate cancer progression: a focus on cancer therapies. Cancer Sci. 2021;112(3):962–9.
209. Matada GSP, Das A, Dhiwar PS, Ghara A. DDR1 and DDR2: a review on signaling pathway and small molecule inhibitors as an anticancer agent. Med Chem Res. 2021;30(3):535–51.
210. Tian Y, Bai F, Zhang D. New target DDR1: a "double-edged sword" in solid tumors. Biochim Biophys Acta Rev Cancer. 2023;1878(1):188829.
211. Bibi S, Zeng W, Zheng P, Mousavi Mehmandousti SM, Jin T. Discoidin domain receptor 1 as a potent therapeutic target in solid tumors. hLife. 2024;2(9):454–66.
212. Zhavoronkov A, Ivanenkov YA, Aliper A, Veselov MS, Aladinskiy VA, Aladinskaya AV, et al. Deep learning enables rapid identification of potent DDR1 kinase inhibitors. Nat Biotechnol. 2019;37(9):1038–40.
213. Cui G, Deng S, Zhang B, Wang M, Lin Z, Lan X, et al. Overcoming the tumor collagen barriers: a multistage drug delivery strategy for DDR1-mediated resistant colorectal cancer therapy. Adv Sci. 2024;11(33):2402107.
214. Theik NWY, Muminovic M, Alvarez-Pinzon AM, Shoreibah A, Hussein AM, Raez LE. NTRK therapy among different types of cancers, review and future perspectives. Int J Mol Sci. 2024;25(4):2366.
215. Barbacid M. The Trk family of neurotrophin receptors. J Neurobiol. 1994;25(11):1386–403.

216. Reichardt LF. Neurotrophin-regulated signalling pathways. Philos Trans R Soc Lond Ser B Biol Sci. 2006;361(1473):1545–64.
217. Jiang T, Wang G, Liu Y, Feng L, Wang M, Liu J, et al. Development of small-molecule tropomyosin receptor kinase (TRK) inhibitors for NTRK fusion cancers. Acta Pharm Sin B. 2021;11(2):355–72.
218. Butti MG, Bongarzone I, Ferraresi G, Mondellini P, Borrello MG, Pierotti MA. A sequence analysis of the genomic regions involved in the rearrangements between TPM3 and NTRK1 genes producing TRK oncogenes in papillary thyroid carcinomas. Genomics. 1995;28(1):15–24.
219. Vaishnavi A, Capelletti M, Le AT, Kako S, Butaney M, Ercan D, et al. Oncogenic and drug-sensitive NTRK1 rearrangements in lung cancer. Nat Med. 2013;19(11):1469–72.
220. Sartore-Bianchi A, Ardini E, Bosotti R, Amatu A, Valtorta E, Somaschini A, et al. Sensitivity to Entrectinib associated with a novel LMNA-NTRK1 gene fusion in metastatic colorectal cancer. J Natl Cancer Inst. 2016;108(1):djv306.
221. Amatu A, Sartore-Bianchi A, Bencardino K, Pizzutilo EG, Tosi F, Siena S. Tropomyosin receptor kinase (TRK) biology and the role of NTRK gene fusions in cancer. Ann Oncol. 2019;30(Suppl 8):viii5–15.
222. Rogers C, Morrissette JJD, Sussman RT. *NTRK* point mutations and their functional consequences. Cancer Genet. 2022;262–263:5–15.
223. Scott LJ. Larotrectinib: first global approval. Drugs. 2019;79(2):201–6.
224. Al-Salama ZT, Keam SJ. Entrectinib: first global approval. Drugs. 2019;79(13):1477–83.
225. Zhao B, Burgess K. TrkC-targeted kinase inhibitors and PROTACs. Mol Pharm. 2019;16(10):4313–8.
226. Soni UK, Jenny L, Hegde RS. IGF-1R targeting in cancer—does sub-cellular localization matter? J Exp Clin Cancer Res. 2023;42(1):273.

Index

A. Sabarwal et al. (eds.), *Receptor Tyrosine Kinases in Cancer*, Cancer Drug Discovery and Development, https://doi.org/10.1007/978-3-031-93894-8

The manufacturer's authorised representative in the EU is Springer Nature Customer Service Centre GmbH, Europaplatz 3, 69115 Heidelberg, Germany. If you have any concerns regarding our products, please contact ProductSafety@springernature.com

Printed and bound by CPI Group (UK) Ltd, Croydon, CR0 4YY
30/09/2026
02225589-0016